This book examines 200 contractual problems which regularly arise on building and engineering projects and provides a detailed explanation of their solutions, citing standard contract conditions and key parts of legal judgements as authority. A succinct summary is provided at the end of each detailed solution.

It covers problems together with their solutions in respect of:

- Procurement matters
- Tenders and bidding
- Design issues
- Letters of intent
- Contractor's programme
- Contractor's float
- Delays
- Concurrent Delays
- Extensions of time
- Liquidated/delay damages
- Unliquidated damages
- Variations
- Loss and expense/additional cost claims
- Acceleration
- Global claims
- Payment
- Damage to the works
- Exclusion clauses
- Retention of title
- Practical completion
- Defect correction
- Adjudication

This book deals with a broad range of construction contracts including JCT Standard Form and Design and Build, New Engineering Contract NEC3, ICE and GC/Works/1.

This book was first published under the title of One Hundred Contractual Problems and Their Solutions, with a second edition entitled One Hundred and Fifty Contractual Problems and their Solutions. This third edition adds 50 new problems and replaces 15 of those in the last edition. Of the remainder half have been the subject of revision.

"Deserves a place on every site and in every office as the standard handbook on contractual problems"
Construction Law Digest

About the Author

Roger Knowles is a chartered surveyor, chartered arbitrator, barrister and CEDR-accredited mediator. He left school at 16 years of age and commenced work as a trainee quantity surveyor, following which he secured his qualifications through distance learning courses. His early career was that of a quantity surveyor, when he gained a great deal of experience of working on construction sites. He founded James R Knowles in 1973, a company of dispute resolution consultants, as a sole practitioner.

Roger built up James R Knowles until it was a world renowned brand and in 1998 became a quoted company on the AIM market. He sold his interests in James R Knowles in 2006, since which time he has acted as an independent disputes resolution consultant. His career has involved representing clients in many parts of the world, including East Asia, the Middle East and throughout the United Kingdom. His work includes the preparation and resolution of construction claims, representing parties in arbitration and adjudication and acting as arbitrator, adjudicator and mediator.

Roger has been involved in training for many years. His specialist subjects include construction law, contract administration, arbitration, adjudication and ADR and he has lectured at over 1,300 seminars at locations all over the world.

Since the mid-1970s, Roger has been a regular contributor to technical journals, with more than 400 published articles to his name. His career has also included representing trade associations and he is a former member of the Joints Contract Tribunal which publishes JCT contracts.

Roger was the founder of Quantity Surveyors International and joint author of the Public Sector Partnering Contract.

200 Contractual Problems and their Solutions

Third Edition

Roger Knowles
FRICS, FCIArb, FQSi, Barrister

WILEY-BLACKWELL
A John Wiley & Sons, Ltd., Publication

This edition first published 2012

Wiley-Blackwell is an imprint of John Wiley & Sons, formed by the merger of Wiley's global Scientific, Technical and Medical business with Blackwell Publishing.

Registered office:
John Wiley & Sons, Ltd, The Atrium, Southern Gate, Chichester, West Sussex, PO19 8SQ, UK

Editorial offices:
9600 Garsington Road, Oxford, OX4 2DQ, UK
The Atrium, Southern Gate, Chichester, West Sussex, PO19 8SQ, UK
2121 State Avenue, Ames, Iowa 50014-8300, USA

For details of our global editorial offices, for customer services and for information about how to apply for permission to reuse the copyright material in this book please see our website at www.wiley.com/wiley-blackwell.

First edition published

Library of Congress Cataloging-in-Publication Data
Knowles, Roger.
200 contractual problems and their solutions/Roger Knowles. – 3rd ed.
p. cm.
Includes bibliographical references and index.
ISBN 978-0-470-65831-4 (hard cover:alk. paper) 1. Contracts–Great Britain–Miscellanea. I. Title. II. Title: Two hundred contractual problems and their solutions.
KD1554.K58 2012
346.4102–dc23

2011035016

A catalogue record for this book is available from the British Library.

Wiley also publishes its books in a variety of electronic formats. Some content that appears in print may not be available in electronic books.

Set in 10/12.5 pt Minion by Toppan Best-set Premedia Limited

1 2012

Contents

Foreword

by Tony Bingham

In the first edition of this book, published in 2000, Roger Knowles discussed 100 contractual problems; then 150 and now 200. Wouldn't you have thought that by now the numbers would be going the other way? In 1993 Sir Michael Latham conducted an inquiry into common contractual problems in construction – 'to formulate possible remedies'. He reflected on his predecessor in 1963, Sir Harold Banwell, treading the same path; and little by way of remedies had arisen. Meanwhile, Roger Knowles did something effective. He recognized that contractual problems in building and civil engineering ran with the territory. Like it or not, they are normal. Construction contracts and contractual problems are everyday stuff; they are there to be managed. Just like, men, materials, plant, the weather and all sorts are to be managed, so too is the contractual side of things. 'Contractual Problems' is the 'the other 50% of building'.

To know the answers, it is ever so useful to pinpoint the questions first. The book is so welcome. Indeed, as I read the manuscript I found I didn't want to stop. That's because a quick glimpse at the next conundrum and the next was immediately relevant to the everyday 'construction disputes' world. Then, I developed an appetite for Roger Knowles' answers. Look, he has been in the midst of these contractual problems for 40 years and more; all of it in building and civil engineering. He is highly qualified, highly experienced and his easy style highly engaging. Take a glimpse at two things: the general list of 22 headings, then the 200 problems. They are the way of things and this book of answers is a practical work to contractual management – and, probably, dispute avoidance.

Tony Bingham
Barrister at Law

3 Paper Buildings
Temple, London
EC4Y 7EU

Preface

Since the last edition of this book, published in 2005, there have been a significant number of changes, in the form of new legislation, revised standard forms of contract and case law, to merit increasing the number of contractual problems from 150 to 200. Of the 150 Problems included in the last edition, 15 have been replaced, as they are no longer relevant. There have also been changes to approximately one-half of the Problems which remain.

The Housing Grants, Construction and Regeneration Act 1996 (Construction Act) and its early interpretation by the courts featured in the last edition of the book. Ten years have elapsed since the Act came into force, during which time the leading lights in the construction industry came to the conclusion that some aspects of the Act weren't working as well as expected, in particular payment to subcontractors and certain matters relating to adjudication. It was considered that changes needed to be made and therefore Parliament was lobbied, with a view to amending legislation being introduced. The result has been the Local Democracy, Economic Development and Construction Act 2009, which applies to all contracts entered into from 1 October 2011. The standard forms of contract have been amended to take account of the Housing Grants, Construction and Regeneration Act 1996 and have been re-issued to include the changes introduced by the Local Democracy, Economic Development and Construction Act 2009.

The Construction Act 1996 has resulted in controversies arising from its wording, which resulted in dispute, some of which have been corrected by the new Act. The payment provisions of the new Act provide for a payment notice to be served in respect of all payments which are to be made, with either the payer or the payee to serve the notice. Where the payee is to give the payment notice, if the payee intends to pay less than the sum included in the notice, then a 'pay less' notice needs to be served. The Construction Act governs only contracts which are in writing or evidenced in writing. Disputes and referrals aplenty have involved courts interpreting what is meant by 'evidenced in writing'. To try and overcome the problems this wording has created, the new Act covers oral contracts as well as those in writing or evidenced in writing. Whilst this solves the problems relating to the interpretation of evidence in writing, it will create an evidential difficulty as to what are the provisions of the oral agreement.

A significant body of case law now exists relating to statutory adjudication, which has been described in the Foreword to *Coulson on Construction Adjudication*, as 'The mass of authorities on adjudication form an impenetrable jungle, through which it is not easy to hack out a path'.

A great deal of controversy has arisen as a result of the case of *Bridgeway Construction* v. *Tolent Construction*, which involved a contract, the terms of which required a subcontractor to pay the contractors' costs of referring a dispute to adjudication even where the subcontractor was successful. The reason for this dispute was the presence of what has become known as a *Tolent* clause in the terms of the contract. This problem is addressed by the new Act and will no longer be enforceable. Some legal commentators, however, have offered the opinion that the provisions in the new Act will have no effect upon a *Tolent* clause.

In an effort to improve cash flow on projects and provide security of payment down the supply chain, Project Bank Accounts are now starting to be employed on some major projects. The JCT has produced standard terms for use where a Project Bank Account is employed.

Legislation which emanates from Europe relating to procurement has caught out a number of local authorities, who have failed to demonstrate the required levels of transparency in their methods of selecting contractors. Failure to observe the procurement requirements can result in either substantial damages having to be paid to unsuccessful contractors or the whole process having to be repeated.

Problems relating to the use of letters of intent are ongoing. Courts regularly advise against their use, but the culture is built into construction industry methodology and unlikely to change. The court, however, has provided good advice on what should be included in letters of intent.

Delay analysis has become a fast-growing skill which has been developed by construction managers. These skills have been demonstrated in a number of court cases and provide evidence of delays which have occurred on construction projects. The courts are now becoming familiar with this process and offering views as to how relevant it is as evidence in legal actions. In particular, concurrent delays, which have been the subject of observations by judges in a number of cases in the past, have been examined again in several recent cases.

Jargon can be confusing to many, but understood by only the few. Terms such as 'Prevention Principle', 'Time of the Essence' and 'Time at Large' are examples of jargon which is frequently used.

Mediation is fast becoming a front-line alternative dispute resolution technique. The standard forms of contract are now catering for mediation and the courts are penalising parties in their awards of costs, if parties refuse to have their disputes referred to mediation.

All these matters are fully explained in this edition of the book.

The Institute of Civil Engineers has, from the outset, been responsible for the publication of ICE conditions, hence the title. The contract, now referred to as the Infrastructure Conditions of Contract has, since August 2011, been published by the Association of Consultant Engineers and the Civil Engineering Contractors' Association.

Roger Knowles

Chapter 1

Procurement

1.1. What are 'entire contracts' and how relevant are they in the construction industry?

1.1.1. Those for whom construction work is undertaken usually require the contract price to be fixed at the outset and not to change throughout the construction of the project. This is to ensure that the final sum paid equates to the contract price. This is often referred to as a 'lump sum fixed price', or in legal circles 'entire contracts'. Most major construction projects are let employing one of the standard forms of contract, which in the main do not provide for the contract price to be a fixed price. Construction work is bedevilled with uncertainties: for example, in many cases some of the work is constructed below ground, where surprises are often encountered during the construction phase. Who, therefore, takes the risk of ground conditions which are more difficult to work with than anticipated at the time the contract was let? The contract, if properly drafted, should provide the answer. If the contractor is expected to take the risk, then it is not unreasonable for a sum to be included in the contract price to cover the risk. Some standard forms of contract, for example the ICE 6th and 7th Editions and FIDIC, make it clear that the contractor's price only includes conditions which could have been reasonably foreseen by an experienced contractor. Therefore, if ground conditions which are met during the construction process are more onerous to work with than could reasonably have been expected to have occurred, then any resultant additional cost incurred by the contractor will be added to the contract price. Some conditions of contract, for example JCT 2011 Standard Building Contract With Quantities, includes for the design to be undertaken by an architect or contract administrator appointed by the employer. Any additional cost incurred by the contractor in overcoming errors in the design will be added to the contract sum. Almost without exception, the standard forms of contract include a clause to the effect that the contract price will be adjusted if additional cost is incurred by the contractor because of changes in legislation which were unknown at the time the contract was entered into. The standard forms of contract in general use therefore cannot be classed as 'entire' contracts.

1.1.2. Contracts which do not provide for any change in the price and are thus fixed-price, are sometimes referred to as entire contracts, which encompass the inclusive price principle. This principle has been expounded in the 11th Edition of *Hudson's Building and*

200 Contractual Problems and their Solutions, Third Edition. Roger Knowles.

Engineering Contracts and states that on a construction project the contractor is entitled to be paid for the work defined in the contract, which is deemed to include all work that is both indispensably and contingently necessary. Indispensable work is all work which, by implication or as a matter of interpretation of the contract as a whole, has to be carried out in order that the final work should comply with the express requirements or descriptions in the contract documents. Contingent work is all work that is necessary to achieve completion of that described, irrespective of the difficulties which may be encountered. According to this principle, any additional work necessary to achieve completion of the work described in the contract documents must be done at the expense of the contractor. This is the situation unless the contract states otherwise.

1.1.3. The inclusive price principle came to the fore in the case of *Safe Homes Ltd* v. *Mr and Mrs Massingham* (2007). Mr Dale, representing Safe Homes Ltd, undertook to construct a new house for the lump sum of £130,000 and to complete it in 17 weeks. The drawings were provided by Mr and Mrs Massingham's architect. Safe Homes claimed an entitlement to extra payment for additional and varied work owing to inadequately defined work, necessary but ill-defined work and compliance with building regulations and other statutory requirements. These claims were rejected by the court on the basis that the contract was an entire lump sum inclusive-price contract. Such contracts are subject to two over-riding principles that are applicable unless varied by the express terms of the contract. These principles are that contractors must, without additional payment, carry out all work necessary to enable the overall scope of work to be completed, even if that work has not been defined in the contract documents, and undertake all work needed to overcome any obstruction or unforeseen eventuality that must be overcome to complete their work.

1.1.4. This case worked in favour of the building owner, as the contractor was required to undertake work which had not been foreseen at tender stage, for no additional payment, on the basis that the contract was a fixed-price entire contract. The case of *SWI* v. *P&I Data Services* (2007) provided the opposite result, where a subcontractor was paid for work which was not undertaken. SWI submitted a quotation to provide work for P&I at the GlaxoSmithKline site in Stevenage for £337,243, in accordance with drawings provided by P&I and tender record sheets produced by SWI. Some of the work was not required to be undertaken, and P&I reduced the price to be paid to reflect the reduced volume of work. It was agreed by both parties that the value of work which was not undertaken amounted to £40,000. The Court of Appeal, however, held that SWI was entitled to payment in full, with no reduction for the work which was not carried out. The contract was a fixed-price lump sum entire contract, with no provision for variations, and therefore the Court of Appeal ordered that the contract price be paid in full, even though some of the work had not been carried out.

1.1.5. Whether the contract is let on a fixed-price basis or one which is subject to price change will depend upon the wording in the contract. To ensure that the rights and obligations of the parties are confined within the wording of the contract, some contracts include an entire agreement clause, also known as 'entire understanding' clauses, 'four corners' clauses or 'zipper' clauses. This clause usually states that the contractor's obligations are fully set out in the contract and no supplementary evidence based upon correspondence, discussions and the like is admissible. The purpose of this type of clause is to eliminate any opportunity for introducing evidence in the form of other documents which may be at variance with the wording of the contract, as a means of enhancing payments.

SUMMARY

Entire contracts are not the norm on major construction projects. Whilst employers like a price which is fixed at the outset and does not fluctuate, most major projects are let using one of the standard forms of contract. These contracts contain a sharing of the risks between the contractor and employer. If one or more of the employer's risks becomes a reality, the contractor may become entitled to an additional payment. For example, unforeseen bad ground conditions on civil engineering contracts; architect or engineer design errors where employer design applies; and changes in legislation usually result in the contractor receiving additional payment.

There are examples of contracts where the inclusive price principle applies, referred to as 'entire contracts'. These contracts are fixed-price in every sense of the word. Where they apply, the contractor is required to undertake all necessary work to ensure practical completion in accordance with the work as described in the contract documents. Any additional work necessary to achieve completion must be carried out at the contractor's expense.

1.2. Do projects where the parties enter into partnership arrangements require a formal contract to be agreed?

1.2.1. Sir John Egan created quite a stir when, in *Rethinking Construction*, he suggested that where parties to a construction project operate on a partnering basis, a formal contract is unnecessary. The exact words Sir John used are:

> Effective partnering does not rest on contracts. Contracts can add significantly to the cost of a project and often add no value to the client. If the relationship between a constructor and employer is soundly based and the parties recognise their mutual dependence then formal contract documents should gradually become obsolete.

What he probably meant by this statement was that on partnering projects agreements reached between the parties do not require to be legally enforceable. There is no doubt that where collaborative working arrangements exist, formal disputes are a rarity. Disputes do arise, but they are usually resolved amicably by representatives of the parties without any involvement by lawyers or other external consultants.

1.2.2. Despite the good relationships which are fostered by partnering, there are examples of disagreements which have been referred to court. In the case of *Baird Textiles Holdings Ltd* v. *Marks and Spencer PLC* (2001), Baird claimed it had lost a sum of £50m as a result of Marks and Spencer breaching a partnering arrangement. Baird had been a supplier of garments to Marks and Spencer for more than 30 years, which represented between

30% and 40% of Baird's annual turnover. There was no guarantee of the number of garments which would be ordered, but over the years there had been a steady but varying amount of business. The parties worked together to ensure that what was not selling well could be returned and new designs jointly developed. The chairman and chief executive of Marks and Spencer stated that for over 70 years the relationship between Marks and Spencer and its suppliers was governed by the principles of partnership. Baird considered that their relationship with Marks and Spencer was a partnership.

1.2.3. Without giving any prior warning, Marks and Spencer gave notice to Baird that, following the end of the current production run, it would not be placing any more business with them. Baird commenced proceedings, claiming that the termination was a breach of an implied contract arising from the manner in which the parties had conducted their business over many years. On the basis of their past relationship, Baird considered it was entitled to be given a minimum of three years' notice if Marks and Spencer intended to terminate the arrangement. The court rejected the arguments made by Baird. It would only recognise an implied contract if it was necessary to do so. Courts are reluctant to make a bargain between the parties which they had not made for themselves.

1.2.4. There have been other cases where the parties have entered into a partnering arrangement, which have ended in disputes which have been referred to the courts. In the case of *BP Exploration Operating Co Ltd* v. *Kvaerner Oil Field Products Ltd* (2004), reliance was placed upon negotiations to establish a partnering relationship. This was not accepted by the courts as evidence for the interpretation of the parties' obligations to procure insurance. The case of *Birse Construction Ltd* v. *St David Ltd* (2000) involved a dispute arising from a contract on which a partnering relationship existed. The matter was referred to the Court of Appeal, as the parties could not agree the terms which applied to the contract. No doubt the parties, in view of the relaxed atmosphere which existed relating to contractual matters, did not see any urgent requirement to enter into a formal contract before work commenced.

1.2.5. The CIC Guide to Project Team Partnering provides the following advice concerning the use of a formal contract on schemes where partnering applies:

> An effective contract can play a central role in partnering, it sets out the common and agreed rules; it states the agreed mechanism for managing the risk and the reward; it lays down the guidelines for resolving disputes. But the central thrust of the new thinking is that the contract should not encourage a self-serving or adversarial state or a battle with other team members for the benefit of one party.

1.2.6. Some contracts have been designed to accommodate partnering and include:

- PPC 2000
- JCT Constructing Excellence
- NEC Partnering Option X12
- Public Sector Partnering Contract – Partnering Agreement

1.2.7. Parties who enter into arrangements where partnering is to take place have an option of either following the advice of Sir John Egan and avoiding the formal contract route, or deciding to accept the guidance of CIC and enter into a formal arrangement. One of the points raised by Sir John is the significant cost of entering into a contract. *Rethinking Construction* was produced before any of the standard forms which are now available for use on partnering projects. There is no downside to the use of a standard form of contract, most of which are fairly simple to complete, with little expense involved. Perhaps the well-worn phrase is still appropriate, which states that parties are best advised to 'sign the form of contract, put it in a drawer and then forget all about it'.

SUMMARY

Advice has been provided by two good authorities as to whether it is advisable, where partnering relationships exist, as to whether or not the parties should enter into a formal contract. Sir John Egan, in *Rethinking Construction*, is of the opinion that a contract involves significant cost and adds no value to the client. On the other hand, the CIC advises that a contract can play a central role in partnering. Since Sir John produced his report, a number of standard forms of contract which cater for partnering have been produced and can be entered into at very little cost, which removes one of Sir John's arguments. He makes the point that a contract is of no value to the client, but omits to mention the benefit or otherwise to the contractor. Partnering relationships are all about people and with changes in personnel, takeovers and mergers, attitudes can change. Sign the contract, put it into a drawer and forget about it, may well be the best advice.

1.3. What is the effect of an agreement to undertake work which is expressed as being 'Subject to Contract'?

1.3.1. The term 'Subject to Contract' originated in agreements for the sale of land by private treaty. Such agreements were unenforceable until a formal contract had been drawn up and agreed by the parties. In recent times the use of the term 'Subject to Contract' has been used in the formation of construction contracts. This development has led to uncertainty as, in some legal cases, it has been held that the document in which the words 'Subject to Contract' appears constitutes a binding contract, whilst in others the reverse is the case.

1.3.2. In the case of *Bennett (Electrical Services)* v. *Inviron* (2007), Bennett was engaged by Inviron to carry out labour-only electrical work. A letter was sent by Inviron to Bennett headed 'Subject to Contract' and stated an intention for the parties to enter into a full subcontract, incorporating Inviron's standard terms. The letter instructed Bennett to proceed with the works required to progress the subcontract and Bennett duly obliged. A formal subcontract was never entered into; however, a dispute arose concerning payment and Bennett commenced adjudication proceedings. The first adjudicator to be appointed refused to proceed, as he considered that he had no jurisdiction, and a second adjudicator was appointed. He proceeded with the adjudication and provided in his decision that Inviron should make a further payment to Bennett. No such payment was made and the matter was referred to court for enforcement. It was held by Judge Wilcox that 'Subject to Contract' had its ordinary meaning, whereby liability did not arise until a full contract had been entered into.

1.3.3. The decision in the case of *Skanska Rashleigh Weatherfoil Ltd* v. *Somerfield Stores Ltd* (2006) tells a different story. Somerfield sent a letter of invitation to tender to Skanska on 19 June 2000, enclosing a draft of the proposed Facilities Management Agreement. Negotiations took place following the sending of the letter, but were incomplete when Somerfield, who were anxious to get the work under way, sent a detailed letter to Skanska dated 17 August 2000 headed 'Subject to Contract'. The key matters included in the letter were:

- We now wish to appoint you to provide us with the Services which are more particularly described in the contract (Ref JRB/2240842 DRAFT3 14th June 2000 Contract) enclosed with the tender . . .
- This appointment is, however strictly subject to contract and to the approval of the board
- In consideration of the above and whilst we are negotiating the terms of the Agreement, you will provide the Services under the terms of the Contract from 28th August 2000 (or such other date as we may advise to you) until 27th October 2000 (the Initial Period), such services to be provided at the prices detailed in the tender return provided by you (as subsequently amended) as more particularly itemised on the attached schedule.

This letter was signed and returned by Skanska. A dispute arose as to the meaning of the term 'provide the Services under the terms of the Contract'. Somerfield argued that the letter headed 'Subject to Contract' contained all the terms of the Facilities Management Agreement and therefore a contract existed between the parties. Skanska was of the view that the purpose of the letter was limited to identifying the nature of the work to be performed and therefore constituted only a temporary arrangement and not the contract for providing services for a three-year period. The judge in the lower court, Mr Justice Ramsey, agreed with Skanska, and Somerfield appealed. The Court of Appeal considered that Mr Justice Ramsey made a wrong decision. Lord Justice Neuberger considered that a contract had come into operation, which comprised the draft Facilities Management Agreement, except where inconsistent with the contents of the 17 August 2000 letter. Lord Justice Neuberger was influenced, in arriving at his decision, by the words 'under the terms of the contract' which appeared in the 17 August 2000 letter.

1.3.4. The meaning of the term 'subject to formal contract' was an issue in the Court of Appeal case of *Stent Foundations* v. *Carillion Construction (Contracts) Ltd* (2000). It was held that a contract, though never formalised, came into being, even though in the letter of intent the words 'subject to formal contract' appeared. The court was of the opinion that, as the parties had agreed all the essential terms, a contract had been properly formed.

SUMMARY

It seems clear that the term 'Subject to Contract', in terms of the sale of land, traditionally meant that, in the absence of a formal contract, no binding agreement arises. The use of this wording in construction projects, however, does not ensure the same degree of certainty. From the legal cases referred to, it is not clear that whether a letter which is headed 'Subject to Contract' does or does not constitute a binding contract depends upon the wording in the letter.

1.4. What is Two-Stage Tendering and how does it operate?

1.4.1. Two-Stage Tendering is best suited to large or complex schemes, as the tendering cost involved is usually greater than for a single stage tendering process. The main aim is to choose a contractor as early as possible. It is usual, where two stage tendering applies, to appoint the contractor during the design period, to enable the professional team to make use of the contractor's expertise. In particular, where an employer design method is adopted, two-stage tendering also enables the contractor to have an input in the planning of the project.

1.4.2. Two-Stage Tendering can be used with any of the standard forms of contract. The method of procurement and the conditions of contract, or standard conditions if applicable, need to be established before inviting first-stage bids. If the employer intends to amend a standard form for use on the project, these amendments should be made known to first-stage bidders as, if they involve a major shift in risk, pricing levels may be affected.

1.4.3. The first stage is the process used for the selection of a contractor and establishing a level of pricing to be used in the calculation of the contract price. Documentation is usually kept to a minimum for reasons of time and cost. In the more traditional method of procurement, where bills of quantities are the norm, approximate bills of quantities will normally be used to arrive at an estimated cost of the works and provide rates on which the contract price can be calculated. Bills of quantities are not as frequently used as in the past, so other methods have been devised. Where bills of quantities are not to be used, a schedule of work, with accompanying rates and prices, which can be employed to enable a fair comparison between competing contractors, is often used. Where the contract is to be let using a cost-reimbursable method, the pricing of overheads and profit and preliminaries should be established at the first stage. If the intention is to make use of a target price, this can form part of the contractor's bid and can be adjusted as the design is developed.

1.4.4. It is important that the design process is frozen at the point at which first-stage bids are received. A comprehensive drawing register, which is properly updated with all revisions, together with a specification or schedule of works, is essential. This will enable the first-stage price to relate back to the design as it had developed up to that point. When arriving at the contract price, the first-stage price or target cost can be adjusted to reflect subsequent changes in design.

1.4.5. It is helpful if the design of one of the early stages of the work, such as the substructure, is completed at the point of first-stage bids. This enables a final price to be established for one of the stages, and possibly an early start on that part of the work can be made.

1.4.6. The situation is different where the procurement involves the contractor providing a full design service. The facility for making a selection mainly on price is more difficult.

In the final analysis, it may be that at this stage the contractor only produces details of the pricing of overheads and profit and the preliminaries, together with an approximate estimate of the overall cost; the intention being that the contractor will be required to design the project down to the estimated price.

1.4.7. Once the tender enquiry documents for the first stage have been completed, a shortlist of interested contactors is invited to submit bids. Once tenders have been returned, one contractor is selected to proceed to the next phase. No contract is awarded at this stage and there is no guarantee that either the work will proceed, or, if it does, that the contractor who is successful at this stage will go on to be appointed. The contractor is very much at risk and can incur substantial costs without there being any obligation on the part of the employer to reimburse those costs. With no binding commitment by either party, the contractor can withdraw at any time, leaving the employer to face probable delay and additional costs.

1.4.8. The first-stage selection process may be purely on the basis of lowest cost; however, it is now common practice for the bidding contractors to be asked to include quality as part of their submission. When making the selection, both price and quality are taken into consideration in making the choice of contractor to proceed to the second stage.

1.4.9. The purpose of stage two is to convert the outline information produced during stage one into the basis for producing a firm contract between the client and contractor as soon as possible. The contractor will usually have an important input into the design process, by making suggestions which relate to the buildability of the project. The contractor may also have an input into drawing up the contract documents and, together with the professional team, arriving at an agreed contract price. The contractor will also have a major input into health and safety issues.

1.4.10. In view of the substantial input which may be required from the contractor at this early stage of the project, there are now in place some standard mini contracts designed to cater for this type of arrangement, which include:

- JCT Pre-Construction Services Agreement
- PPC2000 – Pre-Possession Agreement
- PSPC – Prestart Agreement

These mini-contracts set out the services to be provided and also a payment mechanism.

1.4.11. Examples of the type of services which may be provided by the contractor before a formal contract is entered into include:

- Health and safety
- Risk management
- Value management
- Public consultation
- Design
- Life cycle costing
- Environmental impact

- Establishing an open book accounting system
- Training
- Supporting funding applications
- Selection of specialists whose work may or may not form part of the main contract.

1.4.12. During the period between the first stage and second stage the contractor may be required to undertake accommodation work, such as site clearance, which can be provided for under this early contractual arrangement.

1.4.13. The second stage is complete when a contract for the whole of the project has been drawn up and signed by the parties.

SUMMARY

The purpose of a two-stage tendering procedure is to provide for an early appointment of the contractor. This enables the contractor's knowledge to be used in the design of the work. Buildability and health and safety are two matters where the contractor will normally have a major input. The contractor can be appointed in competition with others, using pricing documents relating to the preliminary design. This pricing information is used during the second stage to build up the contract price for the work.

The contractor can expend a considerable amount of money during the second stage, with no certainty of being appointed. During this period he or she is at risk. To overcome this problem, there have been produced a number of mini contracts, which can be used to engage the contractor during this period, which set out the services to be provided and the method of payment. During this second stage, the contractor can be employed in undertaking accommodation work such as site clearance, using the mini contracts.

1.5. Where tender enquiry documents indicate that an established procedure for selecting contractors will apply, but the employer does not follow the procedure, will an unsuccessful party be entitled to claim damages from the employer?

1.5.1. Many tender enquiries in the private sector provide little, if any, information as to the methods to be adopted in deciding which contractor will be awarded the contract, following receipt of tenders. Contractors requested to submit tenders often assume that, as all tenderers have been through a prior vetting system, it will come down to the bidder submitting the lowest price being appointed. In an effort to regularise the tender selection process, the NJCC produced a Code for Selective Tendering. The Code laid down procedures for both single-stage and two-stage tendering. The use of these codes has, in the past, been used in the UK quite extensively. In order to give the process some credibility, employers or their architects, in the tender enquiry, would state that the tendering procedure was to follow the NJCC Code. The NJCC Code of Procedure has now been superseded by guidelines published by the CIB.

1.5.2. Having stated that the tendering procedure would follow the NJCC Code, could the employer be exposed if he or she proceeded with the selection process completely ignoring the procedure? There may be a situation arising where an unsuccessful bidder could demonstrate that, had the Code been followed, he or she would have been awarded the contract.

1.5.3. This situation has been examined by the courts and an answer provided. In the case in question, *J&A Development* v. *Edina Manufacturing Ltd and Others* (2006), J&A Development was asked by Edina Manufacturing Ltd to submit a tender for the construction of a workshop, offices and associated work at an industrial estate in Lisburn. The tender enquiry documents prepared by ADP Architects and Design Partnership stated that the tendering procedure was to be in accordance with the principles of the Code of Procedure for single-stage selective tendering, published by the NJCC in 1996.

1.5.4. Section 7.1 of the Code of Procedure states:

> . . . good tendering procedure demands that a contractor's price should not be altered without justification. In particular NJCC strongly deplores any practice which seeks to reduce any tender arbitrarily where the tender has been submitted in free competition and no modification to the specification, quantity or conditions under which the work is to be executed, or to be made, or to reduce tenders other than the lowest, to a figure below the lowest tender.

1.5.5. Six tenders were received, of which J&A Development in the sum of £1,074,982 was the lowest. A decision was then made, following discussions between the architect and Edina, that meetings should be held with each of the three contractors who had submitted the lowest tenders, to see if their tender prices could be reduced. The meetings took place, at which the tenderers were invited to reduce their tender amounts. Two of the contractors agreed, but J&A Development refused. Kylen Construction, who had produced the second lowest tender, agreed to reduce its price by £25,000 and was awarded the contract.

1.5.6. J&A Development commenced proceedings against Edina, for damages for breach of contract, on the basis that Edina failed to follow the Code of Procedure. The court decided that J&A Development was entitled to be paid damages. Whilst no contract to construct the work had come into operation, there was a collateral agreement to the effect that the Code would be applied. Clearly, in inviting three tenderers to meetings to discuss reducing the sums tendered, Edina was operating in contravention of the Code. In arriving at its decision, the court took note of the Court of Appeal decision in the case of *Blackpool and Fylde Aero Club Ltd* v. *Blackpool Borough Council* (1990) and the lower court decision in the case of *Fairclough Building Contractor* v. *Borough Council of Port Talbot* (1993).

1.5.7. The court awarded J&A Development damages for breach of the terms of the Code. As they had submitted the lowest price, if the Code had been followed, they would have been awarded the contract. The damages which Edina were ordered to pay were therefore based upon J&A Development's tender costs in the sum of £6,530, plus loss of profit in the sum of £161,247. The amount awarded for loss of profit was reduced by 20% to take account of the availability of other work.

SUMMARY

Where a tender enquiry states that a standard procedure for dealing with tenders, such as those produced by the NJCC or the CIB, will apply, there exists between the employer and tendering contractors a collateral agreement to the effect that this procedure will be applied. If the employer fails to implement the Code, this will amount to a breach of the collateral agreement. Any tendering contractor who can demonstrate a loss, as a result of the failure on the part of the employer to comply with the Code, will be entitled to reimbursement.

1.6. What liability does a tendering contractor have who in its bid names key personnel to be employed on the project, but when work commences replaces some of the named personnel?

1.6.1. It is common practice when submitting bids for work on construction projects for there to be a requirement for key personnel, who will be involved in the project, to be named in the bid document. Will there be any liability if, when the work proceeds, the person named is not employed on the project? This situation occurred in the case of *Fitzroy Robinson* v. *Mentmore Towers* (2010). The case concerns the redevelopment of the 'In and Out Club' in Piccadilly and an associated country house in Buckinghamshire, which was to be converted into a luxurious club with hotel accommodation. Mentmore engaged Fitzroy Robinson to act as the architect on the scheme. The programme indicated that work on the project was scheduled for completion at the end of May 2009. During the summer of 2005, meetings took place between the parties at which the project was discussed. Mr Blake was a key member of the Fitzroy Robinson team. It was made clear both before and in the bid document, dated 20 September 2005, that if Fitzroy Robinson was to be appointed, then Mr Blake would oversee the design and be actively involved in the project. The wording in the bid document stated:

> Team Leader – Mr Blake, or such other individual of comparable standing, ability and experience, as the Employer may at his discretion approve.

1.6.2. Mr Blake tendered his resignation on 17 March 2006. He received an offer to continue in his employment, but refused the offer. Mr Blake, under the terms of his employment, was obliged to work a 12 months' notice period. However, it was not until 5 November 2006 that Fitzroy Robinson informed Mentmore that Mr Blake was due to leave the company. Mr Blake was requested not to inform Mentmore of his intention to leave; in his words, he was sworn to secrecy. He was key to the project and the loss of Mr Blake, Fitzroy Robinson considered, could have affected their appointment.

1.6.3. Fitzroy Robinson's duties were to undertake the design of the project and achieve planning permission. An application for planning consent of the Piccadilly part of the project was submitted on 1 December 2006 and for the Buckinghamshire property in April 2007. Mr Blake left his employment on 16 March 2007 and continued to undertake work on the project on an hourly basis to help facilitate the submission of the planning application for the Buckinghamshire property. It was argued by Mentmore that Fitzroy Robinson was guilty of fraudulent misrepresentation, which involved:

1. The existence of a contract between the parties.
2. The representation – i.e. that Mr Blake would be involved – occurred before the contract was entered into.
3. The representation was made knowing it to be false, or without belief in its truth, or recklessly, careless whether it be true or false.
4. The representation acted as an inducement to Mentmore to enter into the contract.
5. Mentmore had suffered loss.

1.6.4. The court held that Fitzroy Robinson was guilty of fraudulent misrepresentation and that Mentmore was entitled to claim any loss which it incurred as a result. What Fitzroy Robinson was obliged to have done was to advise Mentmore, at the earliest opportunity, of Mr Blake's intending departure, which they failed to do. This notification should have been made after Mr Blake had turned down the offer made by Fitzroy Robinson to remain in post, having received his resignation.

1.6.5. The court, however, decided that Mentmore would be unlikely to have suffered very much loss as a result of Mr Blake's departure. There was no evidence that his departure resulted in delay. In view of the timing of the notice being submitted, it was unlikely that Mentmore would have decided to appoint a different firm of architects. The case seemed to have been a pyrrhic victory for Mentmore.

1.6.6. In this case, Fitzroy Robinson named Mr Blake as being the person who would be employed on the project, knowing full well that he was due to leave his employment with the company. The court held this to amount to fraudulent misrepresentation. This is not always the case, as quite often personnel named in the bid will leave after the contract has been entered into and their name has been submitted in good faith.

1.6.7. Where personnel are named in the contract, it normally states that if they are not available, then a suitably qualified and experienced alternative will be found. This may not prove too difficult to achieve, although the temptation could be to name as an alternative a person who is available and not necessarily as experienced and qualified as the person originally named. The person named may have a high reputation and as a result cannot easily be replaced.

1.6.8. Where a named person is not available and this situation is already known by the party submitting the bid, in like fashion to the Mentmore action, there may be a case brought for fraudulent misrepresentation. Where the naming is in good faith, it will be a matter of breach of contract if the person is named in the contract documents but is substituted, if the substitute is not up to the standard of the named person. However, if a person is named in the bid, or other document which does not become a contract document, it may be a question of breach of warranty.

1.6.9. The most difficult part of any action by a disappointed employer who is deprived of a named person is to demonstrate that he or she has suffered a financial loss as a result. Without being able to demonstrate loss, in like fashion to the Mentmore action, there will be nothing to recover and therefore the award will be a nil recovery. If, for example, the employer replaces the chosen consultant or contractor with another, due to the named person not being available, and this is a reasonable action to take, then any additional price paid by the employer for the work to be carried out by the replacement consultant or contractor would be the basis of the claim.

SUMMARY

Where a contractor states that a named person will be employed on a project, and the person is substituted, there will be no right of action on the part of the employer if the contract allows for a substitute. The wording of the contract, however, will usually state that if the named person is not available, then a substitute of equal experience and qualification will be provided. Where a substitution is made and the substitute turns out not to be of equal experience and qualification to the named person, the employer may have a right of action for breach of contract. Alternatively, if the person is named in a document which is not a contract document, then any action may be based upon a breach of warranty. Should the contractor name a person, knowing that person will not be available, there may be a right of action for fraudulent misrepresentation. For the employer to succeed, however, it is essential for resultant monetary loss to be demonstrated.

1.7. Can a contract which is freely entered into by the parties not be enforced on the grounds that the effect would be commercial nonsense?

1.7.1. The basic rule when interpreting contracts, as explained in *Pioneer Shipping* v. *BTP Tioxide* (1982), is that:

> whilst it seeks to give effect to the intention of the parties, it must give effect to that intention as expressed, that is, it must ascertain the meaning of the words actually used

In arriving at a conclusion, evidence is permitted of the circumstances in which the contractual document was made, any special meaning of words, the customs and certain other matters which may assist the court in arriving at the expressed intention of the parties.

1.7.2. When deciding the rights and responsibilities of the parties to a contract, the usual source of information is the contract entered into by the parties. There are, however, many examples of agreements on matters relating to the contract which were never included in the finally concluded contract. What is the status of these agreements? Are they enforceable, or, as they have not been included in the concluded contract, are they non-binding?

1.7.3. This was the subject of a dispute which was referred to the House of Lords in the case of *Chartbrook Ltd* v. *Persimmon Homes Ltd* (2009). The parties to the dispute involved a developer, the defendant, and a claimant who sold them land for development. The intention was for planning consent to be obtained by the developer for the construction of mixed residential and commercial premises which were to be sold on long leases. The claimant would then be paid for the sale of the land in accordance with a formula. This comprised the Total Land Value, comprising the value of land for housing, commercial and car parking, which caused no problems. There was also a further payment which related to the success of the project, referred to as the Additional Residential Payment (ARP), the interpretation of which was the subject matter of the dispute. The APR was defined as:

> 23.4% of the price achieved for each Residential Unit in excess of the Minimum Guaranteed Residential Unit Value less the Costs and Incentives

The Minimum Guaranteed Residential Value was the Total Land Value divided by the number of flats. However, when a straight calculation was carried out, the net result was that the claimant became entitled to be paid in total £9,168,427.

1.7.4. It was argued by the defendant that the purpose behind the division of the total payment into the Total Land Price and the Additional Residential Payment was to provide a minimum payment for the land and a further payment to allow for the possibility of an increase if the market prices rose and the flats sold for more than expected. The formula adopted by the defendant resulted in a total payment of £5,580,565. In support of their argument the defendant sought to introduce negotiations which were undertaken prior to the contract being entered into, to overturn the wording in the contract. The House of Lords refused to accept this argument and in so doing relied upon the decisions in *Inglis* v. *John Buttery and Co* (1878) and *Prenn* v. *Simmonds* (1971), where it was held that pre-contractual negotiations are inadmissible evidence for the purpose of supporting the construction of a contract.

1.7.5. The House of Lords nonetheless came down on the side of the defendant. It was considered that the wording included in the contract regarding the Additional Residential Payment made no commercial sense. The decision followed the reasoning of the House of Lords in its decision in *Investors Compensation Scheme Ltd* v. *West Bromwich Building Society* (1998). It was held in this case that the principal question to be answered when seeking to interpret a contract is 'what a reasonable person having all the background knowledge which would have been available to the parties would have understood them to mean using the language in the contract'. In this case, Lord Hoffman laid down the following five principles to be used when considering the meaning of a clause in a contract:

- The correct meaning is what the document would convey to a reasonable person with the relevant background knowledge
- The background includes everything in the matrix of fact, i.e. relevant background material
- The law excludes the prior negotiations of the parties
- The meaning of words in a document is not the same as the literal meaning of words, but the meaning that one would reasonably understand against the relevant background

- The rule that the words should be given their natural and ordinary meaning reflects the commonsense proposition that it is not easily accepted that people have made linguistic mistakes, particularly in formal documents.

1.7.6. Whilst the courts do not easily accept that parties have made linguistic mistakes when drawing up a contract, where something has clearly gone wrong with the language which the parties have used, the courts are not required to attribute to the parties an intention which a reasonable person would not have understood them to have had.

1.7.7. This decision allows for contracts to be interpreted in a manner which is not expressed by the wording and may leave the door open to a party having made a bad deal to argue later that the wording makes no commercial sense and would not have been used by a reasonable person. In other words, what was stated in the contract was not what the parties really meant.

SUMMARY

The basic rule, when interpreting contracts, is that a court will enforce the intentions of the parties which are expressed in the contract. If the contract does not include what the parties intended, then those intentions do not form part of the contract. The next point to consider, when interpreting the meaning of a contract, is that the court will take into account what a reasonable person, having all the background knowledge which would have been available to the parties, would have understood the contract to mean, using the language in the contract.

In the case of *Chatbrook Ltd* v. *Persimmon Homes Ltd* (2009), the House of Lords refused to enforce a clause in a contract for the sale of land to be used for house building regarding an enhancement of the price dependent upon the sale price achieved for the houses. This was based upon how a reasonable person would have interpreted the wording in the contract of sale to mean. In this case, the literal interpretation of the wording would have resulted in the seller of the land becoming entitled to uplift of the price, which would be commercial nonsense and out of all proportion with reality.

1.8. Can an architect or engineer be held to have acted negligently for advising a client to use a procurement method which is inappropriate for the project concerned?

1.8.1. Architects and engineers are usually involved in projects from the outset. Early advice given often relates to the procurement methods to be adopted. There is an ever-growing array of alternatives from which to choose. Is it to be the lowest price, or partnering with collaborative working? Will contractor design be appropriate and is this procurement method in the client's best interest? If a JCT contract is to be employed, will the advice regarding which of the JCT alternatives is to be used follow the recommendations included in 'Deciding on the Appropriate JCT Contract', published by the JCT? If the project is mainly of an engineering nature, should the client be advised to use the current version of the standard ICE contract or the Engineering and Construction Contract (NEC 3)?

1.8.2. The possibility always exists for the recommendation to be incorrect. For example, an architect may recommend that a JCT With Quantities form of contract be used. This is a contract which involves the architect producing a full design and the quantity surveyor a complete bill of quantities before the work goes out to tender. This process is time-consuming if it is to be done properly. It is not suitable where time constraints require tenders to be received at a time which would not allow sufficient time for the design to be properly completed and a full bill of quantities prepared. Delays and additional cost may be incurred during the construction period as a result of the architect's drawings not being issued to the contractor on time. Under the circumstances a design and build would have been more suitable.

1.8.3. The appropriate procurement method was the issue in the case of *Plymouth and South West Co-operative Society Ltd* v. *Architecture Structure and Management Ltd* (2006). Plymouth and South West Co-operative Society Ltd (Plymco) wished to develop its flagship store, including the construction of a number of retail units at Derry Cross, Plymouth. Plymco appointed Architecture Structure and Management Ltd (ASM) to undertake the necessary architectural, engineering and quantity surveying services. It was a priority that the cost of the scheme did not exceed £5.5m. ASM produced a budget in the sum of £5.65m and was instructed to make savings to ensure the price fell within the budget. Plymco's board decided to go ahead with the scheme in April 1996 and ASM was appointed shortly thereafter. On 10 October 1996 an agreement for lease was signed by Plymco with Argos, which provided for the completion of the Argos works by 21 April 1997. At this stage Argos was the only tenant Plymco had secured. It was anticipated that the building contract would be let by July or August 1996. One of the problems associated with the scheme was that it was a requirement that the store remained open for business during the construction of the works.

1.8.4. ASM advised letting the building contract by means of a two-stage tender process using the National Joint Consultative Committee for Building's Code, with a view to entering into a JCT 1980 With Approximate Quantities Form of Contract. Competitive tenders were received and Exeter Building Company (EBC) was selected in late October 1996. The contract, however, was not signed until January 1997. The contract sum was £5,036,061; however, due to the tight timescale, 87% of the approximate bill of quantities was provisional and described as 'not detailed save in outline'. The work was completed on time, but the final account was in the region of £7.8 million and, because of the high volume of provisional work, included 7,500 variations.

1.8.5. It was alleged by Plymco that £2 million of the overspend resulted from the procurement method recommended by ASM, which made it impossible to operate effective cost control. It was alleged that ASM should have advised that the work be carried out in two distinct phases, referred to as the Argos first solution: the first phase to comprise the work for Argos, followed by the remainder of the work.

1.8.6. The court held that ASM's overriding obligation was to ensure that the cost of the work did not exceed £5.5 million. ASM had a duty to advise Plymco on the most suitable method of procurement to achieve completion within the financial ceiling. It was the court's view that ASM should also have advised as to what decisions were required to be taken by Plymco and the dates by which they were to be made, but had failed to do so. It was the view of the court that ASM should have advised Plymco to have work carried out in two phases, which would have resulted in a later completion of the works, but with cost certainty. The court was convinced that Plymco would have accepted this advice.

1.8.7. The court experienced some difficulty in arriving at a sum to be paid by ASM to Plymco. The documents which would have greatly assisted in proving Plymco's cost entitlement should have been retained by ASM but were not. The quantity surveying experts retained by each side reached agreement on the basis of comparing what the scheme would have cost, had the Argos first approach been used, with the actual final cost. The difference, which amounted to £1.3m, ASM was obliged to pay to Plymco.

1.8.8. Whilst this case went against the architect, it is often very difficult to demonstrate that, if a different method of procurement had been employed from the one advised, the costs incurred would have been less. In the Plymco case, no doubt ASM was under great pressure to secure the completion of all the work by 21 April 1997. It is easy for the judge in hindsight to say that if ASM had suggested the Argos first option, it would have been accepted by Plymco. ASM however appears to have badly managed the process of securing decisions from Plymco, which were essential for completion of the project, resulting no doubt in delay and additional cost.

1.8.9. There are examples of cases being brought by contractors against professional consultants they have engaged in compiling their tenders. In the case of *Copthorne Hotel* v. *Arup and Associates* (1996) the pre-tender assessment of piling costs was half the actual costs incurred by the contractor, but negligence was not established. In a case relating to advice provided by professional consultants in assisting contractors to secure contracts, the contractor must be able to show that it relied upon the information provided by the consultant, which they often are unable to demonstrate. In *Gable House Estates* v. *Halpern Partnership* (1996) it was shown that the employer would have taken a course of action regardless of the consultant's advice, which meant there was no loss involved.

1.8.10. In cases of this kind, expert evidence plays a large part. In the Plymco case, experts appointed for both sides were of the opinion that, if cost certainty was the objective and if Plymco lacked sufficient experience of this kind of work and the need to make timely decisions, then ASM had not performed its duties in the appropriate manner. The experts, once the court had decided on liability, agreed the *quantum*.

SUMMARY

There will usually be some difficulty in successfully bringing an action against a professional consultant for advising the use of the wrong procurement method. Where it seems apparent that this is the case, it is essential for the employer to be able to demonstrate that additional cost has been incurred as a result of the advice given. In the Plymco case it is clear that, due to the tight timescale, the procurement method recommended by the architect was inappropriate. The court was convinced that the cost of the scheme was the most critical factor and not the time for completion. In the court's view, the architect should have advised the use of a procurement method which would have been more cost certain, but would have resulted in a much later completion date. The judge was of the opinion that it would have been acceptable to the client if given the option at the outset. However, this was very much a matter of speculation on his part.

1.9. Where an unsuccessful tenderer is prevented from adjusting its tender after it has been submitted but before the deadline for submission of tenders has arrived, is the tenderer entitled to compensation?

1.9.1. It is not uncommon for a tender to be submitted and subsequently an error discovered by the tendering organisation, which may or may not have an effect upon the price. Sometimes the error is discovered after the deadline for submission of tenders has passed. However, there are occasions when the error is discovered before the closing date. What are the options available to the tenderer? It is always open to the tenderer to withdraw its tender. Alternatively, a polite request to adjust the tender may be a favoured option. If a request made for an alteration to be made is refused, does the tenderer have any entitlement to compensation?

1.9.2. There are no hard and fast rules concerning this matter. The Public Contracts Regulations 2006, which apply only in the public sector, require all tenderers to be treated equally and in a non-discriminatory manner. This requirement is unlikely to be of much assistance. Often, tender enquiry documents stipulate that where a genuine error has occurred, a tenderer may amend its tender prior to the deadline. To allow a tenderer to adjust its tender, however, could, in certain cases, lead to abuse. A tenderer, having heard details of its competitors' bid prices, may decide it would be in its interest to alter its own price if it were permitted.

1.9.3. The request to amend a tender may not affect the price, but may relate to supporting documentation which is intended to support the bid. For example, tenderers may be required to submit their health and safety records with their bid, but, because of an error, the documentation was omitted.

1.9.4. Where the tender enquiry stipulates that genuine errors may be corrected prior to the latest time for receipt of tenders, there should be no problem in making amendments. What rights, if any, do tenderers have for amending tenders to correct errors when the tender enquiry documents are silent on the matter? There is very little authority which relates to this aspect of tendering.

1.9.5. The case of *J B Leadbetter and Co* v. *Devon County Council* (2009) involved a dispute which arose in connection with the award of a four-year framework agreement in Devon. The tender process was governed by the Public Contracts Regulations 2006. Tenders were to be uploaded electronically to a dedicated website. Only one upload was allowed, as the electronic system had been designed so as not to be capable of accepting additional information, to prevent collusion. An integral part of the tender process was the completion of four case study templates. One of the bidders omitted to upload the case studies with its tender and was allowed to send the documents in hard copy, which they accomplished before the deadline. The claimants accidentally omitted the case studies element when uploading their bid, and as a result they tried to re-upload again, 15 minutes prior to the tender deadline of 3.00 pm, but were unsuccessful as the system allowed only one upload. They sent an email with the case studies attached, but this occurred at 3.26 pm, which was too late. The tender was rejected by the defendant.

1.9.6. The case turned upon whether, in rejecting the tender, the defendant was in breach of regulation No 4 of the Public Contracts Regulations 2006, which required them to treat tenderers equally and in a non-discriminatory way. In addition, it was argued that the defendant, as a general principle of community law, owed the claimant an obligation to act proportionately in relation to the tender, and it had been in breach of that obligation. The specific wording in the Invitation to Tender stated:

> Should a material and genuine error be discovered in the tenderer's submission during the evaluation period by the tender evaluation team, the tenderer will be given the opportunity of confirming their offer or of amending it to correct the error.

1.9.7. It was held that the defendant had not been in breach of the duty of equality and non-discrimination. The court found that it was wrong to describe the claimant's tender as submitted before the deadline as containing an error. It was substantially incomplete, by reason of the omission of the case studies. The wording of the tender enquiry neither obviated the need to submit a complete tender, nor provided a means by which tenderers could supply substantial documents, or substantial sections of documents, after the deadline, so as to complete their tenders.

1.9.8. Proportionality was capable of applying to the implementation of the terms of a procurement process. The exercise of discretionary powers necessarily involved judgment and the court would not intervene unless the decision was unjustifiable. In this case the court considered that it would not intervene in respect of the provisions included in the invitation to tender. There might be circumstances where proportionality would, in exceptional circumstances, require the acceptance of a late submission of the whole, or significant portions, of a tender, most obviously where there was an error on the part of the procuring authority. Generally, even if there is a discretion to accept late submissions, there is no requirement to do so, particularly where, as in this case, it results from a fault on the part of the tenderer.

1.9.9. This case does not specifically deal with the matter relating to an error which the tenderer wishes to correct. The court was of the opinion that it was not a case of correcting an error but a straightforward late submission. This ensured that the court did not have to consider the provision in the tender enquiry document, which allowed genuine errors to be corrected in the period when tenders were being evaluated.

1.9.10. There is no hard and fast rule as to whether genuine errors can be corrected prior to the deadline as of right. In the absence of an express clause in the tender enquiry document to the effect that errors either can or cannot be amended, we will have to await a case on the matter. The court would have to recognise an implied term to this effect if such an entitlement were to exist.

1.9.11. If the employer is in breach of an expressed obligation to allow tenders containing genuine errors to be corrected, or in breach of an implied term, the disadvantaged tenderer would be entitled to recover financial damages in respect of the breach. How they would be calculated depends upon the circumstances. If it could be shown that the disadvantaged tenderer would have been awarded the project, the damages would be based upon wasted tendering costs and loss of profit. If there was uncertainty as to whether the disadvantaged tenderer would have been successful, then the damages would be based on loss of opportunity of making a profit. This figure would be based on how much profit the tenderer would have made on the project, but heavily discounted to take into account the chances of being successful.

SUMMARY

There is little in the form of case law which deals with this problem. What little case law exists demonstrates that courts have little sympathy with tenderers who wish to amend their tenders due to their own shortcomings.

To allow a tenderer to adjust a tender price after submission, due to an error, may lead to abuse of the system, particularly where information concerning prices submitted by competitors becomes known to other tenderers. There seems little to fear of problems arising by allowing tenders to be amended where an error occurred which does not affect price.

Some tender enquiry documents include a specific clause which allows genuine errors to be corrected. In the absence of such a clause, it would require a court to accept that there was an implied term to the effect that errors were capable of being corrected. So far, there is no record of any court recognising such a right.

1.10. What is the difference between Management Contracting and Construction Management?

Management Contracting

1.10.1. Management contracting is appropriate for large-scale projects requiring an early start on site. The design is undertaken on behalf of the employer and this procurement route is ideal where work needs to be started before the design on the project is completed. It therefore is of great assistance where the period available up to completion is restricted. This procurement route is also suitable for projects where the design is sophisticated or innovative, requiring proprietary systems or components designed by specialists.

1.10.2. The management contractor does not carry out any construction work, but manages the project on behalf of the client. It is a cost-reimbursable form of contract, with the management contractor being paid a fee. All the work is undertaken by subcontractors, referred to as works contractors, in distinct works packages, employed by the management contractor. A cost plan is produced at an early stage based upon estimates of the works packages, plus preliminaries and the management contractor's fee.

1.10.3. The employer appoints the architect or contract administrator, CDM co-ordinator, quantity surveyor and any other consultant who may be required. There are two distinct time periods involved. During the first period, which is the preconstruction period, the management contractor should be appointed as early as possible to enable it to have an input into such matters as the design of the project, in particular the buildability aspect; health and safety matters; preparation of the budgets for works packages; and the programme. The fee to be paid to the management contractor is usually agreed at the outset of the preconstruction period.

1.10.4. During the construction period the works packages are put together by the management contractor in conjunction with the employer's professional team. The management contractor will be required to manage, organise and supervise the works contractors, to ensure that the work is carried out in accordance with the requirements of the contract and completed on time.

1.10.5. The management contractor is paid the final cost of all the works packages plus any preliminaries and the fee. The fee is usually a lump sum, as paying the fee as a percentage of the total of the works package is not conducive to keeping the works package costs to a minimum.

1.10.6. The most commonly used standard form of management contract is the JCT Management Building Contract. It is an important concept of this management contract that the consequences of any default on the part of any works contractor do not fall on the management contractor. The management contractor is required to ensure that the work is carried out without defects and on time. However, this requirement does not bite if the only reason for a breach of the obligation is a breach of the works contract by a works contractor. There is no such comfort offered to the management contractor by the Engineering and Construction Contract (NEC 3) Option F Management Contract, which makes it clear that the management contractor is responsible for all work undertaken by the subcontractors.

1.10.7. This type of procurement method is generally regarded as low financial risk from the management contractor's point of view. The client, however, is at greater risk financially than would be the case with a traditional procurement route, where the contractor works for a pre-determined lump sum. Employers can be caught out where the contract runs over a long period and unexpected inflation takes place, which results in the final cost of the project exceeding the cost plan.

Construction Management

1.10.8. Construction management offers an alternative to management contracting and in like manner is suitable for large projects where an early start on site is required. The major difference between construction management and management contracting is that the construction manager acts solely as a manager and is not in contract with the trade contractors, who undertake all of the work.

1.10.9. An additional difference between management contracting and construction management is that the construction manager is a first appointment and will be responsible for selecting the design team, even if they are in contract with the employer.

1.10.10. The employer enters into separate trade contracts with each of the trade contractors who will be carrying out the work. The JCT has produced a standard Construction

Management Appointment and a standard Construction Management Trade Contract for use on construction management projects.

1.10.11. The construction manager acts as an agent on behalf of the employer and manages the trade contractors' work and also the design. It is usually advisable for the employer to select and appoint its own quantity surveyor, who will act independently from the construction manager, to ensure that impartial cost advice is being provided.

1.10.12. In like manner to the management contract, the construction management contract is a cost-reimbursable contract, with the construction manager being paid a fee.

1.10.13. The downside of this procurement route is that, if there is a serious dispute between the construction manager and a trade contractor which cannot be amicably resolved, any formal proceeding must be commenced by the employer against the trade contractor.

1.10.14. In like manner to management contracting, construction management is low financial risk from the construction management constructor's point of view. The financial risk for the employer arising from the two procurement routes is also the same.

SUMMARY

Management contracting and construction management are both suitable for large projects where an early start on site is required. These are cost-reimbursable contracts, with the management contractor and the construction manager being paid a fee. The major difference between the two procurement systems is that in the case of management contracting, all the work is undertaken by works contractors who are subcontracted to the management contractor, whereas with regard to construction management, all the work is carried out by trade contractors who are all contracted to the employer. Both methods are low financial risk from the point of view of the management contractor and construction management contractor, but in times of unpredictable high inflation, the final cost paid by the employer often exceeds the cost plan.

1.11. A public sector project is advertised and tenders invited. Within the advertisement, it is stated that selection will be on the basis of the most advantageous submission. After tenders have been submitted, selection is made employing an evaluation method which has not been revealed to the tenderers. Would the unsuccessful tenderers have any entitlement to compensation and on what basis?

1.11.1. There is a requirement under European law, embedded in the laws applicable in the UK, which applies to most public bodies and publicly funded organisations and provides for equal treatment of all tenderers. Where appointments are based upon the most economically advantageous bid, the contracting authority must specify which criteria from a specified list it will use to determine the most economically advantageous. The principle is designed to be transparent and objective. Unsuccessful tenderers are entitled, where practical, to be informed of their score and the reasons why the successful tenderer was preferred.

1.11.2. Contractors who are not appointed and consider that the regulations have not been followed may challenge the award. There is a 10-day standstill period if the challenge is made electronically and 15 days, if made otherwise. Contracts cannot be entered into during this period.

1.11.3. There have been a number of court cases brought by unsuccessful contractors on the basis that the authority has failed to comply with the regulations. Courts have the power to set aside awards, which can involve expensive re-tendering and/or award damages to unsuccessful tenderers. Often the court will make a decision which is for the award damages to be assessed, leaving the parties to agree the *quantum*. If agreement cannot be reached, the parties would then normally revert back to the court for a decision.

1.11.4. Legal cases where these matters have been the subject of the dispute include:

1.0. *Aquatron Marine* v. *Strathclyde Fire Board* (2007)
This dispute related to a contract for the repair and maintenance of breathing apparatus. The contractor was awarded £110,000, based upon loss of profit, because the authority used different criteria in the evaluation from what appeared in the invitation to tender, which appeared in the *OJEU*.

2.0. *Henry Brothers (Magherafelt) Ltd and Others* v. *Department of Education Northern Ireland* (2007)
The work for which tenders were received involved the provision of major construction works relating to the modernisation of schools. Several unsuccessful contractors brought an action relating to the methods used in the selection process. It was argued that the price evaluation, based merely on a percentage to be added to the prime cost for overheads and profit, was unfair, as it failed to take account of efficiency levels. The court rejected the claim on the basis that the evaluation was transparent, fair and without discrimination.

3.0. *Lettings International* v. *London Borough of Newham* (2008)
The requirements of the authority were for management and other services related to private sector lettings. The authority was held to be at fault for not disclosing in the *OJEU* advertisement that the bids were to be evaluated using weightings and sub-criteria which were not stated.

4.0. *McLaughlin and Harvey* v. *Department of Finance and Personnel Northern Ireland* (2008)
In this case, the method of selection was not disclosed prior to receipt of tenders. It was held by the court that this was not transparent and therefore unfair.

5.0. *McConnell Archive Storage Ltd* v. *Belfast City Council* (2008)
The case related to the selection of a company to undertake a document storage and retrieval service. Following the evaluation process, McConnell was advised that it was the successful bidder. Subsequently, at a debriefing of Morgan, a rival bidder, it became clear that the spreadsheet used in the evaluation process contained an arithmetical error, which, if adjusted, would make Morgan the winner.

Morgan was then appointed, which was contested by McConnell. The court rejected the submission on the grounds that the notification to McConnell was not a contract and could subsequently be changed.

6.0. *Emm G Lianakis AE and Others* v. *Dimos Alexandroupolis and Others* (2008)
This case was heard before the European Court of Justice and involved the Municipal Council of Alexandroupolis, which had sought bids for open planning services. The authority was held liable for introducing the weighting factors and sub-criteria after submission of tenders. This information should have been made known at the outset.

7.0. *Sita UK Ltd* v. *Greater Manchester Waste Disposal Authority* (2010)
Dissatisfied bidders who wish to contest an award must do so under the Regulations within three months. The contract was place by GMWDA for a waste disposal project with VL on 8 April 2009. To comply with the Regulations, Sita should have commenced an action within three months of that date. Sita did not commence proceedings until 27 August 2009, which the court held to be out of time.

8.0. *Mears Ltd* v. *Leeds City Council* (2011)
This case involved capital improvement and refurbishment work for social housing in the Leeds area. Mears claimed that Leeds had been in breach of the Regulations in that they failed to act transparently. It was alleged that Leeds had issued changes to the pricing aspects of the Online Solutions Submission after receiving the tenders. In making an application for an injunction to prevent Leeds from entering into a contract with another bidder, Mears applied for disclosure of certain documents, including the model answers used by those who evaluated the tenders after submission. The court considered that the model answers should be disclosed as being necessary for disposing fairly of the proceedings and determining whether there were criteria, sub-criteria or weightings which had not been made available to tenderers.

9.0. *J Varney and Sons Waste Management Ltd* v. *Hertfordshire County Council* (2011)
In this case Varney tendered unsuccessfully for the operation of 18 Household Waste Recycling Centres in Hertfordshire. It was alleged that the council applied criteria, sub-criteria and weightings which were inconsistent with the information which it had disclosed. In the invitation to tender there was a statement to the effect that the staffing levels proposed by the tenderers would play a significant part in the evaluation of tenders. In submitting its tender, Varney had included for supplying high levels of good-quality staff for each site. When it came to evaluating tenders, staffing levels were given very little significance by the council. The Court of Appeal, however, found against Varney. It was held that it was made clear to tenderers that the basis of the award would be customer satisfaction and price.

10.0. *Traffic Signs and Equipment Ltd* v. *Department for Regional Development and Dept. of Finance and Personnel* (2011)
A decision to award a contract using assessment criteria where 40% of the marks were allocated to quality was found to be unlawful on the basis that the allocation could not be justified.

SUMMARY

There is a requirement under European law, which is embedded in UK laws which apply to most public bodies and publicly funded organisations and provides for equal treatment of all tenderers. Where appointments are based upon the most economically advantageous bid, the contracting authority must specify which criteria from a specified list it will use to determine the most economically advantageous bid. The principle is designed to make the selection process transparent and objective. Unsuccessful tenderers are entitled, where practical, to be informed of their score and the reasons why the successful tenderer was preferred.

Contractors who have unsuccessfully bid for projects covered by the Regulations may wish to consider what actions they may take. Courts now have a number of options they can adopt if a public body fails to comply with the Regulations. Unsuccessful contractors are entitled to be awarded damages where this occurs. If the contractor would have been awarded the contract, had the correct process been followed, as was the situation in the case of *Aquatron Marine* v. *Strathclyde Fire Board* (2007), an award of loss of profit would be appropriate. If there is no certainty that the claimant would have been awarded the contract, then any award would be based upon loss of opportunity which amounts to a discounted profit loss. The court also has power to order the tender process to be repeated.

Chapter 2

Tenders and Bidding

2.1. What is meant by the Battle of the Forms?

2.1.1. The construction industry has a tradition of formal contracts being drawn up and signed before work commences. The use of standard forms of contract such as the JCT, ICE, FIDIC and NEC are examples of formal contracts regularly employed for the purpose. Legally binding contracts, however, may come into being employing less formal means. What is required is an agreement made by the parties involved. This may occur when an unambiguous offer is unconditionally accepted. A further ingredient required is consideration, which means that each party has to contribute something of value which is of benefit to the other party or parties. It is also necessary for the terms under which the work will be carried out to be agreed.

2.1.2. In the construction industry, it is quite common for a supplier of materials, or a subcontractor, to submit a quotation for the supply of goods and/or services. For a binding contract to come into play, what is required is an unconditional acceptance from the contractor. There is no need for a formal contract to be drawn up for the agreement to be binding. Where an invitation to tender is involved, this does not usually form part of the contract, although the intention is that the quotation falls into line with the invitation to tender. If the supplier or subcontractor submits a quotation which does not mirror the invitation to tender, but which is unconditionally accepted by the contractor, the contract will be based upon the quotation and acceptance. The status of the invitation to tender is merely what it purports to be, an invitation to the supplier or subcontractor to submit a quotation.

2.1.3. A struggle often occurs between prospective parties to a contract when conflicting sets of conditions are submitted by the parties to each other; which set of conflicting conditions of contract are to apply? In addition to the standard forms of contract, some commercial organisations will have their own conditions which they will normally wish to use, whilst others will have a set of preferred alterations to the industry standard forms for use as the basis of the contract. What often happens is that each party tries to impose its preferred terms upon the other. This is referred to as the battle of the forms. The classic case which deals with the battle of the forms is *Butler Machine Tools* v. *Ex-cell-o Corporation* (1979), where the plaintiff submitted a quotation for the supply of machine tools, which included terms and conditions stating that the prices quoted

200 Contractual Problems and their Solutions, Third Edition. Roger Knowles.

were variable dependent on inflation. The order, when it arrived, required a fixed price and had a tear-off slip to be signed by the plaintiff as acknowledging receipt of the order and agreeing to the terms included therein. An acknowledgement of order was sent, but the plaintiff, in an accompanying letter, made it clear that the supply of the machine tools would be on the basis of the terms included in the quotation, i.e. a fluctuating price. The quotation and letter accompanying the acknowledgement of the order were on the basis of a fluctuating price, whereas the order and signed acknowledgement referred to a fixed price. The court preferred to consider the order and the signed acknowledgment as being an offer and acceptance and hence the basis of the contract. The plaintiff unfortunately submitted a quotation based upon a fluctuating price, but was obliged to supply the machine tools at the price quoted, but without the benefit of it being subject to inflation. Lord Denning explained the law in the following terms:

> It will be found that in most cases where there is a battle of the forms there is a contract as soon as the last of the forms is sent and received without objection being taken to it . . . the difficulty is to decide which form, or which part or parts of which form is a term or condition of the contract. In some cases the battle is won by the man who fires the last shot.

2.1.4. The subsequent case of *Chichester Joinery Ltd* v. *John Mowlem* (1987) illustrates the point that where there are conflicting conditions of contract, the party who submits its conditions last will usually triumph. The facts relating to this case are:

- Chichester submitted a quotation to Mowlem in November 1984, based upon its own terms and conditions which accompanied the quotation.
- Mowlem sent to Chichester a pro-forma enquiry form which referred to conditions on its reverse, but no conditions appeared. This was probably due to the document being a photocopy.
- The parties discussed the proposed contract at two subsequent meetings.
- On 14 March, Mowlem sent to Chichester a purchase order which stated that the terms and conditions of the purchase order, as set out on the reverse side, are expressly declared to apply to the purchase order. It requested Chichester to sign the acceptance of the order and return it within seven days. Chichester did not sign and return the acceptance.
- On 30 April 1985, Chichester, before commencing deliveries, sent a printed document to Mowlem headed 'Acknowledgement of Order'. The acknowledgement was stated to be 'subject to the conditions overleaf'.

2.1.5. The court held that Chichester's conditions applied to the contract. The original quotation sent by Chichester in November 1984 was an offer. Mowlem did not accept the offer, but instead sent a purchase order which constituted a counter-offer. This counter-offer was not accepted by Chichester, who sent an acknowledgement on 30 April 1985 which amounted to a counter-counter-offer. This was accepted by Mowlem by taking delivery of the joinery without in any way contesting the terms in Chichester's acknowledgement. This is referred to as acceptance by conduct.

SUMMARY

A struggle often occurs between prospective parties to a contract, as to which conditions of contract are to apply. In addition to the standard forms of contract, most commercial organisations will have their own conditions they will normally wish to use, whilst others will have preferred alterations to standard industry forms for use as the basis of the contract. What often happens is that each party tries to impose its preferred terms upon the other. The process may involve a quotation being submitted by one of the parties, which refers to its preferred conditions of contract. If the quotation is unconditionally accepted, the conditions included in the quotation will prevail. Alternatively the quotation may receive, by way of response, an order based upon differing conditions of contract usually referred to as a counter-offer. If there is no adverse response to the order, but work is allowed by both parties to get under way, the terms in the order will prevail. This is referred to as the 'battle of the forms' and, usually, the party who fires the last shot wins the battle.

2.2. If a tender which contains an error is accepted in full knowledge of the error, has the tenderer any redress?

2.2.1. A binding contract will come into being when an unambiguous offer receives an unconditional acceptance. Main contractors' quotations may be unconditionally accepted by building owners and quotations from subcontractors unconditionally accepted by main contractors, to form binding contracts. Where an error appears in a quotation which has been unconditionally accepted, the general rule is that errors cannot be corrected. Does this rule still apply if the person responsible for accepting the quotation is aware of the error at the time of the acceptance?

2.2.2. There is in law an ordinary commercial freedom or discretion to accept or reject a tender, or to negotiate with whoever seems best in the eyes of the person seeking tenders. However, business people are expected to act fairly and can be penalised for indulging in sharp practice. Where the dividing line comes between freedom to contract and a court's decision to intervene on the basis that the contract has come into being as a result of unfair sharp practice on the part of one of the parties is hard to define.

2.2.3. In the case of *Traditional Structures Ltd* v. *HW Construction Ltd* (2009), the claimant was requested by the defendant to provide a quotation for the steelwork and roof cladding at a site in Sutton Coldfield. A specification was provided along with some details concerning the roof cladding. Due to an error on the part of the estimator, the tender which was submitted to the defendant contained only the price for the steelwork; no price had been included for the roof cladding. The following wording appeared in the quotation:

> For the supply and delivery of structural steelwork and claddings erected onto prepared foundations (by others) to form the proposed buildings as detailed above our budget price would be

Steelwork: £37, 573.43 + VAT

Terms: The above prices are net

2.2.4. The papers in the claimant's files made it clear that the price for the roof cladding was not included in the price for the steelwork. The defendant's managing director indicated that as far as he was concerned, the price for the steelwork included the roof cladding. The matter of whether the price included roof cladding was never raised by the defendant, despite the numerous meetings which took place between them during the period from the date of submission of the quotation to the date of the acceptance. In addition, the defendant in an email asked the claimant how long the £37,573.43 plus VAT 'for the floor support beams and roof structure was to remain open'. This was taken to mean the steelwork package without the roof cladding.

2.2.5. The problem became clear when the claimant made an application for payment for the roof cladding and the defendant refused to pay, on the grounds that the price for roof cladding was included in the price for steelwork. The matter was not resolved and was referred to the courts. In finding in favour of the claimant, the judge was convinced that the defendant's managing director was aware that the price for roof cladding had been missed out of the quotation. The managing director failed to ask the questions that an honest man would have asked.

2.2.6. The judge ordered that the contract be rectified to include the price for the roof cladding. He indicated that he considered that the defendant knew about the mistake and had wilfully and recklessly failed to make the sort of enquiries about the roof cladding, prior to accepting the quotation, which a reasonable and honest man would have made.

SUMMARY

Where a quotation containing an error is unconditionally accepted, the general rule is that the tenderer cannot adjust the error. If the error results in a loss, then the tenderer will be required to sustain the loss. Where, however, a quotation is submitted and the recipient is aware that it contains an error, a different result can apply. If the court is satisfied that the person receiving the quotation is aware of the error before acceptance and fails to do what an honest person would do, then the court will normally order rectification of the contract to eliminate the error.

2.3. Where a contractor or subcontractor submits a tender with its own conditions of contract attached, which are neither accepted nor rejected, do these conditions apply if the work is allowed to proceed?

2.3.1. The general rule is that silence never constitutes acceptance. This was well established in the case of *Felthouse* v. *Brindley* (1862), where an uncle, in a letter, offered to buy a horse from his nephew, for £30.15, adding, 'If I hear no more about him I shall consider the horse is mine for £30.15.' No reply was received by the uncle from his nephew. It was held that there was no contract and, therefore, the uncle did not gain ownership of the horse. The reason for the rule is that it is in general undesirable to impose on an offeree the trouble and expense of rejecting an offer which he does not wish to accept.

2.3.2. A recipient of an offer who remains silent can still consent to the offer by his conduct, showing an intention to accept the terms of the offer. An old example of acceptance by conduct is well illustrated in the case of *Brogden* v. *Metropolitan Railway Co* (1877). A railway company submitted to a merchant a draft agreement for the supply of coal. He returned it, marked 'Approved', but also made a number of alterations, to which the railway company did not express assent. Nevertheless, the company accepted deliveries of coal under the draft agreement for two years. It was held that, once the company began to accept these deliveries, there was a contract on the terms of the draft agreement as amended.

2.3.3. Applying this principle to modern procurement methods, a main contractor who, having received a quotation from a subcontractor, allows the subcontractor to commence work without comment, may by its conduct be said to have accepted the quotation.

2.3.4. An example of acceptance by conduct on a construction project is illustrated by the case of *Jean Shaw* v. *James Scott Builders Co* (2010). This case involved the formation of a contract based upon an offer and acceptance by conduct. The Shaws employed the services of an architect, Mr Peter White, to design a new house. The defendant, Scott, was engaged to build the house. A draft contract was produced, but no formal contract. Work commenced, but the project got into difficulties and the architect was replaced by Mr Grime, who appointed Mr Percy, a quantity surveyor, to help sort out matters.

2.3.5. Mr Percy drew up a formal contract and emailed copies to Mr Shaw, Mr Grime and Scott on 23 November 2005. In the email, Mr Percy stated that there was some uncertainty as to the drawings which had been used to produce the original price. Mr Percy asked the parties to let him know if there was any objection or disagreement by close of business on 28 November 2005. Mr Shaw accepted the document, but Scott did not respond. Having heard nothing from Scott, Mr Percy wrote to Mr Shaw to advise him that the document represented a contract between him and Scott.

2.3.6. The work continued until January 2007, when disagreements between the parties began to surface. Mr Shaw wrote to Scott instructing that work should stop. Scott submitted a loss and expense claim, which was disputed. A disagreement occurred as to what constituted the contract. It was held by the court that Mr Percy's contract documents had expanded on the parties' pre-existing agreement. The court held that a party's silence does not imply consent to an offer. However, the surrounding circumstances might lead a court to infer that a party's silence is acceptance by conduct, particularly where the parties have been in negotiations. It concluded that despite Scott's silence when receiving the email from Mr Percy dated 23 November 2005, it should be treated as representing the agreed terms of the contract.

SUMMARY

It is fairly unusual for tenders to be submitted without there being any form of response. The recipient of the tender, however, may be satisfied with the offer and not bother to send a formal communication of acceptance. An instruction to commence work may then follow. The general rule is that silence does not constitute acceptance, and if this rule applied to this situation there would be no contract. However, courts are ready to recognise that even though silence reigns, acceptance by conduct may have occurred. A contractor, on receipt of a subcontractor's quotation, may without accepting or rejecting the quotation, instruct that work be started. This instruction would amount to an acceptance by conduct of the quotation.

2.4. The submission of an unambiguous quotation which receives an unconditional acceptance can normally form the basis of a legally binding contract. If, following the commencement of work, a formal contract is signed which contains conditions which are at variance with those referred to in the quotation and acceptance, which of the competing conditions apply to the work: those in the signed formal contract, or those referred to in the offer and acceptance?

2.4.1. It is not uncommon for work to commence before the contract has been drawn up and signed. Letters of intent are sometimes used as a mechanism for enabling a start to be made on site before a formal contract has been drawn up and signed. Where negotiations are still proceeding and there is no evidence of the existence of a contract, then a formal contract drawn up and signed after work has commenced will normally have retrospective effect. All the work undertaken from the start of the project will then be governed by the terms and conditions of the signed contract, the reason being that this is what the parties expect to happen. In the case of *Trollope and Colls* v. *Atomic Power Construction* (1963), the plaintiff submitted a tender for carrying out certain civil engineering work as subcontractor to the defendant. Negotiations were under way and work started in June 1959 on the basis of a letter of intent. Work continued and the formal contract was finally drawn up and signed on 11 April 1960. A dispute arose between the parties in respect of payment for variations. The matter in dispute was whether the conditions contained in the formal contract governed the work carried out before 11 April 1960.

2.4.2. Mr Justice Megaw, in finding that the contract had retrospective effect and therefore applied from the date work commenced, said:

> So far as I am aware there is no principle of English law which provides that a contract cannot in any circumstances have retrospective effect . . .

2.4.3. It is clear from this case that if there is no contract concluded when work commences, there will be no difficulty in a contract subsequently drawn up and signed having retrospective effect. What is the position if the process of quotation and acceptance results in a legally binding contract coming into effect, but with terms and conditions which vary from those include in the subsequently signed formal contract? It may be that the parties at a later date decided that the terms and conditions referred to in the offer and acceptance process are inappropriate. Certain terms may change in the intervening period, for example the date for completion. There may be subsequent disputes, where the completion dates differ, due to the financial consequences of the changed completion date not being properly reflected in the formal contract. There is no reported case where this has been an issue. It will in all probability come down to the intentions of the parties. When all is said and done, the parties have willingly signed the formal contract and therefore it is not unreasonable to suppose that they were in agreement with all of its terms. If there is a dispute, the party who contends that the some of the conditions in the formal contract do not apply will have an uphill struggle to demonstrate, to the satisfaction of the court, that it was not the intention of the parties that the signed contract should supersede the quotation and acceptance contract.

SUMMARY

The case of *Tollope and Colls* v. *Atomic Power Construction* (1963) established that a contract when drawn up can have retrospective effect. If, therefore, work commences in accordance with an instruction contained in a letter of intent and a formal contract is subsequently signed, all the work will be subject to the terms and conditions contained in the formal contract. A more difficult situation may arise if, at the time work starts, there is a legally binding contract in place based upon a quotation being unconditionally accepted. If this is the case, the inference will be that the parties intended the terms and conditions included in the formal contract to supersede those referred to in the offer and acceptance. It will be for the party who disputes this to prove otherwise.

2.5. Where an employer includes with the tender enquiry documents a site survey which proves misleading, can this be the basis of a claim?

2.5.1. Problems often arise where unforeseen adverse ground conditions occur which add to the contractor's or subcontractor's costs. Who pays the bill? There is no obligation upon the employer to provide information concerning ground conditions. Most of the standard forms of contract make specific reference to the contractor satisfying himself as to the ground conditions he may encounter. If the contract is silent as to the ground conditions, the contractor will normally be deemed to have taken the risk.

2.5.2. Disputes may arise where the employer provides details of the ground conditions which prove to be inaccurate or misleading. Employers like to ensure that they do not take the risk for inaccuracies in the ground information which may lead to the contractor submitting a claim. Some standard forms of contract, however, make it clear that the contractor, in addition to making its own inspection of the site, has based its tender on the information made available by the employer.

2.5.3. The ICE 6th Edition, clause 11(3), deals with the matter in precise terms where it states:

> The contractor shall be deemed to have
>
> (a) based his tender on the information made available by the Employer and on his own inspection and examination all as aforementioned.

It would seem that where the ICE conditions apply, the employer is taking the responsibility for the accuracy of the information he provides at tender stage. The ICE 7th Edition is worded in a slightly different manner, where it states in clause 11(3):

> The Contractor shall be deemed to have . . . based his tender on his own inspection and examination as aforesaid and on all information whether obtainable by him or made available by the Employer . . .

The intention of the revised wording is that the contractor's tender is deemed to be based upon not only information provided by the employer or as a result of his own inspection but also information derived from other sources. This could include information obtained from utilities such as water or gas companies.

2.5.4. The Engineering and Construction Contract (NEC 3) provides in clause 60.1 for the contractor to recover time and cost where physical conditions are encountered which the contractor would have judged at the contract date to have had a small chance of occurring. In judging physical conditions the contractor, in accordance with clause 60.2, is assumed to have taken into account the site information which will normally be provided by the employer.

2.5.5. GC/Works/1, condition 7(1), requires the contractor to have satisfied himself, among other matters, as to the nature of the soil and materials to be excavated. Condition 7(3) goes on to say that if ground conditions are encountered which the contractor did not know of and which he could not reasonably have foreseen (having regard to any information which he had or ought reasonably to have ascertained) he will become entitled to claim extra.

2.5.6. Under the Misrepresentation Act 1967 a subsoil survey which proves to be misleading, though innocently made, could give rise to a claim for damages. A good defence for the employer is, however, available if he can demonstrate that he had reasonable grounds to believe and did believe that the information contained in the subsoil survey was correct.

2.5.7. Employers often include with information provided to contractors at tender stage a statement to the effect that they will have no liability to the contractor if the information proves to be inaccurate and as a result the contractor incurs additional cost. A disclaimer, however, will be of no effect unless it can be shown to be fair and reasonable, having regard to the circumstances which were, or ought reasonably to have been, known or in the contemplation of the parties when the contract was made. Max Abrahamson, in his book *Engineering Law and the ICE Contracts*, says:

> The courts are obviously disinclined to allow a party to make a groundless misrepresentation without accepting liability for the consequences.

In *Howard Marine & Dredging Co Ltd* v. *A Ogden & Sons (Excavations) Ltd* (1978), owners of a barge were held liable to contractors who had hired the barge for construction works on the strength of a misrepresentation of the barge's deadweight, even though the charterparty stated that the charterers'

> acceptance of handing over the vessel shall be conclusive that they have examined the vessel and found her to be in all respects . . . fit for the intended and contemplated use by the charterers and in every other way satisfactory to them.

The defendant's marine manager had said at a meeting that the payload of the barge was 1,600 tonnes, whereas in fact it was only 1,055 tonnes. The mis-statement was based on the manager's recollection of a figure given in Lloyds' Register that was incorrect. He had at some time seen shipping documents which gave a more correct figure, but that had not registered in his mind.

2.5.8. In *Pearson and Son Ltd* v. *Dublin Corporation* (1907), the engineer had shown a wall on the contract drawings in a position which he knew was not correct. There was a clause in the contract to the effect that the contractor would satisfy himself as to the dimensions, levels and nature of all existing works and that the employer did not hold himself responsible for the accuracy of information given. Nevertheless, this was held to be no defence to an action for fraud. Alternatively, the contractor who is misled may have a remedy in tort for negligent mis-statements, on the principle of *Hedley Byrne & Co Ltd* v. *Heller & Partners* (1963).

2.5.9. The question of misrepresentation under the Misrepresentation Act 1967 and negligent mis-statement was the subject of the decision in *Turriff Ltd* v. *Welsh National Water Authority* (1979). An issue in the case related to an alleged misrepresentation in the specification, in which it was stated that satisfactory test units had already been carried out by Trocoll Industries. It was also stated that if competently incorporated into the works, the units could achieve a standard of accuracy within the desired tolerances and also that all the features of the units were proven. These representations were found by the court to be incorrect. The judge found that Turriff had relied upon the statements and had entered into the contract on the understanding that the representations were true. It was found by the court that the employer was unable to prove there were reasonable grounds for believing the representations to be true and therefore, in accordance with the Misrepresentation Act 1967, Turriff was entitled to the damages they had suffered as a result. The court also held that in so far as the representations contained in the specification were statements of opinion and not fact (and thus not within the ambit of misrepresentation), they in any event amounted to negligent mis-statements. In line with the decision in *Hedley Byrne & Co Ltd* v. *Heller & Partners Ltd* (1963), Turriff was entitled to damages.

2.5.10. It is common for contracts to state that the contractor is responsible for satisfying himself as to the conditions of the subsoil and that no claim will be accepted for failure to do so. The ICE 7th Edition states in clause 11 (2) that the contractor is deemed to have inspected and examined the ground and subsoil and hydrological conditions. This type of wording was examined in the Australian case of *Morrison Knudsen International Co Inc* v. *Commonwealth* (1972). The employer had provided the contractor with information concerning ground conditions, which did not form part of the contract. Unfortunately, it failed to disclose the presence of large quantities of cobbles in certain locations, which cost the contractor a significant sum of money to remove. The employer argued that a clause in the contract which required the contractor to satisfy himself of the ground conditions rendered the contractor's claim worthless. The court, however, disagreed in holding that the clauses did not protect the employer from liability for any misleading or erroneous content in the site information, supplied to the contractor prior to the submission of a tender.

2.5.11. Contractors who suffer from incorrect subsoil surveys may be able to demonstrate that the survey became a condition or warranty in the contract, as was the situation in the case of *Bacal Construction* v. *Northampton Development Corporation* (1975). The contractor had submitted as part of his tender sub-structure designs and detailed priced bills of quantities for six selected blocks in selected foundation conditions. The designs and the priced bills of quantities formed part of the contract documents by virtue of an express provision in the contract. The foundation designs had been prepared on the assumption that the soil conditions were as indicated on the relevant borehole data provided by the corporation. During the course of the work, tufa was discovered in several areas of the site, the presence of which required the foundations to be re-designed in those areas and additional works carried out. The contractor claimed that there had thereby been breach of an implied term or warranty by the corporation that the ground conditions would accord with the basis upon which the contractor had designed the foundations. They claimed they were entitled to be compensated by way of damages for breach of that term or warranty. The corporation denied liability, maintaining that no such term or warranty could be implied, but the court found in favour of the contractor.

2.5.12. The liability for inaccurate subsoil information provided by the employer at tender stage was an issue in *Co-Operative Insurance Society* v. *Henry Boot* (2002) in connection with a contract for the reconstruction of Lomond House in Glasgow. In this case the employer, the Co-Operative Insurance Society, had commissioned Terra Tek Ltd to undertake a subsoil survey before tenders were sought. During the excavation work, water and soil flooded into the sub-basement excavations. It was alleged by Henry Boot that the problem stemmed from the inaccuracy of the ground water levels as shown in the subsoil survey. It argued that the method used for dewatering would have been adequate if the information on the subsoil survey had been accurate. The court was asked to decide as a preliminary issue whether the employer was responsible for the information contained in the Terra Tek report. The contract bills, under a heading 'Site Investigation', stated that the report had been issued to the contractor with the tender documents. In addition, drawing A303 expressly stated that the piling design and specification were based upon the Terra Tek report and drawing 303A stated that for prevailing ground conditions refer to Site Investigation Report'. It looked to be game, set and match at this stage. The court took a different view from that expressed on behalf of Henry Boot. Even though the drawings were referred to as contract documents and contained reference to the Terra Tek report, the judge felt that the report itself was not a contract document and did not form a part of the contract. The judge considered that clause 2.2.2.4 in the contract was more persuasive. This clause stated that the contractor is deemed to have inspected and examined the site and its surroundings and to have satisfied himself as to the nature of the site, including the ground and subsoil, before submitting his tender. The judge may have been legally correct, but his decision defied the practicalities of preparing a tender and also common sense. Who would expect a contractor, provided at tender stage with a detailed subsoil survey, to visit the site of an existing building in a street in Glasgow and start digging up the ground to ensure that the information included in the Terra Tek report was correct? The conditions of contract used were the JCT conditions. If the conditions of contract had been ICE 7th Edition, then the decision might have been different. Clause 11(2) indicates that the contractor is deemed to have inspected and examined the site and satisfied himself as to the form and nature of the subsoil, but only in as far as is practicable and reasonable. This should be read with clause 11(3), which states that the contractor is deemed to have based his tender on his own inspection and all information, whether obtained by him or made available by the employer.

2.5.13. It would seem that where an employer provides a site survey to contractors at tender stage they are entitled to rely upon it when calculating their tender price. If the survey proves to be inaccurate and as a result the contractor incurs additional costs, he will usually be able to make out a good case for recovering those costs.

2.5.14. Many contracts are let, however, where the employer provides a subsoil survey but no specific reference is made to the employer taking responsibility. In many instances the specification will include a disclaimer. Despite silence in the contract, or even a disclaimer, if the information is incorrect because of fraud or recklessness by the employer, architect or engineer and the contractor suffers loss as a consequence, he may have a good case for recovering his additional costs. Employers who specifically exclude liability for information provided or who limit the liability for incorrect information will have to show, in accordance with the Unfair Contract Terms Act 1977, that the exclusion or limitation is reasonable. In any event, exclusion clauses will not relieve employers from the results of their negligence unless liability for negligence is expressly excluded.

2.5.15. It may be in the employer's interests, when providing a subsoil survey, to make it clear that the information is intended to show the ground conditions which occur at the location of the boreholes only and on the dates on which they were taken. The contractor should be expressly informed not to assume that the conditions apply anywhere else on the site or at any later period.

2.5.16. It is worth noting that in the case of *Railtrack plc* v. *Pearl Maintenance Services Ltd* (1995) it was held that, as the contract provided expressly for the contractor to ascertain the routes of existing services located below ground, he was liable for damage which occurred to underground services when the work was being carried out.

2.5.17. The decision in the USA case of *T.L. James and Co Inc* v. *Traylor Brothers Inc* (2002) drew attention to the pitfalls that contractors can encounter in not undertaking their own site investigations prior to submitting a tender. Traylor submitted the lowest bid for the construction of a marine terminal, which included dredging, driving piles and constructing a deck and terminal building on the piles. The tender documents warned that the site had 'numerous steel and timber piles removed to the approximate existing mudline'. The documents also advised that a more detailed map was available in the archive rooms of the Port of New Orleans. Bidders were also instructed to undertake a pre-bid site investigation. Traylor neither looked at the map nor undertook more than a cursory site investigation. James, Traylor's dredging subcontractor, encountered a large number of cut-off piles, and Traylor claimed compensation, arguing that the actual number and nature of the piles had not been disclosed and could not have been reasonably foreseen. The United States Court of Appeal, Fifth Circuit, rejected this. The real issue was whether the site conditions could have been anticipated based on the information available at the time the bid was submitted. The court concluded that the conditions could have been anticipated, but were not, because Traylor had failed to conduct an adequate site investigation.

2.5.18. Employers can often protect themselves from claims for additional payment resulting from incorrect information they supply to contractors prior to receipt of tenders by including a 'no reliance' clause in the contract. Such a clause provides that the contractor has not relied upon the information provided by the employer or its agent in entering into the contract. The clause will normally provide protection for the employer, whether or not the contractor has relied upon the information so provided. This type of clause has received legal backing by the courts in such cases as *Howard Marine and Dredging Co* v. *A Ogden and Sons* (1978) and *Emcor Drake and Scull* v. *Edinburgh Royal Venture* (2005), on the basis of evidential estoppel. Contractors may, however, resist such a clause on the basis that despite what the clause states, the employer believed that the contractor had in fact relied upon the information provided. In the case of *Watford Electronics Ltd* v. *Sanderson GFL* (2001), the judge said:

> ... It may be impossible for a party who has made representations, which he intended to be relied upon, to satisfy the court, that he entered into the contract in the belief that a statement by the other party, that he had not relied upon those representations, was true.

SUMMARY

The ICE 6th and 7th Editions conditions make it clear that the contractor's tender is deemed to have been based upon his own inspection of the site and any information provided by the employer. With this in focus, any additional costs resulting from unforeseen physical conditions will be recoverable. In the absence of specific wording in the conditions, the employer may wish to introduce a clause excluding liability, should the site survey prove inaccurate. Whilst it is feasible to exclude liability if for any reason, including negligence, the information proves inaccurate, such an exclusion would have to be reasonable. However, the few cases which deal with this matter tend to lead to the conclusion that exclusion clauses of this nature are unlikely to find favour with the courts.

Employers, however, may wish to include a non-reliance clause in the contract, which states that the contractor has not relied upon the information provided by the employer in putting together his tender. This type of clause, however, may fail if the employer in fact considered that the contractor had relied upon the information provided.

2.6. If, after tenders have been received, the employer decides not to proceed with the work, are there any circumstances under which the contractor/subcontractor can recover the costs associated with tendering or preparatory work, for which no instruction was given?

2.6.1. It is not unusual for contractors or subcontractors to carry out work prior to a contract being let. Often, this is done in contemplation of the contract being entered into to help the client, or to ensure a flying start or keep together key operatives. Sometimes, the contractor having made a start without the benefit of a contract, the employer decides not to proceed with the work. Contractors look to recover payment for their efforts and employers will usually deny any liability.

2.6.2. The case of *Regalian Properties plc* v. *London Docklands Development Corporation* (1991) dealt with this matter. Negotiations began in 1986 for the development of land for housing. A tender in the sum of £18.5m was submitted by Regalian for a licence to build when London Docklands obtained vacant possession. The offer was accepted 'subject to contract' and conditional upon detailed planning consent being obtained. Delays occurred in 1986 and 1987, because London Docklands requested new designs and detailed costings from Regalian. Because of a fall in property prices in 1988, the scheme became uneconomic, the contract was never concluded and the site never developed. Regalian claimed almost £3m which they had paid to their professional consultants in respect of the proposed development. The court rejected the claim. The reasoning was that, where negotiations intended to result in a contract are entered into on express terms that each party is free to withdraw from negotiations at any time, the costs of a party in preparing for the intended contract are incurred at its own risk and it is not entitled to recover them by way of restitution if for any reason no contract results. By the deliberate use of the words 'subject to contract', each party has accepted that if no contract were concluded any resultant loss should lie where it fell.

2.6.3. A different set of circumstances arose in the case of *Marston Construction Co Ltd* v. *Kigass Ltd* (1989). A factory belonging to Kigass was destroyed in a fire in August 1986. Kigass thought that the proceeds from its insurance policy would cover the costs of rebuilding and accordingly invited tenders for a design and build contract for this work. Marston submitted a tender. It was believed by Kigass that the terms of its insurance policy required that the rebuilding work be performed as quickly as possible, so Marston was invited to a meeting in December 1986 to discuss its tender. At this meeting, it was made clear to Marston that no contract would be concluded until the insurance money was available, but both Marston and Kigass firmly believed that the money would be paid and that the contract would go ahead. Marston received an assurance that it would be awarded the contract (subject to the insurance payment), but did not receive an assurance that it would be paid for the costs of the preparatory work. Marston carried out the preparatory work, but no contract was signed because the insurance money was insufficient to meet the costs of rebuilding. It was established that no contract between the parties existed and there was no express request for preparatory work to be carried out. However, the judge found in favour of the contractor on the basis that Kigass had expressly requested that a small amount of design work be carried out and that there was an implied request to carry out preparatory work in general.

2.6.4. A leading case on the subject matter is *William Lacey (Hounslow) Ltd* v. *Davis* (1957), where it was found that the contractor was entitled to be paid for preparatory work. The proper inference from the facts in this case is not that this work [i.e., the preparatory work] was done in the hope that this building might possibly be reconstructed and that the plaintiff company might obtain the contract, but that it was done under a mutual belief and understanding that this building was being reconstructed and that the plaintiff company would obtain the contract.

2.6.5. Payment is due in restitution if the contractor, in the absence of a contract, is instructed to carry out work. In the case of *British Steel Corporation* v. *Cleveland Bridge & Engineering Co Ltd* (1981), the judge held:

> Both parties confidently expected a formal contract to eventuate. In those circumstances, to expedite performance under the contract, one requested the other to expedite the contract work, and the other complied with that request. If thereafter, as anticipated, a contract was entered into, the work done as requested will be treated as having been performed under that contract; if, contrary to that expectation, no contract is entered into, then the performance of the work is not referable to any contract the terms of which can be ascertained, and the law simply imposes an obligation on the party who made the request to pay a reasonable sum for such work as has been done pursuant to that request, such an obligation sounding in quasi contract, or, as we now say, restitution.

2.6.6. A situation may occur where work is undertaken on a speculative basis – for example, preliminary design work – and it is agreed that payment will be made if the construction work on the project goes ahead. In the case of *Dinkha Latchen* v. *General Mediterranean Holdings SA* (2003) an architect undertook preliminary design work in connection with a hotel and also a tennis club, both of which were in Tangiers, on the basis that no payment would be made until a building permit was obtained. The work began to escalate, with the client instructing the architect to do a great deal of design work and to visit Tangiers to engage local architects. It was the view of the court that there must come a point in the relationship when each party, had they addressed the question, would have recognised that there was no longer any intention that further work would be unremunerated. It was held by the court that at some point between May 1994 and February 1995 the conduct of the parties was such as to give rise to an intention that any further work would be remunerated.

SUMMARY

The basic rule is that if a contractor carries out work prior to a contract being entered into he does so at his own risk. However, the court may order payment if it can be shown that the work was expressly requested or there was an implied requirement that the work be carried out. There may also be an obligation to make payment, if a benefit is derived from the work which was undertaken.

2.7. What is a tender contract and will it assist a contractor/subcontractor who submits a valid tender which is ignored in seeking compensation?

2.7.1. Contractors and subcontractors often fear that tenders they submitted were not considered. The reasons can be varied. For example, the successful bidder of an earlier stage of the work is almost certain to be awarded the work in a subsequent phase; or the employer may appoint a subsidiary who will do the work, alternative tenders being invited merely to put pressure on the subsidiary to reduce the price.

2.7.2. In *Blackpool & Fylde Aero Club Ltd* v. *Blackpool Borough Council* (1990), the council invited tenders for a concession to operate pleasure flights from the airport. Tenders were to be received by 12 noon on 17 March 1983. The letter box was supposed to be emptied by 12 noon each day, but this was not always the case. The club's tender, although delivered on time, was rejected as being late. The club maintained that the council had warranted that if a tender was returned by the deadline it would be considered and sought damages in contract for breach of that warranty, and in negligence for the breach of the duty that it claimed was owed. It was held that the form of the invitation to tender was such that, provided an invitee submitted his tender by the deadline, he was entitled, under contract, to be sure that his tender would be considered with the others. The court as a result found in favour or the aero club. A tenderer whose offer is in the correct form and submitted in time, is entitled, not as a matter of expectation, but of contractual right, to be sure that his tender will be opened and considered with all other conforming tenders, or at least that his tender will be considered if others are. This is sometimes referred to as a tender contract. A contractor or subcontractor whose properly submitted tender is not considered could levy a claim for damages, which would normally be the abortive tender costs. Where the tender was submitted on a design and construct basis, this could prove expensive. There may also be a possibility that the contractor or subcontractor could successfully frame a claim based upon loss of opportunity.

2.7.3. In the case of *Fairclough Building Ltd* v. *Borough Council of Port Talbot* (1992) a different set of circumstances arose. The Borough Council of Port Talbot decided to have a new civic centre constructed and advertised for construction companies to apply for inclusion on the selective tendering list for the project. Fairclough Building Ltd in March 1983 applied to be included on the list. Mr George was a construction director of Fairclough, whose name appeared on Fairclough's letter of application. His wife, Mrs George, had been employed by the council since November 1982 as a senior assistant architect and the Borough Engineer was aware of the connection between Mrs George and Fairclough for some time, as she had disclosed this at the time that the council employed her. Fairclough was invited to tender under the NJCC Two-Stage Tendering Code of Procedure. Mrs George, now the Principal Architect, was to be involved in reviewing the tenders and wrote to the Borough Engineer reminding him of her connection with Fairclough. The council considered the position and, having obtained counsel's opinion about how to proceed, decided to remove Fairclough from the select tender list for the project. Fairclough brought proceedings for breach of contract. The

Court of Appeal held that, under the circumstances, the council had only two alternatives. One was to remove Fairclough from the tender list, and the other was to remove Mrs George altogether. In removing Fairclough from the list, the council had acted reasonably. The council had no obligation to permit Fairclough to remain on the selected list and were not in breach of contract.

2.7.4. In giving its decision, the Court of Appeal distinguished *Blackpool & Fylde Aero Club Ltd* v. *Blackpool Borough Council.* Blackpool and Fylde Aero Club did in fact submit a tender by posting it, by hand, in the council's letter box before the deadline. However, the council's staff failed to empty the letter box properly, with the result that the tender was considered to have been delivered late, and rejected. In Fairclough, the council had perhaps been in error in shortlisting Fairclough in the circumstances of the connection, but had acted reasonably in removing them from the select tender list in the light of Mrs George's involvement.

2.7.5. Employers in the public sector must comply with the provisions of the Public Contracts Regulations 2006 where the value is above the financial threshold. In addition, they may find themselves at odds with any obligations derived from an implied tender contract. In the case of *Lettings International* v. *London Borough of Newham* (2008) the authority was held to have acted in breach of the Public Contracts Regulations 2006 and of an implied contract when conducting a tender procedure for contracts for procurement, maintenance and management of private-sector leased accommodation, through a failure to disclose weightings for the award criteria and irregularities in the scoring criteria.

2.7.6. The New Zealand case of *Pratt Contractors Ltd* v. *Transit New Zealand* (2003) dealt with a claim from a contractor whose tender was rejected, despite being the lowest by over £1m. An appeal from a decision of the Court of Appeal in New Zealand was referred to the Privy Council in London. The reason for not awarding the contract to the lowest bidder was that the appointment of the successful contractor was based upon a 'weighted attribute method'. This system adopted a formula by which marks were given to quality attributes as well as price. The Privy Council, in reaching a decision against the contractor, laid down a few ground rules to be applied when selecting a contractor. It was the duty of the employer to comply with any procedures set out in the tender enquiry. There was no obligation, however, for the employer to follow any of its own internal procedures. An obligation exists of good faith and fair dealings, but this does not extend beyond a requirement for members to express views honestly held. This duty of fairness did not require the employer to appoint people who approached the task of tender selection with no preconceived views about the tenderers.

SUMMARY

A tenderer who submits a tender in the correct form has a contractual right to have his tender opened and considered, but circumstances may arise in particular cases, where conflicting duties make it reasonable for properly submitted tenders not to be considered. This follows a number of legal cases which have established that, when submitting a tender, the employer and tendering contractor often enter into an implied tender contract. The terms of these tender contracts may vary, depending upon the circumstances.

2.8. If an architect/engineer, acting as employer's agent in a Design and Construct contract, approves the contractor's drawings and subsequently errors are found, will the architect/engineer have any liability?

2.8.1. A matter which needs to be addressed at the outset is whether the employer's agent should involve himself in approving contractor's drawings? Often, the employer's agent is a quantity surveyor or other non-design specialist without the appropriate expertise.

2.8.2. The standard forms of contract deal with the matter of the contractor submitting drawings to the employer in several different ways.

- JCT Design and Build 2011 Edition: requires the contractor to submit to the employer copies of the contractor's design documents, as and when necessary from time to time, in accordance with the contractor's design submission procedure set out in Schedule 1, or as otherwise stated in the contract documents. No reference is made to any action to be taken by the employer or its agent on receipt of the drawings.
- GC/Works 1/Design and Build: Condition 10A forbids the contractor from commencing work until the drawings have been submitted and they have been examined by the project manager, who has either confirmed that he has no questions to raise in connection with the drawings, or that such questions have been raised and answered to his satisfaction.
- ICE Design and Construct: Clause 6(2)(a) requires the contractor to submit drawings to the employer's representative and not to commence work until the employer's representative consents thereto.

2.8.3. One standard form which makes reference to approval of contractor's drawings is MF/1, which refers in clause 16.1 to the engineer approving drawings. If, however, an employer's agent, whether he be architect or engineer, approves the contractor's drawings which are subsequently shown to include an error, he may be liable. It would in the first instance be necessary to identify, in the conditions of employment, what his responsibilities were. Most consultancy agreements require the consultant to exercise due skill and care in carrying out his duties.

2.8.4. *George Fischer (GB) Ltd* v. *Multi Design Consultants Roofdec Ltd, Severfield Reece and Davis Langdon and Everest* (1998) is a case which, among other matters, examined the obligations of the employer's representative. George Fischer was the employer under an amended JCT With Contractor's Design. Multi Construction were main contractors and Davis Langdon and Everest both quantity surveyors and employer's representative. The project included the construction of the employer's UK head office. From the outset the roof leaked and, despite some reduction of the problems following the taping over of the end lap joints, the leaking continued. The main contractor, Multi Construction Ltd, became insolvent and went into liquidation. A claim was made against Davis Langdon and Everest as a result of the roof leaks. The case against them was that, under their contract with George Fischer, they had an obligation to approve all working drawings. They were also, it was alleged, obliged to make visits to the site to ensure that the work was being carried out in accordance with the drawings and specification. George Fischer claimed the problems with the roof would not have occurred had Davis Langdon and Everest carried out their duties as required by their contract. A further difficulty arose in that, under the terms of George Fischer's contract with Multi Construction, Davis Langdon and Everest were obliged to issue a certificate of practical completion, which had the effect of releasing the bond. Davis Langdon and Everest's defence was that they were not obliged to approve all working drawings. With regard to site visits, they contended that access was unsafe, which prevented them making these visits and even if such site visits were made, the defective formation of the lap joints would not have been seen, as workmen usually provide work of appropriate quality when being observed by the employer's representative. Their defence to the claim resulting from the issue of the certificate of practical completion was that in fact the certificate they issued was not one of practical completion but a substantial completion certificate, accompanied by a two-page document of incomplete work under a heading of reserved matters. The judge was not impressed with Davis Langdon and Everest's defence. He considered that the contract made it clear that they were obliged to approve working drawings. The reasons given for not inspecting the work were dismissed. Davis Langdon and Everest's certificate he considered to be a certificate of practical completion as, apart from the final certificate, this was the only certificate of completion referred to in the contract. The moral of the story is that those who work as employer's representatives should ensure that the wording of their contracts with the employers is crystal clear as to the duties they are required to undertake and, for the avoidance of any doubt, equally clear as to the duties that are not required.

SUMMARY

If the employer's agent approves contractor's drawings he may have a liability. Most consultancy agreements require the consultant to exercise due skill and care in carrying out his duties. If, due to a failure to exercise due skill and care, an error remains unnoticed, then the employer's agent may have a liability. As with any claim based on allegations of negligence, the employer will have to demonstrate that the errors resulted in additional cost. It is important for the employer's agent's conditions of engagement to spell out whether or not he is required to approve drawings.

2.9. Does an employer have any liability for not sending a subsoil survey which is in his possession to tendering contractors, the absence of which leads a successful contractor to significantly underprice the risk of bad ground?

2.9.1. Some contracts specifically require employers to disclose information in their possession which is relevant. The FIDIC 1999 Edition in clause 4.10 requires the employer to make available to the contractor all relevant data in the employer's possession on sub-surface and hydrological conditions. By way of contrast, the ICE 7th Edition, clause 11(1), provides for the contractor to take into account in the price only the information concerning the nature of ground and subsoil and hydrological conditions made available by the employer.

2.9.2. In the absence of specific obligations written into the contract, the law in the UK is vague on whether there is a legal obligation on an employer to disclose to a contractor relevant information which is in his possession. It has been argued that a failure to disclose information concerning ground conditions which may affect the contractor's tender price could amount to a negligent misstatement on the part of the employer. This line of argument was used in the Australian case of *Dillingham Construction Pty Ltd* v. *Downs* (1972), where it was recognised that the employer could owe the contractor a duty of care and this would include disclosure of relevant information. The decision went against the contractor, as it was held that there had been no reliance by the contractor on the employer providing the information.

2.9.3. There have been decisions of the Canadian courts which support the view that a duty of disclosure exists. In the case of *Quebec (Commission Hydroelectrique)* v. *Banque de Montreal* (1992), it was held that the law imposes a positive obligation to provide information in cases where one party is in a vulnerable position. In the case of *Opron Construction Co* v. *Alberta* (1994), the court took into account lack of time, the opportunities available for the tenderer to acquire the information, whether the information was indispensable and the degree of technicality of the data.

2.9.4. In the USA, there is authority for the proposition that, at least in government contracts, when the government agency is in possession of information which may be relevant to the work to be undertaken by the contractor, there is a duty to fully disclose the information to the contractor: *D Federico Co* v. *Bedford Redevelopment Authority* (1983).

2.9.5. The Misrepresentation Act 1967 may come to the aid of contractors who have made no financial provision for bad ground conditions, for which an allowance could have been made if the employer had issued information in its possession. It is possible for a liability to arise under the statute in respect of a failure to provide correct or relevant information, or where only a part of the information is provided. This was an issue in the decision of *Howard Marine and Dredging* v. *Ogden* (1978). In this case it was held that Howard Marine's manager was negligent in stating that the payload of a barge was 1,600 tonnes, when in fact it was only 1,055 tonnes. Howard Marine was held to be liable for the manager's misrepresentation under section2(1) of the Misrepresentation Act 1967.

SUMMARY

The terms of the contract may require the employer to disclose information concerning ground conditions to the contractor. In the absence of a contractual obligation, the law in the UK is unclear. In Canada, it has been held that the employer's obligation to disclose will depend upon the time available, the opportunities for the tenderer to acquire the information, whether the information was indispensable and the degree of technicality of the data. The law in the USA requires government agencies to disclose relevant information to contractors where government projects are involved. There has been no legal decision in the UK on this matter; however, in the event of such a case being heard, the court may consider that a failure to disclose amounts to a negligent misstatement or a liability under the Misrepresentation Act 1967. Alternatively, the courts may chose to follow the legal decisions arrived at in the Canadian courts and in courts of the USA.

2.10. If a subcontractor submits a lump sum estimate to a contractor to carry out the subcontract work and it is unconditionally accepted, can he later change the price on the basis that the lump sum was only an estimate?

2.10.1. The *Building Contract Dictionary* by Chappell, Marshall, Powell-Smith and Cavender, 3rd edition, published by Blackwell Publishing, provides two possible meanings for the word 'estimate':

- 'Colloquially and in the industry it means "probable cost" and is then a judged amount, approximate rather than precise';
- 'A contractor's estimate, in contrast, may, dependent upon its terms, amount to a firm offer, and if this is so, its acceptance by the employer will result in a binding contract'.

2.10.2. With two meanings attached to the word, using 'estimate' in the submission of an offer to carry out construction work can be ambiguous. In the case of *Crowshaw* v. *Pritchard and Renwick* (1899) a contractor submitted a quotation for construction work in the following terms:

> ESTIMATE – Our estimate to carry out the sundry alterations to the above premises, according to the drawings and specifications, amounts to £1,230.

2.10.3. The quotation was accepted, but the contractor refused to proceed with the work. He contended that submitting an 'estimate' for carrying out the work was not intended to amount to an offer capable of acceptance. It was held by the court that the submission was a firm offer which had been accepted.

2.10.4. The case of *Sykes* v. *Packham* (2011) involved a house refurbishment contract. The contractor submitted a price of £88,830 plus VAT, referred to as an 'estimate of cost' for undertaking the work, which he alleged was an estimate. He considered that an entitlement existed to be paid in accordance with the costs involved in carrying out the work. The house-owner considered the price to be a fixed price. The judge had his own opinion, which was that the builder had an entitlement to be paid a reasonable price for the work undertaken, which was *not* 'cost plus'. In deciding what was a reasonable price, the judge explained that he took a 'rough justice' approach. He knocked 20% off the wages bill for time wasted and a sum of £175 per day was allowed for the builder's expenses. All the subcontractors' accounts were allowed, plus the cost of materials. A mark-up of 20% was allowed on the subcontractors' accounts, 15% on materials and 25% on the operatives' wages.

2.10.5. With these decisions in mind, it is important for contractors and subcontractors to appreciate that, if the intention in submitting an estimate is that it is not intended to be a firm offer to carry out the work, this should be made clear. If, on the other hand, it is intended to be a firm offer but the price is subject to subsequent adjustment, then again this must be explained.

SUMMARY

The word 'estimate' has two meanings. It may refer to a probable cost or approximate sum which can later be adjusted, or a firm price which is fixed. When submitting a quotation, the term 'estimate' should either not be used at all, or qualified to provide its precise meaning.

2.11. Where a tender enquiry requires tenders to remain open for acceptance for a specific period of time, can a contractor or subcontractor who has submitted a tender as required withdraw the tender before the period expires, without incurring a financial liability?

2.11.1. It is a well-established principle of English law that an offer can be withdrawn at any time up to its being unconditionally accepted. This rule applies even if the offeror undertakes to leave the offer open for a specified period of time. In the case of *Routledge* v. *Grant* (1828), the defendant offered to buy a house, giving the plaintiff six weeks to provide an answer. It was held that the offer could be withdrawn within the six-week period without incurring any liability. The situation would be different if consideration, e.g. a payment, were made to the offeror in return for keeping the offer open.

2.11.2. On international projects, contractors are often required to provide a tender bond. If the tender is withdrawn before the period for acceptance has expired, the employer will be entitled to levy a claim against the bondsman.

2.11.3. The law in Hong Kong appears to differ from UK law. In the case of *City Polytechnic of Hong Kong* v. *Blue Cross (Asia Pacific) Insurance HCA* (1999), the plaintiff invited tenders from several insurance companies to provide health insurance for its staff. The tenders were required to remain open for acceptance for a period of three months. Blue Cross submitted the lowest tender, but withdrew it before the three months period elapsed. The Polytechnic accepted another and more expensive insurer's tender and claimed the difference back from Blue Cross. At first instance, the claim was rejected. On appeal, however, the claim was successful on the grounds that there existed an implied contract that the tender would remain open for acceptance for a period of three months. Blue Cross was in breach of the implied contract and was obliged to pay damages.

SUMMARY

Under English law, an offer can be withdrawn at any time up to its being unconditionally accepted. This applies even if the offeror agrees to keep the offer open for a fixed period of time. Where some form of consideration, e.g. a payment, is provided to the offeror in return for keeping the offer open, it cannot be withdrawn until the period for acceptance expires. In Hong Kong it has been held that an undertaking to keep an offer open for acceptance for a given period of time represents an implied contract. A withdrawal of the offer before the period comes to an end constitutes a breach of contract.

Chapter 3
Design

3.1. What is the difference between a fitness for purpose responsibility and an obligation to exercise reasonable skill and care?

3.1.1. Clients, when appointing a designer, whether architect, engineer, contractor or subcontractor, expect the building or structure to operate when complete in the manner envisaged when the appointment was made. If the building or structure fails to meet the client's expectations, there are often questions asked of the designer and/or contractor as to whether the problem resulted from a failure on their part to meet their contractual obligations. These obligations will normally take the form of implied or express terms in the conditions of appointment or the terms of the contract under which the work was carried out. In the absence of an express term in the contract for providing a design service, there will be an implied term that the designer will use reasonable skill and care. The standard is not that of the hypothetical 'reasonable man' of ordinary prudence and intelligence, but a higher standard related to his professed expertise. This was laid down in *Bolam* v. *Friern Hospital Management Committee* (1957) by Mr Justice McNair, in stating:

> Where you get a situation which involves the use of some special skill or competence, then the test whether there has been negligence or not is not the test of the man on top of a Clapham omnibus, because he has not got this special skill. The test is the standard of the ordinary skilled man exercising and professing to have that special skill. A man need not possess the highest expert skill at the risk of being found negligent. It is well-established law that it is sufficient if he exercises the ordinary skill of an ordinary competent man exercising that particular art.

3.1.2. The case of *London Fire and Emergency Planning Authority* v. *Halcrow Gilbert Associates* (2007) dealt with damage and liability in respect of a fire which occurred at a training facility. It seems that artificial smoke was distributed through ducts to the various rooms in which training exercises took place. Mineral oil in the smoke coalesced on the outsides of the ducts and formed droplets, which then leaked out into the ducts and contaminated the insulation, which then caught fire. As this was a new design, the employer alleged that Halcrow should have carried out investigations as to the risks involved, which might have resulted in steps being taken which could have prevented the fire from

200 Contractual Problems and their Solutions, Third Edition. Roger Knowles.

starting. The defendant indicated that he expected a fine film of oil to develop and not the droplets which led to the fire. The judge, in finding in favour of the defendant, considered that the conclusions that a film of oil would form was reasonable and that he had not failed to exercise reasonable skill and care in failing to identifying the likely problem of droplets forming. In this case the defendant was found to have exercised reasonable skill and care, with no liability, but nonetheless failed to produce a product which was fit for its purpose.

3.1.3. A person who professes to have a greater expertise than in fact he possesses will be judged on the basis of his pretended skills. In *Wimpey Construction UK Ltd* v. *DV Poole* (1984), a case where, unusually, the plaintiffs were attempting to prove their own negligence, they attempted to convince the judge that a higher standard was appropriate to the case under consideration. They put forward two 'glosses', as the judge referred to them:

> First, that if the client deliberately obtains and pays for someone with specially high skill, the Bolam test is not sufficient.
>
> Second, that the professional person has a duty to exercise reasonable care in the light of his actual knowledge, not the lesser knowledge of the ordinary competent practitioner.

As regards the first gloss, the judge felt obliged to reject it in favour of the *Bolam* test. However, the judge accepted the second gloss, not as a qualification of the *Bolam* test, but as a direct application of the principle in *Donoghue* v. *Stevenson* (1932). This requires reasonable care to be taken to avoid acts or omissions which one can reasonably foresee would be likely to injure a neighbour.

3.1.4. Another important aspect of reasonable skill and care is what is generally referred to as the 'state of the art' defence. Briefly, what this means is that a designer is only expected to design in conformity with the accepted standards of the time. These standards will generally consist of Codes of Practice, British Standards or other authoritative published information.

3.1.5. Unlike a professional designer, such as an architect, where a contractor or subcontractor undertakes design work or production of working drawings, there is, in the absence of an express term in the contract, an obligation to produce a product fit for its purpose. This is in marked contrast to a professional designer's implied obligation of reasonable skill and care. The duty to produce a building fit for its purpose is an absolute duty, independent of negligence. It is a duty which is greater than that imposed upon an architect employed solely to design, who would only be liable (in the absence of an express provision) if he were negligent. Express provisions to the contrary will obviously negate any implied terms. The contractor's position is best illustrated by the following extracts from leading cases:

Independent Broadcasting Authority v. *EMI Electronics Limited* (1980):

> In the absence of a clear, contractual indication to the contrary, I see no reason why [a contractor] who in the course of his business contracts to design, supply and erect a television aerial mast is not under an obligation to ensure that it is reasonably fit for the purpose for which he

> knows it is intended to be used. The Court of Appeal held that this was the contractual obligation in this case and I agree with them. The critical question of fact is whether he for whom the mast was designed relied upon the skill of the supplier to design and supply a mast fit for the known purpose for which it was required . . . In the absence of any terms (express or to be implied) negativing the obligation, one who contracts to design an article for any purpose made known to him undertakes that the design is reasonably fit for the purpose.

Greaves Contractors Limited v. *Baynham Meikle & Partners* (1975):

> Now as between the building owners and the contractors, it is plain that the owners made known to the contractors the purpose for which the building was required, so as to show that they relied on the contractors' skill and judgment. It was, therefore, the duty of the contractors to see that the finished work was reasonably fit for the purpose for which they knew it was required.

In the circumstances of this case, the designers were also held to have a liability to ensure that their design was fit for its purpose.

Young and Marten v. *McManus Childs* (1969):

> I think that the true view is that a person contracting to do work and supply materials warrants that the materials that he uses will be of good quality and reasonably fit for the purpose for which he is using them unless the circumstances of the contract are such as to exclude any such warranty.

3.1.6. The House of Lords' decision in *Slater* v. *Finning* (1996) held that no liability lies where a party is not aware of the particular purpose for which the goods are intended, or where the proposed use deviates from the normal use. The principle, as expressed by Lord Keith, was:

> As a matter of principle . . . it may be said that where a buyer purchases goods from a seller who deals in goods of that description there is no breach of the implied condition of fitness where the failure of the goods to meet the intended purpose arises from an abnormal feature or idiosyncrasy not made known to the seller by the buyer or in the circumstances of the use of the goods by the buyer. That is the case whether or not the buyer is himself aware of the abnormal feature or idiosyncrasy.

Lord Steyn provided a useful example of the application of this decision in the construction industry in saying:

> If a contractor in England buys pipes from a dealer for use in a pipe-laying project the seller would normally assume that the pipes need merely to be suitable to withstand conditions in our moderate climate. If the contractor wishes to use the pipes in arctic conditions for a Siberian project, an implied condition that the pipes would be fit to withstand such extreme weather conditions could only be imputed to the seller if the buyer specifically made that purpose known to the seller.

In the case of *J Murphy and Sons Ltd* v. *Johnston Precast Ltd* (2008), Johnston was engaged by Murphy to supply a length of glass-reinforced plastic pipe in a tunnel and surrounded it with foam concrete. Due to alkaline attack, the pipe in the void was unable to withstand the pressure exerted by the concrete and the pipe burst. It was held that, whilst Johnston had an obligation to supply a pipe which was fit for its purpose, they were not made aware of the fact that it was to be surrounded by foam concrete.

3.1.7. In the case of *PSC Freyssinet Ltd* v. *Byrne Brothers (Formwork) Ltd* (1996), the court had to decide liability where a design failure occurred due to a lack of provision for early thermal movement. The defendant, Byrne Brothers, was a subcontractor for the design and construction of the car park superstructures at the Lakeside Shopping Complex in Thurrock, Essex. PSC was employed by Byrne Brothers to design and install post-tensioned reinforcement and grouting. Whilst PSC owed a fitness for purpose obligation, it was not responsible for the design of the whole beam, as to do so would require a consideration of its relation to the entire structure and not merely the subframe. The court considered it absurd to hold PSC to a 'fitness for purpose' term when their work might be affected by information supplied by a third party, namely the architect.

3.1.8. JCT Design and Build Contract places the following design responsibility upon the contractor:

> the Contractor shall have in respect of any inadequacy of such design the like liability to the Employer, whether under statute or otherwise, as would an architect or as the case may be other appropriate professional designer . . .

GC/Works/1 1998 imposes a different responsibility. Condition 10, Alternative B states:

> The Contractor warrants to the Employer that any Works . . . will be fit for their purposes, as made known to the Contractor by the Contract.

ICE Design and Construct, clause 8(2), requires the contractor in carrying out his design responsibility to 'exercise all reasonable skill and care'. It can be seen that some of the standard forms reduce the contractor's 'fitness for purpose' obligation, which the law would normally imply, to the less onerous task of exercising reasonable skill and care. The main reason for this is the difficulty contractors have in obtaining insurance cover for a fitness for purpose obligation.

SUMMARY

In the absence of an express term in the conditions of contract, a designer, whether architect, engineer or other designer, will have an implied obligation to carry out his design obligation employing reasonable skill and care. The test is whether the level of skill provided is the standard of the ordinary skilled person exercising and professing to have that skill.

Where a contractor or subcontractor undertakes a design responsibility in conjunction with an obligation to construct the works there is, in the absence of an express term in the contract, an implied obligation to produce a design which is reasonably fit for its purpose. This is an absolute duty and any failure of the design solution will place a responsibility upon the design and construct contractor or subcontractor, whether or not the problem results from negligence. For this obligation to arise, the contractor or subcontractor, at the time the contract was entered into, must be aware of the purpose for which the facility is to be employed.

Some of the standard forms reduce the contractor's 'fitness for purpose' obligations which the law would normally imply to the less onerous 'reasonable skill and care'.

3.2. Where a contractor/subcontractor's drawings are 'approved', 'checked', 'inspected', etc. by the architect/engineer and subsequently an error is discovered, who bears the cost – the contractor, subcontractor or employer? If the employer bears the cost, can he recover the sum involved from the architect/engineer?

3.2.1. In general terms, when an employer appoints an architect or engineer to design a building or work of a civil engineering nature, he is entitled to expect the architect or engineer to be responsible for all design work. This basic principle was established in the case of *Moresk Cleaners Ltd* v. *Thomas Henwood Hicks* (1966). The plaintiffs were launderers and dry cleaners who appointed the defendant architect to undertake the design work of an extension to their laundry. Instead of designing all the work himself, the architect arranged for the contractor to design the structure. The employer brought an action for defective design against the architect, who argued that his terms of engagement entitled him to delegate the design of the structure to the contractor. It was held that an architect has no power whatever to delegate his duty to anybody else. Sir Walter Carter QC had this to say:

> [Counsel for the architect] in a very powerful argument, asks me to say alternatively that the architect had implied authority to act as agent for the building owner to employ the contractor to design the structure and to find that he did just this. I am quite unable to accept that submission. In my opinion he had no implied authority to employ the contractor to design the building. If he wished to take that course, it was essential that he should obtain the permission of the building owner before that was done.

3.2.2. Nevertheless, the architect or engineer in his terms of engagement may include a term which permits him to use a specialist contractor, subcontractor or supplier to design any part of the works, leaving the architect or engineer with no responsibility if the design work undertaken by others contains a fault, but the employer has to agree to this. Where a part of the design work is carried out by a subcontractor or supplier in accordance with an express term in the architect's or engineer's conditions of appointment, it is in the employer's interest to obtain some form of design warranty from the subcontractor or supplier. The employer would then be able to seek to recover any loss or damage resulting from design faults by the subcontractor or supplier on the basis of the warranty.

3.2.3. If, however, an architect or engineer (having excluded his responsibility for a subcontractor's design in the terms of his appointment) approves, checks or inspects a subcontractor's drawing, does he then take on any responsibility for a failure of the design? It is essential for the architect or engineer to make it clear to both employer and subcontractor exactly what he is doing with the drawings. If he is checking the design carried out by the subcontractor or supplier, he may find that, even though the terms of his appointment exclude liability, he may have adopted a post-contract amendment to the conditions and with it responsibility. The employer will be left to bring an action against either the architect/engineer or the subcontractor who carried out the design. An unfortunate aspect of English law is that both may be held to be jointly and severally liable. In other words, the employer can extract the full amount of his loss or damage from either party. This can be useful to the employer if a subcontractor carried out the design and subsequently became insolvent, leaving a well-insured architect who had checked the design to stand the full amount of the loss. Alternatively, the employer may decide to sue both, leaving the court to allocate his loss or damage between the joint defendants, after he has been paid in full by one or other of them.

3.2.4. If the architect/engineer is not checking the design, then he must make it very clear what he is doing. Ideally, it should be set out in the architect's/engineer's terms of appointment precisely what his duties are with regard to design work undertaken by a contractor, subcontractor or supplier. Should the employer commence an action against the architect/engineer alone, then, under the Civil Liability (Contribution) Act 1978, the architect/engineer may seek a contribution from the contractor, subcontractor or supplier whose design was faulty. In the event of the employer deciding to sue the contractor, subcontractor or supplier alone they, likewise, may seek a contribution from the architect/engineer.

3.2.5. The fact that an engineer receives drawings does not in itself imply that he has any liability for errors in design. In *J Sainsbury plc* v. *Broadway Malyan* (1998) a claim for defective design was settled out of court. The problem related to the design of a wall between a store area and retail area. Due to the low level of fire protection, fire spread and caused substantial damage. The architect attempted to off-load some of the liability upon an engineer to whom the drawings had been sent for comment. It was held that, if the architect wanted to get the structural engineer's advice on fire protection, he needed to say so. Simply to transmit the drawings for comment, without specifying any area in which comment was requested, was not sufficient to imply any obligation.

3.2.6. A different slant was placed upon acceptance of drawings by the engineer in the case of *Shanks & McEwan (Contractors) Ltd* v. *Strathclyde Regional Council* (1994), which arose out of the construction of a tunnel for a sewer. A method of construction was employed using compressed air to minimise water seepage. The tunnel and shaft segments, in compliance with the specification, were designed by a supplier to the main contractor. The main contractor was to be responsible for the adequacy of the design insofar as it was relevant to his operations, but it was also a requirement of the specification that design calculations were to be submitted to the engineer. In the course of construction, fine cracks appeared in the prefabricated tunnel segments because of a design fault. The engineer was prepared to accept the work, subject to the segments being made reasonably watertight and confirmed the same in a letter to the contractor dated 21 September

1990. Clause 8(2) of the ICE 5th Edition, which governed the contract, states that the contractor shall not be responsible for the design of the permanent works. There seemed to be a conflict between clause 8(2) and the specification, which placed responsibility for the design of the tunnel segments onto the contractor. The contractor levied a claim for the cost of the repair work. It was the view of the Court of Session in Scotland that, following acceptance by the engineer of the design of the segments, the contractor was entitled to expect that the approved design would not crack. The letter from the engineer dated 21 September 1990, which accepted repair work to the segments, was held to be a variation and therefore the contractor was entitled to be paid for that work.

3.2.7. The employer's ability to recover from the engineer any costs incurred because of design error on the part of the contractor or subcontractor will depend upon a number of factors. If the design faults lie with the contractor or subcontractor, it is to those who caused the error that the employer would normally address his claim. If the employer is unable to recover from the contractor or subcontractor, for example because of insolvency, he may wish to turn his attentions to the engineer. The ability to recover will depend upon the terms of the engineer's appointment. If the matter is referred to court, all involved in the design process will normally be joined into the action. In *London Underground* v. *Kenchington Ford* (1998), the design of a diaphragm wall at the Jubilee Line station of Canning Town became the subject of a dispute. The diaphragm wall was designed by Cementation Bachy (the contractors). London Underground argued that Kenchington Ford (the engineer) had failed to realise that there had been a mistake in computation made by Cementation Bachy and consequently the diaphragm wall was designed too deep and hence over-expensive. The error had resulted from Cementation Bachy misinterpreting the load shown on the drawing. The contract stated that Cementation Bachy would be responsible for design errors, whether approved by the engineer or not. Kenchington Ford was under a duty to London Underground to provide services, which included the correction of any errors, ambiguities or omissions. The judge concluded that Kenchington Ford should have checked and discovered the error, and as they had not, this constituted a breach of duty. In *George Fischer (GB) Ltd* v. *Multi Design Consultants Roofdec Ltd, Severfield Reece and Davis Langdon and Everest* (1998), a complex multiparty action, the employer's representative was held to be partly liable in respect of the design error. The employer's representative's conditions of appointment obliged him to approve all working drawings. Following judgment in favour of the employer, the parties agreed on the sum payable as damages. Multi Design Consultants, who carried out the design function, were liable in the sum of £940,000, with the liability of the employer's representative, Davis Langdon and Everest, being £807,388.

SUMMARY

The approval of a contractor's or subcontractor's drawings by the architect or engineer, will not usually relieve the contractor or subcontractor from liability. Employers who incur costs due to this type of error will normally commence an action against both the contractor/subcontractor who prepared the drawings and the architect/engineer who gave his approval. The court will decide on the apportionment of blame. Where the employer incurs cost due to errors in the contractor/subcontractor's design, these costs may be recovered from the engineer/architect if a duty to check the drawing was expressly or impliedly provided for in the conditions of appointment and the errors result from a failure to carry out the checking properly.

3.3. Who is responsible for co-ordinating design? Can a main contractor be legitimately given this responsibility, even though he has no design responsibility?

3.3.1. When an employer appoints an architect/engineer to design a building or work of a civil engineering nature, he is entitled to expect the architect/engineer to be responsible for all design work. This basic principle was established in *Moresk Cleaners Ltd* v. *Thomas Henwood Hicks* (1966).

3.3.2. This being the case, the architect/engineer will also be responsible for co-ordinating design, unless there is an express term in the contract to the contrary.

3.3.3. Specifications for mechanical and electrical work and other specialist disciplines often refer to the subcontractor being responsible for design co-ordination. This will not absolve the architect from his design responsibilities, expressed or implied, in the conditions of engagement. If the specification which refers to a subcontractor being responsible for design co-ordination becomes a main contract document, then the employer may bring an action against the main contractor for breach in respect of any loss or damage resulting from poor design co-ordination. Any liability on the part of the main contractor would be recoverable from the subcontractor under the terms of the subcontract. Alternatively, design co-ordination may be specifically referred to in a design warranty entered into by the subcontractor, in which case the employer may commence an action for breach of warranty against the subcontractor for faulty co-ordination.

3.3.4. Where the contractor is required to design a part of the work only, it will be the architect's responsibility to ensure that the contractor's design is properly co-ordinated with his own design work.

3.3.5. The main contractor's responsibility for design co-ordination will be dependent upon the terms of the contract. Design by contractors, employing either a full design and construct procedure or a partial design and construct, is on the increase. Even without a design responsibility, the terms of the main contract may impose a responsibility upon the main contractor to undertake design co-ordination. However, it is unlikely that, in the absence of express terms in a main contract or subcontract, an obligation to co-ordinate design will rest on the main contractor or subcontractor.

3.3.6. If the contractor is required to co-ordinate design work, an express clause must be included in the contract which is fully descriptive of the co-ordinating activities required of the contractor. A brief term which states that the contractor is responsible for co-ordinating the work of all subcontractors, including design, would not be adequate. A much more descriptive clause is necessary. This clause should indicate which trades are involved and expressly state that all costs and losses resulting from a failure properly to co-ordinate the subcontractors' design and working drawings will be borne by the main contractor.

3.3.7. Where there is no reference to a design obligation in the main contract, it is unlikely that the main contractor will become liable for any defective design by a subcontractor: *Norta* v. *John Sisk* (1971).

3.3.8. The problem often starts with the appointment of the architect or consulting engineer. It is essential that his conditions of appointment spell out clearly the duties which he is required to undertake. Clarity in the terms of the main contract and subcontract are also essential.

3.3.9. Building Information Modelling is now being used on some of the larger projects. As this system allows a three-dimensional perspective of the project, design coordination is much easier and the likelihood of clashes of services becomes less likely.

SUMMARY

The architect/engineer will normally be responsible for design co-ordination, except where the contractor is appointed on a design and build basis. It is possible for an architect to disclaim the responsibility for design co-ordination in his conditions of engagement with the employer and place the burden upon the contractor's shoulders. For a main contractor to take on a responsibility for design co-ordination will require a fully descriptive clause in the main contract conditions of contract.

3.4. Can a contractor be held responsible for a design error where the employer appoints an architect and no provision exists in the contract for the contractor to undertake any design responsibility?

3.4.1. It is commonplace for a contractor to have placed upon him by the terms of contract, a full design responsibility. Some contracts provide for parts only of the work to be designed by the contractor. If under the contract the employer appoints an architect, whose duty it is to prepare all the drawings, with no reference being made to a contractor's design responsibility, can a situation ever arise where the contractor finds himself liable for a design fault?

3.4.2. In the case of *Edward Lindenberg* v. *Joe Canning, Jerome Contracting Ltd* (1992) the plaintiff engaged the defendant builder for some conversion work on a block of flats. During the work, load-bearing walls in the cellar were demolished, which caused damage in the flat above. The plaintiff sued the defendants for breach of contract and/or negligence, seeking repayment of the sums he was forced to pay the building owners under an indemnity. The plaintiff alleged that Canning was in breach of an implied term that he would proceed in a good and workmanlike manner and that he had negligently demolished the load-bearing walls without providing temporary or permanent support. It was held:

(1) As there was no express agreement between the parties, Canning was entitled to be paid on a *quantum meruit* basis for labour and materials.
(2) There was an implied term that the defendant would undertake the work in a good and workmanlike manner and exercise the care expected of a competent builder. He had been supplied with plans, prepared by the plaintiff's surveyor, which supposedly indicated which walls were non-load-bearing. However, as a builder, he should have known that since they were nine-inch walls, they were in fact load-bearing. As he took 'much less care than was to be expected of an ordinary competent builder', he was in breach of contract but not liable in negligence.
(3) The plaintiff was entitled to recover £7,484 (representing the amount he had to reimburse the building owner, plus professional fees), less a sum for contributory negligence.
(4) The plaintiff had been guilty of contributory negligence through his agents, in that Canning had been given plans which wrongly showed which walls were non-load-bearing; oral instruction had been given to demolish walls and no instructions had been given regarding the provision of supports. The liability was attributed at 75% to the plaintiff and 25% to the defendant. The plaintiff's damages were reduced accordingly to £1,871.
(5) Canning was entitled to a *quantum meruit* payment, assessed at £4,893. As this was less than the £7,000 which the plaintiff had advanced to him, Canning was liable to repay the difference.

This case illustrates that the contractor, where the design is faulty, can take on a design responsibility if a reasonably competent contractor would have identified the error.

3.4.3. It is possible for a contractor to have imposed upon him variations to the contract where the work in the variation imposes a design responsibility.

SUMMARY

The fact that the employer employs an architect and the main contract makes no reference to the contractor's design responsibility does not mean that the contractor cannot become responsible for design errors. In the *Joe Canning* case the drawings incorrectly showed which walls were load-bearing. The contractor was, nevertheless, held to be liable in breach of contract for taking much less care than an ordinary competent builder in demolishing the walls which turned out to be load-bearing. It is also possible, although unusual, for a contractor to be issued with a variation which includes a design responsibility.

3.5 Can a main contractor be responsible if a nominated/named subcontractor's design is defective?

3.5.1. Whether a main contractor is responsible for a nominated or named subcontractor's design error is usually decided following a careful study of the contract documents. It is common practice for the architect or engineer to arrange for specialist work to be designed by a subcontractor, who is then either nominated or named in the contract documents. Often, the main contractor has no involvement whatsoever in the design of the specialist work.

3.5.2. The matter is catered for in the ICE 6th and 7th Editions at clause 58(3), which states:

> If in connection with any Provisional Sum or Prime Cost Item the services to be provided include any matter of design or specification of any part of the Permanent Works or of any equipment or plant to be incorporated therein such requirement shall be expressly stated in the Contract and shall be included in any Nominated Sub-contract. The obligation of the Contractor in respect thereof shall be only that which has been expressly stated in accordance with this sub-clause.

The ICE contracts therefore make it crystal clear where the contractor's responsibility lies with regard to the design of a nominated subcontractor's work.

3.5.3. JCT 98, in clause 35.21, in like manner to the ICE 6th and 7th Editions, makes it clear that the contractor is not responsible for design work undertaken by a nominated subcontractor. There is no provision for the appointment of nominated subcontractors in JCT 2011.

3.5.4. In the case of *Norta* v. *John Sisk* (1977), the Irish Supreme Court had to decide the contractor's liability for a nominated subcontractor's design error, where the conditions of the main contract made no reference to design responsibility. The claimant entered into a contract to construct a factory for making wallpaper. Prior to the receipt of tenders from main contractors, the claimant approved a quotation from Hoesch Export for the design, supply and erection of the superstructure of the factory, including roof lights. Hoesch Export became nominated subcontractors to John Sisk, the appointed main contractor. Following practical completion, the roof began to leak, because of faulty design of the roof lights. The claimant sought to recover his losses from the main contractor, John Sisk. No reference was made in the main contract to John Sisk having any design responsibility. It was argued on behalf of the claimant that a design obligation was implied into the main contract. The Irish Supreme Court held that no such term could be implied into the main contract and therefore John Sisk had no liability.

3.5.5. JCT 98 includes for performance specified work. Clause 42 provides for performance specified work to be included in the contract by means of the employer indicating the performance he requires from such work. Before carrying out the work, the contractor must produce a contractor's statement in sufficient form and detail adequately to explain the contractor's proposals. The contractor will be responsible for any fault in the contractor's statement, which may include design work by subcontractors if the fault results from a failure to exercise reasonable skill and care. There is no provision in JCT 2011 for performance specified work.

3.5.6. The main contractor will be responsible for all design work, including that of subcontractors, where design and construct conditions apply, e.g. a JCT Design and Build Contract.

3.5.7. Many non-standard forms of contract or amendments to standard forms make it clear that the main contractor is responsible to the employer for all the nominated subcontractors' work, including design.

SUMMARY

The main contractor will be responsible for design faults in a named or nominated subcontractor's work if there is a clear statement to that effect in the main contract. In the absence of an express obligation, an employer would have to show that such an obligation was implied. This may prove difficult, if the subcontractor's design work was developed through a liaison between the subcontractor and architect/engineer direct, particularly if this took place without any involvement by the contractor. To protect himself against loss due to subcontractors' design faults, it is advisable for the employer to enter into a design warranty direct with the subcontractor. Most of the commonly used standard forms of contract make it clear that the main contractor is not responsible for a nominated subcontractor or nominated supplier's defective design. Where the contract is placed on a design and construct basis, the contractor will be responsible for all design work undertaken by named or nominated subcontractors, unless the main contract states otherwise.

3.6. Must a contractor notify an architect/engineer of defects in his design?

3.6.1. Human errors occur on a regular basis, including design errors by architects and engineers. Contractors may from time to time suspect that a design error has occurred. If this be the case, does the contractor have an obligation to draw attention to the design error?

3.6.2. The case of *Equitable Debenture Assets Corporation Ltd* v. *William Moss and Others* (1984) involved a building where the curtain wall leaked, due to defective design undertaken by a subcontractor. Unfortunately, the subcontractor went into liquidation and the employer brought an action against the architect and main contractor. In finding against the main contractor, the court held that a term should be implied into the contract that the contractor is required to report design defects known to him.

3.6.3. The case of *Victoria University of Manchester* v. *Hugh Wilson and Others* (1984) dealt with a problem of ceramic tiles falling off the exterior face of a building at Manchester University. The cause was a combination of poor design and poor workmanship. With regard to design defects, it was held that the contractor had a duty under an implied term of JCT 63, on which the contract was based, to warn of design defects which they believed to exist. However, there was no obligation on the part of the main contractor to undertake a close scrutiny of the architect's drawings. Judge John Newey said:

> The contractor's duty to warn the architect of defects which they believe existed in the architect's design, did not in my view require them to make a critical survey of the drawings, bills and specifications looking meticulously for mistakes.

3.6.4. A more recent decision is *University of Glasgow* v. *Whitfield and Laing* (1988), which called into question the decisions in *Equitable Debenture Assets Corporation* and *Victoria University*. In this case, it was alleged that the contractor owed an implied duty to the architect to warn of design faults. However Judge Bowsher had this to say when holding that the contractor had no duty to the architect to warn of defects:

> Mr Gaitskell on behalf of the defendant relies on the decisions of Judge Newey QC in *Equitable Debenture Assets Corporation* v. *William Moss* (1984) and *Victoria University of Manchester* v. *Wilson* (1984). On analysis it is clear that both cases were concerned with a duty of a contractor to warn the employer, not a duty owed by the contractor to warn the architect. References to a duty to give a warning to the architect were in both cases references to a duty to warn the architect as agent of the employer. It is clear from page 163 of the report of the *Victoria University of Manchester* case that the learned judge considered that both decisions were founded on implied contract between the contractor and the building owner. In each case, the learned judge cited *Duncan* v. *Blundell* (1820) and *Brunswick Construction Limited* v. *Nowlan* (1974). It is plain from the citation from the *Brunswick Construction* case that the learned judge had in mind the situation where the contractor knew that the owner placed reliance on him in the matter of design. It seems to me that the decisions in *EDAC* v. *Moss and Victoria University of Manchester* can stand with more recent decisions if they are read as cases where there was a special relationship between the parties, but not otherwise, and bearing in mind the difficulties in analysing the meaning of the words 'special relationship' and 'reliance' demonstrated by Robert Goff LJ in *Muirhead* v. *Industrial Tank Limited* (1986). On the facts of the present case it is not necessary to resolve those difficulties.

3.6.5. In *Edward Lindenberg* v. *Joe Canning, Jerome Contracting Ltd* (1992) (see 3.4.2), the contractor was held liable to make a contribution to the cost of remedial works resulting from the demolition of load-bearing walls. The walls were shown on the architect's drawings as non-load-bearing.

3.6.6. The opinion expressed by Judge Bowsher, to the effect that the employer and contractor must have a special relationship before an obligation to warn of design defects arises, does not seem to have been followed in subsequent cases. In *CGA Brown* v. *Carr* (2006), the judge, in respect of a defect in the roof design, was of the view that the builder should have discovered the problem, which was inherent in the design, before commencing construction. A reasonably competent builder, he thought, should have reported the defect to the client. In *J Murphy and Sons Ltd* v. *Johnston Precast Ltd* (2008), the judge expressed his opinion that a duty to warn arose when there was knowledge of a problem, or where there should reasonably have been knowledge.

3.6.7. The subject of a contractor's duty to warn occurred in *Plant Construction plc* v. *Clive Adams Associates and Another* (1999), heard before the Court of Appeal. Ford appointed a company trading under the name of Plant to install two engine mounts in a research and development centre at the Ford Research Engineering Centre. The substructure and underpinning of a roof was subcontracted to JMH. A variation was issued involving the design of the temporary works by a representative of Ford. The design was defective and collapse occurred. JMH was not responsible for the design, but it was held that they had an implied obligation to exercise reasonable skill and care. An experienced contractor such as JMH would have an obligation to warn of errors in design which were obviously dangerous and defective. The decision left open the situation where the design is obviously defective, but not dangerous.

3.6.8. Some forms of contract, for example JCT Design and Build, require the contractor to notify the employer of any discrepancies arising from the Employer's Requirements, Contractor's Proposals and instructions issued by the employer, as required by the conditions of contract.

SUMMARY

It was held in the *University of Glasgow* v. *Whitfield and Laing* case that in the absence of express provisions, the contractor may have an implied duty to the employer to warn of design faults, but only where a special relationship exists between them. There would otherwise appear to be no obligation in the absence of an express term in the contract.

This decision is difficult to comprehend. If correct, a contractor knowing of a design error could carry out construction work without obligation. It is hard to anticipate any subsequent cases following this decision. The decision in the *Equitable Debenture* case is to be preferred, where it was held that an implied term exists in construction contracts that contractors should report design defects known to them. In the case of *Plant Construction plc* v. *Clive Adams* (1999) it was held that a contractor would have an obligation to warn of errors in design which were obviously dangerous and defective.

It is suggested that contractors do have an implied obligation to notify the architect/engineer of suspected errors in the design. This does not, however, extend to the contractor being obliged to make a careful study of the drawings, in an attempt to identify errors.

3.7. Where an architect/engineer includes a new product in his design following advice from a manufacturer and the product proves to be unsuitable, is the architect/engineer liable to the employer for his losses?

3.7.1. Engineers and architects often have difficulty in providing appropriate design solutions to suit planning constraints, environmental considerations and the client's financial position. Manufacturers often make claims that a new product will meet the architect's/engineer's requirements. In the absence of a track record the architect/engineer is seen to be taking a risk in specifying the new product. If, having made checks concerning the manufacturing process and having sought whatever advice is available, the architect/engineer specifies the product, what liability does the architect have to the client if the product proves unsatisfactory?

3.7.2. The case of *Victoria University of Manchester* v. *Hugh Wilson and Others* (1984) arose out of a major development for the plaintiffs, erected in two phases between 1968 and 1976. The first defendants were the architects for the development, the second defendants the main contractors and the third defendants nominated subcontractors. The architects' design called for a building of reinforced concrete (which was not waterproof) to be clad partly in red Accrington bricks and partly in ceramic tiles. In due course, many of the tiles fell off and the University adopted a remedial plan which involved the erection of brick cladding with a cavity between bricks and tiles and with the brick walls attached to the structure by steel ties. It was held that the architect was liable as his design was defective. With regard to the use of untried materials, Judge John Newey had this to say:

> For architects to use untried, or relatively untried materials or techniques cannot in itself be wrong, as otherwise the construction industry can never make any progress. I think, however, that architects who are venturing into the untried or little tried would be wise to warn their clients specifically of what they are doing and to obtain their express approval.

3.7.3. In *Richard Roberts Holdings Ltd* v. *Douglas Smith Stimson Partnership* (1988) a tank lining failed. The employer brought an action against the architect for negligence. The architect's defence was that he had no legal liability, as the employer knew that he had no knowledge of linings. It was held, again by Judge John Newey, that:

> The architects were employed for the design of the whole scheme of which the linings were an integral part. The architects did not know about linings, but part of their expertise as architect was to be able to collect information about materials of which they lacked knowledge and/or experience and to form a view about them. If the architects felt that they could not form a reliable judgment about a lining for a tank they should have informed the employer of that fact and advised them to take other advice . . .

SUMMARY

Where an engineer/architect includes a new product in his design, the employer should be informed at the outset. Failure to advise the employer could leave the engineer/ architect exposed to a liability for negligence, should the new product fail.

3.8. Where an architect/engineer is required by the conditions of the contract to approve, or accepts a contractor or subcontractor's drawings, how long can he take before an entitlement to an extension of time arises?

3.8.1. It is quite common for contractors or subcontractors to be required to produce drawings in respect of their installation. Well-drafted specifications will normally provide for an approval or acceptance system. The system will set out the roles to be played by architect/ engineer and contractor or subcontractor up to the stage of approval or acceptance of the drawings. Usually, a timescale will be included which will indicate the maximum time within which the drawings must be approved or accepted or queries raised. Time will normally be allowed for answering queries with final approval or acceptance, again within a timescale. If the architect or engineer fails to approve, accept or query a contractor or subcontractor's drawing within the timescale, and as a result the completion date for the project is delayed, there is usually an entitlement to an extension of time. If there is no provision for extending time, where delays are caused by late approval or acceptance, then time becomes at large and the contractor or subcontractor's obligation is to complete within a reasonable time.

3.8.2. GC/Works/1 Design and Build requires the contractor to ensure that the programme allows reasonable periods of time for the provision of information from the employer.

3.8.3. Contractors and subcontractors will often indicate on the face of the drawing a period of time within which approval is sought.

3.8.4. Where there is no timescale in the procedures within which the architect/engineer is required to approve or accept or query a contractor's or subcontractor's drawing, or perhaps there is no formal procedure provided for approvals in the specification, the court will normally hold that such a term will be implied to give the contract business efficacy. A clause will usually be implied to the effect that approval by the architect/engineer must be given or any query raised within a reasonable time. What is a reasonable time will depend upon the circumstances of each case and would include such matters as any time allowed on the contractor's or subcontractor's programme; the rate of progress of the work; and the date fixed for completion.

SUMMARY

Ideally, the contract will indicate what period of time is to be allowed for drawing approval; alternatively, the contractor's programme should address the point. If there is no provision in the contract, then it will be implied that a reasonable period will be allowed.

3.9. Where is the line to be drawn between an architect/engineer's duty to design the works or a system and a contractor or subcontractor's obligation to produce working shop or installation drawings?

3.9.1. Where a contract such as JCT 2011, ICE 6th or 7th Editions, MF/1 or GC/Works is employed, the duty to design the works rests with the architect/engineer. However, provision is made in these contracts for some or all of the design work to be prepared by the contractor. Many bespoke engineering contracts require the contractor to be responsible for the detailed design of the plant and of the works in accordance with the specification. Specifications are often written to the effect that specialist engineering subcontractors will be obliged to produce shop or working drawings. There is no hard and fast rule as to where the architect's/engineer's obligations cease and those of the contractor or subcontractor begins. It will be a matter for a decision to be made in each and every case.

3.9.2. In *H. Fairweather & Co* v. *London Borough of Wandsworth* (1987), a subcontract was let using the now out-of-date NFBTE/FASS nominated subcontract, often referred to as the Green Form. The description of the works set out in the appendix to that form was to 'carry out the installation and testing of the underground heat distribution system, as described in [the specification]'. The specification had two provisions. Clause 1.15 made it the subcontractor's responsibility to provide the installation drawings and they were also 'responsible for providing all installation drawings in good time to meet the agreed programme for the works'. Section 3(b) of the technical specification also required detailed drawings to be prepared and supplied by the subcontractor. Before entering into the nominated subcontract, Fairweathers had written to the architect in an endeavour to disclaim 'any responsibility for the design work that may be undertaken by your nominated subcontractor'. They also asked for 'a suitable indemnity against defects in design work carried out by the nominated subcontractor'. The architect's reply drew attention to the provisions of clause 1.15 and pointed out that these did not 'require [them] to assume responsibility for the design of the system'. Fairweathers did not take the matter further and entered into the subcontract. The arbitrator found that the installation drawings were not design drawings. The judge agreed with him, although he had not seen the drawings. It does not appear that there was any dispute about responsibility for the content of the installation drawings and it would seem from this case that one cannot deduce that 'installation drawings' in general do not embody any 'design'. The architect had made it clear that the installation drawings were to be provided so as to meet the requirements of the programme and that the subcontractors were not responsible for the design of the system. However, in the course of preparing a detailed design for the installation of a system, decisions are taken of a design nature by the person responsible for the preparation of the drawings. In the absence of a clear contrary indication, the responsible contractor, subcontractor or supplier will be held liable.

3.9.3. It is not always obvious where the line is to be drawn between design or conceptual design and shop or working drawings. What is the purpose of the shop or working drawings? Some may argue that the intention is that the contractor's or subcontractor's duty is to fill in the gaps left in the design or conceptual design drawings. Others may argue that the purpose of shop or working drawings is to convert design information into a format to enable the materials to be manufactured and fixed.

3.9.4. It is essential, if a named or nominated subcontractor is to produce shop or working drawings, for the contract to stipulate in clear terms what is meant by these terms.

SUMMARY

It would seem that it is almost impossible to produce a dividing line to differentiate between design drawings and working, shop, or installation drawings. Each case would have to be judged on its merits. A reasonable interpretation is that the purpose of shop or working drawings is to convert design information into a format to enable the materials to be manufactured and fixed. It is advisable for the contract to stipulate in clear terms what is meant by these terms.

3.10. Where an item of work has been properly provided for in the Employer's Requirements but is missing from the Contractor's Proposals, can the contractor claim extra payment for doing the work, on the grounds that it was never included in the contract price?

3.10.1. If we were living in a perfect world, then all contract documents would be fault-free. Unfortunately, human beings are often known to err and, as a result, discrepancies are apt to appear between the employer's requirements and contractor's proposals.

3.10.2 The recitals to the JCT Design and Build Contract state:

> the Employer wishes to have the design and construction of the following work carried out . . . and the Employer has supplied the Contractor with documents showing and describing or otherwise stating his requirements (Employer's Requirements).

In response to the Employer's Requirements the contractor has supplied to the employer documents showing and describing the contractor's proposals for the design and construction of the works (Contractor's Proposals)

The contractor's obligations are expressed in the following terms:

> The Contractor shall carry out and complete the Works in a proper and workmanlike manner and in accordance with the Contract Documents

The Contract Documents are defined in the contract as comprising:

> the Agreement and these Conditions together with the Employer's Requirements, the Contractor's Proposals and the Contract Sum Analysis.

3.10.3. A difficult situation arises if there is a discrepancy between the employer's requirements and the contractor's proposals. This is a common occurrence in practice: for example, the employer's requirements may call for engineering bricks below the damp proof course, whereas the contractor's proposals allow for semi-engineering bricks, either type of brick being fit for the purpose. It is clear, however, that an instruction would have to be issued as to which of the alternatives is to apply. The contract is silent as to how this type of discrepancy is to be dealt with, but a clue as to how the situation can be resolved is contained in the third recital, which states:

> The Employer has examined the Contractor's Proposals and subject to the conditions is satisfied that they appear to meet the Employer's Requirements

It is arguable that, as the employer has declared that he has examined and is satisfied with the contractor's proposals, any discrepancy between the employer's requirements and contractor's proposals which comes to light after the contract has been entered into should be interpreted in the contractor's favour. There is, however, no authority for this argument. An amendment to the wording of the third recital should be made to indicate which takes precedence.

3.10.4. The GC/Works/1 Design and Build contract is reasonably clear as to which of the employer's requirements or the contractor's proposals takes precedence. Condition 2(2) states:

> In the case of discrepancy between the Employer's Requirements and either the Contractor's Proposals or the Pricing Document, the Employer's Requirements will prevail without adjustment to the Contract Sum.

Further references to discrepancies are made in condition 10A, which states:

> To demonstrate compliance with the Employer's Requirements the contractor shall ensure that relevant work will be the subject of a Design Document.

Condition 10A(7) develops the theme further, by stating that:

> In case of any discrepancy between Employer's Requirements and Design Documents the Employer's Requirements shall prevail, without any adjustment to the Contract Sum.

'Design documents' are defined as any drawing, plan, sketch, calculation, specification or any other document prepared in connection with design by the contractor. The intention is to catch any document, whether prepared prior to the submission of the tender or subsequently prepared by the contractor for design purposes. All these documents will be subsequent to the employer's requirements.

3.10.5. The ICE Design and Construct Conditions are also clear as to the priority of those key documents, in that clause 5(b) states:

> If in the light of the several documents forming the Contract there remain ambiguities or discrepancies between the Employer's Requirements and the Contractor's Submission the Employer's Requirements shall prevail.

3.10.6. The Engineering and Construction Contract (NEC 3) is completely silent on the matter. It will therefore be a matter of proper provision being included in the Works Information.

3.10.7. A court may take the view that, whilst the Employer's Requirements and Contractor's Proposals are silent with regard to a particular item of work, the requirement to have the work undertaken was obvious (*Williams* v *Fitzmaurice* (1858)). For example, a door will always require ironmongery and a house will require flooring. This being the case, it should have been included for in the contractor's price.

SUMMARY

Unfortunately, the JCT Design and Construct Contract does not address the difficulty, which may arise where there is a conflict between the employer's requirements and the contractor's proposals. The contract is silent as to which will take precedence. It is likely, however, that a court would hold that the contractor's proposals take precedence as the recitals indicate that:

> the Employer has examined the Contractor's Proposals . . . and is satisfied that they appear to meet the Employer's Requirements.

The ICE Design and Construct and GC Works/1 Design and Build contracts make it clear that the employer's requirements will take precedence over the contractor's proposals. The Engineering and Construction Contract (NEC 3) contract is silent on the matter.

3.11. Is the contractor entitled to payment for design in full when the design work has been completed, or should payment for design costs be spread over the value of work as and when it is carried out?

3.11.1. Contracts which are well drafted will usually be precise as to how much is to be paid or the manner in which payment is to be calculated and the timing of the payment. Design and construct contracts are no exception, and so the contract should be clear as to when payment for both the design function and construction of the works is to be made.

3.11.2. Contracts such as GC/Works/1 Design and Build provide for milestone payments. This being the case, the milestone payment chart should make it clear when payment for design is to be made. In considering the make-up of each payment, consideration should be given to the contractor's pre-contract and post-contract design costs. Provision for payment of the pre-contract design costs should be included in the first milestone. The post-contract costs should be costed in accordance with a design programme and allocated to the appropriate milestone. Condition 48B provides for mobilisation payments if stated in the abstract of particulars. The calculation of this payment would normally include pre-contract design costs. Where milestone payment and mobilisation payment provisions do not apply, payment of the pre-contract design costs should be included in the first advance on account. Design costs should be included in subsequent advances on account to accord with the progress of the post-contract design.

3.11.3. JCT Design and Build Contract is similar to GC/Works/1. Payment method Alternative A provides for stage payments. The analysis of stage payments included in the contract particulars should make it clear in which stage the pre-contract and post-contract design costs will be paid. If Alternative A does not apply, Alternative B comes into operation. In this case, payment of the pre-contract design costs should be included in the first interim payment and the remaining design costs to be included in subsequent payments to suit the progress of the design. In like manner to GC/Works/1, JCT Design and Construct Contract provides an option for an advance payment to be made. Such advance payment would normally include pre-contract design costs.

3.11.4. ICE Design and Construct makes provision in clause 60(2)(a) for a payment schedule to be included in the contract. This schedule should make it clear as to when payment for pre-contract and post-contract design costs are to be made. If there is no schedule, design costs should be dealt with in the same manner as Alternative B of the JCT Design and Build Contract. No provision is made for advance payment or mobilisation payment.

SUMMARY

Payments should reflect the fact that design costs comprise pre-contract design costs and post-contract design costs. Where stage or milestone payments apply, these costs should be properly allocated to the appropriate stage or milestone. The first stage payment should include for pre-contract design costs. If there is no provision for stage or milestone payments, the first interim payment should include all of the pre-contract design costs. The post-contract design costs should be included in subsequent interim payments, to suit the progress of the design.

3.12. On a design and construct project, where the architect is novated from the employer to the contractor, is there any impediment upon the contractor's ability to recover from the architect loss he suffers because of architect design errors which occurred during his employment by the employer?

3.12.1. It has become a common practice for employers wishing to enter into a design and construct contract to start off the process by appointing an architect themselves. The intention is for the architect to be involved in the planning application and to work up the design to a state where tenders from contractors can be sought. When the contractor is appointed, the architect, by way of a novation agreement, becomes a part of the contractor's team. Under the novation agreement the contractor takes responsibility for the work carried out by the architect both pre- and post-contract.

3.12.2. The wording of the novation agreement provides for the contractor to stand in the employer's shoes with regard to negligence on the part of the architect. Any right of redress vested in the employer regarding the negligence of the architect in the pre-contract stage is transferred to the contractor. Contractors have derived comfort from this arrangement. They considered that any loss incurred as a result of an architect's pre-contract error could be recovered from the architect.

3.12.3. In the case of *Blyth and Blyth* v. *Carillion Construction Ltd* (2001), the architect was responsible for design errors in the pre-contract stage, which resulted in the contractor incurring additional cost. The novation agreement allowed the contractor to pursue claims against the architect which would have been available to the employer. In other words, whatever loss the employer would have incurred resulting from the design errors was recoverable by the contractor. The design errors affected the contractor's price, but would not have involved the employer in any additional cost. The contractor therefore recovered nothing. This will have come as a shock to regular design and construct contractors. Serious rewriting of novation agreements was obviously necessary as a result of this decision.

3.12.4. The CIC/Nov Agr novation agreement published in 2004 by the Construction Industry Council deals with this problem. In clause 4(a), the consultant warrants to the contractor that all services provided to the employer have been performed in accordance with the terms of the original appointment.

SUMMARY

Under a design and construct contract, the contractor takes responsibility for all design work carried out both pre- and post-contract. Often, however, the employer engages an architect to produce a preliminary design and secure planning consent. The intention of a novation agreement is to transfer the architect's design obligation in the pre-contract stage from the employer to the contractor. Comfort can be drawn by the contractor from the novation agreement concerning design errors due to negligence by the architect which occurred in the pre-contract stage. Whilst the contractor can be held responsible to the employer for the design errors, redress by the contractor can be sought from the architect. Unfortunately, due to the particular wording of many novation agreements, as the employer is unlikely to suffer loss due to an architect's negligence as the risk has been transferred to the contractor, he in turn can recover nothing from the architect.

The CIC/Nov Agr novation agreement published in 2004 by the Construction Industry Council deals with this problem. In clause 4(a), the consultant warrants to the contractor that all services provided to the employer have been performed in accordance with the terms of the original appointment.

Chapter 4

Letters of Intent

4.1. Mr Justice Clarke, in the case of *RTS Flexible Systems Ltd* v. *Molkerei Alios Muller* (2010), said 'This case is another example of the perils of proceeding with work under a letter of intent'. What did he mean?

4.1.1. Fast-track construction methods often overtake the procedures for drawing up a formal contract, which in many instances lacks the necessary urgency. Instructions to commence work on site are therefore often given before a contract has been entered into. Parties to construction contracts are, at the outset, often imbued with good relationships and feel comfortable about a start being made in the absence of a formal contract. Usually, a letter is despatched providing the instruction to commence and including words to the effect that the formal contract is to follow in due course. This type of letter is often referred to as a 'letter of intent'. There is no set format for a letter of intent, the drafter making up his own mind as to its contents.

4.1.2. Difficulties can often arise where the work proceeds and no contract is ever entered into. The parties are often in disagreement as to their respective rights and obligations. There have, in the past few years, been numerous referrals, at great expense, to courts for those rights and obligations to be established. The courts are left to examine the precise wording contained in letters of intent and to pick over the meaning, often with large sums at stake. It is not surprising, therefore, that some judges have condemned the use of the letter of intent, because of the uncertainty as to the intentions of the parties as expressed therein.

4.1.3. In the case of *EFDC Group* v. *Brunel University* (2006), Judge Lloyd defined letters of intent as follows:

> Some are merely expressions of hope; others are firmer but make it clear that no legal consequences ensue; others presage a contract and may be tantamount to an agreement 'subject to contract'; others are contracts falling short of the full-blown contract that is contemplated; others are in reality that contract in all but name. There can be no precise assumptions, such as looking to see if words such as 'letter of intent' have or have not been used. The phrase 'letter of intent' is not a term of art. Its meaning and effect depend on the circumstances of each case.

200 Contractual Problems and their Solutions, Third Edition. Roger Knowles.

4.1.4. Mr Justice Akenhead highlighted the dangers of using letters of intent in the case of *Diamond Build Ltd* v. *Clapham Park Homes Ltd* (2008), when he said:

> This is yet another case which relates to a letter of intent on a construction project . . . The case illustrates the dangers posed by letters of intent which are not followed up promptly by the parties progressing the formal contract anticipated by them at the letter of intent stage.

4.1.5. If, despite advice to the contrary, the parties are set upon using a letter of intent, there is judicial guidance as to when it is safe to send a letter of intent. Judge Coulson, in the case of *Robert Cunningham, Catherine Good, Geland E Corporation Ltd* v. *Colett and Farmer* (2006), made the following suggestions as to when a letter of intent may be appropriate:

> It seems to me, a letter of intent can be appropriate in circumstances where:
>
> - The contract work scope and price are either agreed or there is a clear mechanism in place for such work scope and price to be agreed.
> - The contract terms are (or are very likely to be) agreed.
> - The start and finish dates and the contract programme are broadly agreed.
> - There are good reasons to start work in advance of the finalisation of all the contract documents.
>
> In those circumstances I can see that, if the employer wants the work to start on site promptly and the contractor is also keen to commence work, then a careful letter of intent can be appropriate.

4.1.6. The main difficulty with letters of intent is that there is no standard format and no universal agreement as to their purpose, other than to provide a mechanism to allow work to commence. Each letter of intent is drafted on an *ad hoc* basis, leaving a great deal of uncertainty as to its meaning. When the formal contract fails to materialise and the parties get into dispute, there is often extreme difficulty in sorting out the respective parties' rights and obligations. Many of these cases have come before the courts, examples in addition to those already referred to comprise:

- *British Steel Corporation* v. *Cleveland Bridge Ltd* (1984).
- *Kitson Insulating Contractors Ltd* v. *Balfour Beatty Buildings Ltd* (1989).
- *Durabella Ltd* v. *J Jarvis and Sons Ltd* (2001).
- *AC Controls Ltd* v. *BBC* (2002).
- *Metropolitan Special Projects Ltd* v. *Margold Services Ltd* (2001).
- *Shimizu Europe Ltd* v. *LBJ Fabrications Ltd* (2003).
- *Tesco Stores Ltd* v. *Costain Construction Ltd* (2003).
- *Twinsec Ltd* v. *GSE Building and Civil Engineering Ltd* (2003).
- *ABB Engineering and Construction Pty Ltd* v. *Abigroup Construction Ltd* (2003).
- *Harvey Shopfitters* v. *ADI* (2004).
- *Westminster Building Company* v. *Buckingham* (2004).
- *Mowlem PLC* v. *Stena Line Ports Ltd* (2004).

- *Bryen and Langley Ltd* v. *Martin Boston* (2005).
- *Skanska Rashleigh Ltd* v. *Somerfield Stores Ltd* (2006).
- *Bennett Electrical Services Ltd* v. *Inviron Ltd* (2007).
- *Diamond Build Ltd* v. *Clapham Park Homes* (2008).
- *Whittle Movers Ltd* v. *Hollywood Express Ltd* (2009).

4.1.7. In view of the problems which letters of intent can create relating to the uncertainty of the rights and obligations of the parties, it is hardly surprising that in the case of *RTS Flexible Systems Ltd* v. *Molkerei Alois Muller Gmbh* (2010), Lord Clarke in the Supreme Court had these words to say concerning letters of intent:

> The different decisions in the courts below and the arguments in this court demonstrate the perils of beginning work without agreeing the precise basis upon which it is to be done. The moral of the story is to agree first and to start work later

SUMMARY

The scale of the problem that can arise from starting work without a formal contract but merely a letter of intent can be judged by the number of cases coming before the courts relating to projects which have commenced in this manner. The absence of an agreement of all the major terms and conditions which will apply can create serious problems for the parties. In probably the vast majority of cases, following the sending of a letter of intent a formal contract is produced at a later stage. However, all too often, where work is commenced on the basis of a letter of intent, in the absence of proper agreement, the likelihood of disputes arising is high. The main difficulty is that letters of intent come in all shapes and sizes and can be very uncertain in the wording as to the rights and obligations of the parties. This led Lord Clarke in the case of *RTS Flexible Systems* v. *Milkerei Alois Muller* to say, 'This case is another example of the perils of proceeding with work under a letter of intent'. The problem lies not so much with the letter of intent, with its objective of getting work started, but in doing so when crucial contractual matters are still to be agreed, and a failure to quickly follow up with a formal contract.

4.2. What risk is a contractor taking if it receives a letter of intent which places a cap on expenditure, but carries out work in excess of the cap?

4.2.1. There may be circumstances where the employer is anxious to have work commenced, possibly to avoid losing financial backing, but before all crucial matters have been agreed. One option is to instruct the contractor to commence work using a letter of intent, but to cap the expenditure. The intention of this process is for the employer to be in a position to replace the contractor if agreement is not reached.

4.2.2. What is the position of the contractor if, when work is stopped, the expenditure has exceeded the cap? Is payment limited to the amount of the cap, or is the contractor entitled to a payment which reflects the amount of work undertaken? In the case of *Diamond Build Ltd* v. *Clapham Park Homes Ltd* (2008) a tender was submitted in respect of the refurbishment of a number of properties. A letter of intent was sent by way of response, instructing the contractor to get on with the work with due diligence and stating a date for completion. The letter of intent went on to state that if no contract was executed, Clapham Park would reimburse Diamond Build, up to a maximum of £250,000, for its reasonable costs up to the date on which a decision was made not to proceed. It was stated in the tender enquiry that a JCT contract would be used, executed as a deed. A contract was drawn up and sent to Diamond Build, but was not signed. After work had progressed for some time, Clapham Park became dissatisfied with the work being carried out by Diamond Build and terminated the arrangement. Clapham was prepared to pay only the amount of the cap. Diamond Build argued that they were not working on the basis of the letter of intent, as they had been sent the contract documents which governed the work they had undertaken. As the letter of intent no longer applied, the cap was not relevant.

4.2.3. The court disagreed, concluding that there *was* a contract in existence as set out in the letter of intent, which provided start and finish dates and a sum of £250,000 to be spent. It was the opinion of the judge that there was no reason why the cap should not be enforced, even though Diamond Build claimed the value of the work far exceeded the cap. The judge suggested that Diamond could have protected itself by agreeing caps with its subcontractors, to ensure that its total liability did not exceed the cap. Alternatively, Diamond Build should have requested Clapham Park to increase the cap.

4.2.4. The case of *AC Controls Ltd* v. *British Broadcasting Corporation* (2002) produced a different result. BBC required the installation of an essential software system which would control, monitor and record access to and from 57 BBC properties and was looking to AC Controls Ltd to carry out the work. The intention was for a formal MF/1 contract to be entered into. The BBC was anxious for a start to be made and therefore a letter of intent was sent, to enable AC Controls Ltd to start on the design work. A financial cap of £500,000 was indicated in the letter of intent. Work was carried out by AC Controls Ltd which well exceeded the amount of the cap. The BBC insisted that their liability extended only to the capped figure of £500,000. It was held by the court that the BBC was entitled to determine the arrangement when the cap was reached and, as they failed to do so, AC Controls Ltd was entitled to be paid a reasonable sum for the work undertaken.

4.2.5. In the case of *Eugena Ltd* v. *Gelande Corporation* (2004), the claimant contractor was asked to submit a tender to undertake work for the defendant. The intention was that a JCT contract would be entered into. The defendant was anxious for work to get under way and so sent to Eugena Ltd a letter of intent, instructing that work be commenced. A cheque for £40,000 was enclosed with the letter of intent, together with an instruction that capped expenditure at £50,000. Negotiations with regard to a formal contract were entered into but not concluded. Nothing was finalised and so it was agreed that work would cease. Eugena Ltd submitted a final invoice in the sum of £76,000, but Gelande Corporation refused to pay any more than the £50,000 capped figure. It was argued on behalf of the claimant that it was entitled to be paid £76,000, which was the value of the work for which the defendant had received benefit. It was held that Eugena Ltd was entitled to be paid the sum of £50,000 included in the cap, and no more.

4.2.6. The circumstances in the case of *Mowlem PLC* v. *Stena Line Ports Ltd* (2004), which related to the construction of a new ferry terminal at Holyhead in Anglesey, were somewhat unusual. In all, 14 letters of intent, each of which included a cap on expenditure, were sent by Stena Line to Mowlem. The last letter of intent was dated 4 July 2003 and included a cap of £10m. Due to the discovery of rock and instructions from the employer to undertake additional work, expenditure exceeded the £10m cap. The letter of intent expressly required Mowlem to carry out work up to 18 July 2003. Mowlem argued that Stena allowed them to carry on working after this date and, in addition, as work had been carried out over and above what had been envisaged, a payment in excess of the cap should be made. The court found against Mowlem, as it was considered that each of the letters of intent were themselves contracts. The last letter provided for Mowlem to carry out the works, including any post 4 July 2003 variations, for a payment of its reasonable costs up to a maximum of £10m. Mowlem could have requested a further letter of intent to cover the work required, over and above the work included within the £10m cap, but failed to do so.

SUMMARY

It can be seen from these judgments that there is no hard and fast rule as to the financial entitlements of a contractor who undertakes work which results in a financial cap included in a letter of intent being exceeded. Some judges take the view that where a cap occurs, the employer must order a stop to the work when the cap has been reached; otherwise there is a liability to pay extra. This is a most impractical view, as the employer is not in a position to know when the cap has been reached. It is a much more reasonable approach to expect the contractor to seek an increase, when expenditure is approaching the cap, which is the view taken by judges in the majority of cases involving this problem which have reached the court.

4.3. When work is undertaken in accordance with a letter of intent without a contract being entered into, on what basis is the contractor or subcontractor entitled to be paid for the work carried out?

4.3.1. It is important for the wording contained in a letter of intent to indicate in precise terms how any sums to be paid to a contractor or subcontractor are to be calculated for any work undertaken, in the event of a contract not being concluded.

4.3.2. In the case of *Robertson Group (Construction) Ltd* v. *Amey-Miller Joint Venture* (2005), a letter of intent was sent relating to extensive refurbishment work at the Royal High School, Edinburgh, under a PFI arrangement. The wording of the letter of intent regarding payment stated:

> Should a formal contract fail to be entered into for any reason other than the default of negligence of Robertson Construction then all direct costs and directly incurred losses shall be underwritten and reimbursed by the Joint Venture.

4.3.3. No contract was ever entered into and a dispute arose concerning an interpretation of the words 'all direct costs and losses'. The dispute related to whether the wording should be interpreted to include profit. The court took the view that, in the commercial world, there is an expectation, or at least a hope, of returning a profit. A failure to make a profit would amount to a loss; therefore, as the letter of intent safeguards the claimant from making a loss, then there existed an entitlement to be paid a profit.

4.3.4. A difficulty arose in the case of *ERDC Group Ltd* v. *Brunel University* (2006) relating to the mode of payment which applied after a letter of intent expired. ERDC was required to construct a new sports facility for the University. Because of planning delays, the University decided to have work started by way of a letter of intent. The letter of intent, which instructed the contractor to undertake a limited amount of work, indicated that payment would be made in accordance with the terms and conditions of the JCT With Contractor's Design Form of Contract. This letter of intent was followed by two more, both of which made reference to payment in accordance with the JCT Form of Contract. Work which was instructed to be carried out in accordance with the last letter of intent expired on 1 September 2002. One of the matters in dispute related to how work which had been carried out after 1 September 2002 should be valued, as no contract was ever entered into.

4.3.5. ERDC submitted eight applications for payment in accordance with the provisions of the JCT Contract. However, in the absence of a concluded contract, ERDC argued that they were entitled to be paid on a *quantum meruit* basis for all of the work, employing the recorded costs. This was disputed by the University. Judge Lloyd concluded that, in the absence of a formal contract, based upon what was stated in the letters of intent, there was a contract in place for work undertaken prior to 1 September 2002, with the terms and conditions of the JCT Form of Contract applying. The more difficult question was the basis for payment of the work carried out after 1 September 2002. Because the letter of intent expired on 1 September 2002, it could not be said that work was carried out under a contract after this date. The court agreed that work undertaken after this date should be valued on a *quantum meruit* basis; however, ERDC's costs should not be employed as the basis for the valuation. It would not be right to have a part of the work valued in accordance with the JCT Form of Contract, with the remainder on a cost-reimbursable basis. The judge decided that work carried out after 1 September should be valued on the same basis as work executed prior to this date, namely, in accordance with the JCT Form of Contract, which was thus deemed to be a *quantum meruit*.

4.3.6. In the case of *Skanska Ransleigh Weatherfoil* v. *Somerfield Ltd* (2006) the lower court and Court of Appeal were at odds as to the meaning of the words:

> while we are negotiating the terms of the agreement, you will provide the services under the terms of the contract from 28th August 2000 until 27th October 2000.

The lower court considered these words were limited to the actual provision of the services and did not include all the remainder of the terms included in the Facilities

Management Agreement, such as payment, which was the subject of the negotiations. The Court of Appeal disagreed and held that the wording was intended to mean all the terms in the Facilities Management Agreement, including the payment terms.

4.3.7. The letter of intent may be fairly clear in that, in the event of a contract not materialising, for example, it may indicate that the contractor will be paid 'reasonably ascertainable costs', being the wording of the letters of intent which were used in the case of *Bryn and Langley Ltd* v. *Martin Boston* (2005). Difficulties can arise and argument ensues regarding what are to be regarded as reasonable costs. The question of profit has already been dealt with; but what about reworking due to faults in the workmanship, or allegations of inefficiency or lack of competitive bids for subcontracted work?

4.3.8. Many letters of intent fail to address the question of payment at all. It is well established that a fair price would be paid in the absence of a contract. The term *quantum meruit* is often used to denote a fair price. But what constitutes a fair price? In the case of *EDRC Group Ltd* v. *Brunel University* (2006), the claimant considered a *quantum meruit* payment should be based on costs, whereas the court decided that contract rates should be applied.

SUMMARY

Letters of intent should be avoided, if at all possible. Those involved where letters of intent are employed, should always make provision in the letter of intent as to what should happen if no contract comes into being. Of crucial importance is the basis for payment in respect of work undertaken in accordance with the requirement of a letter of intent, if no contract comes into being. The payment entitlement should be as comprehensive as possible, to avoid disputes at a later date. Where the letter of intent is silent as to payment in the event of no contract being concluded, the contractor will be entitled to payment of a fair and reasonable amount to reflect the work undertaken. Unfortunately, what constitutes a fair and reasonable payment can be a matter of conjecture.

4.4. Under what circumstances, if any, could a letter of intent be regarded as a concluded contract?

4.4.1. Letters of intent can lead to confusion as to the rights of the parties. In the case of *Cunningham* v. *Collett Farmer* (2006), Judge Coulson said that letters of intent should only be used where agreement had been reached as to the work, scope and price, conditions which are to apply and also the start and finish dates. If all these matters have been set out in a letter of intent, it could be argued that a contract exists between the parties. In the case of *British Steel Corporation Ltd* v. *Cleveland Bridge Ltd* (1984), the court had to deal with the question as to whether a particular letter of intent created a contract. In the context of the case, Lord Justice Goff said:

> Now the question is whether, in a case such as the present one any contract has come into existence must depend on a true construction of the relevant communications which have

passed between the parties and the effect (if any) of their action pursuant to those communications. There can be no hard and fast answer to the question whether a letter of intent will give rise to a binding agreement; everything must depend on the circumstances of the particular case.

4.4.2. The case of *Kitsons Insulation Contractors Ltd* v. *Balfour Beatty Ltd* (1989) relates to a dispute arising from the construction of Phase I development at the White City for the BBC. Kitsons submitted a tender in the sum of £1,100,000, in October 1987, for the modular toilet facilities and accessories, which included a payment schedule. Balfour Beatty sent a letter of intent to Kitsons in March 1988. The general gist of the letter was that Balfour Beatty intended to enter into a subcontract with Kitsons, using the standard subcontract DOM/2 1981 edition, amended to suit Balfour Beatty's particular requirements, which were to be forwarded to Kitsons in due course. Kitsons signed the letter of intent, returned it to Balfour Beatty and then commenced work. In August 1988 Balfour Beatty drew up and sent a formal contract to Kitsons. An accompanying letter indicated that, until Kitsons signed and returned the contract, no payment would be made. Kitsons refused to sign the contract, as it contained a number of significant changes from those set out in the letter of intent and did not include the payment schedule. The court held that no contract had been concluded, as there were a number of important matters not agreed which were not covered by the letter of intent, such as the method of payment, and were too significant for a contract to come into place.

4.4.3. In the case of *Mitsui Babcock Engineering Ltd* v. *John Brown Engineering Ltd* (1996), a dispute arose as to whether a contract had come into being. In October 1992 Mitsui Babcock Engineering sent a proposed contract to John Brown Engineering Ltd, which was subject to the MF/1 conditions. Clause 35 of the conditions, which was headed 'performance tests', was struck out and a marginal note 'to be discussed and agreed' written in. The document was signed by both parties, but there was no agreement concerning performance tests and liquidated damages for failure to achieve the performance tests. Despite the parties' inability to reach agreement on these matters, the court held that there was, nonetheless, a binding contract.

4.4.4. In some cases, a letter of intent may represent only one of the documents which make up a contract. The case of *Twintec Ltd* v. *GSE Building and Civil Engineering Ltd* (2003) involved a complex situation involving a quotation, a letter of intent, an acceptance letter and a meeting, which when read together formed the basis of the contract.

4.4.5. In the case of *Mowlem PLC* v. *Stena Line Ltd* (2004), there was a series of letters of intent; each one superseded the previous one. The court held that each letter of intent constituted a binding contract.

4.4.6. As can be seen in the *Mowlem* case, the contract which results from the letter of intent may not extend to the full project, but merely a part. These mini contracts were referred to in the case of *British Steel Corporation Ltd* v. *Cleveland B ridge Ltd* (1984) as 'if' contracts, on the basis that payment will be made if the work in the letter of intent is carried out. A contract for the full project may or may not subsequently be forthcoming.

SUMMARY

There is no straightforward answer to the question as to whether a particular letter of intent is the basis of a binding contract. The courts will look at the communications, both written and oral, between the parties to ascertain whether or not a contract has been formed. Correspondence between the parties will be examined to see if agreement has been reached on all the major issues, in particular the scope of works and the payment terms. The court, in reaching its decision, will also take into account how the parties have conducted themselves.

4.5. What are the advantages and disadvantages to an employer and contractor in work being commenced on the basis of a letter of intent?

4.5.1. The main purpose of commencing work on the basis of a letter of intent is to enable a start to be made before a formal contract has been drawn up and signed by the parties. Unfortunately, there is no standard format for letters of intent and, therefore, they come in all shapes and sizes. There have been numerous legal cases where the wording of a letter of intent has been the subject matter of a dispute. It is clear that one of the problems relating to letters of intent is that there is often uncertainty as to the respective rights and obligations of the parties which have been created by a letter of intent. At an early stage, contractors often require subcontractors to commence work and major orders for materials need to be placed, if completion dates are to be achieved. There is also the matter of appointing personnel for the project. Such actions will normally involve major financial commitments and require an undertaking from the employer to reimburse the contractor if the work is terminated without a binding contract being entered into. Contractors need to be assured that before commencing these expensive undertakings that there is a commitment on the part of the employer to make payment in respect of these matters. Letters of intent should therefore be scrutinised by contractors to ensure they adequately provide for these costs to be reimbursed.

4.5.2. Whilst the employer may have gained an advantage, usually commercially, from the early start, it could leave him vulnerable. If the contractor refuses to sign the formal contract when it is presented, the employer is left with a dilemma. Does the employer allow the contractor to continue with the work or find a replacement? In allowing the contractor to continue, it is unlikely that there will be any agreement as to the payment mechanism for the whole project. The letter of intent may merely state that the contractor will be reimbursed its costs until a formal contract has been concluded. Attempting to agree as to what constitutes costs can often create difficulties. The contractor, if allowed to complete the project in the absence of a concluded contract, will be entitled to receive payment on a *quantum meruit* basis; in other words, a payment which is fair and reasonable. Reaching an agreement of a fair and reasonable price for the work could cause a problem. Should the payment be based upon the contractor's costs or its tender price?

There is no hard and fast rule to answer this question. This dilemma can often provide the contractor with a negotiating advantage, as in the absence of a binding contract contractors often become more ambitious as to the terms under which they are prepared to carry on and complete the works.

4.5.3. If, on the other hand, the employer decides to replace the contractor, it is unlikely that the contractor will go willingly and disputes as to the quality of work and payment will probably occur. Where a project has been commenced by one contractor and completed by another and, subsequently, faults arise in the workmanship or materials, it may not be clear in whose work those faults originate. Again, this could lead to serious disputes.

4.5.4. Contractors are usually delighted to be requested to commence work on a project. This usually means that the competition has been eliminated and the job is theirs. Commencement of work on a project by means of a letter of intent is usually seen by both parties as a temporary arrangement, which will be shortlived, until the contract is signed up. These short periods often go from a few days or weeks to many months. Payment often becomes a problem, in the absence of precise amounts which will be paid at agreed stages. An undertaking to reimburse a contractor in accordance with its costs often leads to disagreements. In the case of *Kitsons Insulation Contractors Ltd* v. *Balfour Beatty Buildings Ltd* (1989), work commenced on the basis of a letter of intent. With work progressing, Kitsons' request for payment was met with the response from Balfour Beatty that payment would not be forthcoming until they signed a formal contract which included payment and other provisions to which Kitsons had not given its agreement.

4.5.5. An advantage to each party is that if, following commencement, events occur which render the contract undesirable, either party can withdraw at any time.

SUMMARY

The main advantage of the use of a letter of intent is to allow work to commence earlier than would have been the case, if delayed until a formal contract was signed. Once the formal contract is signed, there should be few problems relating to the contractual relationship between the parties. However, it is not uncommon for the parties never to sign a formal contract. This being the case, the parties often have difficulties in establishing their rights and responsibilities. Disputes, as a result, are a common feature of letters of intent.

4.6. Could an architect, engineer or project manager be negligent for recommending to an employer that a letter of intent be used?

4.6.1. On many projects, the completion date triggers off a chain reaction of events which can have major financial consequences for an employer. Delays to the project in the design stage can often threaten to jeopardise the start; and hence, the completion date. Architects, engineers and project managers are under pressure to gain time to overcome delays, and the use of a letter of intent is often seen as a solution. However, in view of the many cases referred to earlier in this section, the use of a letter of intent is fraught with danger. What would be the position if an architect, engineer or project manager recommended the use of a letter of intent which resulted in a major dispute and created more expenditure than it saved? Would an action in negligence on the part of the employer be likely to succeed?

4.6.2. The case of *Cunningham* v. *Collett and Farmer* (2006) involved the sending of a letter of intent, which resulted in the project running into serious difficulties. The employer as a result brought an action against the architect in negligence for the recovery of his costs and losses. The claimants, Mr Cunningham and Ms Good, employed the defendants to act as architect in respect of the refurbishment of property in Hertfordshire. Two tenders were received for the work, but each exceeded the budget figure of £500,000. The defendant recommended that the contract be awarded to Eugena, whose price was £605,722 subject to certain changes, which included a reduction in the work-scope and hence price. The defendant recommended the use of a letter of intent to allow work to start, to be followed by the formal contract as soon as possible thereafter.

4.6.3. A number of delays took place, including the sending of the letter of intent. No contract was ever entered into. The defendant ceased work on the grounds that his fees had not been paid. The matter was referred to adjudication and the adjudicator decided in favour of the defendant, but the claimant refused to make payment. A court action was commenced to enforce the adjudicator's decision. The claimants in turn commenced an action against the defendant for breach of contract. As part of their action, the claimants alleged that the defendant was negligent in recommending the use of a letter of intent.

4.6.4. The claimant alleged that the defendant should have advised them that a letter of intent was 'not appropriate at all and/or premature in all the circumstances'. In pursuing their case, the claimants submitted, on the basis of expert evidence, that a letter of intent was not an acceptable way to commence any building project.

4.6.5. By way of response, Judge Coulson had this to say:

> I agree . . . that letters of intent are used too often in the construction industry as a way of avoiding, or at least putting off, potentially difficult questions as to the final make-up of the contract and the contract documents. There is no doubt that, sometimes they are issued in the hope that once the work is underway potentially difficult contract issues will somehow resolve themselves. They are plainly not appropriate in such circumstances. But having said all that, I do not agree that letters of intent are as a matter of principle, always or almost always inappropriate. There will be times when a letter of intent is the best way of ensuring that the works can start promptly with a clear timetable both for the finalisation of the contract formalities, and for the carrying out of the works themselves.

4.6.6. Judge Coulson went on to say that letters of intent could be appropriate where the scope of works, the price and contract terms and conditions had been agreed, as well as the start and finish dates. In the case in question, Judge Coulson found that agreement had been reached on these matters and, therefore, the architect had not been negligent.

SUMMARY

The letter of intent is a very commonly used method of getting work commenced in the absence of a formal contract. Judge Coulson, in the one reported case relating to an action in negligence against an architect for recommending the use of a letter of intent, said that recommending the use of a letter of intent was not in itself an act of negligence. A letter of intent, he said, would be appropriate where the scope of the work, the price and contract terms and conditions had been agreed, together with the start and finish date. The implication being that, in the absence of agreement on these matters, the use of a letter of intent was inappropriate. A recommendation to use a letter of intent without there being agreement on these matters could therefore amount to an act of negligence. Most experienced Architects, Engineers and Project Managers, however, will be careful to point out to their clients the advantages and shortcomings in the use of a letter of intent and also the views of Judge Coulson, leaving the final decision to the employer as to whether or not to use a letter of intent.

Chapter 5
Programme

5.1. Where a contractor submits a programme which is approved or accepted by the architect/engineer, is he obliged to follow it or can he amend it at his own discretion?

5.1.1. The programme is usually intended to be a flexible document. If the contractor gets behind, say, because of the insolvency of a subcontractor, he would normally expect to revise the programme in an attempt to make up lost time. For this reason, programmes are rarely listed as contract documents. It is the requirement of most contracts that obligations provided for in contract documents must be carried out to the letter. With a programme containing some hundred or more activities, compliance with the start and finish date for each without the possibility of revision would be impractical. For this reason, programmes should not be contract documents.

5.1.2. An exception to the general convention that the programme is not a contract document is GC/Works/1, where, in the definitions section, the programme is listed as a contract document. This could make life difficult for a contractor undertaking work on a contract where these conditions apply. There is an obligation to carry out work in accordance with the contract documents, which would include the programme. The usual intention in contracts such as JCT is for the programme to be flexible, in order to give the contractor maximum scope to make changes on a regular basis, when he causes delays, to ensure that completion on time is achieved.

5.1.3. Nevertheless, some forms of contract will not permit the contractor to amend its programme, once accepted, without approval. For example, GC/Works/1, condition 33(2), states:

> the contractor may at any time submit for the PM's agreement proposals for the amendment of the Programme.

This is an essential provision, in the light of the programme being a contract document. MF/1, clause 14.4, is worded along similar lines, to the effect that the engineer's consent is required before the contractor can make any material change to the programme.

5.1.4. Clause 14(4) of the ICE 6th and 7th Editions empowers the engineer to require the contractor to produce a revised programme, if progress of the work does not

200 Contractual Problems and their Solutions, Third Edition. Roger Knowles.

conform with the accepted programme. The revised programme must show the modifications to the accepted programme to ensure completion on time. That apart, there is no restriction placed upon the contractor who wishes to revise his accepted programme.

5.1.5. The Engineering and Construction Contract (formerly the NEC), clause 32.1, calls on the contractor to show on each revised programme

- the actual progress achieved on each operation and its effect upon the timing of the remaining work;
- the effects of implemented compensation events and of notified early warning matters;
- how the Contractor plans to deal with any delays and to correct notified Defects; and
- any other changes which the Contractor proposes to make to the Accepted programme.

Clause 32.2 goes on to add that the contractor is required to submit a revised programme to the project manager for acceptance.

5.1.6. Other forms of contract, for example JCT 2011, do not expressly require the contractor to seek approval to the amendment of his programme. However, if amendments are made without approval, the architect may, however, feel under no obligation to issue drawings to meet the revised programme.

SUMMARY

Some forms of contract require the contractor to seek approval or acceptance before amending his programme; for example, GC/Works/1, MF/1 and the Engineering and Construction Contract. In the absence of an express requirement to seek approval to amend, the contractor can revise his programme as he wishes. An architect or engineer, however, who has not been asked to approve or accept an amended programme may feel under no obligation to issue drawings in good time to enable the contractor to comply with the revised programme.

5.2. Is a subcontractor obliged to follow a main contractor's programme?

5.2.1. Most standard forms of contract provide for the contractor to produce a programme. A failure on the part of the contractor to produce the programme amounts to a breach of contract. It is unusual, however, for a contract to state expressly that a contractor must follow the programme. An exception is GC/Works/1, which states in condition 34(1) that the contractor

> shall . . . proceed with diligence and in accordance with the Programme or as may be instructed by the PM . . .

5.2.2. The contract programme is rarely classified in a contract as a contract document. If it were so, then contractors would be required to carry out work strictly in accordance with the programme. This could prove very exacting and, in many instances, impossible.

5.2.3. The situation with subcontractors is similar to that of a main contractor. An example of the obligation of a subcontractor with regard to a main contractor's programme occurred in *Pigott Foundations* v. *Shepherd Construction* (1994). Pigott was employed as a domestic subcontractor to design and construct bored piling on a new 14-storey office block. The main contract was JCT 80, the subcontract DOM/1, 1980 edition, and Shepherd Construction the main contractor. Pigott's subcontract provided for work to be carried out in eight weeks. Piling work commenced on 26 June 1989. However, the bulk of the work had been completed, with the exception of a few piles, by 20 October 1989. Pigott then left site and returned in April 1990 to complete the remaining piles. After commencement, work had proceeded at a slow pace, with only one pile completed in the first week and further difficulties arose due to piling work, which was alleged to be defective. It was not clear whether this was due to faulty design or bad workmanship. Pigott claimed that the difficulties arose as a result of ground conditions. A solution to the problem was reached, which involved installing additional piles. It was decided in this case that where DOM/1 conditions apply, a subcontractor is not required to comply with the main contractor's programme. Clause 11.1 of the subcontract required Pigott to complete the subcontract reasonably in accordance with the progress of the work, and the judge, not surprisingly, decided that this did not mean that the subcontractor must carry out its work in accordance with the main contractor's programme.

5.2.4. If there exists an obligation for a subcontractor to carry out work to suit a main contractor's programme, it can be a two-edged sword for the main contractor. Such a requirement would place an obligation upon the main contractor to provide access to enable the subcontractor to carry out the subcontract work in accordance with the main contractor's programme. Contractors often experience difficulties in this respect, as happened in the case of *Kitson Sheet Metal Ltd* v. *Matthew Hall Mechanical & Electrical Engineers Ltd* (1989). The court had to decide whether Kitsons, the subcontractors, were entitled under the contract to work to the programme and whether any written order requiring departure from it constituted a variation. Clause 18(1) of the subcontract required the subcontractor to commence when instructed and to proceed with due diligence and without any delay and to complete the works within the periods stated in the sixth schedule. Unfortunately, the sixth schedule did not contain any time periods and made no reference to the programme, but instead required the subcontract works to be carried out in accordance with the instructions of Matthew Hall site management, to enable the mechanical installation to be handed over by 18 March 1985. It was held that the parties must have recognised the likelihood of delays and of different trades getting in each other's way and that the prospects of working to programme were small. Provided Matthew Hall had done their best to make areas available for work, they were not in breach of contract, even if Kitsons were brought to a complete stop. Kitsons were therefore unable to recover the additional costs due to a substantial overrun on the contractor's programme. A similar situation occurred in the case of *Martin Grant & Co Ltd* v. *Sir Lindsay Parkinson & Co Ltd* (1984). Again, the court held that there was no entitlement for the subcontractor to claim extra due to delays to the main contract programme.

SUMMARY

A subcontractor is not required to follow a main contractor's programme unless provided for expressly in the terms of the subcontract; equally, the main contractor is not obliged to grant access, etc. to enable the subcontractor to do so.

5.3. Who owns float time in the contractor's programme, the architect/engineer or the contractor?

5.3.1. Most prudent contractors will allow some form of contingency in their programme. Risk analysis is becoming a frontline skill in construction projects. More of the risk and, hence, uncertainty is being placed upon contractors. Bad ground, strikes, weather conditions, shortages of labour and materials are now regularly allocated in the contract as the contractor's risk. Contractors and their subcontractors often make mistakes, which have to be corrected. The contractor, therefore, will be unwise not to make provision in his programme for these uncertainties. A prudent contractor will always include an element of float in his programme to accommodate these variables.

5.3.2. The question, however, is this; if the contractor has clearly programmed an activity to take longer than is estimated to complete, can the employer take advantage of the float time free of cost? This might prove useful if the architect/engineer is late issuing drawings, or delays have been caused by the employer himself. It may be argued that float will not be on the critical path and so the employer using it will not cause any delay or disruption. Hence, the contractor will not become entitled to compensation.

5.3.3. Nevertheless, Keith Pickavance, in his book *Delay and Disruption in Construction Contracts*, at page 335 makes reference to a case heard before the Armed Services Board of Contract Appeals in the USA (Heat Exchanges (1963)). Here, it was held that the contractor's original cushion of time (which was not necessary for performance) should still be preserved when granting an extension of time for employer caused design delays. In an earlier case, the Army Corporation of Engineers' Board of Contract Appeals had recognised the contractor's right to reprogramme, thereby giving him the benefit of the float. American courts also took the line on a management dispute that total float may be used to programme jobs for all contractors; free float belongs to one contractor for programming any one activity. Neither total float nor free float should be used for changes, that is, variations, as expressed in *Natken and Co* v. *George A Fuller & Co* (1972).

5.3.4. The matter of float was in evidence in the case of *Ascon Contracting Ltd* v. *McAlpine Construction* (1999). McAlpine was the main contractor for the construction of a five-storey building known as Villiers Development in Douglas, Isle of Man, near to the sea front. Ascon was appointed as subcontractor for constructing the reinforced concrete floor slabs, basement perimeter walls and upright columns between floors. The subcontract period was 29 weeks, commencing on 28 August 1996, with completion by 5 March

1997. Practical completion of the subcontract work was not achieved until 9 May 1997, nine weeks late. No extension of time was granted. Ascon submitted a claim for an extension of time and payment of £337,918. McAlpine counterclaimed for £175,000 liquidated damages paid to the employer, plus its own loss and expense. McAlpine's claim against Ascon in respect of liquidated damages alleged to have been paid to the employer in the sum of £175, 000, left the judge unimpressed. A dispute between McAlpine and the employer had been compromised by a final payment of £9,475, inclusive of all McAlpine's claims, and explicitly no deduction was made for liquidated damages. Part of McAlpine's case against Ascon was that, had all subcontractors started and finished on time and McAlpine executed their own work on time, practical completion would have been achieved 5 weeks early. McAlpine's argument was that the five weeks' float was for their benefit, to absorb their own delays. As the five weeks had been used by Ascon and other subcontractors, McAlpine claimed they were entitled to recover their lost benefit. The judge rejected this argument. He considered the float to be of value in the sense that delays could be accommodated in the float time. This would avoid an overrun to the contract period and, hence, any liability to pay liquidated damages to the employer. The judge went on to say that McAlpine, whilst accepting the benefit against the employer, could not claim against the subcontractors. The judge seems to be taking the view that if float time is available, it can be used on a first come, first served basis.

5.3.5. It would seem that in this country, it is unlikely for an arbitrator to award an extension of time if the employer's delay did not affect the completion date. Judge Lloyd, in the case of *Royal Brompton Hospital* v. *Hammond* (2002), appeared to be of the view that the project owns the float, when he held that a contract administrator is bound to take any unused float into account, and stated that this is because an extension of time is only granted if completion would otherwise be delayed beyond the then current completion date. He went on to recognise the potential unfairness that this would cause to contractors in the event of a delay caused by the employer taking place before one caused by the contractor. Where this occurs, the contract administrator should inform the contractor that if he, the contract administrator, causes a delay later in the project, an extension of time will be granted for a period which will not exceed the period of the float.

5.3.6. The Society of Construction Law Delay and Disruption Protocol 2002 expresses an opinion concerning float time which is not dissimilar to the judgment in the *Ascon* case, where it states:

> unless there is express provision to the contrary in the contract, where there is remaining float in the programme at the time of an Employer Risk Event, an EOT should only be granted to the extent that the Employer Delay is predicted to reduce to below zero the total float on the activity paths affected by the Employer Delay.

What is being said is that if a contractor becomes entitled to an extension of time due to an employer's delay, any remaining float time must be taken up first and only if there is then an anticipated delay to completion will an extension of time be relevant. Any contractor delays which occur after all float has been used and which affect the completion date will create a liability on the part of the contractor to pay liquidated damages to the employer.

5.3.7. The Engineering and Construction Contract (NEC) contract requires the contractor to show on his programme the contractual completion date, the planned completion date and the float. When considering an entitlement to an extension of time, a delay to the completion date is assessed as 'the length of time that, due to the compensation event, planned completion is later than planned completion on the accepted programme'. In the NEC Guidance Notes it states that this method of assessment means that any terminal float, resulting from an early planned completion date, will be preserved. This appears to mean that the contractor is entitled to retain the float.

SUMMARY

There is no hard and fast rule as to who owns float, but it would seem that, as a contractor will normally include float in his programme to accommodate his risk items which cannot be accurately predetermined in terms of time involvement and also to provide time for correcting mistakes, then the float belongs to him and the employer or architect/engineer cannot object if later reprogramming by the contractor absorbs it. This is supported by legal decisions made in the USA. There is, however, a conflicting view expressed in the Society of Construction Law Delay and Disruption Protocol 2002, which states that float belongs to the project and may be used on a first come, first served basis. Legal decisions in the UK support the view expressed in the Protocol.

5.4. What is the effect of making the programme a contract document?

5.4.1. The standard forms in general use require the contractor to produce a programme. JCT 2011, clause 2.9 requires the contractor to produce his master programme as soon as possible after the execution of the contract. Details to be included in the programme, if required, such as the critical path must be set out in the contract documents. GC/Works/1, where the programme is included as a contract document, is more precisely worded where it states in condition 33(1):

> The Contractor warrants that the Programme shows the sequence in which the Contractor proposes to execute the Works, details of any temporary work, method of work, labour and plant proposed to be employed, and events, which, in his opinion, are critical to the satisfactory completion of the Works; that the Programme is achievable, conforms with the requirements of the Contract, permits effective monitoring of progress, and allows reasonable periods of time for the provision of information required from the Employer; and that the Programme is based on a period for the execution of the Works to the Date or Dates of Completion.

ICE 6th and 7th Editions, in clause 14, require the contractor to submit a programme to the Engineer for approval within 21 days of the award of the contract, but no reference is made to the type of programme required.

5.4.2. It is usually considered to be unwise for a progamme to be given the status of a contract document. Where a programme is given such a status, the contractor is required to comply with it to the letter. All flexibility, which is the key to catching up when progress gets behind, disappears. This apart, most standard contracts allow the contractor to revise its programme, but this could result in a great deal of time spent in undertaking reprogramming exercises.

5.4.3. An example of a problem for the employer when a programme is given the status of a contract document arose in the case of *Yorkshire Water Authority* v. *Sir Alfred McAlpine & Son (Northern) Ltd* (1985). The plaintiff invited tenders for a tunnel at Grimwith Reservoir. The contract was to incorporate the ICE 5th Edition and clause 107 of the specification stated:

> Programme of Work: In addition to the requirement of clause 14 of the conditions of contract, the contractor shall supply with his tender a programme in bar chart or critical path analysis form sufficiently detailed to show that he has taken note of the following requirements and that the estimated rates of progress for each section of the work are realistic in comparison with the labour and plant figures entered in the Schedule of Labour, Plant and Sub-Contractors . . .

The defendant submitted a tender in the standard ICE form accompanied by a bar chart and method statement. The method statement was approved at a meeting and, two months later, the defendant's tender was accepted by letter. A formal agreement was signed incorporating, inter alia, the tender, the minutes of the meeting, the approved method statement and the plaintiff's letter of acceptance. The method statement had followed the tender documents in providing for the construction of the works upstream. The contractors maintained that in the event it was impossible to do so and, after a delay, the work proceeded downstream. The contractors contended that, in the circumstances, they were entitled to a variation order under clause 51(1) of the ICE conditions. The dispute was referred to arbitration. The arbitrator made an interim award in favour of the contractor and the employer appealed. It was held by the court:

(1) The method statement was not the programme submitted under clause 14 of the contract.
(2) The incorporation of the method statement into the contract imposed on the contractors an obligation to follow it, in so far as it was legally or physically possible to do so.
(3) The method statement therefore became the specified method of construction, so that if the variation which took place was necessary for completion of the works, because of impossibility within clause 13(1), the contractors were entitled to a variation order under clause 51 and payment under clause 51(2) and 52.

5.4.4. A similar situation arose in *English Industrial Estates Corporation* v. *Kier Construction Ltd and Others* (1992). The instructions to tenderers provided that the tender should be accompanied by a full and detailed programme indicating the tenderer's proposed work sequence, together with a brief description of the arrangements and methods of demolition and construction which the tenderer proposed to adopt for the carrying out of the contract works. In due course, Kier prepared a method statement and enclosed it with their tender. It provided that details for on-site crushings of suitable demolition arisings, removal of unsuitable arisings and excavation material, and filling excavations with suitable crushed arisings. Disputes arose as to whether or not the contractors, through their subcontractors, were entitled to export off site materials excavated which they thought uneconomic to crush. This dispute, *inter alia*, was referred to an arbitrator, whose decision was upheld by the court. His award was that Kier's method statement was a contract document, but nevertheless the wording was such that under it, the contractor remained free:

> to decide which arisings it was uneconomic to crush; and to import replacement fill in place of hard arisings he chose not tocrush; and to export uncrushed hard arisings.

5.4.5. In *Havant Borough Council* v. *South Coast Shipping Company Ltd* (1996) a method statement was given the status of a contract document. The contractor was unable to follow the method statement, because of a court injunction which restricted the hours of working. To overcome the problem (which involved excessive noise), the contractor worked a different system from that provided for in the method statement. The court held that this constituted a variation and he was entitled to be paid.

5.4.6. Programmes provided by many contractors are drawn up with claims in mind. Often, unrealistic timescales are given within which drawings and instructions for the expenditure of prime cost and provisional sums are to be produced. Architects and engineers need to be very watchful at the time programmes are submitted to ensure that proper warning is given to the contractor that the programme is unrealistic, with regard to the timing of the architect/engineer's functions. Unfortunately, there are no sanctions provided in the contract where a contractor neglects to provide a programme as required. In the absence of express wording, the architect/engineer would be in breach in refusing to allow a contractor who had not produced a programme to start work. The exception is the Engineering and Construction Contract (NEC), which provides in clause 50.3 for only 25% of payments due to be made until the contractor fulfils his obligation to provide a programme.

5.4.7. As is evident from the aforementioned cases, it can be unwise to give a programme the status of a contract document. The net result, if this be the case, is that it ties the hands of both main contractor and architect/engineer in the manner in which they carry out their duties. It can also result in the employer having to pay additional sums with no tangible benefit.

SUMMARY

If a programme is given the status of a contract document, the contractor will be obliged to follow it to the letter. Should events not due to the contractor's own negligence arise which make it impossible to follow the programme, then it is likely that a court would award the contractor a variation and additional cost necessary to overcome the problem.

As is evident from the aforementioned cases, it can be unwise to give a programme the status of a contract document. The net result, if this be the case, is that it ties the hands of both main contractor and architect/engineer in the manner in which they carry out their duties. It can also result in the employer having to pay additional sums with no tangible benefit.

Chapter 6
Delays and Delay Analysis

6.1. If work is delayed due to two or more competing causes of delay, often referred to as concurrent delays, one of which is the responsibility of the contractor/subcontractor or a neutral event and the other is a result of some fault of the architect, engineer or employer, is there an entitlement to an extension of time and loss and expense?

6.1.1. Delays can be excusable delays, which may be either due to fault by the employer or his agents, such as late access, or delay in issuing drawings, or on the other hand delays due to a neutral event, such as excessively adverse weather or *force majeure*. Inexcusable delays are those which are due to fault on the part of the contractor or his agents. When it occurs, the contractor is usually entitled to an extension of time when delay to completion is caused by an excusable delay, but not when the delay is inexcusable.

6.1.2. Most of the standard forms of contract provide for financial recovery in addition to an extension of time, where events stated in the contract occur which give rise to additional cost: for example, late issue of instructions from the architect under a JCT contract. A great amount of confusion and muddled thinking has been experienced by judges in trying to unravel the effect in terms of additional time and cost entitlements under the standard forms of contract where concurrent delays have occurred. Does 'concurrent' refer to delays which start and finish on the same dates, or does it provide for some element of overlap? Do different rules apply in respect of extension of time entitlements from additional cost claims which arise from concurrent delays? These sort of questions have been examined in some detail by the courts in attempting to ascertain the entitlement of the parties where concurrent delays occur. Decisions of the judges as to what constitutes a concurrent delay have not been consistent.

6.1.3. Lord Osborne, in the case of *City Inn Ltd* v. *Shepherd Construction* (2010), offered the following advice as to what constitute concurrent delays, which he considered could apply to any of the following:

- The delays occur in a way in which they have common features.
- The delays share a common start and finish date.

200 Contractual Problems and their Solutions, Third Edition. Roger Knowles.

- The delays share either a common start or finish date.
- For a part of the time, the delays overlap.

6.1.4. The case of *Wells* v. *Army and Navy Co-operative Society* (1903) is an early case which provides assistance relating to concurrent delays, where the judge stated:

> . . . the fact that delay has been caused by matters for which the contractor is responsible will not deprive the contractor of his right to claim an extension of time for delay caused by a relevant event.

Peak Construction (Liverpool) Ltd v. *McKinney Foundations Ltd* (1976) was a helpful decision, in that judge Salmon LJ was of the view that an extension of time should be available in cases where delay has been caused partly by the fault of the contractor and partly by the fault of the employer.

6.1.5. Judge Dyson, in the case of *Henry Boot Construction (UK) Ltd* v. *Malmaison Hotel (Manchester) Ltd* (1999), was even more generous to contractors when, without dissent, he included the following matters which had, surprisingly, been agreed by both parties:

> . . . it was agreed that if there are two concurrent causes of delay, one of which is a relevant event, and the other is not, then the contractor is entitled to an extension of time for the period of delays caused by the relevant event notwithstanding the concurrent effect of the other event. Thus, to take a simple example, if no work is possible on a site for a week not only because of exceptionally inclement weather (a relevant event), but also because the contractor has a shortage of labour (not a relevant event), and if the failure to work during that week is likely to delay the work beyond the completion date by one week, then if he considers it fair and reasonable to do so, the architect is required to grant an extension of time of one week. He cannot refuse to do so, on the grounds that the delay would have occurred in any event by reason of the shortage of labour.

6.1.6. Arguments as to a contractor's or subcontractor's entitlements where two competing causes of delay occur which affect the completion date have been addressed in *Keating on Building Contracts*, 6th edition, which suggests that the law is unclear on this matter, but has offered assistance at 8.015 by suggesting as a solution:

> Where a contractor claims payment under the contract, e.g. for delay resulting from variation instructions and there is a competing cause of delay. The competing cause of delay could be . . . no one's fault e.g. bad weather or . . . the contractor's own delay in breach of contract.
>
> the claimant succeeds . . . if he establishes that the cause on which he relies, was the effective dominant cause of delay.

Keating goes on to state, in 8.018, that there is authority – of varying weights – for a number of disjointed propositions, which include:

> The Devlin approach: 'If a breach of contract is one of two causes of a loss, both causes co-operating and both of approximately equal efficacy, the breach is sufficient to carry judgment for the loss.

This would apply where, for example, there were two competing causes of delay which entitled a contractor to an extension of time, one a neutral event, such as excessively adverse weather, and the other being a breach such as late issue of instructions by the architect. Following the Devlin approach, the contractor would be entitled to extra time and loss and expense due to the late issue of instructions.

> The dominant cause approach: 'If there are two causes, one the contractual responsibility of the defendant and the other the contractual responsibility of the plaintiff, the plaintiff succeeds if he establishes that the cause for which the defendant is responsible is the effective, dominant cause.

Which cause is dominant is a question of fact, which is not solved by the mere point of order in time, but is to be decided by applying commonsense standards. In the case of *H Fairweather and Co Ltd* v. *London Borough of Wandsworth* (1987), 'dominant' was said to have a number of meanings, such as 'ruling, prevailing and most influential'.

> The burden of proof approach: 'If part of the damages is shown to be due to a breach of contract by the plaintiff, the claimant must show how much of the damage is caused otherwise than by his breach of contract, failing which he can recover nominal damages only.

An example would be a delay caused by the contractor having to correct defective work, occurring at the same time as a delay caused by the employer. To succeed, the contractor would have to demonstrate that his losses were due to the employer's actions and not his own.

6.1.7. The 'dominant cause of delay' theory was rejected by the court in the case of *H Fairweather & Co Ltd* v. *London Borough of Wandsworth* (1987). H Fairweather and Co Ltd were the main contractors for the erection of 478 dwellings for the London Borough of Wandsworth, employing JCT 63 conditions. Long delays occurred and liability for those delays was referred to arbitration. With regard to the delays the architect granted an extension of time of 81 weeks under clause 23(d), by reason of strikes and combination of workmen. The *quantum* of extension was not challenged, but Fairweather contended before the arbitrator that 18 of those 81 weeks should be reallocated under clause 23(e) or (f). The reasoning behind the contention was that only if there was such a reallocation could Fairweather ever recover direct loss and expense under clause 11(6) in respect of those weeks reallocated to clause 23(e), or clause 24(1)(a) in respect of those weeks reallocated to clause 23(f). The arbitrator's reasoning is to be found in sections 6.11 and 6.12 of his interim award, which state as follows:

> It is possible to envisage circumstances where an event occurs on site which causes delay to the completion of the works and which could be ascribed to more than one of the eleven specified reasons, but there is no mechanism in the conditions for allocating an extension between different heads, so the extension must be granted in respect of the dominant reason.
>
> I accept the respondent's contention, that faced with the events of this contract, nobody would say that the delays which occurred in 1978 and 1979 were caused by reason of the Architect's instructions given in 1975 to 1977. I hold that the dominant cause of the delay was the strikes

> and combination of workmen and accordingly the Architect was correct in granting his extension under condition 23(d).

In a later paragraph of his award, paragraph 6.14, the arbitrator declared:

> For the sake of clarity I declare that this extension does not carry with it any right to claim direct loss and/or expense.

The arbitrator's award was the subject of an appeal. The judge in the case disagreed with the arbitrator's ruling that the extension of time should relate to the dominant cause of delay. He said in his judgment:

> 'Dominant' has a number of meanings: 'Ruling, prevailing, most influential'. On the assumption that condition 23 is not solely concerned with liquidated or ascertained damages but also triggers and conditions a right for a contractor to recover direct loss and expense, where applicable, under condition 24, then an architect and in his turn an arbitrator has the task of allocating, when the facts require it, the extension of time to the various heads. I do not consider that the dominant test is correct.

He went on to say:

> The arbitrator erred in law in his award in relation to the strikes in holding... that as between the various heads under clause 23 pursuant to which an extension of time may be granted, the extension should be granted in respect of the dominant reason, whereas he should have held that where the reasons for delay correspond with more than one head, the extension may be granted under either or both heads.

This decision is in contrast to Keating's 'Dominant Cause' theory.

6.1.8. The dominant cause approach, however, received the following support in *Leyland Shipping* v. *Norwich Union* (1918).

> Which cause is dominant is a question of fact which is not solved by mere point of order in time, but is to be decided applying common sense standards.

6.1.9. Where an employer delays the contractor, he will not be entitled to deduct liquidated damages, even though the contractor is also in default: *Wells* v. *Army & Navy Co-operative Society* (1903). However, just because the employer is prevented from deducting liquidating damages in respect of a delay because of the concurrency of delays does not mean the contractor is entitled to recover financial loss. With this in mind, Keith Pickavance, in the second edition of his book *Delay and Disruption in Construction Contracts*, at p. 352, states:

> Lastly, and this is a legal conceptual problem, the rules which apply to recovery of actual damages for delay are not the same rules that apply to the relief of liquidated damages for delay. If C's progress on the critical path has been interfered with by D's act of prevention, then

> C must be given sufficient time to accommodate the effects of that and be relieved for LADs [liquidated and ascertained damages] for a commensurate period.
>
> On the other hand if, during the period of disruption to progress or prolongation for which an EOT [extension of time] has been granted, the predominant cause of C's loss and expense is disruption, or prolongation caused by a neutral event or his own malfeasance (for which he bears the risk), then he will not be able to recover damages for the compensable event unless he can separate those costs flowing from the compensable event from those costs which are at his own risk.

In other words, if two delays are running in parallel, one cause being the contractor's default and the other a breach by the employer, an extension of time should be awarded to the contractor, but no monetary reimbursement.

6.1.10. A simplistic approach sometimes taken is the 'first past the post' approach. This adopts the logic that, where delays are running in parallel, the cause of delay which occurs first in terms of time will be used for adjustment of the contract period. Other causes of delay will be ignored unless they affect the completion date and continue on after the first cause has ceased to have any delaying affect. In this case only the latter part of that second cause of delay will be relevant to the calculation of an extension of time. For example, delays may run in parallel because of the late issue of drawings and inclement weather. If the late issue of drawings causes a delay commencing on 1 February and inclement weather makes work impossible from 14 February, the late issue of drawings is the 'first past the post' and will take precedence over inclement weather until the drawings are issued. If drawings are issued on 21 February, but the inclement weather continues until 28 February, the contract completion date will be adjusted in respect of weather for the latter period.

The Australian Building Works Contract, JCC-E 1994, published by the Joint Contracts Committee, under clause 10.11, 'Predominant Causes of Delay', applies the 'first past the post' approach to delays caused by the proprietor's default.

6.1.11. It has been suggested that, when delays are concurrent, some form of apportionment may be appropriate. The matter of allocating or apportionment was referred to in *Keating on Building Contracts*, 6th edition, p. 213:

> Where the loss or damage suffered by the plaintiff results partly from his own conduct and partly from the defendant's breach of contract, it is correct in principle for the damages to be apportioned.

This seems to be at variance with Keating's own 'burden of proof' approach, referred to above. In the case of *Tennant Radiant Heat Ltd* v. *Warrington Development Corporation* (1987), a warehouse roof collapsed because of an accumulation of rainwater, due to blocked outlets. The landlord had a liability for damages and the tenant the responsibility for repair, both of which could have caused the collapse. The court, under the circumstances, apportioned the loss between both of the matters which could have contributed to the cause of the collapse. Lord Justice Dillon explained the rationale as follows:

> The problem which this court faces, on the claim and counterclaim alike, is in my judgment a problem of causation of damages. On the claim, the question is how far the damage to its goods which the lessee has suffered was caused by the corporation's negligence, notwithstanding the lessee's own breach of covenant. On the counterclaim, the question is how far the damage to the corporation's building which the corporation has suffered was caused by the lessee's breach of covenant, notwithstanding the corporation's own negligence. The effect is that on each question, apportionment is permissible.

6.1.12. In the case of *John Doyle Ltd* v. *Laing Management (Scotland) Ltd* (2004) the court had another opportunity of giving its opinions concerning concurrent delays. It seemed to favour the 'dominant cause' approach. However, if the dominant cause approach could not be applied, then it considered apportionment might be appropriate. The court, in explaining the significance of dominant cause and apportionment, stated as follows:

> '[14] . . . it may be possible to identify a causal link between particular events for which the employer is responsible and individual items of loss. On occasion that may be possible, where it can be established that a group of events, for which the employer is responsible, are causally linked with a group of heads of loss, provided that the loss has no other significant cause. In determining what is a significant cause, the 'dominant cause' approach described in the following paragraph is of relevance. Determining a causal link between particular events and particular heads of loss may be of particular importance, where the loss results from mere delay, as against disruption; in cases of mere delay such losses as the need to maintain the site establishment for an extended time can often readily be attributed to particular events, such as the late provision of information or design changes . . .
>
> [15] . . . the question of causation must be treated by the application of common sense to the logical principles of causation . . . In this connection, it is frequently possible to say that an item of loss has been caused by a particular event notwithstanding that other events played a part in its occurrence. In such cases, if an event or events for which the employer is responsible can be described as the dominant cause of an item of loss, that will be sufficient to establish liability, notwithstanding the existence of other causes that are to some degree at least concurrent. . . .
>
> [16] . . . even if it cannot be said that events for which the employer is responsible are the dominant cause of the loss, it may be possible to apportion the loss between the cause for which the employer is responsible and other causes. In such a case it is obviously necessary that the event or events for which the employer is responsible should be a material cause of the loss. Provided that condition is met, however, we are of the opinion that apportionment of loss between the different causes is possible in an appropriate case. Such a procedure may be appropriate in the case where the causes of the loss are truly concurrent, in the sense that both operate together at the same time to produce a single consequence.

6.1.13. The decision in the case of *City Inn* v. *Shepherd Construction* (2010) seems to approve of both the dominant cause and apportionment methods, where Lord Osborne stated:

> In the fourth place, if a dominant cause can be identified as the cause of some particular delay in the completion of the works, effect will be given to that, but leaving out of account any cause

> or causes which are not material. Depending on whether or not the dominant cause is a relevant event, the claim for extension of time will or will not succeed.
>
> In the fifth place, where a situation exists in which two causes are operative, one being a relevant event and the other some event for which the contractor is to be taken to be responsible and neither of which could be described as the dominant cause, the claim for extension of time will not necessarily fail. In such a situation, which could as a matter of language, be described as one of concurrent causes, in a broad sense... it will be open to the decision maker, whether the architect, or other tribunal, approaching the issue in a fair and reasonable way, to apportion the delay in the completion of the works occasioned thereby as between the relevant event and the other event.

6.1.14. A 'but for' test has been developed, which can be advantageous to contractors. The argument often runs that 'but for' the architect's instruction, the delay would not have occurred. In the Australian case of *Quinn* v. *Burch Brooks (Builders) Ltd* (1966), auditors failed to identify that a company was in a substantial loss position. It was argued that, but for the auditor's errors, the loss-making company would have ceased trading and subsequent trading losses would therefore have been avoided. In the UK case of *Turner Page Music Ltd* v. *Torres Design Associates Ltd* (1997) an action was brought against an architect for negligence, when there was a severe financial overrun on a construction project. The employer subsequently sold the building and alleged that, but for the architect's negligence, the sale would not have been necessary. The courts in both cases, however, rejected the 'but for' argument.

6.1.15. The Society of Construction Law Delay and Disruption Protocol 2002 presents a view which is similar to the 'burden of proof approach' referred to in 3.5.2, when it states:

> 1.10.4 Where an Employer Risk Event and a Contractor Risk Event have concurrent effect, the Contractor may not recover compensation in respect of the Employer Risk Event unless it can separate the loss and/or expense that flows from the Employer Risk Event from that which flows from the Contractor Risk Event. If it would have incurred the additional costs in any event as a result of Contractor Delays, the Contractor will not be entitled to recover those additional costs. In most cases this will mean that the Contractor will be entitled to compensation only for any period by which the Employer Delay exceeds the duration of the Contractor Delay.
>
> 1.10.5 The loss and/or expense flowing from an Employer Delay cannot usually be distinguished from that flowing from Contractor Delay without the following:
>
> 1.10.5.1 an as-planned programme showing how the Contractor reasonably intended to carry out the work and the as-planned critical path;
>
> 1.10.5.2 an as-built programme demonstrating the work and sequence actually carried out and the as-built critical path;
>
> 1.10.5.3 the identification of activities and periods of time that were not part of the original scope;
>
> 1.10.5.4 the identification of those activities and periods of time that were not part of the original scope and that are also at the Contractor's risk as to cost; and
>
> 1.10.5.5 the identification of costs attributable to the two preceding sub-sections.

This advice, however, only applies to financial reimbursement of additional costs resulting from delays.

SUMMARY

There is no hard and fast rule concerning which delay takes precedence, where concurrent delays are affecting the completion date. The courts have been inconsistent with regard to this matter. There is judicial approval for stating that when one delay is the responsibility of the employer and the other the contractor's fault or a neutral event, the contractor becomes entitled to an extension of time, but not necessarily additional cost. There is support for the dominant cause of delay to be the subject of an extension of time, whilst apportionment has met with some success. The 'first past the post' and the 'burden of proof' approach also have their supporters.

Each case has to be judged on its own merits, when judgments concerning extensions of time are decided. One rule which is not the subject of disagreement is that the architect, engineer or other person responsible for making extension of time decisions must act in a fair and reasonable manner.

6.2. Will a claim for an extension of time and the recovery of loss and expense which does not precisely detail the period of delay and the amount claimed in respect of each claim matter causing delay (i.e. a failure to link cause and effect), sometimes referred to as a global claim, fail?

6.2.1. The proper manner of presenting a claim before a court or arbitrator is to link the cause of delay and extra cost with the effect. For example, if the architect or engineer is six weeks late in issuing the drawings for the foundations (cause), the date for completion of the work may, as a consequence, be delayed by six weeks (effect). In recent times, contractors and subcontractors have been ever willing to shortcut the need to link cause and effect by the use of the global claim. All causes of delay and additional cost under the 'global claim' method, are lumped together and one overall delay and monetary claim given as a consequence. The usual requirement to link each cause of delay and monetary claim with the effect is ignored. The preparation of claims on a global basis has been the subject of considerable controversy for a number of years. Cost savings can arise because of not having to retain copious records, analyse them meticulously and spend time on research. It is, however, a difficult and developing area of law, the subject of a great deal of court scrutiny.

6.2.2. In support of the global claim, contractors and subcontractors draw comfort from the dicta in a number of legal cases. In *J. Crosby & Sons Ltd* v. *Portland Urban & District Council* (1967), the contract overran by 46 weeks. The arbitrator held that the contractor was entitled to compensation in respect of 31 weeks of the overall delay, and he awarded the contractor a lump sum by way of compensation, rather than giving individual periods of delay against the nine delaying matters. By way of justification, the arbitrator, in his findings, said:

> The result, in terms of delay and disorganisation, of each of the matters referred to above was a continuing one. As each matter occurred its consequences were added to the cumulative consequences of the matters which had preceded it. The delay and disorganisation which ultimately resulted was cumulative and attributable to the combined effect of all these matters. It is therefore impracticable, if not impossible, to assess the additional expense caused by delay and disorganisation due to any one of these matters in isolation from the other matters.

The respondent contested that the arbitrator was wrong in providing a lump-sum delay of 31 weeks, without giving individual amounts in respect of each head of claim. Mr Justice Donaldson, however, agreed with the arbitrator, saying:

> I can see no reason why he (the arbitrator) should not recognise the realities of the situation and make individual awards in respect of those parts of individual items of claim which can be dealt with in isolation and a supplementary award in respect of the remainder of these claims as a composite whole.

A similar award occurred in *London Borough of Merton* v. *Stanley Hugh Leach* (1985), where Mr Justice Vinelott said:

> The loss or expense attributable to each head of claim cannot in reality be separated.

6.2.3. This type of claim is now referred to as a 'global' or 'rolled up' claim. These decisions were thrown into question by *Wharf Properties Ltd and Another* v. *Eric Cumine Associates and Others* (1991). In this case the plaintiff made no attempt to link the cause with the effect in respect of a claim by the employer against his architect for failure properly to manage, control, co-ordinate, supervise and administer the work of the contractors and subcontractors, as a result of which the project was delayed. Six specific periods of delay were involved, but the statement of claim did not show how they were caused by the defendant's breaches. The plaintiff pleaded that, because of the complexity of the project, the interrelationship and very large number of delaying and disruptive factors and their inevitable knock-on effects and so on, it was impossible at the pleadings stage to identify and isolate individual delays in the manner the defendant required and that this would not be known until the trial. The defendant succeeded in an application to strike out the statement of claim. The Court of Appeal in Hong Kong decided that the pleadings were hopelessly embarrassing as they stood (some seven years after the action began) and an unparticularised pleading in such a form should not be allowed to stand. The matter was, nevertheless, referred to the Privy Council in view of the apparently differing view taken by the courts in *Crosby* and *London Borough of Merton*. The Privy Council, however, rejected the assertion that these two decisions justified an unparticularised pleading. Lord Oliver said:

> Those cases establish no more than this, that in cases where the full extent of extra cost incurred through delay depend upon a complex interaction between the consequence of various events,

so that it may be difficult to make an accurate apportionment of the total extra costs, it may be proper for an arbitrator to make individual financial awards in respect of claims which can conveniently be dealt with in isolation and a supplementary award in respect of the financial consequences of the remainder as a composite whole. This has, however, no bearing upon the obligations of a plaintiff to plead his case with such particularity as is sufficient to alert the opposite party to the case which is going to be made against him at the trial. [The defendants] are concerned at this stage not so much with quantification of the financial consequences – the point with which the two cases referred to were concerned – but with the specification of the factual consequences of the breaches pleaded in terms of periods of delay. The failure even to attempt to specify any discernible nexus between the wrong alleged and the consequent delay provides, to use [counsel's] phrase 'no agenda' for the trial.

6.2.4. The editors of *Building Law Reports*, vol. 52, at p. 6 say, by way of observation:

It must therefore follow from the decision of the Privy Council in *Wharf Properties* v. *Eric Cumine Associates* that *Crosby* and *Merton* are to be confined to matters of *quantum* and then only where it is impossible and impracticable to trace the loss back to the event. The two cases are not authority for the proposition that a claimant can avoid providing a proper factual description of the consequences of the various events upon which reliance is placed before attempting to quantify what those consequences were to him. Thus, taking the example before the Privy Council, it seems that it will in future be necessary for a plaintiff to be quite specific as to the delay which it is alleged was caused by an event such as a breach of contract, or an instruction giving rise to a variation. This in turn will mean that those responsible for the preparation and presentation of claims of this kind will need to work hard with those who have first-hand knowledge of the events, so as to provide an adequate description of them. Equally, it will mean that proper records will need to be kept, or good use will have to be made of existing records to provide the necessary detail. It will no longer be possible to call in an outsider who will simply list all the possible causes of complaint and then by use of a series of chosen 'weasel' words try to avoid having to give details of the consequences of those events before proceeding to show how great the hole was in the pocket of the claimant. There must be, as the Privy Council points out an 'agenda' for the trial: there must be a discernible connection between the wrong and, where delays are relied on, the consequent delay.

6.2.5. The Scott Schedule was originally designed to set out relevant points concerning defects in a manner which would easily be digested. It can, however, be adapted to suit any particular form of dispute, where a great many disputed facts are involved. In the case of *Imperial Chemical Industries* v. *Bovis Construction Ltd and Others* (1992), Judge Fox-Andrews QC ordered the plaintiff to serve a Scott Schedule containing:

- the alleged complaint;
- the defendant against whom the claim was made;
- which clause in the contract had been breached; and
- alleged failure consequences of such breach.

The Scott Schedule is not, however, a formula that can be applied to every case.

6.2.6. The whole subject came under review by the Court of Appeal in *GMTC Tools & Equipment Ltd* v. *Yuasa Warwick Machinery Ltd* (1994). This case, relating to a defective computer-controlled precision lathe (to be used in the manufacture of blanks, which in turn were machined to become rotary cutters), had nothing to do with construction work, but nonetheless the principles on which the decision was made will apply. The lathe did not operate as intended and the plaintiff prepared and submitted a claim based on the number of management hours involved in dealing with the problem and the number of hours during which the lathe was inoperable. Difficulties arose when the defendants sought further and better particulars of the claim. The judge ordered that a Scott Schedule should be drawn up providing detailed information attempting to link the cause (the malfunctioning of the lathe, which caused down time) with its effect (the wasted management hours and the purchase of blanks to replace lost production). Following attempts to re-amend the Scott Schedule, the matter came before the Court of Appeal after the plaintiff's failure to comply with an Unless Order.

The Court of Appeal was sympathetic to the plaintiff's situation. Lord Justice Leggatt said the defendant's argument presupposed that the plaintiff's production process was so flexible and instantaneously reactive to a period of down time that it would be possible to link each incident of down time with the purchase of a precise number of blanks to replace lost production. The opinions of the defendant were not accepted. It was the view of Lord Justice Leggatt that a judge is not entitled to prescribe the way in which the *quantum* of damage is pleaded and proved. No judge, he said, is entitled to require a party to establish causation and loss by a particular method. Lord Justice Leggatt, by way of conclusion, said:

> I have come to the clear conclusion that the plaintiff should be permitted to formulate their claims for damages as they wish, and not be forced into a strait-jacket of the judge's or their opponent's choosing.

6.2.7. A further review of global claims occurred in *British Airways Pension Trustees Ltd* v. *Sir Robert McAlpine & Sons* (1994), where a dispute arose out of the development of a site in Croydon. There were defects in the work alleged to be due to faults by the architect, the contractor and others involved in the construction of the project and it was argued that the diminution in value of the property because of the defects was £3.1 m, which formed the basis of the claim, plus the cost of investigating the defects. The defendants requested that further and better particulars be provided in respect of the claim. They asked to be given detailed information as to how much of the diminution in the value of the property could be attributed to each and every defect. For example, if two windows were defective, how did it affect the price paid for the property? They justified this type of question by asserting that until such details are given, the defendant does not know the case to be answered and so faces an unfair hearing. On behalf of the plaintiff, it was argued that all the defects had been identified and therefore, due to their existence, the project was worth £3.1 m less than it would have been without the default. An application was made by the defendants to strike out the claim and dismiss the action, as they had been seriously prejudiced through a failure on the part of the plaintiffs properly to particularise their claim. Judge Fox-Andrews ordered that the claim be struck out and the action dismissed. However, this decision was overruled by the Court of Appeal; Lord Justice Saville in summing up said:

> The basic purpose of pleadings is to enable the opposing party to know what case is being made in sufficient detail to enable that party properly to answer it. To my mind, it seems that, in recent years, there has been a tendency to forget this basic purpose and to seek particularisation even when it is not really required. This is not only costly in itself, but is calculated to lead to delay and to interlocutory battles in which the parties and the courts pore over endless pages of pleadings to see whether or not some particular points have or have not been raised or answered, when in truth each party knows perfectly well what case is made by the other and is able properly to prepare to deal with it. Pleadings are not a game to be played at the expense of citizens nor an end in themselves, but a means to the end, and that end is to give each party a fair hearing.

This attitude is a precursor of the sweeping Woolf reforms which, from April 1999, rewrote the court procedural rules to give judges a much more proactive role in the management of cases and cut down interlocutory delaying tactics of the type described by Lord Justice Saville.

6.2.8. From the decision in *Amec Building Ltd* v. *Cadmus Investment Co Ltd* (1996), it seems that courts in the future will judge each case on its merits, without laying down principles as to whether global claims will or will not be accepted. In this case, the judge's remarks as to the arbitrator's approach to global claims seem relevant:

> Certainly, it seems to me that there is no substance in the complaint that the arbitrator had set his face against global claims and that, thereby, prejudiced Amec. What appears to have happened, is that, upon justifiable complaint of lack of particularity, the arbitrator insisted upon an allocation of the overall claim to particular heads which was attempted by Amec and, when these matters were investigated by the accountants and in evidence and cross-examination, it clearly became quite clear to the arbitrator that there were occasions of duplications, matters compensated elsewhere and a general lack of particularisation. In those circumstances, it seems to be what the arbitrator concluded was that the plaintiff had not proved the costs incurred were due to the fault by Cadmus. . . . As is clear from the careful judgment of the arbitrator, he proceeded to analyse each of the claims made by Amec and decided each upon the evidence that was before him.

6.2.9. In *Inserco Ltd* v. *Honeywell Control Systems* (1996), the court made an award based upon a global claim. The judge's comments make interesting reading:

> Inserco's pleaded case provided sufficient agenda for the trial and the issues are about quantification. Both *Crosby* [*Crosby* v. *Portland Urban District Council* (1977)] and *Merton* [*London Borough of Merton* v. *Stanley Hugh Leach* (1985)] concerned the application of contractual clauses. However, I see no reason in principle why I should not follow the same approach in the assessment of the amounts to which Inserco may be entitled. There is here, as in *Crosby*, an extremely complex interaction between the consequences of the various breaches, variations and additional works and, in my judgement, it is 'impossible to make an accurate apportionment of the total extra cost between the several causative events'. I do not think that even an artificial apportionment could be made – it would certainly be extremely contrived – even in relation to the few occasions where figures could be put on time etc. . . . It is not possible to disentangle the various elements of Inserco's claims from each other. In my view, the cases show that it is legitimate to make a global award of a sum of money in the circumstances of

this somewhat unusual case, which will encompass the total costs recoverable under the February agreement, the effect of the various breaches which would be recoverable as damages, or which entitle Inserco to have their total cost assessed, to take account of such circumstances, and the reasonable value of the additional works similarly so assessed.

6.2.10. In *How Engineering Services Ltd* v. *Linder Ceilings and Partitions plc* (1999) the arbitrator, Mr Jupp, awarded a sum in respect of loss and expense based upon a global assessment. The dispute arose out of two contracts, the Atrium and the Station. In finding for the claimant, the arbitrator accepted the claimant's costs as set out in the points of claim and arrived at a figure of £130,346. From this he deducted £4,186 in respect of work carried out prior to the receipt of the notice, £32,611 in respect of realignment of ceilings, which was treated as a variation order, and £3,155 for remedial work. The award was thus based on a total cost claim calculated on a global basis, against which the defendant appealed on the basis that the arbitrator had not ascertained the sum as required by the arbitration clause. It was the view of the court that, in some cases, the facts are not always clear. Different tribunals would reach different conclusions and an arbitrator is entitled to assess loss and expense in the same way as a court assessing damages. The court upheld the arbitrator's award.

6.2.11. In the case of *Berhards Rugby Landscapes Ltd* v. *Stockley Park Consortium* (1997), the plaintiff landscape contractor entered into an agreement under seal with Trust Securities Holdings (TSH) for the construction of a golf course on a landfill site, under a contract which incorporated the ICE 5th Edition conditions. The work was subject to delay, and detailed and lengthy claims were submitted. It was alleged by the defendant that the claims were bound to fail because of a number of reasons, one of which was that they contained global delay claims for variations. It was held by the court that the global claim was a total cost claim. The plaintiff had quantified its alleged loss by subtracting the expected cost of the works from the final costs. Such a claim was permissible, if it was impractical to disentangle that part of the loss attributable to each head of claim and the situation had not been caused by the plaintiff's conduct. In such circumstances, the inference was that the employer's breaches had led to additional costs and that the cause nexus was to be inferred rather than demonstrated.

6.2.12. The case of *John Doyle Construction Ltd* v. *Laing Management (Scotland) Ltd* (2004) was heard in the Scottish Inner House, Court of Session, and dealt with whether, in principle, global claims were bound to fail. Laing Management was the management contractor for the construction of a new corporate headquarters for Scottish Widows in Edinburgh. A number of works packages were contracted out to John Doyle. An action was brought by John Doyle, seeking an extension of time of 22 weeks, together with loss and expense. John Doyle admitted that, despite their best efforts, it was impossible to identify cause and effect in respect of each item which caused delay and disruption and that was why the claim had been prepared on a global basis. It was argued on behalf of Laing that the basis of the global claim was that all the additional costs incurred by John Doyle resulted from the delay and disruption caused by Laing. This being the case, if one of the events relied upon by Doyle was shown not to have been caused by Laing, then the case would be undermined. The judge was sympathetic to this view when he said:

> . . . if a global claim is to succeed, whether it is a total cost claim or not the contractor must eliminate from the causes of his loss and expense all matters that are not the responsibility of the employer.

The judge felt, however, that the whole matter should be treated with common sense. He considered global claims to be a risky business, but nonetheless allowed proceedings to continue.

6.2.13. It is often argued by contractors that a requirement for strict particularisation places upon them a heavy evidential burden, which defendants know is difficult if not impossible to produce. Defendants in so doing are therefore seeking to profit from their own wrongs. The response by Judge Wilcox in the case of *Skanska Construction UK Ltd* v. *Eggar (Barony) Ltd* (2004) to this assertion was:

> Whilst it is unattractive that the party who has created the difficulty should be its beneficiary, nonetheless the evidential burden upon the claimant to adequately prove its case should not be diluted. The logic of a global claim demands that all events which contribute to causing the global loss are events for which the defendant is liable. If causal events include events for which the defendant bears no liability, the effect of upholding a global claim is to impose on the defendant a liability in part not legally his.

6.2.14. It is not unknown for contractors to submit a claim based upon the total costs for the project, sometimes referred to as a total cost claim. This occurred in the case of *Petromec Inc* v. *Petroleo Brasileiro SA* (2007), where the claimant was contracted to construct an oil rig. Because of a change in specification, the claimant submitted a claim being the difference between the total cost of constructing the rig to the new specification, less what the cost would have been if there had been no specification change. The court held that it was necessary to demonstrate that the costs were reasonable; it was not sufficient merely to assert what costs had been incurred; and therefore the claim failed. It is usually unwise to submit a claim based upon the total cost of the project. A claim submitted in this manner assumes that there was no loss due to underpricing, correcting poor workmanship, or general inefficiency. Most involved in the construction industry would consider that a situation such as this rarely arises.

6.2.15. The case of *London Underground Ltd* v. *Citylink* (2007) is another example of a global claim which finished up in court. The claim was rejected by both the adjudicator and the arbitrator. It was the court's view that a tribunal could accept a lesser claim, but only if the surviving claim was implicit in the case and it was fair to do so without seeking further submissions.

6.2.16. The Society of Construction Law Delay and Disruption Protocol 2002 provides the following view concerning global claims:

> 1.14.1 The not uncommon practice of contractors making composite or global claims, without substantiating cause and effect is discouraged by the Protocol and rarely accepted by the courts.
>
> 1.14.2 If the Contractor has made and maintained accurate and complete records, the Contractor should be able to establish the causal link between the Employer Risk Event and the resultant loss and/or expense suffered, without the need to make a global claim.

1.14.3 In what should only be rare cases where the financial consequences of the various causes of compensation are impossible to distinguish, so that an accurate apportionment of the compensation claimed cannot be made between the several causative events, then in this rare situation it is acceptable to quantify individually those items of the claim which can be dealt with in isolation and claim compensation for the remainder as a composite whole.

1.14.4 The Contractor will nevertheless need to set out the details of the Employer Risk events relied on and the compensation claimed with sufficient particularity so that the employer knows the case that is being made against it.

SUMMARY

The complexity of contemporary claims sometimes needs to be dealt with by a 'global' approach, but this is not a *carte blanche* for the plaintiffs to put in any figure. Detail needs to be provided where it is available and the contractor can be asked and/or ordered to produce information in a specific form, e.g. a Scott Schedule. However, demands for particulars are not to be used as a delaying tactic or as an end in themselves. Will a claim which fails to link cause and effect, fail? The answer must be that it depends upon the circumstances. It is unlikely, however, that the claim will be rejected entirely by the courts merely on the grounds that it has been prepared on a global basis. It must be demonstrated, however, that the items which form the basis of the global claim cannot be separated one from another. The claimant must ensure that problems of the claimant's own making are eliminated from the claim. A total cost claim which is based upon the total cost of the project is likely to fail, as it ignores any element of under-pricing, cost of correcting poor workmanship, or general inefficiencies. Finally, the rules of common sense will normally be applied when a tribunal is dealing with any claim which is calculated on a global basis.

6.3. What is meant by a contractor or subcontractor having to 'use constantly his best endeavours to prevent delay'; does it differ from 'reasonable endeavours'?

6.3.1. Many contracts require a contractor or subcontractor to use constantly his best endeavours to prevent delay. For example, JCT 2011, clause 2.28.6.1, states:

> the Contractor shall use constantly his best endeavours to prevent delay in the progress of the Works.

'Best endeavours' means that all steps to achieve the objective must be taken. *Keating on Building Contracts*, 5th edition, at page 575, has this to say with regard to the wording as it appears in the JCT forms of contract:

> This proviso is an important qualification of the right to an extension of time. Thus, for example, in some cases it might be the contractor's duty to reprogramme the works either to

prevent or to reduce delays. How far the contractor must take the other steps depends upon the circumstances of each case, but it is thought that the proviso does not contemplate the expenditure of substantial sums of money.

6.3.2. The wording of I Chem E, clause 14.3, is a little different, where it states that the contractor shall at all times use his best endeavours to minimise any delay in the performance of his obligations under the contract.

6.3.3. GC/Works/1, in condition 36(6), states that the contractor must endeavour to prevent delays and to minimise unavoidable delays.

6.3.4. In the case of *IBM UK Ltd* v. *Rockware Glass Ltd* (1980), Rockware agreed to sell IBM some land for development, and the sale was conditional upon planning permission being obtained, with a further proviso that IBM 'will make an application for planning permission and use its best endeavours to obtain the same'. The local authority refused planning permission. IBM did not appeal against that decision to the Secretary of State. The parties disagreed on whether, by not appealing, IBM had failed to use its best endeavours to obtain planning permission. The project was a substantial one, in which the purchase price of the land alone was £6,250,000. It was accepted that making an appeal to the Secretary of State would cost a significant amount of money. The court said that, taking into account the background to the matter and the amount of money involved, it was not likely that the parties would have considered a refusal of planning permission at a local level to be the end of the matter, but that they must have had in mind the prospect of an appeal to the Secretary of State. The test of 'best endeavours' which was approved was that the purchasers of the land were bound to take all those steps in their power which are capable of producing the desired results, namely the obtaining of planning permission, being steps which a prudent, determined and reasonable owner, acting in his own interests and desiring to achieve that result, would take. It was expressly stated that the criterion was not that of someone who was under a contractual obligation, but someone who was considering his own interests. Whilst it seems clear a contractor or subcontractor may be required to expend some money to meet the obligation to 'use constantly his best endeavours' to prevent delay, the intention is not to expend large sums, particularly where the delay has been caused by the engineer or architect.

6.3.5. In the case of *Victor Stanley Hawkins* v. *Pender Bros Pty Queensland* (1994), it was held that the term 'best endeavours' should be construed objectively. The test as to whether it had been fulfilled would be that of prudence and reasonableness.

6.3.6. Two cases have involved the court in having to decide the meaning of best endeavours. In *Midland Land Reclamation Ltd* v. *Warren Energy Ltd* (1997), the judge in deciding the case said:

> I reject the submission made on behalf of the defendant that a best endeavours obligation is the next best thing to an absolute obligation or guarantee.

In *Terrell* v. *Maby Todd & Co* (1952), the judge held that a 'best endeavours' obligation only required a party to do what was commercially practicable and what it could reasonably do in the circumstances.

6.3.7. In an article in *Building* (10 September 1999), Neil White explained:

> A best endeavours clause means that you do what a reasonable person would to achieve an objective – it is not a guarantee, it may be overruled by conflicting obligations and it doesn't apply to intangible outcomes such as an agreement.

In the final analysis, a contractor will be expected to do what is commercially practicable and what it could reasonably do in the circumstances.

6.3.8. There is a difference between the meaning of 'best endeavours' and 'reasonable endeavours'. In the case of *Rhodia* v. *Huntsman* (2007), Huntsman agreed to buy Rhodia's chemical business. As part of the sale agreement, the parties undertook to use their 'reasonable endeavours' to ensure that all supply contracts were transferred from Rhodia to Huntsman, which included an energy supply contract with a firm called Cogen. Some of the energy supply contracts were to be transferred to subsidiaries of Huntsman. Unfortunately, Cogen proved difficult and insisted upon a parent company guarantee being provided by Huntsman; otherwise, it would not agree to the novation of its energy supply contract from Rhodia to Huntsman. The novation did not take place, as Huntsman refused to provide a parent company guarantee. Rhodia finished up with a liability in the sum of £14.8 m for energy use, which had to be paid to Cogen. This sum was claimed by Rhodia from Huntsman, on the basis that it had failed to use reasonable endeavours by refusing to provide a parent company guarantee. The court found that Huntsman had, in not providing a company guarantee, failed to use 'reasonable endeavours'. The judge considered that there was a difference between 'reasonable endeavours' and 'best endeavours'. He said:

> An obligation to use reasonable endeavours to achieve the aim probably only requires a party to take one reasonable course, not all of them, whereas an obligation to use best endeavours probably requires a party to take all reasonable courses he can.

Reasonable endeavours, the judge considered, would not involve a party doing something which was contrary to its commercial interests.

SUMMARY

Where a contract requires the contractor to use his 'best endeavours' to prevent delay, he is expected to keep the effect of any matters which could cause delay down to a minimum, or to eliminate them if possible. This would include an obligation to reprogramme, if appropriate. If the delay is the contractor's responsibility, he may consider it financially more advantageous in the absence of an obligation to use best endeavours, to allow the work to overrun the contract period and to pay liquidated damages. This will be particularly relevant if the liquidated damages figure is modest. Some standard forms of contract, for example JCT 2011, allow the employer to determine the contractor's employment if he fails to proceed regular and diligently with the works, which would make it unwise for the contractor to fail to use his best endeavours to prevent delay. If the delay is the responsibility of the architect/engineer or employer, the contractor is not required to expend substantial amounts of his own money to reduce the delay.

The courts have differentiated between 'reasonable endeavours' and 'best endeavours'. An obligation to use 'reasonable endeavours' probably requires a party to take one of a range of possible reasonable courses of action, whereas an obligation to use 'best endeavours' probably requires a party to go further and to pursue all reasonable alternatives.

6.4. What is meant by 'Time is of the Essence'?

6.4.1. The term 'time is of the essence' appears from time to time in bespoke conditions and alterations to standard conditions which apply on construction projects. This term, however, does not appear in any commonly used standard conditions. Those who use the term often do not fully understand its purpose. Where time is of the essence, is a condition of the contract and a date for completion is stated, a failure to complete by the stated date can result in the contract being terminated. In the Australian case of *Carr* v. *Berriman Pty Ltd* (1953), an owner on a construction project was under an obligation to make the site available on a stated date. The site was not made available on that date, and the contractor claimed that time was of the essence and, hence, he was entitled to terminate the contract and claim for his loss of profit. The judge had this to say with regard to 'time being of the essence':

> Where a contract contains a promise to do a particular thing on or before a specified day, time may or may not be of the essence. If it is, the promisee is entitled to rescind, but he may elect not to exercise the right and an election will be inferred from any conduct which is consistent with the contract remaining in being.

6.4.2. Time may be of the essence because there is a clause to that effect in the contract. Alternatively, it may be clear from an examination of the contract documents, as a matter of presumed intention of the parties. Where there has been unreasonable delay in completing within the required time or by the contract completion date, time may become of the essence because of the inconvenienced party serving notice of a revised date for completion. If the revised date is not met, then time may be of the essence. This was summed up in the case of *United Scientific Holdings* v. *Burnley Council* (1978), where it was stated:

> Time will not be considered to be of the essence unless:
> - The parties expressly stipulate that conditions as to time must be strictly complied with, or
> - The nature of the subject-matter of the contract or the surrounding circumstances show that time should be considered to be of the essence, or
> - A party who has been subjected to unreasonable delay, gives notice to the party in default, making time of the essence.

6.4.3. In construction contracts, time is not of the essence unless made so in the manner already described. This seems clear, as most construction contracts contain a liquidated damages clause as a means of compensating the employer if work finishes late. There are, however, rare examples of contracts which contain liquidated damages and extension of time provisions, where time is made of the essence: *Peak Construction* v. *McKinney Foundations* (1971).

6.4.4. It is stated in Para 9.014 of *Hudson on Building and Engineering Contracts*, 11th Edition:

> However it has been a characteristic of poor draftsmanship of recent years, in nearly all commercial fields that 'of the essence' wording is frequently used to 'overegg the pudding' in relation to obviously inappropriate contractual obligations, and should not be accorded the same weight as when used in a more considered way.

This was illustrated in the case of *Clear Homes* v. *Sarcon (no 177) Ltd* (2010), where the defendant agreed to construct an apartment block in Belfast, with Gilbert Ash as the contractor. The claimant signed two documents committing it to purchasing an apartment block, with the defendant completing the work by 31 May 2009. Work was not completed on time, because of a combination of poor weather conditions and problems with the foundations. The claimant was notified that a revised completion date would be October/November 2009. The contract, in clause 8, provided for a reasonable extension of time to be given if work was delayed by, *inter alia*, bad weather and causes outside the claimant's control. Clause 23, however, stated:

> In relation to the time limits specified in this agreement time shall be of the essence.

6.4.5. The ruling of the court was that the agreement represented the parties' intentions, to the effect that time would not be of the essence. The reference to time limits in clause 23, did not, and was not intended to, apply to the completion dates.

6.4.6. An example of a case where the court found that time *was* of the essence is *Charles Rickards Ltd* v. *Oppenheim* (1950). In this case, a car builder undertook to build a body onto a Rolls Royce chassis. The car was to be ready by a specified date, which was held to be the essence of the contract. The date passed but the work was not completed. The purchaser continued to request delivery and new dates were promised and accepted, but still delivery was not forthcoming. Finally, the purchaser gave notice that he would not accept delivery after a specified date. When the car was not delivered by this date, the purchaser cancelled the order. The court held that time had become of the essence when the purchaser served the notice. Under the circumstances, the purchaser was entitled to cancel the order when the car was not delivered by the date for delivery stated in the notice.

SUMMARY

Time will be of the essence if one of the following applies:

- The parties expressly stipulate that conditions as to time must be strictly complied with; or

- The nature of the subject matter of the contract or the surrounding circumstances show that time should be considered to be of the essence; or
- A party who has been subjected to unreasonable delay gives notice to the party in default, making time of the essence.

If time is, or becomes of the essence, the injured party may terminate the contract, if the date for completion is not met. The injured party will then be entitled to claim damages for breach of contract.

6.5. Where delays to completion of the works have occurred and disputes arise as to the appropriate extension of time which should be granted, is the employment of a computer-based critical path analysis essential to establish the true entitlement?

6.5.1. The standard forms of contract regularly used on construction projects, are very clear as to who is responsible for making decisions regarding the granting of extensions of time. In the case of most JCT contracts, the architect or contract administrator is responsible for making the decision. Engineers do likewise on ICE contracts and the project manager does so if the contract has been let using the NEC. It needs to be recognised that the skill of those charged with the duty of deciding the extent, if any, to which a contractor may have an entitlement to an extension of time may vary. There is no procedure laid down in the contract which must be followed in arriving at a decision regarding extensions of time. The courts have therefore taken a very general view as to what is required when granting an extension of time. In the case of *City Inn Ltd* v. *Shepherd Construction* (2007), Lord Drummond Young said:

> What is required of clause 25 is that the Architect should exercise his judgement to determine the extent to which completion has been delayed by relevant events. The Architect must make a determination on a fair and reasonable basis.

6.5.2. In the case of *Balfour Beatty Building Ltd* v. *Chestermount Properties Ltd* (1993), Coleman J made the following observations, which are somewhat obvious to most charged with the duty of making decisions regarding entitlements to extensions of time:

> The underlying objective is to arrive at the aggregate period of time within which the contract works, as ultimately defined, ought to have been completed, having regard to the incidence of non-contractor's risk events and to calculate the excess time, if any, over that period, which the contractor took to complete the works.... the Architect's objective must be to assess whether any of the relevant events has caused delay to the progress of the works and if so how much. He must then apply the result of his assessment of the amount of the delay caused by the relevant event by extending the contract period for completion of the works by a like amount and this he does by means of postponing the completion date.

6.5.3. The judge in the case of *John Barker Construction Ltd* v. *London Portman Hotel* (1996) adversely criticised the architect on the grounds that, when arriving at a decision regarding the contractor's entitlement to an extension of time, he had not carried out a logical analysis in a methodical way, of the impact of the relevant events on the contractor's programme and only made an impressionable, rather than a calculated, assessment. The judge said:

> [The Architect] did not carry out a logical analysis in a methodical way of the impact which the relevant matters were likely to have on the plaintiff's planned programme. He made an impressionistic rather than a calculated assessment of the time which he thought was reasonable for the various items individually and overall.

6.5.4. In the case of *Royal Brompton Hospital NHS Trust* v. *Frederick Alexander Hammond and Others* (2001), HHJ Richard Seymour stated:

> In order to make an assessment of whether a particular occurrence has affected the ultimate completion of the work, rather than just a particular operation, it is desirable to consider what operations, at the time the event with which one is concerned happens are critical to the forward progress of the work as a whole.

To be able to comply with this judge's requirements would involve producing a critical path analysis.

6.5.5. Judge Toulmin, in the case of *Mirant Asia-Pacific Construction (Hong Kong) Ltd* v. *Ove Arup and Partners International Ltd* (2007), which was heard in the TCC, defined a critical path analysis as:

> . . . the sequence of activities through a project network from start to finish, the sum of whose durations determine the overall project duration.

6.5.6. Judge Humphrey Lloyd was strident in his comment regarding the evidence required to establish an entitlement to an extension of time, which supports the use of a critical path analysis, when he said:

> By now one would have thought that it is well understood that on a contract of this kind, in order to attack on the facts, a clause 24 certificate for non-completion (or an extension of time determined under clause 25), the foundation must be the original programme (if capable of justification and substantiation to show its validity and reliability as a contractual starting point) and its success will similarly depend on the soundness of its revisions on the occurrence of every event so as to be able to provide a satisfactory and convincing demonstration of cause and effect. A valid critical path (or paths) has to be established both initially and at every later material point since it (or they) will almost certainly change. Some means has also to be established for demonstrating the effect of concurrent or parallel delays or other matters for which the employer will not be responsible under the contract.

To say the least, the judge appears to be going over the top. Few architects, contract administrators, engineers or project managers would be likely to go to these extremes when making decisions concerning extensions of time.

6.5.7. In the case of *City Inn Ltd* v. *Shepherd Construction* (2007), Lord Drummond Young made the following observations concerning the use of a critical path analysis, which are at odds with the views of Mr Justice Lloyd:

> In my opinion the pursuers clearly went too far in suggesting that an expert could only give a meaningful opinion on the basis of an as-built critical path analysis... The major difficulty it seems to me is that in the type of programme used to carry out a critical path analysis any significant error in the information that is fed into the programme is liable to invalidate the entire programme.

6.5.8. Further words of wisdom were expressed by Lord Osborne in a later hearing of the case *City Inn Ltd* v. *Shepherd Construction* (2010), when he said:

> In the second place the decision as to whether the relevant event possesses such causative effect is an issue of fact which is to be resolved, not by the application of philosophical principles but rather by the application of common sense.

SUMMARY

Lord Drummond Young and His Honour Mr Justice Lloyd are not in accord regarding the value of a critical path analysis in deciding the correct extension of time which should be awarded. Lord Drummond Young is not convinced that the use of a critical path analysis is essential. He considered that, where the information on which the critical path has been built contains a significant error, it would invalidate the entire programme; to use the old adage 'rubbish in, rubbish out'. His Honour Mr Justice Lloyd, however, expressed the view that the use of a critical path analysis has to be established in deciding entitlements to extensions of time on complex projects. In assessing these diametrically opposite views it is necessary to take into account the fact that judges are not experts with regard to construction processes. Their knowledge is based upon the information gleaned from cases brought before them. It is therefore not surprising that two judges should fail to have a similar view on a key matter such as the use of a critical path analysis.

If a dispute arises in respect of an extension of time award and the matter is referred to arbitration or is the subject of litigation, opinions as to the contractor's entitlement are often submitted by expert delay analysts using techniques which are often not available to the architect, contract administrator, engineer or project manager when making the decisions. It is not uncommon for each side to appoint its own delay analyst and quite usual for the experts to disagree with the extension of time which has been granted. Having done so, they proceed to express opposing views as to the correct extension of time which they consider should have been granted. The judge or arbitrator is then left to decide which of the experts' opinions he prefers, or whether the architect, contract administrator, engineer or project manager was right all along. This is a far cry from the situation in which the architect, contract administrator, engineer or project manager was placed at the time the decision was made. It must also be borne in mind that these decisions are often required by the terms of the contract to be made within a tight timescale. For example, the architect appointed with regard to a JCT contract has 12 weeks from receiving details from the contractor to arriving at a decision.

It is clear that there is no one system which should apply to analysing delays and making decisions as to the contractor's entitlement to extensions of time. It is not an exact science and in all probability, if three experts were presented with the same set of facts concerning delays to a project and asked to determine the contractor's entitlement to an extension of time, they would produce three completely different answers. Whoever makes the decision must be able to demonstrate that there has been a full examination of the facts, based upon what actually happened and establish the effect of the delaying events on the progress and completion of the work. Overriding all this is the requirement that common sense must prevail. A cynic, however, may wish to add the rider that common sense is not all that common.

Chapter 7

Extensions of Time

7.1. Does a contractor or subcontractor lose entitlements to extensions of time if he fails to submit the appropriate notices and details required by the contract?

7.1.1. Most of the standard forms of main contract and subcontract require the contractor and subcontractor to give notice when delays occur to the progress or completion of the works. A question often asked is whether, in the absence of notice, the contractor or subcontractor loses his rights to have the completion date extended. In other words, is the service of notice a condition precedent to the right to an extension of time?

7.1.2. The matter was considered by the House of Lords in the case of *Bremer Handelsgesellschaft mbh* v. *Vanden Avenne-Izegem* (1978), which arose out of a dispute over the sale of soya bean meal. Lord Salmon, referring to how the rights of the parties were affected by the lack of a proper notice, had this to say:

> In the event of shipment proving impossible during the contract period, the second sentence of clause 21 requires the seller to advise the buyers without delay of the impossibility and the reasons for it. It has been argued by the buyers, that this is a condition precedent to the sellers' rights under that clause. I do not accept this argument. Had it been a condition precedent, I should have expected the clause to state the precise time within which the notice was to be served and to have made plain by express language, that unless the notice was served within the time, the sellers would lose their rights under the clause.

7.1.3. From what Lord Salmon has said it seems clear that, for notice to be a condition precedent to a right for more time, the wording of the clause would need to be such that a failure to serve notice would result in loss of rights. The situation of lack of notice was examined in the decision in *Stanley Hugh Leach* v. *London Borough of Merton* (1985) in relation to JCT 63, where Mr Justice Vinelott summarised the position as follows:

> The case for Merton is that the architect is under no duty to consider or form an opinion on the question whether completion of the works is likely to have been, or has been, delayed for any of the reasons set out in clause 23 unless, and until, the contractor has given notice of the cause of a delay that has become 'reasonably apparent' or, as it has been put in argument, that

200 Contractual Problems and their Solutions, Third Edition. Roger Knowles.

the giving of notice by the contractor is a condition precedent, which must be satisfied before there is any duty on the part of the architect to consider and form an opinion on these matters.

I think the answer to Merton's contention is to be found in a comparison of the circumstances in which a contractor is required to give notice, on the one hand, and the circumstances in which the architect is required to form an opinion, on the other hand. The first part of clause 23 looks to a situation in which it is apparent to the contractor that the progress of the works is delayed, that is, due to an event known to the contractor, which has resulted, or will inevitably result in delay. The second part looks to a situation in which the architect has formed an opinion that completion is likely to be, or has been delayed, beyond the date for completion. It is possible that the architect might know of events (in particular 'delay on the part of artists, tradesmen or others engaged by the employer in executing work not forming part of this contract') which is likely to cause delay in completion, but which has not caused an actual or prospective delay in the progress of the work which is apparent to the contractor. If the architect is of the opinion that because of an event falling within sub-paragraphs (a) to (k) progress of the work is likely to be delayed beyond the original, or any substituted completion date, he must estimate the delay and make an appropriate extension to the date for completion. He owes that duty not only to the contractor, but also to the building owner. It is pointed out in a passage from *Keating on Building Contracts*, 4th Edition at page 346, which is cited by the arbitrator, that if the architect wrongly assumes that a notice by the contractor is a condition precedent to the performance of the duty of the architect to form an opinion and take appropriate steps:

. . . and in consequence refuses to perform such duties, the Employer loses his right to liquidated damages. It may therefore be against the Employer's interests for an Architect not to consider a cause of delay of which late notice is given, or of which he has knowledge despite lack of notice.

7.1.4. In *Maidenhead Electrical Services* v. *Johnson Controls* (1996) the terms of the contract laid down that any claim for an extension of time had to be made within ten days of the event for which the claim arises. It was held that a failure to comply with the notice provisions did not render a claim invalid.

7.1.5. The GC/Works/1 contract is somewhat out of line with the other standard forms of main contract in that it refers, in condition 28, to the contractor's written notice being a condition precedent to a right to an extension of time, unless otherwise directed by the Authority. However, the later GC/Works/1 (1998) version does not provide for the delay notice required by condition 36 to be a condition precedent to an entitlement to an extension of time.

7.1.6. JCT 2011 and the ICE 6th and 7th Editions make no reference to a delay notice being a condition precedent to the contractor's entitlement to an extension of time.

7.1.7. The Engineering and Construction Contract (NEC 3), however, takes a different stance. Clause 61.3 requires the contractor to notify the project manager of a compensation event as a condition precedent to an entitlement to an extension of time, where it states:

if the contractor does not notify a compensation event within eight weeks of becoming aware of the event, he is not entitled to a change in the Prices, the Completion Date or a Key Date . . .

7.1.8. The MF/1 and IChemE conditions make no reference to a delay notice being a condition precedent to an extension of time.

7.1.9. The interpretation of the various subcontracts runs in parallel with the main contracts. An exception is the CECA blue form subcontract, for use with the ICE main contract. Clause 6(2) stipulates that it is a condition precedent to the subcontractor's right to an extension of time for a notice to be served within 14 days of a delay first occurring, for which the subcontractor considers himself entitled to extra time.

7.1.10. The Australian case *Turner Corporation Ltd (Receiver and Manager Appointed)* v. *Austotal Pty Ltd* (1998) dealt with the situation of a delay caused by the employer, where the conditions of contract required a written delay notice as a condition precedent to an extension of time. The lack of notice lost the contractor the right to an extension of time. The judge stated:

> If the builder, having a right to claim an extension of time, fails to do so, it cannot claim that the act of prevention, which would have entitled it to an extension of time for practical completion, resulted in its inability to complete by that time. A party to a contract cannot rely on preventing conduct of the other party, where it failed to exercise a contractual right, which would have negated the effect of that preventing conduct.

7.1.11. Following the *Turner Corporation* case, the Australian courts seem to have undergone a change of heart. The principle of prevention became relevant to extensions of time, following the Australian case of *Gaymark Investments Pty Ltd* v. *Walter Construction Group Ltd* (1999). In basic terms, the prevention principle states that one cannot benefit from one's own errors. Where an employer includes in a contract a clause which states that a written notice from the contractor is a condition precedent to the right to an extension, it could come into conflict with the prevention principles. A delay which is caused by the employer is obviously an employer's breach of contract. If the contract denies the contractor an entitlement to an extension of time due to a failure to serve a written notice, this could leave the employer with the benefit of being paid liquidated and ascertained damages which result from its own delays. In the *Gaymark* case, the contract required the contractor to serve a written notice within 14 days of the delay arising. The clause went on to stipulate that the contractor would not be entitled to an extension of time if it failed to serve a proper written notice. The court found in favour of the contractor, even though there had been no written notice served, as it considered it would be unmeritorious to award the employer liquidated and ascertained damages where the delays were of its own making.

7.1.12. Another Australian case, *Abigroup Contractors* v. *Peninsula Balmain* (2002) also involved the workings of an extension of time clause. The wording of the contract made the service of a written notice a condition precedent to the contractor's right to an extension of time. No notice was given, but because the contract gave the superintendent power to grant an extension of time, whether or not a notice had been served, it was held that he was wrong in not exercising the discretion and making an appropriate award.

7.1.13. A situation may arise where the contractor is delayed by the employer and loses rights to an extension of time, due to the lack of a written notice which is expressed in the contract to be a condition precedent. The employer will then become entitled to deduct liquidated damages. The delay, however, may result in the contractor incurring additional cost, which he will still be entitled to recover from the employer, either under the terms of the contract or as a common law damages claim. The odd situation will have then arisen whereby the contractor loses a right to an extension of time, pays liquidated damages to the employer, but is reimbursed the delay costs he incurs.

7.1.14. The case of *City Inn* v. *Shepherd Construction* (2003) arose out of the construction of a hotel in Bristol. A dispute occurred concerning the contractor's entitlement to an extension of time. The contract was JCT 80, but with special amendments inserted at the behest of the employer. One of the amendments dealt with architect's instructions and required the contractor to serve a written notice with regard to an entitlement to an extension of time. The contract stated that the architect could dispense with the contractor's obligations concerning written notice, but in the absence of such a waiver the special amendment stipulated that the contractor would not become entitled to any extension of time if he failed to serve the written notice. Due to a failure on the part of the contractor to serve proper notice following the issue of an architect's instruction, the employer considered he had a right to deduct liquidated and ascertained damages at the rate of £30,000 per week, as stated in the contract. It was argued by the contractor that, even though £30,000 per week represented a reasonable pre-estimate of the employer's loss, the removal of an entitlement to an extension of time due to a failure to serve notice was in itself a penalty. The loss which the employer would incur as a result of a lack of written notice would still be the same. Therefore, a failure by the contractor to serve a written notice had no financial effect upon the employer. The argument was not accepted by the court. If the contractor elects not to comply with the notice requirement, it can properly be regarded as a breach of contract. Should the contractor decide to serve a written notice, then the employer provides an indemnity to the contractor against payment of liquidated and ascertained damages, but only if a written notice is provided. A failure by the contractor to serve a written notice also prevents the architect from reviewing the instruction. For these reasons, the court found against the contractor. The English courts are more likely to follow this decision, which arose from the Scottish courts, and would seem unlikely to adopt the principles established in the Australian *Gaymark* case.

7.1.15. The case of *Steria Ltd* v. *Sigma Wireless Communications Ltd* (2007) arose as a result of the provision of a new computerised system for the fire and ambulance services in the Republic of Ireland. The project was known as a Computer Assisted Mobilising Project (CAMP). CAMP East was the name given to the employer on the project. Sigma was the main contractor in connection with the project and Steria the subcontractor. A subcontract was let for the work using a heavily amended MF/1. A dispute arose between Sigma and Steria relating to the release of the final balance of retention, amounting to 153,786 Euros, which was due for payment at the end of the defects liability period. Sigma asserted that Steria had caused delay in completing the sub-contract works, giving them an entitlement to payment or set-off of 307,573 Euros in respect of liquidated damages and general damages in a sum in excess of 380,000 Euros for losses incurred as a result of delay. The dispute was referred to the Technology and Construction Court.

7.1.16. Clause 6.1 of the conditions of contract includes the following wording:

> . . . if by reason of any circumstance which entitles the contractor to an extension of time for completion of the works under the main contract, or by reason of a variation to the sub-contract works, or by reason of breach by the contractor, the sub-contractor shall be delayed in the execution of the sub-contract works, then in any such case, provided the sub-contractor shall have given within a reasonable period, written notice to the contractor of the circumstances giving rise to the delay, the time for completion hereunder shall be extended by such period as may in all the circumstances be justified . . .

The court provides a clear ruling of what is required concerning notice to ensure compliance with the provisions of this clause, namely:

1. The notice must make it clear that it is a request for an extension of time under clause 6.1.
2. The notice must explain how and why the relevant circumstances have caused delay.
3. The notice must give an assessment of the delay.
4. The notice must be given within a reasonable period.

The court also stated that the written notice must emanate from Steria. A note in the minutes of a site meeting prepared by the main contractor, or the employer, to the effect that the subcontract works had been delayed and the reason for the delay, would not in itself amount to a good notice under clause 6.1. It was successfully argued by Sigma that the wording of Clause 6.1 made the sending of a proper written notice by Steria a condition precedent to the right to an extension of time. In other words, in the event of non-compliance with clause 6.1, Steria would lose the right to an extension of time. The judge considered that the wording provided a right to an extension of time:

> . . . provided that the sub-contractor shall have given within a reasonable time written notice to the contractor of the circumstances giving rise to the delay.

This wording makes the provision of a notice a condition precedent to the right to an extension of time. This case is important, because the judge decided that the wording of clause 6.1 makes the giving of a written notice a condition precedent, even though those specific words are not included in the clause.

7.1.17. We, as a country, seem to be obsessed by the wording in the small print. This is well illustrated by the decision in the case of *Education 4 Ayrshire Ltd* v. *South Ayrshire Council* (2009). This case arose out of a PPP arrangement for schools in Ayrshire. All the work was subcontracted to Carillion. The contract under which the dispute arose was not one of the standard construction contracts, but one designed for PPP projects. A contract was entered into for the design and construction work at six schools. The work included the removal of asbestos. A survey was carried out and the contract provided for the removal of the asbestos identified by the survey. Once work got under way, the contractor encountered more asbestos than was shown by the survey. Removing the additional asbestos caused delay and additional cost. The contractor claimed the additional work had caused a delay of 16 weeks and additional cost amounting to £815,792 and looked to the contract for a remedy. The contract notice requirements were extremely strict and required the contractor to serve a notice within 20 business days after he became aware of the event giving rise to delay and additional cost. This notice was required to include the reasons for the delay and also a claim for any extension of time and payment of compensation. The contract made it crystal clear that in the event of a failure to comply with this requirement, the contractor would not become entitled to more time or money. The contract stipulated that the notice must be sent to the Chief Executive of the authority by first class post, fax, or by hand.

7.1.18. A dispute arose in respect of the notice of delay. The contractor became aware of the additional asbestos and the likely effect on the completion date on 6 April 2007. A notice was submitted on 2 May 2007, which just met the time scale. However, the notice reported the delay and went on to express an intention to submit a claim. This did not meet the contractual obligations, which required the actual claim to be submitted. The court had no hesitation in finding against the contractor. He had failed to comply with the contract provisions with regard to what should be included in the notice.

SUMMARY

A failure on the part of a contractor or subcontractor to serve a proper delay notice will not result in the loss of rights to an extension of time, unless the conditions of contract expressly state that the service of a notice is a condition precedent to such rights.

7.2. What is the Prevention Principle – does it provide a contractor with assistance in avoiding the payment of liquidated or delay damages where he fails to serve a delay notice, which the contract states is a condition precedent to the granting of an extension of time?

7.2.1. The Prevention Principle was defined in the case of *Barque Quilpe Ltd* v. *Brown* (1904), as follows:

> There is an implied contract by each party that he will not do anything to prevent the other party from performing a contract or to delay him in performing it. I agree that generally such a term is by law imported into every contract.

In the more recent Canadian case of *Perini Pacific Ltd* v. *Great Vancouver Sewerage and Drainage District* (1996), it was said:

> Since the earliest times it has been clear that a party to a contract is exonerated from performance of a contract when that performance is prevented or rendered impossible by the wrongful act of the other party.

7.2.2. The principle can be traced back in construction contracts as far as *Holme* v. *Guppy* (1838). This case involved the construction of a brewery in Liverpool. The employer failed to provide access to the site on the contractual date and this was regarded as an act of prevention. This has led to the general view that when acts of the employer prevent the contractor from achieving completion by the date of completion, if there is no contractual machinery to extend the date for completion:

- The contractor is excused from completing by the date for completion.
- There is no date from which to calculate liquidated or delay damages.
- Time is left at large and the contractor is obliged to complete within a reasonable time.
- Whilst the employer will be unable to levy liquidated or delay damages, unliquidated damages may be charged, in other words, such costs and losses which the employer can demonstrate resulted from the failure on the part of the contractor to complete within a reasonable time.

Contracts which have been drafted with no provision for granting the contractor an entitlement to extensions of time due to delays caused by the employer or his architect or engineer may attract the Prevention Principle.

7.2.3. The Prevention Principle has been used in recent cases where the contract provisions stipulate that the service of a delay notice on the part of the contractor is a condition precedent to the granting of an extension of time. Loss of the right to an extension of time due to the absence of a timely delay notice resulting in the application of liquidated or delay damages, has been met by the argument that such deduction falls foul of the Prevention Principle. If delays are caused by the employer, it has been argued that it cannot be right for such delays to be used to entitle the employer to deduct damages on the ground of lack of a delay notice. If the employer were allowed to deduct liquidated damage in respect of delays which he or she had caused, it would be difficult to relate to the decision in the case of *Alghussein Establishment* v. *Eton College* (1988), where it was stated:

> It has been said as a matter of construction, unless the contract clearly provides to the contrary, it will be presumed that it was not the intention of the parties that either should be entitled to rely on his own breach of duty to avoid the contract, or bring it to an end or to obtain a benefit under it.

7.2.4. There is conflicting case law on the matter of the application of the Prevention Principle, as it is applied where a contractor fails to serve a delay notice which the contract states to be a condition precedent to the contractor's right to an extension of time. In the Australian case of *Turner Corporation Ltd (Receiver and Manager Appointed)* v. *Austotal Pty* (1998), the judge found against the principle applying, when he stated:

> If a builder, having a right to claim an extension of time, fails to do so, it cannot claim that the act of prevention which would have entitled it to an extension of time for practical completion

resulted in its ability to complete by that time. A party to a contract cannot rely on preventing-conduct of the other party where it failed to exercise a contractual right which would have negated the effect of that preventing conduct.

7.2.5. The Prevention Principle was, however, argued successfully in Australia in the later case of *Gaymark Investments Pty Ltd* v. *Walter Construction Ltd* (1999). In this case, the contract conditions provided for the contractor to serve a written notice as a condition precedent to an entitlement to have the completion date adjusted. There were several actions on the part of the employer which resulted in delays to completion. In some instances the contractor complied with the notice provisions, whilst in others he failed to do so. The matter was referred to arbitration, where it was found that the Prevention Principle applied and thus the employer had no right to deduct liquidated damages. On referral to the court, the decision of the arbitrator was upheld. The court considered that if the employer were allowed to deduct liquidated damages, it would receive an entirely unmerited award in respect of delays of its own making.

7.2.6. In the Australian case of *Wunda Products Australia P/L* v. *Kyren P/L* (2010), the judge was of the view that if the contract contains a provision enabling the contractor to gain an extension of time in consequence of the employer's delay, it will have the effect of nullifying the Prevention Principle.

7.2.7. The Prevention Principle has been referred to in the following UK cases

- *Multiplex* v. *Honeywell* (2007).
- *Steria Ltd* v. *Sigma Wireless* (2007).
- *City Inn* v. *Shepherd Construction Ltd* (2007).

The Prevention Principle, however, has not as yet been adopted in the UK as overcoming a condition precedent notice provision, but it seems that it is likely to make further appearances in the courts. It leaves open the debate as to whether a contract which states that a delay notice is a condition precedent has the general effect of excluding the Prevention Principle.

SUMMARY

The Prevention Principle stipulates that a party to a contract cannot insist on compliance with its provisions, if the reason for the non compliance is some action or omission of the party insisting on compliance. For example, if the employer delays the contractor, it cannot insist on completion by the contractor on time. This argument has been used where a contract requires the contractor as a condition precedent to the right to an extension of time to serve a delay notice and it fails to do so. To apply liquidated damages in these circumstances would, it is argued, be a breach of the Prevention Principle. There have been two cases in Australia which address this matter, one of which supported the view that allowing an employer to levy liquidated damages in respect of employer-created delays would be a breach of the Prevention Principle. The other case supported the opposite view. In the UK courts the Prevention Principle has been discussed, but no firm decision has been made as to its application in respect of delay notices.

7.3. Are minutes of site meetings considered by the courts to be adequate notices of delay required by extension of time clauses?

7.3.1. Most standard forms of construction contracts provide in the extension of time clauses for the contractor and subcontractor to serve delay notices. ICE 6th and 7th Editions, in accordance with clause 44(1), require the contractor, within 28 days after the cause of delay has arisen, or as soon thereafter as is reasonable, to deliver to the engineer full and detailed particulars in justification of any claim to an extension of time. JCT 2011, in accordance with clause 2.27.1, requires the contractor to provide a written notice, together with particulars of the expected effects and an estimate of the expected delay, with regard to any relevant event. GC/Works 1/, condition 36, includes a requirement for the contractor to serve a notice requesting an extension of time, which shall include grounds for the request. Engineering and Construction Contract (NEC 3) in clause 61.3 requires the contractor to notify the project manager of an event which has happened, or which he expects to happen, if he believes that the event is a compensation event.

7.3.2. Before giving a definitive answer to the problem, it is necessary to decide whether a failure to serve a notice will lose the contractor or subcontractor his entitlement to an extension of time. The circumstance whereby a written notice becomes a condition precedent can be found in 7.1, above. If a formal notice is expressed as a condition precedent to the contractor's right to an extension of time, the question is often asked whether minutes of site meetings can constitute good notice.

7.3.3. In the Scottish case of *John L. Haley Ltd* v. *Dumfries & Galloway Regional Council* (1988), the court had to decide whether site meetings minutes constituted a good notice. The contract in this matter was let employing the JCT 63 standard form of contract, under which the claimants undertook certain building works to a school, with the contract period set at 78 weeks. This was overrun by 31 weeks, in respect of which a six-week extension of time was granted. When further extensions requests were refused, the matter was referred to arbitration. The claimants argued that they were entitled to an extension as the cause of delay fell within clauses 23(e), (f) and (h). The respondents maintained that the claimants were not entitled to an extension as they had not given written notice of the delay as required under clause 23. The arbiter, following proof before answer, had granted a four-week extension on the basis of site meeting minutes. At the respondent's instigation, the arbiter stated a case for the opinion of the Court of Session as to whether the minutes constituted a good notice under clause 23. The court held that the minutes did not constitute good notice. Unfortunately, the claimants had conceded that notice was a condition precedent to an entitlement to an extension of time and lost their case.

7.3.4. Nevertheless the parties to a contract often agree at the outset that the obligation to submit written notices will be waived and delays will be recorded in the site meeting minutes instead. Where this occurs, the employer would be estopped from denying the contractor an entitlement to an extension of time through lack of written notice.

7.3.5. The court had to decide in the case of *Steria Ltd* v. *Sigma Wireless Communications Ltd* (2007), details of which are provided in 7.1, whether site meeting minutes provided an adequate notice to comply with the contract requirements. The court stated that the written notice must emanate from Steria. A note in the minutes of a site meeting prepared by the main contractor or the employer, to the effect that the subcontract works had been delayed and the reason for the delay, would not in itself amount to a good notice under clause 6.1. It left unanswered as to whether site meeting minutes written by the claimant, or for that matter a written progress report, would suffice. If a court were to hold that the minutes or progress report constituted good notice, they would be required to include the level of detail needed to satisfy the requirements of the conditions of contract relating to the delay notice.

SUMMARY

Whether site meetings minutes constitute a good delay notice will depend upon the precise wording of the contract. It would seem, however, following the Scottish decision of *John L. Haley Ltd* v. *Dumfries & Galloway Regional Council* (1988) that in the case of the majority of the standard forms of contract, the site meeting minutes will not constitute good notice, unless the parties specifically amend the contract in this respect. The decision in the case of *Steria Ltd* v. *Sigma Wireless Communications Ltd* (2007) held that the notice must emanate from the claimant, and therefore minutes of a meeting which were not written by Steria did not constitute good notice. It left unanswered whether minutes written by the claimant would qualify.

7.4. Can an architect/engineer grant an extension of time after the date for completion has passed?

7.4.1. It is desirable for extensions of time to be granted at such times that contractors always know in advance the date for completion to which they are working. Some contracts will be precise as to when an architect/engineer must deal with extensions of time. JCT 2011, clause 2.28.2, is precise in stating:

> Whether or not an extension is given, the Architect/Contract Administrator shall notify the contractor of his decision in respect of any notice under clause 2.27 as soon as is reasonably practicable and in any event within 12 weeks of receipt of the required particulars. Where the period from receipt to the completion date is less than 12 weeks, he shall endeavour to do so prior to the completion date.

The ICE 7th Edition states under clause 44(5):

> The Engineer shall within 28 days [14 days in the 6th Edition] of the issue of the Certificate of Substantial Completion for the Works or any Section thereof review all the circumstances of the kind referred to in sub-clause (1) of this Clause and shall finally determine and certify

> to the Contractor with a copy to the Employer the overall extension of time (if any) to which he considers the Contractor entitled . . .

GC/Works/1, condition 36, requires the project manager to grant extensions of time 'as soon as possible and in any event within 42 days from the date any such notice is received . . . '

7.4.2. The question of the timing of an extension of time award has been discussed in the following legal cases, where the contracts were not precise as to the timing of an extension of time award.

7.4.3. In *Miller* v. *London County Council* (1934), the express wording of the contract provided:

> It shall be lawful for the Engineer, if he thinks fit, to grant from time to time, and at any time or times, by writing under his hand such extension of time for completion of the work and that either prospectively or retrospectively and to assign such other time or times for completion as to him may seem reasonable.

It was held that the words 'either prospectively or retrospectively' did not give the engineer power to fix a new date for completion after the completion of the works.

7.4.4. In *Amalgamated Building Contractors Ltd* v. *Waltham Holy Cross Urban District Council* (1952) the wording in the contract, which was the then current RIBA contract, provided in clause 18 that 'the Architect shall make a fair and reasonable extension of time for completion of the works'. Lord Denning, with regard to the time within which the architect would be required to make a decision, had this to say:

> The contractors say that the words in clause 18 mean that the architect must give the contractors a date at which they can aim in the future, and that he cannot give a date which has passed, I do not agree with this contention. It is only necessary to take a few practical illustrations to see that the architect as a matter of business, must be able to give an extension even though it is retrospective – in such a case, seeing that the cause of the delay operates until the last moment, when the works are completed, it must follow that the architect can give a certificate after they are completed . . .

Lord Denning distinguished the decision in *Miller* v. *LCC* in the following terms:

> These practical illustrations show that the parties must have intended that the architect should be able to give a certificate which is retrospective, even after the works are completed . . . *Miller* v. *London County Council* (1934) is distinguishable. I regard that case as turning on the very special wording of the clause which enables the engineer 'to assign such other time or times for completion as to him may seem reasonable'. Those words, as Mr Justice du Parcq said, were not apt to refer to the fixing of a new date for completion *ex post facto*. I would also observe that on principle there is a distinction between cases where the cause of delay is due to some act or default of the building owner, such as not giving possession of the site in due time, or ordering extras, or something of that kind. When such things happen the contract time may well cease to bind the contractors, because the building owner cannot insist on a condition if it is his own fault that the condition has not been fulfilled.

SUMMARY

The timing of the architect/engineer's decisions concerning the granting of an extension of time may be stipulated in the contract. This being the case, the architect/engineer will be required to comply with the requirements. If no time is laid down in the contract for the decision to be made, there seems no impediment to the Architect or Engineer making a decision after the date for completion has passed.

7.5. If the architect/engineer issues a variation after the extended completion date but before practical completion, should an extension of time be granted employing the date the variation is issued or by adding the net period of delay resulting from the variation to the existing completion date? Alternatively, does the issue of a variation at this time render time at large?

7.5.1. Architects and engineers should issue variations for extra work at times appropriate to the progress of the works. Often, where delays have occurred, variations are issued after the date has passed when work should have been completed. Contractors frequently argue that, due to the timing of the variation, an extension of time should be granted up to the date the variation was issued plus adequate time to carry out the extra work. A contrary argument is often made: that the issue of a variation after the completion date in the contract has passed which causes delay does not rank for an extension of time but leaves time at large, requiring the contractor to finish in a reasonable time.

7.5.2. The case *Balfour Beatty Building Ltd* v. *Chestermount Properties Ltd* (1993), heard before Mr Justice Colman of the Commercial Court, arose out of an appeal against an award of Christopher Willis, a well-known and respected arbitrator, and deals with the subject matter in question.

7.5.3. The works, employing JCT 80, comprised the construction of the shell and core of an office block. Work commenced in September 1987, the completion date being 17 April 1989, later extended to 9 May 1989. As work was not completed by this date, a certificate of non-completion was issued by the architect under clause 24.1. By January 1990 the work had still not been completed. During the period 12 February 1990 to 12 July 1990, the architect issued instructions for the carrying out of fit-out works as a variation to the contract. Practical completion of the shell and core was achieved on 12 October 1990 with the fit-out works not finished until 25 February 1991. The architect issued two extensions of time, to give a revised completion date of 24 November 1989. The variations with regard to the fit-out works were issued after the revised completion date but prior to practical completion, during a period of culpable delay. The architect then revised the non-completion certificate to reflect the extended completion date.

7.5.4. The contractor argued that the effect of the issue of variations during a period of culpable delay was to render time at large, leaving the contractor to complete within a reasonable time. This being the case, the employer would lose his right to levy liquidated damages. Alternatively, the contractor contended that the architect should have granted an extension of time on a gross basis. In this case it was argued that the fit-out work should have taken 54 weeks, this period to be added to 12 February 1990, when the fit-out variation was issued. It was the employer's contention that the correct approach should be a net extension of time, that is to say, one which calculated the revised completion date by taking the date currently fixed for completion and adding to it the 18 weeks that the architect considered to be fair and reasonable for completing the fit-out work. The main plank in support of the contractor's argument was that if the net method was adopted, the extended completion date would expire before the variation giving rise to the extension had been instructed, which was logically and physically impossible. If the contractor's line were followed, it would provide him with a windfall which swept up his delays. While recognising this, the contractor considered the problem resulted from the employer's own voluntary conduct in requiring a variation during a period of culpable delay.

7.5.5. Mr Justice Colman did not agree with the contractor and found in favour of the employer, providing the following reasons:

> When the architect reviews extensions of time under clause 25.3.3.2 following practical completion, he is entitled to reduce the extended contract period to take account of omissions. These may have been issued during a period of culpable delay. It would, therefore, be illogical for the architect to have to deal with additions differently to the way he deals with omissions.
>
> The objective of clause 25.3.1 is for the architect to assess whether any of the relevant events have caused a delay and if so by how much. He must then apply the result of his assessment to give a revised completion date. It would need clear words in the contract to allow the architect to depart from a requirement to postpone the completion date by the period of delay caused by the relevant event.

Mr Justice Colman finally concluded by saying:

> In the case of a variation which increases the works, the fair and reasonable adjustment required to be made to the period for completion may involve movement of this completion date to a point in time which may fall before the issue of the variation instruction.

7.5.6. This decision is likely to apply to a contract where the conditions of contract are JCT 2011, as the wording relating to extension of time entitlements is similar to that in JCT 98. However, a disagreement could take place where the Engineering and Construction Contract (NEC 3) conditions apply. It is the contractor's duty to submit quotations for additional work, which are required to include proposed adjustments to the time for completion and the contract sum. The quotation is required to be submitted for agreement by the project manager in advance of instructions being issued for the work to be undertaken. It is feasible that, in submitting the quotation, the contractor would apply the gross method, as did Balfour Beatty, as few contractors bother to research legal cases when seeking extensions of time. The project manager may, on the other hand, apply the net method.

7.5.7. This decision is unlikely to apply to ICE 6th and 7th Editions, where under clause 47(6) liquidated damages are suspended during a period of delay resulting from variations, a clause 12 situation, or any other delaying event outside the control of the contractor.

SUMMARY

Where an architect/engineer issues a variation after the contract completion date has passed, but before practical completion, it is appropriate where resultant delays occur for an extension of time to be granted. Such extension of time will be calculated by extending the completion date by the net period of delay. This is unlikely to apply to ICE 6th and 7th Editions, which provide for the suspension of liquidated damages during a period of delay caused by variations which are issued after the contract date for completion has passed.

7.6. When an architect/engineer is considering a contractor's application for an extension of time, can he reduce the period to which the contractor is entitled to reflect time saved by work omitted?

7.6.1. Architects and engineers when issuing variations which omit work often consider that, where the variation shows a saving in time, they are entitled to reduce the contract period or reissue an extension of time already granted but showing a shorter period.

7.6.2. Some contracts deal with this question. For example, JCT 2011 states in clause 2.28.6.3:

> No decision of the Architect/Contract Administrator under clause 2.28.4 or clause 2.28.5.2 shall fix a Completion Date for the Works or any Section earlier than the relevant Date for Completion.

From the wording in this clause it is clear that the architect cannot reduce the contract period, irrespective of how many omissions he issues. Where, however, practical completion takes place after the completion date in the contract the situation is different, as the architect or contract administrator may, in accordance with clause 2.28.5.2:

> . . . fix a Completion Date earlier than that previously fixed, if in his opinion that is fair and reasonable having regard to any instruction for Relevant Omissions issued after the last occasion on which a new Completion Date was fixed for the Works or Section.

7.6.3. The ICE 6th and 7th Editions, in clause 44(5), state in the following terms that the engineer cannot decrease an extension of time already granted:

> No such final review of the circumstances shall result in a decrease in any extension of time already granted by the Engineer pursuant to sub-clauses (3) or (4) of this Clause.

There is no reference to taking into account time saved by omissions when granting extensions of time, but this may be implicit in 'all the circumstances known to him' which the engineer considers in assessing the delay under 44(2)(a).

7.6.4. GC/Works/1 states, under condition 36(4):

> The PM shall not in a final decision, withdraw or reduce any interim extension of time already awarded, except to take account of any authorised omission from the Works or any relevant Section that he has not already allowed for in an interim decision.In this case, the project manager cannot reduce the contract period, but can subsequently reduce an extension of time already granted to take account of omissions.

7.6.5. GC/Works 1/Edition 2 states, under condition 28:

> In determining what extension of time the contractor is entitled to the Authority shall be entitled to take into account the effect of any authorised omissions from the Works.

7.6.6. MF/1, IChemE and the Engineering and Construction Contract (NEC3) make no reference to taking into account omissions when extensions of time are considered. It would seem reasonable, however, for the engineer or supervising officer to take into account omissions when making a decision concerning an extension of time. The contractor, however, having been given a period within which to carry out the work, e.g. the contract period or an extended contract, should not see the period subsequently reduced in the absence of express provision in the contract.

SUMMARY

Some forms of contract deal specifically with the question. In the absence of specific wording in the contract, it is unlikely that a court would accept that an architect or engineer has power to reduce the contract period or any extension of time already granted. Nevertheless, the contractor should continue to proceed diligently with the work, despite there being surplus time due to omissions, since such periods of surplus time can be taken into account, if the need arises, when further extensions of time are being considered.

7.7. Where a contractor's progress is behind programme, will he be entitled to an extension of time where progress and completion is affected by exceptionally adverse weather, but would not have been so affected if work had been on programme?

7.7.1. Many standard forms of contract, for example JCT 2011 clause 2.29.8, ICE 7th Edition clause 44(1)(d), and FIDIC 1999 Edition clause 8.4(c), include provisions which allow the contractor an extension of time for completion of the work where affected by exceptionally adverse weather or climatic conditions. In the case of *Walter Lawrence and Son*

Ltd v. *Commercial Union Properties (UK) Ltd* (1984), the progress and completion of the work was delayed due to exceptionally adverse weather conditions. The progress of the work was behind programme and it was argued by the architect that, had work been carried out in accordance with the programme, the weather would not have affected the progress or completion of the works.

7.7.2. The court did not accept the architect's interpretation of the contract. It is clear from the wording of the contract, which was JCT 63, that if the delaying event, which in this case was exceptionally adverse weather, affects progress and, as a consequence, the completion date, then a right to an extension of time arises. There is no reference in the extension of time clause in the contract to the programme and, therefore, the contractor's entitlement should not be affected if he is behind programme. The wording of JCT 2011, ICE 7th Edition and FIDIC 1999 Edition is similar to that of JCT 63 and, if applied to the circumstances which prevailed in this case, would produce a similar result.

SUMMARY

If a contractor's progress is behind programme, it will not affect an entitlement to an extension of time where completion is delayed due to exceptionally adverse weather conditions. The governing factor is whether the exceptionally adverse weather delayed the actual progress and in turn the completion date.

7.8. Some standard forms of contract, such as the JCT contracts, provide for extensions of time where work is delayed due to '*force majeure*'. What is *force majeure*?

7.8.1. Some contracts provide a definition as to the meaning of *force majeure*. The standard forms of contract produced by the JCT use the term *force majeure*, but include no such definition. The presence in JCT contacts of *force majeure* is important, as it is one of the relevant events which provides contractors and subcontractors with an entitlement to extensions of time for completion. By way of contrast, the ICE conditions make no reference to *force majeure*.

7.8.2. The FIDIC conditions of contract are one of the few standard conditions which include a definition of *force majeure*. It is defined as an exceptional event or circumstance:

- Which is beyond a party's control;
- Which such party could not reasonably have provided against before entering into the contract;
- Which having arisen, such party could not reasonably have avoided or overcome;
- Which was not substantially attributable to the other party.

7.8.3. *Force majeure* does not have a precise meaning, nor does it give rise to any legal doctrine under the laws which apply in the UK; it is a legal concept developed under French law.

Under the French Civil Code, *force majeure* is a good defence to a claim for breach of contract; however, to succeed, convincing evidence must be produced to demonstrate that:

- Performance was impossible;
- The event was unforeseen and unavoidable.

7.8.4. In the absence of a definition in the contract and the lack of a legal doctrine under English law, the parties to a contract which includes a *force majeure* clause are left to search through case law for assistance. From the few cases where *force majeure* has been interpreted in the English courts, it is considered that *force majeure* involves matters which are outside the control of the parties. It includes Act of God and also events such as strikes and breakdown of machinery. The edict by the Government in the UK in the 1970s that electricity supplies would be limited during a coal miners' strike was generally regarded as *force majeure*. Another example is the effect on air transport which occurred due to a volcano which erupted in Iceland in 2010. A pandemic was predicted to occur because of swine flu in 2009 and 2010 which, had it occurred, would in all probability have been *force majeure*.

7.8.5. It was held in the case of *Matsoukis* v. *Priestman* (1915), which related to the application of a liquidated damages clause in a shipbuilding contract, that the universal coal strike and breakdown of machinery amounted to *force majeure*, but not bad weather. In the case of *Hackney Borough Council* v. *Dore* (1922), the defendant was responsible for supplying electricity. A clause in the contract provided a get-out if the supply was interrupted by *force majeure*. The court had to decide whether the clause came in aid of the defendant where the supply was interrupted because of two of the workmen refusing to do the work necessary to maintain the supply. It was argued that if the men were dismissed, which would have overcome the problem, there would have been a strike. The Court of Appeal dismissed this argument. It took the view that *force majeure* applied only to strikes which actually proceeded; it did not apply to fear, no matter how reasonable, of a threatened action. When interpreting a *force majeure* clause it was said in the case of *Lebeaupin* v. *Crispin* (1920) that:

> a *force majeure* clause should be construed in each case with close attention to the words which precede or follow it and with due regard to the nature and general terms of the contract . . .

7.8.6. In the case of *Trandin Aviation Holdings Ltd* v. *Aero Toy Store LLC* (2010), the court had to decide whether a change in economic circumstances fell within the definition of *force majeure*. The case arose out of a contract to sell a jet aircraft. The price for the aircraft was \$31.75 m. A down payment of \$3 m was made at the outset, with the balance due on delivery of the aircraft. When construction of the aircraft had been completed, the defendant refused to take delivery or to pay the balance of money which was due. The claimant determined the contract with the result that, under the terms of the contract, the deposit was not returned to the defendant. The contract made it clear that the deposit represented liquidated damages, with no obligation on the part of the claimant to make a repayment.

7.8.7. The contract contained a *force majeure* clause, which stated that neither party would be liable to the other for a number of reasons, including Act of God and any other cause beyond the seller's control. It was argued by the defendant that the unanticipated unforeseeable and cataclysmic downward spiral of the world's financial markets triggered the operation of the *force majeure* clause. The judge held that it is well established in English law that a change in economic/market circumstances affecting the profitability of a contract, or the ease with which the parties' obligations can be performed, is not generally regarded as being a *force majeure* event. The court gave as an example the facts and decision in the case of *Thames Valley Power Ltd* v. *Total Gas and Power Ltd* (2006).

7.8.8. How *force majeure* should be interpreted under the EU Regulations was the subject of the decision in *Dairyvale Foods Ltd* v. *Intervention Board of Agriculture Produce* (1982), where the court had to consider whether industrial action came within the definition of *force majeure*. It was held that to be effective, the occurrence had to be:

- An external event beyond the control of the party relying on it, and
- Have consequences which could not be avoided.

7.8.9. In the Singaporean case of *RDC Concrete PTE Ltd* v. *Sato Kogoyo (S) Pte Ltd* (2007), the court held that the non- or short supply of concrete between 18 November 2004 and 5 April 2005 did not constitute *force majeure*.

7.8.10. Just prior to Christmas 2010, torrential rains fell in Queensland, Australia, which caused major flooding. In fact, an area the size of France was badly affected by the floods, to such an extent that it virtually brought the coal industry, which supplies 60% of the global coal consumption, to a halt. The mining companies relied upon *force majeure* to save them from legal actions being brought by their customers. Rio Tinto, which has major coal interests in the region, invoked the *force majeure* clause in its contracts, and issued the following press release:

> The severe monsoonal rain, on top of the significant rainfalls in November and December, has had an adverse impact on mining operations and has cut access roads and rail networks.

SUMMARY

Where *force majeure* is included in a contract which excuses one of the parties from performance, its meaning should ideally be defined in the contract. A rare example of a standard form which includes a definition of *force majeure* is the FIDIC conditions. If there is no definition in the contract, then in the event of a dispute, the court will be left to decide whether the events which have prevented performance are in fact a *force majeure*. Under UK law, there is no legal doctrine which applies to *force majeure*; the courts are therefore left to their own devices as to whether an event constitutes a *force majeure*. Acts of God, strikes and breakdown of machinery have been held to be a *force majeure*. Shortages of materials and the downward spiral of world financial markets have been considered not to be *force majeure*. Events such as a pandemic and interruption of air traffic due to a volcano eruption, whilst not tested in court, are considered to be further examples of a *force majeure*.

Chapter 8
Liquidated/Delay Damages

8.1. What is the difference between liquidated damages and a penalty?

8.1.1. The terms liquidated damages and penalty are often interpreted incorrectly. Liquidated damages are enforceable, whereas a penalty is not. Lord Dunedin had this to say of liquidated damages in the case of *Dunlop Pneumatic Tyre Co Ltd* v. *New Garage & Motor Co Ltd* (1915):

> The essence of a penalty is payment of money stipulated as *in terrorem* of the offending party; the essence of liquidated damages is a genuine covenanted pre-estimate of damage.

Lord Dunedin also went on to say: if the sum is 'extravagant and unconscionable in amount in comparison with the greatest loss that could conceivably be proved to have followed the breach', it will be regarded as a penalty and unenforceable.

8.1.2. In the case of *Public Works Commissioner* v. *Hills* (1966), in deciding what constituted a penalty, the judge said:

> The question whether a sum stipulated is a penalty or liquidated damages is a question of construction to be decided upon the terms and inherent circumstances of each particular contract, judged of as at the time of the making of the contract, not as at the time of the breach. Where the stipulated liquidated damages are held to be a penalty, the employer will be able to recover only the amount of unliquidated damages he can prove.

8.1.3. It does not affect the situation as to whether the sum included in the contract is referred to as a penalty or liquidated and ascertained damages. *Keating on Building Contracts*, 5th edition at page 225, states:

> Though the parties to a contract who use the words 'penalty' and 'liquidated damages' may *prima facie* be supposed to mean what they say, yet the expression used is not conclusive. The court must find out whether the payment stipulated is in truth a penalty or liquidated damages.

8.1.4. In the case of *Jeancharm Ltd* v. *Barnet Football Club Ltd* (2003), Jeancharm undertook to supply football kits to Barnet. If Barnet paid late, they were required under the terms of the contract to pay interest at a rate of 5% per week. Jeancharm were obliged to pay

200 Contractual Problems and their Solutions, Third Edition. Roger Knowles.

20 p per garment for each day's delay in delivery. Claims were levied by both parties and the judge in the lower court ordered Barnet to pay Jeancharm the sum of £5,000. It was argued by Barnet before the Court of Appeal that 5% per week equated to 260% per annum and was a penalty. It was argued by Jeancharm that in deciding whether 5% per week was a penalty, the court should examine fully the risk borne by both parties, which would include the 20 p per garment per day for late delivery. The Court of Appeal held that on any basis, 260% per annum interest should be regarded as a penalty.

8.1.5. Lord Justice Peter Gibson in the *Jeancharm* case considered there were four principles relating to liquidated damages which differentiated them from a penalty:

(1) The court had to look at the substance of the matter, rather than the form of words used, in order to identify the parties' intentions.
(2) The essence of a penalty would be distinct from a genuine pre-estimate of loss. In other words, the provision would not be construed as a penalty where a genuine pre-estimate had been carried out.
(3) The question of whether such a clause was to be treated as a penalty was a question of construction or interpretation of the contract at the time of making the contract, not at the time of the breach.
(4) If the amount was extravagant or unconscionable in comparison with the greatest loss that could be perceived, the clause would be a penalty.

8.1.6. There seems a growing tendency, where disputes occur regarding whether a provision in the contract which has been entered into by the parties stands or is set aside, for the court to support the *status quo*. The Hong Kong case of *Phillips Hong Kong Ltd* v. *Attorney General of Hong Kong* (1993) seems to have set the tone, where Lord Woolf in giving advice to the Privy Council in connection with an appeal, said

> It is now evident that the power to strike down a penalty clause is a blatant interference with the freedom of contract and is designed for the sole purpose of providing relief against oppression for the party having to pay the stipulated sum. There is no place where there is no oppression.

8.1.7. Mr Justice Jackson in the case of *Alfred McAlpine* v. *Tilebox* (2005), following the contemporary thinking regarding penalty clauses, said:

> Because the rule about penalties is an anomaly within the law of contract, the courts are predisposed, where possible, to uphold contractual terms which fix the level of damages for breach. This predisposition is even stronger in the case of commercial contracts freely entered into between parties of comparable bargaining power.

8.1.8. In the case of *Azimut-Benetti Spa* v. *Healey* (2010), a dispute arose with regard to a clause in a Yacht Construction Contract which stated that in the event of the contract being wrongly terminated, the purchaser was required to pay a sum equivalent to 20% of the purchase price. When the contract was terminated, the claim for payment of the 20% was resisted on the grounds that it represented a penalty and not liquidated damages.

The court, in finding that the 20% was not a penalty, considered that where both parties had the benefit of expert representation in the conclusion of the contract and the terms, including the liquidated damages clause, were freely entered into, such clauses will generally be enforced. Further, it was held that in a commercial contract of this kind, what the parties have agreed should normally be upheld.

8.1.9. Whilst the intention of liquidated damages is that of a genuine estimate of the likely level of the loss, when viewed at the time the contract was entered into, courts seem to be prepared to look at actual loss in providing assistance in arriving at a decision as to whether the sum should be enforced. In the case of *Lordsvale Finance plc* v. *Bank of Zambia* (1996), Colman J had this to say on the matter:

> whether a provision is to be treated as a penalty is a matter of construction to be resolved by asking whether at the time the contract was entered into the predominant contractual function of the provision was to deter a party from breaching the contract or to compensate the innocent party for the breach. That the contractual function is deterrent rather than compensatory can be deduced by comparing the amount that would be payable on breach with the loss that might be sustained if the breach occurred.

In other words, by comparing the estimated loss which formed the basis of the sum included in the contract with the actual loss sustained.

8.1.10. It is instructive to note that, of the many cases which have come before the courts where a liquidated damages sum included in a contract has been challenged as being a penalty, few have been successful.

SUMMARY

Liquidated and ascertained damages are a reasonable pre-estimate of the losses the employer is likely to incur if work is completed late. Such a sum is enforceable if, due to his own default, the contractor completes work late. A penalty, on the other hand, is a sum included in the contract which is intended to penalise the contractor and is far greater than the employer's estimated loss. Such a sum would be unenforceable. The courts have, however, over many years shown a reluctance to overturn a figure included in the contract which was freely negotiated by the parties.

8.2. If the employer suffers no loss as a result of a contractor's delay to completion, is he still entitled to deduct liquidated damages?

8.2.1. Contractors often argue that if they can show that when delays to completion occurred the employer suffered no loss, or a substantially reduced loss, then the liquidated damages expressed in the contract will not become payable.

8.2.2. The essence of liquidated damages is that they are a genuine covenanted pre-estimate of loss: *Clydebank Engineering and Shipbuilding Co* v. *Don Jos, Yzquierdo & Castaneda* (1905). This case arose out of a contract for the construction of four torpedo boats, which included a clause under which a sum of £500 per week would be payable in the event of the ships being delivered late. The ships were delivered late and a claim in the sum of £67,500 was levied by the Spaniards against Clydebank Shipping Co. One of the arguments employed by the defendant was that, after the date the ships were due for delivery, but before actual delivery, a large part of the Spanish navy was destroyed by the American fleet. Had the ships been delivered on time, the likelihood was that they would also have been sunk. Clydebank had therefore done the Spaniards a favour in delivering the boats late. Lord Halsbury, in rejecting this argument, said:

> I do not think I have ever heard such an argument before, nor do I think that I am likely to hear it again. Nothing could be more absurd and to give effect to it would be a striking example of defective jurisprudence.

8.2.3. In the case of *BFI Group of Companies Ltd* v. *DCB Integration Systems Ltd* (1987), a contract had been let using the JCT Minor Works Form to alter and refurbish offices and workshops. A dispute arose concerning liquidated damages and was referred to arbitration. The arbitrator held that there had been a delay in completion, but declined to award liquidated damages on the grounds that the employer had suffered no resulting loss. The work had been delayed and an extension of time awarded. However, by the time the extended date for completion arrived, work was not complete, as the roller shutters had still to be installed. None the less, BFI used the time awaiting the arrival of the roller shutter doors to fit out the premises, which normally would not have commenced until all the work had been completed. In bringing forward the fitting out BFI managed to avoid any loss. An appeal was lodged against the arbitrator's award and heard by Judge John Davies QC. He decided that the liquidated damages clause automatically came into play when the contractor completed late without a contractual justification and the employer was not required to demonstrate that he had suffered loss. The arbitrator was wrong in law in refusing to award payment of liquidated damages.

8.2.4. It was said by Lord Woolf in the Hong Kong case of *Philips Hong Kong Ltd* v. *The Attorney General of Hong Kong* (1993):

> It is to [the parties'] advantage, that they should be able to know with a reasonable degree of certainty the extent of their liability and the risk which they run as a result of entering into the contract. This is particularly true in the case of building and engineering contracts. In the case of those contracts provision for liquidated damages should enable the employer to know the extent to which he is protected in the event of the contractor failing to perform his obligations.
>
> Liquidated damages are therefore a reasonable pre-estimate of the loss the employer anticipates he will suffer if the contractor completes late. Its advantage is that the contractors know in advance the extent of risks they are taking and employers do not have the expense and difficulty of proving their loss item by item.

8.2.5. In the case of *Bovis Construction* v. *Whatling* (1995), it was held that a clause such as a liquidated damages clause which limits liability should state clearly and unambiguously the scope of the limitation and should also be construed with a degree of strictness.

SUMMARY

It was made clear by the decision in *BFI Group of Companies* v. *DCB Integration Systems Ltd* (1987) that an employer may, where they are provided for in the contract, deduct liquidated damages even though in the event he has suffered no loss.

8.3. If a delay is caused by the employer for which there is no specific entitlement to an extension of time expressed in the extension of time clause, will this result in the employer losing his right to levy liquidated damages?

8.3.1. One purpose for including an extension of time clause in a contract between employer and contractor is to provide a mechanism for adjusting the completion date where delays which affect completion are caused by the architect, engineer or employer and so preserve the employer's right to deduct liquidated damages in the event of further delay through the fault of the contractor.

8.3.2. There have been a number of cases which have come before the courts where contractors have completed late due to a combination of delays caused by the employer and delays due to fault on the part of the contractor. The employer will experience difficulties in deducting liquidated damages, as, in the absence of a clause giving the architect the power to grant an extension of time resulting from employer delays, the liquidated damages entitlement in respect of delays caused by the contractor falls away. This is because there is no mechanism written into the contract for adjusting the completion date as a result of employer delays. Time becomes at large, and the contractor's obligation is to finish in a reasonable time. What is a reasonable time may be the subject of dispute, and even if that can be agreed the employer's entitlement will be to recover the financial losses he can prove resulted from the contractor failing to complete within a reasonable time.

8.3.3. In the case of *Dodd* v. *Churton* (1897), the contractor was delayed by two weeks due to extra work ordered by the employer. The contractor, due to his own failings, also delayed completion by 25 weeks. Unfortunately for the employer, there was no provision in the contract whereby an extension of time could be awarded due to delays caused by additional work. The court held that time had become at large as a result of the delays caused by the ordering of the additional work, and thus the employer lost the right to deduct liquidated damages as a result of the 25-week delay caused by the contractor.

8.3.4. The case of *Peak Construction (Liverpool) Ltd* v. *McKinney Foundations Ltd* (1970), which is often quoted with regard to contracts which do not provide for extensions of time where late completion results from employer delays, relates to the construction of multistorey flats in Kirby, which is on the outskirts of Liverpool. Delays occurred, because of defective piling, but also as a result of procrastination on the part of Liverpool Corporation in making decisions. There was no provision in the contract for extending the completion time because of delays caused by the employer, which resulted in Liverpool Corporation being unable to deduct liquidated damages as a result of the delays caused by the defective piling. Lord Justice Pillimore summed up the situation succinctly when he said:

> I would re-state the position because I think it needs to be stated quite simply. As I understand it, a clause providing for liquidated damages [clause 22] is closely linked with a clause which provides for an extension of time [clause 23]. The reason for that is that when the parties agree that if there is delay the contractor is to be liable, they envisage that the delay shall be the fault of the contractor and, of course, the agreement is designed to save the employer from having to prove the actual damage which he has suffered. It follows, once the clause is understood in that way, that if part of the delay is due to the fault of the employer, then the clause becomes unworkable if only because there is no fixed date from which to calculate that for which the contractor is responsible and for which he must pay liquidated damages. However, the problem can be cured if allowance can be made for that part of the delay caused by the actions of the employer, and it is for this purpose that recourse is had to the clause dealing with extension of time. If there is a clause which provides for extension of the contractor's time in the circumstances which happen, and if the appropriate extension is certified by the architect, then the delay due to the fault of the contractor is disentangled from that due to the fault of the employer and a date is fixed from which the liquidated damages can be calculated.

8.3.5. In *Rapid Building Group* v. *Ealing Family Housing* (1984), the employer granted possession late because of the presence of squatters on the site. The terms of the contract were JCT 63, which makes no provision for extending the completion date due to delay in granting possession of the site. Work was completed 43 weeks late, of which three weeks was the result of the presence of squatters. For the reasons given in the *Peak Construction (Liverpool)* v. *McKinney Foundations Ltd* (1970) case, Ealing Family Housing was unable to recover liquidated damages because of the late possession.

8.3.6. To ensure that all delays by employer, architect or engineer are properly catered for in the extension of time clause, fully comprehensive wording is required. JCT 2011, in clause 2.29.6, provides for extensions of time to be awarded where delays have occurred due to 'any impediment, prevention or default, whether by act or omission, by the Employer, the Architect/Contract Administrator, the Quantity Surveyor or any of the Employer's Persons…'. GC Works/1 states, under condition 36(2):

> The PM shall award an extension of time under paragraph (1) only if he is satisfied that the delay, or likely delay, is or will be due to
>
> (b) the act, neglect or default of the employer, the PM or any other person for whom the Employer is responsible.

The ICE 7th Edition, clause 44(1)(e), covers all delays caused by the employer, the wording being 'any delay impediment prevention or default by the Employer'. The Engineering and Construction Contract (NEC3) provides for the project manager to grant extensions of time in respect of discrete causes of delay by the employer, such as is stated in clause 60.1(2), which deals with the late granting of access. Clause 60.1(18) is a sweep-up clause, which covers all breaches of contract which are not one of those otherwise provided for in the contract.

SUMMARY

If the contract does not provide grounds for extending the completion date because of employer's delays, employers who cause delay to completion will lose their right to deduct liquidated damages in respect of the contractor's delays. Effective contracts avoid this by having a fully comprehensive extension of time entitlement which covers all defaults on the part of the employer and those for whom he is responsible.

8.4. Are liquidated damages which are calculated using a formula or based upon a percentage of the contract sum enforceable?

8.4.1. It is not uncommon for contracts to include liquidated damages which have been calculated in accordance with a formula or a percentage of the contract sum. MF/1 provides in clause 34.1 for liquidated damages to be expressed as a percentage of the contract value. A limit to the contractor's liability is achieved by making provision for the amount of damages to be capped.

8.4.2. It has been argued that as the definition of liquidated damages is a genuine covenanted pre-estimate of loss (*Dunlop Pneumatic Tyre Co Ltd* v. *New Garage & Motor Co Ltd* (1915)), damages based upon a formula which includes an estimate only of the value of the work cannot be a genuine pre-estimate of damage. However, Lord Dunedin in the *Dunlop* case made the following important comment:

> It is no obstacle to the sum stipulated being a genuine pre-estimate of damages that the consequences of the breach are such as to make precise pre-estimation almost an impossibility. On the contrary, that is just the situation when it is probable that pre-estimated damage was the true bargain between the parties.

8.4.3. Certain projects, for example a new road, a school or church, would present the parties to a contract with a difficult task in pre-estimating the loss for late completion. These are the types of projects which Lord Dunedin probably had in mind. In *Robophone Facilities* v. *Blank* (1966), Lord Justice Diplock said:

> And the more difficult it is likely [to be] to prove and assess the loss which a party will suffer in the event of a breach, the greater the advantages to both parties of fixing by the terms of the contract itself an easily ascertainable sum to be paid in that event.

Lord Woolf, in the same case, said:

> The court has to be careful not to set too stringent a standard and bear in mind what the parties have agreed should normally be upheld. Any other approach will lead to undesirable uncertainty especially in commercial contracts.

8.4.4. In the case of *J.F. Finnegan* v. *Community Housing* (1993), the liquidated damages were to be calculated using the following formula:

$$\frac{\text{Estimated Cost of the Scheme} \times \text{Housing Corp. Lending Rate} \times 85\%}{52}$$

The court held that the formula was a genuine attempt to estimate in advance the loss the defendant would suffer from late completion. Judge Carr concluded by saying:

> I find that the formula used was justified at the time the parties entered into the contract.

8.4.5. The courts have been reluctant to strike down liquidated damages where the contract has been freely agreed between the parties. In the case of *Steria Ltd* v. *Sigma Wireless Communications Ltd* (2008), His Honour Judge Stephen Davies said:

> ...the question is a broad and general question and that in commercial contracts the courts should exercise great caution before striking down a clause as penal.

8.4.6. From what has been said by the judges in various legal cases involving disputes concerning liquidated and ascertained damages, it is unlikely that liquidated damages expressed as a percentage of the contract sum will not be enforced. It would, of course, depend upon all the circumstances and the percentage used.

SUMMARY

It would seem that, provided:

- the calculation of an accurate estimate of liquidated damages is difficult or impossible;
- the intention is not to penalise the contractor for completing late, but to compensate the employer;
- the formula is a genuine attempt to estimate in advance the employer's loss;

a court will not refuse to enforce liquidated damages which have been calculated by the use of a formula or applying a percentage to the contract sum.

8.5. If the architect or engineer fails to grant an extension of time within a timescale laid down in the contract, will this prevent the employer from levying liquidated damages?

8.5.1. Many modern contracts, such as JCT 2011 and ICE 7th Edition, lay down timescales within which extension of time awards are to be decided or decisions promulgated. In JCT 2011, clause 2.28.2 states:

> Whether or not an extension of time is given, the Architect/Contract Administrator shall give his decision in respect of any notice under clause 2.27 as soon as is reasonably practical and in any event within 12 weeks of receipt of receipt of particulars.

The ICE 7th Edition requires the engineer, under clause 44(5), to make a final decision regarding decisions concerning extensions of time within 28 days of the issue of the certificate of substantial completion. GC/Works/1, in condition 36(1), provides for the project manager to make a decision within 42 days of receiving the contractor's written notice. Will a failure by the architect, contract administrator or engineer to comply with these timescales be fatal to the employer's right to deduct liquidated damages?

8.5.2. There have been two legal cases where this question has been considered, *Temloc* v. *Errill Properties Ltd* (1987) and *Aoki Corp* v. *Lippoland (Singapore) Pte Ltd* (1994).

8.5.3. The case of *Temloc* v. *Errill* arose out of a contract let using JCT 80. By the terms of this contract, the architect is required to make decisions concerning extensions of time within a set timescale. With regard to the effect on the employer's entitlements should the architect fail to give his decision within the timescale, Lord Justice Croom-Johnson in the Court of Appeal had this to say:

> In my view, even if the provision of clause 25.3.3 [requirement for the architect to review extensions of time within 12 weeks of practical completion] is applicable, it is directory only as to time and is not something which would invalidate the calculation and payment of liquidated damages. The whole right of recovery of liquidated damages under clause 24 does not depend on whether the architect, over whom the contractor has no control, has given his certificate by the stipulated day.

8.5.4. A similar matter was the subject of the decision in *Aoki Corp* v. *Lippoland (Singapore) Pte Ltd* (1994). Clause 23.2 of the SIA Conditions of Contract makes it a condition precedent that the contractor notifies the architect of any event, direction or instruction which the contractor considers entitles him to an extension of time. The architect is then required to respond in writing within one month. As soon as possible after the delay has ceased to operate and it is possible to decide the length of the extension, the architect will notify the contractor of his award. If the contractor fails to complete the work by the completion date, or extended completion date, the architect must issue a delay certificate as soon as the latest date for completion has passed. The contractor notified the architect of delays, but the architect failed to notify the contractor of whether in principle an entitlement to an extension of time existed. Eventually, the architect, without giving his decision in principle, refused all requests for extension except one, for which he allowed 15 days. The employer deducted liquidated damages. It was held:

- A decision by the architect on the principle of the contractor's right to an extension was not a condition precedent to a valid determination of the contractor's entitlement. The contractor could, however, claim damages as a result of the architect's failure to make a decision, which might include the cost of increasing the labour force.
- There is no rule that delay in the issue of the delay certificate after the date for completion or the latest extended date for completion renders the delay certificate invalid.

8.5.5. The Engineering and Construction Contract (NEC 3), unlike the JCT, ICE and GC/Works contracts, provides the remedy for a failure on the part of the project manager to reach a decision within a timescale laid down in the contract. Clause 62.3 gives the project manager the power to instruct the contractor to submit a quotation in respect of time and money with regard to a compensation event. The contractor has three weeks in which to submit the quotation, and the project manager two weeks to respond. If the project manager fails to respond within the timescale and the contractor hasn't agreed to any extension of time, the contractor may notify the project manager to this effect. If there is no response to the contractor's notification by the project manager within two weeks, then the quotation is treated as being accepted. It should be stressed that the latter two-week period is triggered by the contractor's notification of a lack of response by the project manager to the quotation.

8.5.6. It would seem that failure by the architect, contract administrator or engineer to make a decision concerning extensions of time within a timescale laid down in the contract is not fatal to the employer's right to deduct liquidated damages. This is not the case with the Engineering and Construction Contract (NEC 3), where a failure on the part of the project manager to respond to the contractor's quotation in respect of a compensation event within the timescale laid down in the contract could be fatal to the employer's right to deduct liquidated damages.

SUMMARY

Unfortunately, contracts such as JCT 2011 which provide a timescale within which the architect must grant an extension of time do not state what effect a failure to comply with the timescale will have upon the employer's right to deduct liquidated and ascertained damages. However the decisions in *Temloc* v. *Errill Properties Ltd* and *Aoki Corp* v. *Lippoland* suggest that, provided a proper decision is made by the architect at some stage concerning extensions of time, a failure to meet the deadline will not affect the employer's rights.

This is not the case with the Engineering and Construction Contract (NEC 3), where a failure on the part of the project manager to respond to the contractor's quotation in respect of a compensation event within the timescale laid down in the contract could be fatal to the employer's right to deduct liquidated damages.

8.6. If the contractor delays completion but no effective non-completion certificate is issued by the architect under a JCT contract, will this mean that the employer loses his right to deduct liquidated damages?

8.6.1. JCT 2011 is an example of a contract which makes reference, under clause 2.31, to the architect issuing a certificate when the contractor fails to complete on time. The question frequently asked is: whether, in the absence of the architect's certificate, the employer remains entitled to deduct liquidated damages where the contractor finishes late.

8.6.2. This procedure was the subject of a decision of the High Court in the case of *A Bell & Son (Paddington) Ltd* v. *CBF Residential Care & Housing Association* (1989). A Bell, the contractors, entered into a contract with CBF Residential Care for the construction of an extension to a nursing home. The contract was JCT 80 Private Edition with Quantities, with a date for completion of 28 February 1986. Liquidated damages for late completion were stated to be £700 per week.

8.6.3. Work commenced on time, but completion was not achieved by 28 February 1986. The contractor served a delay notice and the architect granted an extension of time to provide a new completion date of 25 March 1986. Completion was not, however, achieved by this new date. At this stage the employer, CBF Residential Care, considered that as completion was late, liquidated damages would be due. JCT 80 requires the architect to issue a certificate of non-completion and provides for the employer to write to the contractor indicating an intention to deduct liquidated damages. Both architect and employer complied with this procedure. However, the architect subsequently had second thoughts and granted two further extensions of time. The first extended the completion date to 14 April 1986; the second further extended the date to 21 April 1986. Unfortunately, the contractor did not complete the work until 18 July 1986, when the architect issued a certificate of practical completion. A long delay then occurred, before the architect, on 3 December 1987, granted another extension of time extending the completion date to 20 May 1986. There was still, however, a shortfall between the date of 20 May 1986, by which time the contractor should have completed and 18 July 1986, when practical completion was achieved.

8.6.4. The architect issued a final certificate on 25 February 1988, but the balance due was reduced by £4,900 in respect of liquidated damages, before payment. It was argued by the contractor that liquidated damages should not have been deducted, as the procedures required by JCT 80 had not been properly complied with. Following his first granting of an extension of time, showing a revised date for completion of 14 April 1986, the architect had issued a non-completion certificate, indicating that the contractor had failed to achieve this date, but following the grant of further extensions of time, the non-completion certificate was not re-issued to reflect the revised dates for completion. It was argued that it should have been and, in the absence of a properly re-issued non-completion certificate, the employer lost the right to deduct liquidated damages. The architect had issued a final certificate and was hence *functus officio*, and therefore no longer had authority to administer the contract. It was therefore too late to re-issue the non-completion certificate. In finding in favour of the contractor and ordering that the £4,900 be paid, plus interest and costs, the court held:

> Construing clause 24.1 strictly and in accordance with its plain and ordinary meaning, it demands the issue of a certificate when a contractor had not completed by 'the completion date...' I think that when a new completion date is fixed, if the contractor has not completed by it, a certificate to that effect must be issued, and it is irrelevant whether a certificate has been issued in relation to an earlier, now superseded completion date...
>
> Construing clause 24.2.1 in a similar manner to clause 24.1, since the giving of a notice is made subject to the issue of a certificate of non-completion, if the certificate is superseded, then logically the notice should fall with it. If a new completion date is fixed, any notice given by the employer before it is at an end.

Accordingly, the condition precedent to the permissible deduction of liquidated damages, i.e. the issue of an architect's non-completion certificate, had not been fulfilled and the employer therefore lost the right to deduct liquidated damages.

8.6.5. The matter of a non-completion certificate was again referred to in *J.F. Finnegan* v. *Community Housing* (1993), when it was held that a written notice from the employer under JCT 80 is a condition precedent to the right to deduct liquidated damages.

8.6.6. The wording has been tightened up in JCT 2011, where it states in clause 2.31:

> If a new Completion Date is fixed after the issue of such a certificate, such fixing shall cancel that certificate and the Architect/Contract Administrator shall where necessary issue a further certificate.

8.6.7. The requirement for a non-completion certificate to be issued is peculiar to the JCT family of contracts. No such requirement is to be found in the ICE, GC/Works/1 or the Engineering and Construction Contract (NEC 3).

SUMMARY

The employer will lose the right to deduct liquidated damages where JCT 2011 applies, if the architect/contract administrator fails to issue a proper non-completion certificate under clause 2.31.

8.7. Can a subcontractor who finishes late have passed down to him liquidated damages fixed under the main contract which are completely out of proportion to the subcontract value?

8.7.1. The question presumes that the subcontractor has a contractual obligation to finish within a timescale and is in breach of the obligation if he completes late. Where a subcontractor is in breach, he will have a liability to pay damages to the main contractor. The general principles covering damages for breach of contract are explained in *Hadley* v. *Baxendale* (1854) and later fully considered in *Victoria Laundry (Windsor) Ltd* v. *Newman Industries* (1949).

8.7.2. Briefly, the injured party is entitled to recover any loss likely to arise, in the usual course of things, from the breach, plus such other loss as was in the contemplation of the parties at the time the contract was made and which is likely to result from the breach. The contractor, as injured party, is entitled to levy a claim for damages against a subcontractor who completes late. These damages should include only those losses which under normal circumstances are likely to arise and are within the contemplation of both parties. In all probability, a court would hold that the contractor's claim should include his own additional costs, plus any legitimate claims received from the employer and other subcontractors who have suffered financially as a result of the subcontractor's late completion. If the normal standard forms of contract are employed, the employer will levy a claim for liquidated damages against the main contractor, if the main contract completion is delayed due to a default on the part of a subcontractor. Under normal circumstances, these liquidated damages will form a part of the main contractor's claim against the defaulting subcontractor, irrespective of the value of the subcontract works.

8.7.3. Where the sum for liquidated damages under the main contract could be classed as out of the ordinary and therefore not within the contemplation of the subcontractor, it may be argued that the subcontractor is obliged to reimburse the main contractor only that element of the employer's liquidated damages which is normal and usual. Two problems arise out of this type of argument. Firstly, the question of what we mean as normal and usual; and secondly, if the sum for liquidated damages is so out of the ordinary, it might be regarded as a penalty and unenforceable against the main contractor.

8.7.4. Usually, main contractors will send to subcontractors, with the tender enquiry, details of the main contract, including the sum stated therein for liquidated damages. This procedure prevents subcontractors from arguing that the sum was outside their contemplation when they entered into the subcontract.

8.7.5. One way out for subcontractors is to include in the subcontract an amount for liquidated damages which provides a cap on their liabilities. In *M.J. Gleeson plc* v. *Taylor Woodrow Construction Ltd* (1989), Taylor Woodrow were management contractors for work at the Imperial War Museum and entered into a subcontract with Gleeson. The management contract provided for liquidated damages at £400 per day and clause 32 of the subcontract provided for liquidated damages at the same rate. Clause 11(2) of the subcontract also provided that if the subcontractor failed to complete on time the subcontractor should pay:

> a sum equivalent to any direct loss or damage or expense suffered or incurred by [the management contractor] and caused by the failure of the subcontractor. Such loss or damage shall be deemed for the purpose of this condition to include for any loss or damage suffered or incurred by the authority for which the management contractor is or may be liable under the management contract or any loss or damage suffered or incurred by any other subcontractor for which the management contractor is or may be liable under the relevant subcontract.

8.7.6. Gleeson finished late and they received from Taylor Woodrow a letter as follows:

> We formally give you notice of our intention under clause 41 to recover monies due to ourselves caused by your failure to complete the works on time and disruption caused to the following subcontractors. The following sums of money are calculated in accordance with clause 11(2) for actual costs we have incurred or may be liable under the management contract.

Then followed a summary of account, showing deductions of £36,400 for liquidated damages, being £400 per day from 31 May 1987 to 31 August 1987, and £95,360 in respect of 'set-off' claims from ten other subcontractors.

8.7.7. Gleeson applied for summary judgment under Order 14 in respect of the retained sum of £95,360 and were successful. Judge Davies found that Taylor Woodrow had no defence, when he said:

> On the evidence before me, therefore, Taylor Woodrow's course of action against Gleeson in respect of set-offs is for delay in completion. It follows that it is included in the set-off for

liquidated damages, and to allow it to stand would result in what can be metaphorically described as a double deduction.

Taylor Woodrow was therefore prevented from claiming both the liquidated damages and the losses suffered in respect of claims received from other contractors. The intention of liquidated damages is that they are comprehensive and intended to cover all costs and losses which result from delay to completion.

SUMMARY

Subcontractors who, in breach of their subcontract, complete late will be liable to pay the resultant loss and damage incurred by the contractor. These will normally include any liability the main contractor has to pay liquidated damages to the employer which result from the delay. This procedure will apply irrespective of the value of the subcontract works.

It is open to the subcontractor to argue, if the main contract liquidated damages are extremely high, that the sum involved was outside his contemplation at the time the contract was entered into. To forestall this type of argument main contractors, usually within the tender enquiry documents, will set out details of the main contract, including the sum included for liquidated damages.

8.8. What is meant by 'time at large'? How does it affect the employer's entitlement to levy liquidated damages for late completion?

8.8.1. 'Time at large' means there is no time fixed for completion, or the time set for completion no longer applies. Agreements for work to be carried out are often entered into without a completion period being stated. Letters of intent often contain instructions to commence work without a completion date being agreed. In these cases time is said to be 'at large'.

8.8.2. Contractors can find themselves trapped into contracts where the time allowed for completion is too short and the amount of money to which they are entitled is insufficient to meet their additional costs. In these circumstances they may turn to alternative means of rectifying the situation, other than the normal claims for extensions of time and additional payment.

8.8.3. For some time, contractors have used the 'time at large' argument in an attempt to avoid paying liquidated damages. Their normal approach is to say that the contract period has either never been established or that, due to delays caused by the employer for which there is no express provision in the contract for extending the completion date, time becomes at large. This being the case, the contractor's obligation is merely to finish within a reasonable time.

8.8.4. The contractor successfully used this argument in the case of *Peak Construction (Liverpool) Ltd* v. *McKinney Foundations Ltd*, heard before the Court of Appeal in 1970. It was held that, as delays on the part of the City Council in approving remedial works to the piling were not catered for in the extension of time provisions, the right to liquidated and ascertained damages was lost and time became at large. The Corporation was left with an entitlement to claim such common law damages as it was able to prove, which resulted from the contractor failing to complete within a reasonable time.

8.8.5. The case of *Rapid Building Group* v. *Ealing Family Housing*, heard before the Court of Appeal in 1984, involved a contract let using JCT 63. Unfortunately, due to the presence of squatters on the site, the contractor was given possession late. There is no provision in JCT 63 for extensions of time for late possession. The contractor was therefore able to argue successfully that time became at large. The obligation was altered to completing within a reasonable time and the employer lost its rights to levy liquidated and ascertained damages.

8.8.6. In the case of *Inserco Ltd* v. *Honeywell Control Systems* (1996), Inserco contracted to complete all work by 1 April 1991. Because of additional and revised work and lack of proper access and information, Inserco was prevented from completing on time. There was no provision in the contract for extending the completion date and time was held to be at large.

8.8.7. The construction of Wembley Stadium by Muliplex (Construction) UK Ltd resulted in a number of headline-grabbing disputes, some of which were referred to court. One such case, *Multiplex (Construction) UK Ltd* v. *Honeywell Control Systems* (2007), involved an argument concerning time at large. The dispute related to clause 4.2, which dealt with variation proposals in the subcontract and clause 11, which provided a mechanism to extend the completion date. Multiplex issued a number of revised programmes for the subcontract works, ostensibly under clause 4.2, but it was argued by Honeywell that the extension of time mechanism in the subcontract did not provide for extensions of time resulting from revised programmes. Time, it was argued as a result, became at large. The judge disagreed; he considered that clause 11 provided Multiplex with the power to grant extensions of time to take account of revised programmes; hence the arguments put forward on behalf of Honeywell concerning time at large fell to the ground.

8.8.8. If time does become 'at large', the contractor's obligation is to complete within a reasonable time. What is a reasonable time is a question of fact: *Fisher* v. *Ford* (1840). Calculating a reasonable time is not an easy matter and would depend on the circumstances of each case. As *Emden's Building Contracts*, 8th edition, puts it in Volume 1 at page 177:

> Where a reasonable time for completion becomes substituted for a time specified in the contract . . . then in order to ascertain what is a reasonable time, the whole circumstances must be taken into consideration and not merely those existing at the time of the making of the contract.

8.8.9. The judge in the case of *Hick* v. *Raymond and Reid* (1893), which is often referred to when the question of what is a reasonable time arises, stated:

> ...invariably been held to mean that the party upon whom it is incumbent, duly fulfils his obligation, notwithstanding protracted delay, so long as such delay is attributable to causes beyond his control and he has neither acted negligently nor unreasonably.

8.8.10. Vincent Powell-Smith, in his book *Problems in Construction Claims* at page 78 has this to say concerning 'time at large':

> If for some reason time under a building contract becomes 'at large', the Employer can give the contractor reasonable notice to complete within a fixed reasonable time, thus making time of the essence again: *Taylor* v. *Brown* (1839). However, if the contractor does not complete by the new date, the Employer's right to liquidated damages does not revive, and he would be left to pursue his remedy of general damages at common law.

8.8.11. Time may also become at large where the architect or engineer fails properly to administer the extension of time clause as required by the contract. An example would be where an architect or engineer fails to make any award where a proper entitlement exists. If the employer interferes with the extension of time process to the detriment of the contractor, time would again become at large.

SUMMARY

Time is at large when a contract is entered into with no period of time fixed for completion. Where this occurs, the contractor's obligation is to complete work within a reasonable time. There may also be circumstances which arise, rendering a completion period fixed by the contract as no longer operable, again rendering time at large. An example is where a delay is caused by the employer and the terms of the contract make no provision for extending the completion date due to delays by the employer.

A failure on the part of the architect, contract administrator or engineer to administer the extension of time clause as required by the contract, or interference on the part of the employer in the extension of time process, to the detriment of the contractor, would also render time at large. Where time becomes at large, the contractor's obligation is to complete the work within a reasonable time.

8.9. Can a contractor challenge the liquidated damages figure included in a contract as being a penalty and unenforceable after the contract is signed? If so, will it be a matter for the employer to prove the figure to be a reasonable pre-estimate of anticipated loss?

8.9.1. A golden rule when interpreting contracts is that both parties are bound by the terms of the contract into which they enter. With this in mind, can a contractor, having signed the contract which includes a sum for liquidated damages, later challenge the figure as being a penalty? Liquidated damages are not usually challenged on the grounds that they represent a penalty until they are levied, or there is a threat to have them levied.

8.9.2. *Bramhall and Ogden* v. *Sheffield City Council* (1985) is an example of a challenge to the liquidated damages included in the contract which only arose after they had been levied. This case arose out of a contract which was let by Sheffield City Council to Bramhall and Ogden, for the construction of 123 dwellings, employing a JCT 63 standard form of contract. A completion date of 6 December 1976 was written into the contract, together with liquidated damages in the sum of £20 per week for each uncompleted dwelling. Various extensions of time were granted by the architect, up to 4 May 1977; however, the last of the houses was not taken into occupation by the council until 29 November 1977.

8.9.3. The council, in respect of the dwellings which were completed after 4 May 1977, withheld £26,150 by applying the liquidated damages sum of £20 per dwelling per week. This was challenged by the contractor, who argued that the wording of clause 22 in the contract referred to the contractor's failure to complete the works. The liquidated damages provision applied to a failure to complete individual dwellings. The works would include external works as well as the dwellings. Counsel for the contractor expressed the opinion that the council was wrong in failing to use the sectional completion supplement. This is a standard supplement for use with JCT 63 which allows for separate amounts in respect of liquidated damages to be applied to the sections of the work specified in the contract which are to be completed on different dates.

8.9.4. The court found in favour of the contractor, when Judge Lewis Hawser said:

> It seems to me, therefore, that, in the absence of any provision for sectional completion in this contract, the respondents were not entitled to claim or deduct liquidated damages as provided in the contract.

This might not have been the end of the matter, as the judge remarked that the council would still be entitled to claim damages for breach of contract. This would have been a more exacting task, because of the need to prove all sums claimed, rather than merely levying liquidated damages.

8.9.5. The decision in the case of *Jeancharm Ltd* v. *Barnet Football Club Ltd* (2003) is another example of a late challenge to the liquidated damages figure included in a contract which also proved to be successful. Jeancharm undertook to supply football kits to Barnet. If Barnet paid late, they were required under the terms of the contract to pay interest at a rate of 5% per week. Jeancharm were obliged to pay 20 p per garment for each day's delay in delivery. Claims were levied by both parties and the judge in the lower court ordered Barnet to pay Jeancharm the sum of £5,000. It was argued by Barnet before the Court of Appeal that 5% per week equated to 260% per annum and was a penalty. It was argued by Jeancharm that in deciding whether 5% per week was a penalty the court should examine fully the risk borne by both parties, which would include the 20 p per garment per day for late delivery. The Court of Appeal held that on any basis, 260% per annum interest should be regarded as a penalty.

8.9.6. In the case of *Philips Hong Kong* v. *The Attorney General of Hong Kong* (1993), the contractor again made a late challenge to the liquidated damages figure on the grounds that it was a penalty.

The case involved a contract in which the liquidated damages varied between HK$77,818 per day and HK$60,655 per day, depending on how much of the works had been handed over. Again, there was no difficulty in leaving it until the end of the day to make the challenge. However, the contractor was unable to demonstrate that the liquidated damages clause was a penalty and so unenforceable. In arriving at a decision, the court was influenced by the decision in *Robophone Facilities Ltd* v. *Blank* (1966), where Lord Justice Diplock stated that:

> The onus of showing that a stipulation is a penalty clause lies upon the party who is sued upon it.

In other words, the contractor facing a claim for liquidated damages which he challenges as being a penalty is put to proof that his allegation is correct. It is not for the employer to prove that the liquidated damages figure is a reasonable pre-estimate of loss.

SUMMARY

A contractor who enters into a contract which contains a liquidated damages figure can, at a later stage, challenge the amount as being a penalty and unenforceable. However, where he makes such a challenge, it is up to him to demonstrate that the amount is a penalty and not a reasonable pre-estimate of the employer's loss. It is not for the employer to justify the figure.

8.10. If liquidated damages to be enforceable must be a reasonable pre-estimate of loss, how can public bodies or organisations financed out of the public purse be capable of suffering loss?

8.10.1. It is an established principle that, for liquidated damages to be enforceable, the sum claimed should be a genuine pre-estimate of anticipated loss. From this, contractors often argue that organisations such as health authorities, education trusts and the like are financed from the public purse and as such can never suffer loss. This argument, if successful, would be unsavoury to any right-minded person.

8.10.2. This argument was rejected by the judge in the case of *Multiplex Constructions Pty Ltd* v. *Abgarus Ltd* (1992), who had this to say with regard to non commercial losses:

> Thirdly, if the arguments addressed by the builder are correct in relation to works of a public nature, such as dams or major road works, where traditionally such public works do not yield a cash flow, or any cost of capital incurred in the works in for instance in the case of a dam related to a water supply, to be recouped over a defined period of time at a defined interest rate, delay in completion of construction would simply defer commencement of that recoupment period such that it could be said, on one view that the delay caused the proprietor no loss. Conceptually I do not think it is correct to say that public works, because they may not yield a cash flow, cannot result in damages to the state or public authority if delay in construction occurs. Whilst the example may be peripheral to the one being here considered, it demonstrates that, at least in some instances, an appropriate measure of liquidated damages is the cost of capital tied up for the period of delay. I regard it as an inadequate answer, in the case of a public work, to say that if the work were delayed say six months, no damage is suffered, and no liquidated damages could be validly agreed, because there was no delay in receipt of

cash flow, and there was mere deferment of a planned recoupment of capital and interest costs over time.

8.10.3. The effect of public money was at issue in the case of *Design 5* v. *Keniston Housing Association Ltd* (1986). The plaintiffs were a firm of architects who brought an action for unpaid fees and the defendant a registered housing association. It was argued, by way of defence, that failures on the part of the plaintiffs in their design, supervision and contract administration had resulted in an increase in expenditure amounting to £14 m. In answer to the counterclaim, the architects maintained that, whether or not they had been at fault, and whether or not the costs of the scheme had been increased, the housing association had suffered no loss. This was because the housing association was entitled to receive Housing Association Grant (HAG) from the Department of the Environment in a sum equal to the actual cost of the scheme, regardless of any fault by the architects. Judge Smout, in finding for the housing association, said:

> It is pertinent to note that the general rule, that only nominal damages can be awarded where there has been a wrong but no loss, has never been absolute. Various exceptions are as old as the rule itself, others have developed piecemeal. In this respect, it is sufficient to echo the comments expressed in the argument of the defendants, namely that the purpose of Housing Association Grants is to provide housing for the needy, and not to be used to relieve professional advisers from the financial consequence of breach of contract and negligence.

The housing association did not get a 'windfall', as the Housing Association grant was reduced to take account of the money which the claimant was obliged to pay to the housing association.

8.10.4. For public sector projects, losses have traditionally been calculated under three headings:

- Notional interest on capital employed;
- Additional supervision/administration costs;
- Additional accommodation costs.

A regularly used formula on local authority work to provide a weekly liquidated damages rate is:

$$\frac{85\% \times \text{estimated cost of the scheme} \times \text{interest rate}}{52}$$

This formula was approved by the court in the case of *JF Finnegan Ltd* v. *Community Housing Association Ltd* (1993).

SUMMARY

It would seem that it is not open to a contractor to argue that, as the employer is publicly funded and hence incurs no loss when delay to completion occurs, liquidated damages are not appropriate. The courts have proved to be robust in enforcing liquidated damage provisions on public sector contracts.

8.11. If liquidated damages become unenforceable and hence an entitlement to unliquidated damages arises, can the unliquidated damages be greater than the liquidated damages?

8.11.1. An extension of time clause serves two purposes. In the first instance, it enables a contractor to be relieved from the obligation to complete on time, if events occur which would otherwise be at his risk, for example excessively adverse weather, as provided for in JCT 2011. Further, an extension of time clause provides a mechanism for adjusting the completion date to take account of delays caused by the architect, contract administrator, engineer, employer and those employed or engaged by the employer. In the event of there being no such provision, where delays occur because of fault on the part of the employer, or those for whom he is responsible, time becomes at large and the right to levy liquidated damages for any delays caused by the contractor is lost. This is not a satisfactory situation. In *Rapid Building* v. *Ealing Family Housing* (1994), Lord Justice Lloyd commented:

> Like Lord Justice Philimore in *Peak Construction (Liverpool)* v. *McKinney Foundations Ltd* (1969) I was somewhat startled to be told in the course of the argument that if any part of the delay was caused by the employer, no matter how slight, then the liquidated damages clause in the contract, clause 22 becomes inoperative.

In *Peak Construction*, it was held:

> If the Employer is in any way responsible for the failure to achieve the completion date, he can recover no liquidated damages at all and is left to prove such general damages as he may have suffered.

8.11.2. Where liquidated damages become unenforceable it will be for the employer to prove such general or unliquidated damages as he is claiming. This is in contrast to liquidated damages, which require no proof to be enforceable. The question then arises as to whether, should the general damages which the claimant is able to prove, exceed the liquidated damages included in the contract, will payment become due for the greater amount so proved?

8.11.3. Lord Justice Lloyd, in the case of *Rapid Building* v. *Ealing Family Housing* (1984), stood on the fence on the matter, when he said:

> Counsel has argued that although the liquidated damages clause has ceased for the reasons I have mentioned earlier, to be applicable, nevertheless the defendants will not be entitled to recover more than the amount they would have recovered under Clause 22 if Clause 22 had continued to be applicable. Even if this is right, as to which I say nothing…

8.11.4. There is little in the way of case law concerning this matter. In the old case, *Wall* v. *Rederiaktiebolaget Luggude* (1915), it was held that, where a liquidated damages figure was held to be inappropriate, the unliquidated damages which were proved to have been incurred could be levied in full, even though they exceeded the amount of liquidated damages. In contrast, a more recent Canadian case, *Elsley* v. *Collins Insurance Agency Ltd* (1978), provided an opposite view. The question of whether a general damages claim is limited to the amount of liquidated damages included in the contract was referred to in *Cellulose Acetate Silk Co Ltd* v. *Widnes Foundry* (1933), but was left open.

8.11.5. As the case law on this subject is sparse and inconsistent, it may be relevant to apply an old rule to the situation, namely that a party cannot benefit from his own breach. In *Alghussein Establishment* v. *Eton College* (1988), it was held:

> It has been said that, as a matter of construction, unless the contract clearly provides to the contrary it will be presumed that it was not the intention of the parties that either should be entitled to rely on his own breach of duty to avoid the contract or bring it to an end or to obtain a benefit under it.

It would seem that in the light of this decision, an employer who caused a delay for which there was no provision for an extension of time, and so rendered time at large, should not be able to recover in respect of the contractor's delays general damages which exceed the liquidated damages stated in the contract.

SUMMARY

There is little consistent authority on this point, but it seems unlikely that unliquidated damages would be held to be enforceable to an extent that they exceed the amount of liquidated damages which were written into the contract.

8.12. Where a contract includes a single liquidated damages amount for failing to complete the whole of the works by the completion date, what entitlement does the employer have to claim from the contractor who has failed to complete parts of the work by the milestone dates written into the contract?

8.12.1. Most standard forms of contract provide for a date by which all work should be completed. If the contractor fails to complete the work by that date, then he is in breach of contract. JCT 2011, ICE 7th Edition, FIDIC 1999 Edition and GC/Works/1, together with most other standard forms of contract, provide the employer with a right to deduct a liquidated damages sum written into the contract for each day or week the contractor is late completing. An exception to the norm can be found in the Engineering and

Construction Contract (NEC 3). Where this contract is used, liquidated damages for delay are not provided for unless Option J – Delay Damages, applies. Where no provision for liquidated damages is included in the contract, the contractor, if he completes late, will be liable for unliquidated damages. For these to apply, the employer will be required to produce proof of the losses he has sustained as a result of the contractor's failure to complete on time.

8.12.2. The employer may, in addition to including a date by which all of the work must be completed, provide for the handover of the work in stages. Handover dates are usually provided and some standard forms of contract include special conditions which apply. JCT 2011, for example, makes provision in the contract particulars for a sectional completion, whilst ICE 7th Edition includes in the Form of Tender provision for completion dates for sections of the work, sometimes referred to as milestone dates. With this type of arrangement, separate liquidated damages amounts may be inserted into the contract in respect of each section. A failure on the part of the contractor to finish work by the section completion date will leave him open to a claim from the employer for the recovery of liquidated damages.

8.12.3. Where the contract provides for section completion dates but there are no liquidated damages stated, then if the contractor completes a section (or more than one section) late, he will be liable to pay the employer unliquidated damages.

8.12.4. It is sometimes the case that there is a liquidated damages sum written into the contract for a failure on the part of the contractor to complete the whole of the works, but no liquidated damages included for failing to achieve the sectional completion dates. Under these circumstances, if the contractor is late in completing the whole of the works, he will be obliged to pay to the employer liquidated damages. Failure to complete by the sectional completion dates will leave him exposed to a claim from the employer to pay unliquidated damages.

8.12.5. In the case of *E Turner and Sons Ltd* v. *Mathind Ltd* (1986), where a JCT 63 form of contract was employed, the bills of quantities included for phased completion. The sectional completion supplement, which was designed for this purpose, was not used. However, a sum of £1,000 per week was provided in the contract in respect of a failure on the part of the contractor to complete the whole of the works by the completion date. The contractor failed to complete the sections by the dates stated in the bills of quantities and the employer attempted to impose liquidated damages, using a *pro rata* application to the £1,000 per week included in the contract. It was held by the Court of Appeal that a sum based upon a *pro rata* application of the sum included in the contract could not be imposed. The contractor had an obligation to reimburse the employer any damages which could be proven to have been incurred by the employer as a result of the contractor's failure to complete by the sectional completion dates.

8.12.6. If the employer uses a standard form of contract correctly with regard to sectional completion dates and the associated liquidated damages, when claiming liquidated damages for failure on the part of the contractor to complete by milestone dates, the employer may be challenged to demonstrate whether, at the time the contract was entered into, they represented a reasonable pre-estimate. In some cases, it may be impossible to show that very much loss could be expected to result from a failure to meet the milestone dates. This being the case, it would cause the employer some difficulty, but would not provide him with an insurmountable problem.

SUMMARY

Contractors often include a single liquidated damages figure in respect of the whole of the works, but no liquidated damages figures for the milestone completion dates. The employer will be entitled to levy a claim for liquidated damages if the contractor fails to complete the whole of the works by the completion date, and for unliquidated damages for a failure to complete by the milestone dates. However, the employer will be required to prove the losses he has incurred due to the contractor's failure to complete by the milestone dates.

8.13. Is it possible to include in a subcontract an all-embracing sum for liquidated and ascertained damages for delay to completion?

8.13.1. It is common practice for a daily or weekly rate to be included in a standard main form of contract in respect of liquidated damages. These damages become payable by the main contractor to the employer in the event of a delay to completion of the works. The sum for liquidated damages must be a reasonable pre-estimate of the loss the employer is expected to incur should the work be completed late.

8.13.2. It is rare, however, to find a standard form of subcontract which includes provision for the deduction of liquidated damages should the subcontractor complete late. The Public Sector Partnering Contract Option 8 is, however, an exception. The normal procedure is for the main contractor to levy a claim against the subcontractor in respect of actual loss incurred. The delay by the subcontractor may result in delay to the completion of the main contract works, resulting in the employer levying liquidated damages against the main contractor. The main contractor would then pass down the liquidated damages claim to the subcontractor. The contractor may also incur costs of his own resulting from the subcontractor's delay, which may include the cost of accommodation, supervisory staff and plant retained on site during the delay period. These costs will usually be added to the employer's liquidated claim damages and charged against the subcontractor.

8.13.3. One of the difficulties for the subcontractor is that, when pricing for the work, he has no idea what the financial liability will be in the event of his late completion. If the subcontract delay does not affect the main contract completion date, the main contractor's claim may be modest. Alternatively, a substantial claim can be expected if the subcontractor's delay affects the main contract completion date. This type of uncertainty can be avoided if a liquidated and ascertained damage amount is included in the subcontract. The contractor, however, may be in a dilemma in calculating the daily or weekly sum. Should it be assumed that any delay by the subcontractor will automatically affect the main contract completion date and therefore the amount when calculated ought to include the liquidated damages in the main contract? If this is the case, subcontractors may be put off from submitting a tender by the size of the resultant amount.

8.13.4. The main contractor may decide to include the liquidated damages from the main contract when calculating the amount to be included in the subcontract. This figure could then be included in several subcontracts, as delays on the part of any of those subcontractors could affect the main contract completion date. If, in the final analysis, many weeks' delay to each of these subcontracts occurred which did not affect the main contract completion date, the claims from the main contractor would result in a windfall profit. Would courts be prepared to enforce claims of this nature? The view may be taken that it would be unreasonable to assume that all subcontract delays would affect the main contract completion date. Therefore, only those subcontracts in respect of work which was on the critical path current at the time the subcontract was entered into should provide for a subcontract liquidated damage amount which includes the liquidated damages in the main contract.

8.13.5. In the case of *M J Gleeson plc* v. *Taylor Woodrow Construction Ltd* (1989), a sum of £400 per day was included in a subcontract on a project which employed a management contract procurement method in respect of liquidated damages. Clause 11(2) of the subcontract also provided that if the subcontractor failed to complete on time the subcontractor should pay to the management contractor:

> a sum equivalent to any direct loss or damage or expense suffered or incurred by [the management contractor] and caused by the failure of the subcontractor. Such loss or damage shall be deemed for the purpose of this condition to include for any loss or damage suffered or incurred by the authority for which the management contractor is or may be liable under the management contract or any loss or damage suffered or incurred by any other subcontractor for which the management contractor is or may be liable under the relevant subcontract.

Work was completed late and Taylor Woodrow applied a sum for liquidated damages which totalled £36,400, plus a further sum of £95,360 in respect of claims received from ten other subcontractors, who it was alleged were delayed by M J Gleeson. The court held that the sum for liquidated damages was all-embracing and that further sums in respect of claims from other subcontractors were deemed to be included in the £400 daily rate.

8.13.6. Despite the uncertainties which could affect a proper estimate of the losses which a contractor may incur should a subcontractor complete late, there seems no reason why provision could not be made in a subcontract to accommodate a liquidated damages clause in respect of delays to completion.

SUMMARY

There should be no impediment to the inclusion of a sum for liquidated and ascertained damages in subcontracts. The Public Sector Partnering Contract, Option 8 Subcontract, includes for such a provision. The subcontract procurement arrangement in the case of *M J Gleeson plc* v. *Taylor Woodrow Construction Ltd* illustrates the point.

8.14. Are liquidated damages payable in respect of delays which occur after a contractor's employment has been terminated but before practical completion?

8.14.1. Where a contractor's employment is terminated, delays in completing the work are almost inevitable. At the time the contractor's employment has been terminated, the work may already be in delay. The time taken to find a replacement and have the work completed will invariably result in practical completion not being achieved before the contract completion date. This being the case, is the employer entitled to levy liquidated damages against the original contractor? Until recently there has been no legal authority on the subject. Legal commentators have taken the view that the employer, whilst being entitled to deduct liquidated damages in respect of delays which occur before termination, cannot levy liquidated damages in respect of delays which occur after termination, but before practical completion. Any delays which occur after termination would be the subject of a claim by the employer for general damages. It was considered that a contractor who has been replaced following termination loses any opportunity to influence completion, so therefore it would be unfair to levy liquidated damages for the period between termination and practical completion.

8.14.2. The decision of Mr Justice Coulson in the case of *Hall* v. *van der Heiden* (2010) has led to a change of thinking in respect of this matter. Ms Hall entered into a contract with the defendant for the refurbishment of her flat, using a JCT Minor Works contract. The contract completion date was 22 September 2007 and liquidated damages were included in the contract at a rate of £700 per week. Work was not completed by this date, a dispute arose concerning payment for work completed, and the contractor left the site on 22 January 2008. At this stage there was a substantial amount of work which was incomplete and some of the completed work was defective. The contractor's employment was terminated on 12 March 2008 and the incomplete work and defects rectification were undertaken by a replacement contractor. Practical completion was finally achieved on 17 May 2008. An extension of time of six weeks had been granted to the contractor, leaving an overrun of 29 weeks. Ms Hall levied a claim against the original contractor for the additional cost of having the work completed and defects corrected by a replacement contractor, together with liquidated damages for the 29 weeks period of late completion.

8.14.3. Mr Justice Coulson held that the contractor was not entitled to suspend work and therefore found in favour of Ms Hall. It was his view that liability for liquidated damages does not end upon termination for a breach, as to do so would reward the defendant for its own breach. One factor which may have influenced the decision was that the contractor was not represented at the hearing, as he had dismissed his legal team a week earlier. He did not appear in person, nor did his expert witnesses. Nonetheless, this is the only case where the matter has been the subject of a legal decision and, as such, represents the law as it currently stands. Whether it will be followed in any subsequent proceedings, time alone will tell.

SUMMARY

Legal commentators have taken the view that the employer, whilst being entitled to deduct liquidated damages in respect of delays which occur before termination, cannot levy liquidated damages in respect of delays which occur after termination, but before practical completion. It was considered that a contractor who has been replaced following termination loses any opportunity to influence completion, so therefore it would be unfair to levy liquidated damages for the period between termination and practical completion. Mr Justice Coulson, in the case of *Hall* v. *van der Heiden* (2010), did not follow this line of thinking. It was his expressed opinion that liability for liquidated damages does not end when a contractor's employment is terminated, as, if it were otherwise, the contractor would be rewarded for its own default. Whether this decision will be followed in any subsequent case, time alone will tell.

8.15. What are the problems associated with applying liquidated damages where provision has been made in the contract for sectional completion?

8.15.1. There are often good reasons why the employer wishes to have the work completed in phases or sections. The main benefit to the employer is that he is able to secure occupation of phases or sections of the work at an early stage, rather than have to wait for the whole project to be completed. There may be financial advantages, such as allowing access for tenants who will commence paying rent.

8.15.2. Where the employer requires the work to be completed in phases or sections, it is essential that these requirements are clearly stated in the contract documents. The commonly used standard forms of contract, such as the JCT 2011, Engineering and Construction Contract (NEC 3), ICE 7th Edition and GC/Works/1, all make provision for the work to be completed in phases or sections, which allow for early completion and handover of various parts of the project. Separate sums in respect of liquidated or delay damages need to be provided for each of the phases or sections.

8.15.3. The basic rules relating to liquidated damages will still apply, in that the sum included for each of the phases or sections must represent a reasonable estimate of the loss the employer will suffer if completion of the phase or section is late. This may be a relatively simple exercise if the phases or sections are to be handed over early to allow access for tenants, in which case the liquidated or delay damages will be based upon the loss of rental. However, if the purpose of the early access is for the convenience of the employer, for example to transfer some of its staff from existing premises to new, there may be no financial loss involved if the phases or sections are not completed on time. This being the case, liquidated damages, to be enforceable, would have to be modest.

8.15.4. Consideration will also have to be given to the effect of delays on the dates included in the contract for phased or sectional handover. The separate dates for each phase or section will have to be extended in a similar fashion to the completion date for the whole of the works, if affected by a delay which merits an extension of time. Each phase or section will also have its own defects correction period. This will obviously add to the administrative burden of the contractor and architect, or whoever is charged by the contract with the duty of making decisions on the contractor's entitlements.

8.15.5. Disputes have arisen concerning the knock-on effects of delays in earlier phases or sections on those which come later, and how the extension of time entitlement is affected. The case of *Liberty Mercian Ltd* v. *Dean and Dyball Construction* (2008) arose out of a dispute concerning the construction of four retail units at Park Avenue, Aberystwyth, which were for use as a store for Somerfield. Work was to be completed in five sections, with a separate sum in respect of liquidated damages for each one; the contract employed was JCT 1998. Dean and Dyball were eight weeks in delay in completing section 1 and received a four-week extension of time only. They were therefore exposed to liquidated damages for the four weeks overrun. The delay to section 1 had a knock-on effect on all of the following sections and therefore the architect granted four weeks' extension of time for each one. This meant that there was a delay of four weeks to sections 2 to 5, for which no entitlement to an extension of time existed.

8.15.6. Dean and Dyball argued that, because section 1 had been delayed by eight weeks, this resulted in possession of sections 2 to 5 being delayed by eight weeks and hence an extension of time of eight weeks for each section should be granted. The court did not accept this line of reasoning, as it considered that four weeks of the delay to possession of sections 2 to 5 resulted from the culpable delay on the part of Dean and Dyball in section 1.

8.15.7. It was also argued on behalf of Dean and Dyball that the liquidated damages as expressed in the contract amounted to a penalty and as such were unenforceable. Dean and Dyball argued that they were repeatedly penalised for the same delay by the deduction of liquidated damages in each of the remaining sections, due to the delay to section 1. It was also argued that the contract was void for uncertainty, as there was no provision in the contract which addressed the impact of delayed completion in earlier sections upon the remaining sections of the works. The court rejected all the arguments presented on behalf of Dean and Dyball and allowed the liquidated damages to stand.

8.15.8. It is not uncommon, where the standard forms of contract are not used, for the contract merely to stipulate dates on which various sections or phases of the work are to be completed, no provision being made for separate extensions of time and liquidated damages to apply in respect of the separate sections or phases. This situation occurred in the case of *Turner* v. *Mathind* (1986), where there was a clear requirement for the work to be completed in phases. However, there was no provision for liquidated damages to be deducted in the event of the contractor failing to complete by the phased completion dates. Only one amount for liquidated damages was included and that applied only if the whole of the works was not completed on time. It was suggested on behalf of the employer that the sum included in the contract for liquidated damages should be applied *pro rata*, but this was rejected by the court. In theory, the employer could apply general damages for a failure on the part of the contractor to complete by the phased dates. However, as there are unlikely to be any extension of time provisions which apply to the phased dates, any delays to these dates because of default by the employer or those for whom he is responsible would render time at large. The effect would be to replace an obligation to complete by the phased dates with an obligation on the part of the contractor to finish within a reasonable time.

SUMMARY

For the Employer to be able to properly enjoy the benefits of a contract which provides for phased or sectional completion, the contract must be comprehensively drafted. It must be clear that a failure to complete by the phased or sectional completion dates will attract liquidated or delay damages, with proper provisions for the granting of extensions of time where appropriate. The standard forms of contract, such as JCT 2005, ICE 7th Edition, NEC 3 and GC/Works 1, all contain provisions of this nature. Disputes have arisen as to how knock-on delays in the earlier phases or sections which affect later phases or sections should be dealt with. This has now been clarified in the case of *Liberty Mercian Ltd* v. *Dean and Dyball.*

Problems in enforcement will exist for the employer where the contract contains dates for phased or sectional completion with no provision for liquidated damages to be applied. Where this is the case, the employer will be required to demonstrate the losses he has incurred as a result of the contractor failing to meet these dates.

8.16. Do liquidated damages provide a complete remedy for delays to a contract?

8.16.1. The question is occasionally raised as to whether liquidated damages are an exhaustive remedy for delays to the completion of a contract. As a rule, delays result from slow progress which affects the date for completion. However, other factors may come into play which affect the contractor's ability to finish on time. If, for example, handover is delayed because of correcting a defect, or there is a fire caused by negligence where no extension of time can be secured, is the employer's entitlement limited to the recovery of liquidated damages, or if its costs well exceed the rate for liquidated damages, can a greater sum be claimed? Where a cap has been provided in the contract regarding liquidated damages, once the cap has been reached can the employer claim unliquidated damages for the delays thereafter?

8.16.2. *Keating on Construction Contracts*, 8th Edition, para 9-006, has this to say regarding whether liquidated damages are a complete remedy for delay:

> It is suggested that the solution in most (if not all) cases, the clause is clearly expressed to be, or as a matter of proper construction, appears to be a complete remedy for delayed completion, then it matters not why the contractor failed to complete by the due date (provided only the cause of delay was not a matter giving rise to an entitlement to extension of time, or was due to employer's default). The fact that the delay is due to a breach of contract by the contractor

as opposed to merely going slow cannot affect the nature or quality of the loss which the liquidated damages is intended to compensate.

8.16.3. In the case of *Temloc* v. *Errill Properties Ltd* (1987), where liquidated damages had been stated as nil, the Court of Appeal was of the view that liquidated damages exhausted the employer's entitlement where the contractor was responsible for causing delays to the completion of the works. Lord Nourse, as part of his judgment, said:

> I think it clear, both as a matter of construction and one of common sense, that if...the parties complete the relevant parts of the appendix,...then that constitutes an exhaustive agreement as to damages which are...payable by the contractor in the event of his failure to complete the works on time.

8.16.4. In the case of *Pigott Foundations Ltd* v. *Shepherd Construction Ltd* (1993), the meaning of the following clause, which appeared in the subcontract to which the parties were contracted, was the subject of dispute:

> With regard to B9, it was agreed that damages would only apply in the event of Pigott's not completing within the 10 weeks and any sum would be limited to £40,000 (max) at the rate of £10,000 per week.

It was argued by Shepherd, the main contractor, that the sum of £10,000 per week related to the liquidated damages in the main contract, which was intended to be passed on to Piggott, if charged to Shepherd by the employer as a result of delays caused by Pigott. There would be other costs and losses incurred by Shepherd as a result of the delay, in addition to liquidated damages passed down by the employer. These additional costs and losses Shepherd sought to recover from Pigott.

8.16.5. The judge was not impressed by the case which was presented by Shepherd and in finding in favour of Pigott, said:

> The effect of a provision for the payment of liquidated damages for delay in a building contract has been considered in a number of recent authorities, from which it is clear that not only does such a clause have the effect of imposing a liability upon the party who is responsible for the delay, to pay damages at the stated rate, but also it has the effect of precluding the other party to the contract from seeking to avoid the limitation on any amount of damages contained in a liquidated damages clause, by claiming damages for delay or disruption arising from delay in completing the works as damages for the breach of some other provision of the contract

8.16.6. In the case of *Surrey Heath Borough Council* v. *Lovell Construction Ltd* (1998), it was decided that liquidated damages were an exhaustive remedy for delay where a building has been damaged by fire.

8.16.7. The case of *Biffa Waste Services Ltd* v. *Mashinenfabrik Ernst Hese GmbH* (2008) resulted from a contract to construct a recycling plant in Leicester in connection with the collection of waste. Part of the plant comprised a large rotating drum which contained steel balls weighing 50 tonnes. During the commissioning and testing a fire broke out, which was alleged to have been the result of negligence. The losses relating to delays resulting from the fire were far greater than what could be recovered by way of liquidated damages. It was argued by Biffa that, as the claim was for breach of contract which caused delay, their entitlement to recover financial losses was not limited to the sum for liquidated damages included in the contract. The judge, in finding that the liquidated damages figure was exhaustive of Biffa's entitlements, said:

> Further I do not consider that the provisions of clause 47.1 can be construed to draw a distinction between a 'simple' failure to complete and a failure to complete caused by the breach of another obligation under the Design and Build Deed. First, I do not consider that it is possible to draw a distinction between a 'simple' failure to complete and a failure to complete caused by breach of another obligation.

8.16.8. In the case of *M J Gleeson plc* v. *Taylor Woodrow Construction Ltd* (1989), the subcontractor Gleeson finished late. Taylor Woodrow, the main contractor, claimed from Gleeson a sum calculated using the £400 per day liquidated damages written into the subcontract, plus a sum of £95,360 in respect of claims they had received from ten subcontractors relating to the delays caused by Gleeson. The court rejected the claims in respect of the subcontractors' work, as the liquidated damages fully catered for all losses incurred by Taylor Woodrow in respect of late completion by Gleeson.

SUMMARY

It is clear from what is stated in *Keating on Construction Contracts* and the decided legal cases that liquidated damages represent a comprehensive entitlement to the recovery of loss due to late completion. There is no legal facility for differentiating between delay caused by the contractor's slow progress, delay caused by a breach of contract such as correcting defective work, or the consequences of a fire caused by negligence. In all cases the employer's only remedy, if the work is completed late, is to levy liquidated damages.

Chapter 9
Variations

9.1. Where a contractor/subcontractor submits a quotation for extra work which is accepted, is the accepted quotation deemed to include for any resultant delay costs?

9.1.1. It is not uncommon, where a major variation is contemplated, for the architect or engineer to call for a quotation from the contractor or subcontractor. Often, a lump sum is quoted for the extra or varied work. Where quotations are requested for extra or varied work, it may be argued that a separate contract is contemplated. This being the case, in the absence of any wording to the contrary, if the quotation is accepted the contractor/subcontractor will be required to do everything necessary to carry out the extra or varied work. If the extra or varied work disrupts or delays the contract work, it seems reasonable that the quotation should include for any resultant costs which are foreseeable. Over and above such additional costs, the contractor/subcontractor, if this argument is sustainable, will be entitled to payment of additional unforeseen costs. If this argument is correct and the extra or varied work is considered to be a separate contract and causes delay, it could be contended that that time becomes at large. This would mean that the contractor's obligation is to complete within a reasonable time and that liquidated damages are no longer enforceable, even if the contractor is responsible for further delays due to his own default. This subject is dealt with in more depth in Problem No 8.9. Courts, however, seem nowadays to be endeavouring to resolve disputes within the context of the contract. A more likely interpretation is that the quotation and acceptance represents agreement of a financial entitlement which the contractor/subcontractor has under a clause or clauses in the contract.

9.1.2. Some of the standard forms provide for contractors to quote for variations. JCT 2011, clause 5.3.1, provides for the architect or contract administrator to request the contractor to submit a quotation for varied work in accordance with Schedule 2. Clause 1.2.2 of Schedule 2 requires the contractor to include in the quotation for any adjustment to time and clause 1.2.3 for any additions to the price in respect of loss and expense.

9.1.3. The Engineering and Construction Contract (NEC3), in clause 62, provides a detailed procedure for the submission and acceptance of quotations for compensation events and states that such quotations comprise:

200 Contractual Problems and their Solutions, Third Edition. Roger Knowles.

proposed changes to the Prices and any delay to the Completion Date assessed by the Contractor.

The project manager has to reply within two weeks of the submission and will either:

- instruct the contractor to submit a revised quotation,
- accept the quotation,
- give notice that the proposed instruction or proposed changed decision will not be given, or
- give notice that the project manager will make his own assessment.

The project manager may extend the time for submission of the quotation in his reply.

9.1.4. Where the conditions are silent on the matter, a contractor or subcontractor could refuse to submit a quotation.

9.1.5. In the case of the ICE 6th Edition, there is no requirement for the contractor to submit a quotation for varied or additional work. However, a logical argument is that any agreement between engineer and contractor in respect of a quotation relates to the evaluation of a variation under clause 52.1. It needs to be made clear in the quotation, that the price either includes or excludes any additional time requirement and extra costs related to delay. However, it seems unlikely that a quotation would be held to include any adjustment of rates which the engineer has power to fix, where the varied works renders contract rates inappropriate or inapplicable in respect of other parts of the work. The ICE 7th Edition, under clause 53, empowers the engineer to request the contractor to submit a quotation for a proposed variation and the contractor is specifically required to include in his quotation an estimate of any time implications and the cost associated with such delay.

SUMMARY

It may be argued that, in the absence of a clause in the contract which caters for quotations being submitted, an agreement for the contractor to undertake additional or varied work constitutes a separate contract. This being the case, the contractor/subcontractor would be required to include for everything necessary to carry out the extra or varied work. In all probability, it would be expected that foreseeable delay and additional cost to the contract works would be catered for in the quotation. Any unforeseeable additional delay and additional costs would be the subject of a separate claim. A more plausible argument is that the quotation is in respect of an entitlement which the contractor/subcontractor has under one or more of the clauses in the contract, such as the variations clause. In the case of the ICE 6th Edition a quotation may relate to the valuation of a variation under clause 52.1, with no specific reference to resultant additional cost. Where JCT 2011 applies, it is clear that a quotation for varied work must include for any adjustment to time and direct loss and expense. The Engineering and Construction Contract (NEC 3) provides for all variations to the works to be treated as compensation events, for which quotations for the additional works, including any changes to time and all additional financial adjustments which flow from the variation, must be submitted by the contractor, for acceptance by the project manager, before an instruction to carry out the work is issued. Contractors and subcontractors, when submitting quotations for varied or extra works, should make it expressly clear to what extent the quotation includes for additional time and loss and expense resulting from delay and disruption.

9.2. Can a contractor/subcontractor be forced to carry out a variation after practical completion?

9.2.1. Once practical completion has been achieved, it is convenient to have the contractor available to carry out extra work during the defects period. This may be, for example, to correct a design fault or to comply with the occupier's requirements. In many instances this may be inconvenient for contractors, or they may consider that, as they have been ill-served by the employer's quantity surveyor, they have no wish to do any more work. On the other hand, the contractor is familiar with the project and would normally be the most suitable organisation to carry out the work.

9.2.2. Once practical completion has been achieved, however, the contractor has no obligation to carry out varied work instructed by the architect/engineer. An exception would be where the contract expressly provides for variations being issued after practical completion.

9.2.3. *Hudson's Building and Engineering Contracts*, 11th edition, at paragraph 4.182, states:

> The last of the above mentioned cases [*SJ and MM Price Ltd* v. *Milner* (1968)] though somewhat inadequately reported supports the view that variations as well as original contract work cannot be instructed after practical completion of the remainder of the work in the absence of express provision, unless of course the contractor is willing to carry them out.

SUMMARY

The contractor is not obliged to carry out variations where the instruction is issued after practical completion, unless there is a clause in the contract which gives the architect/engineer power to issue an instruction of this nature.

9.3. Where work is omitted from the contract by way of a variation, can a contractor or subcontractor claim for loss of profit?

9.3.1. Contractors often argue that where work is omitted from their contract they lose an opportunity of earning the profit element which was built into the value of work omitted. This being the case, they claim from the employer the loss they allege to have been suffered.

9.3.2. Whether or not the contractor is entitled to the loss of profit is not clear cut. ICE 7th Edition deals with the evaluation of variations in the following clauses:

52(3) – new work to be valued at rates and prices set out in the contract

52(4) – the engineer has power to change rates or prices for any item of work in the contract which are rendered inapplicable due to the varied work.

The equivalent clauses in ICE 6th Edition are 52(1) and (2). These clauses are not too helpful in answering the question as to whether loss of profit should be paid where work is omitted. In the case of *Mitsui Construction Co Ltd* v. *The Attorney General of Hong Kong* (1986), the court seemed to give the engineer wide scope when exercising his powers under the equivalent of clause 52(4) of the ICE conditions to adjust contract rates. A reasonable argument may be that the adjusted rate should take account of lost profit.

9.3.3. Where JCT 2011 applies, clause 5.6.1.1 deals with variations and stipulates that they will be valued at bill rates where work is of a similar character to, is executed under similar conditions as and does not significantly change the quantities. If there is a significant change of quantities, then the contractor may become entitled under clause 5.6.1.2 to a variation to the rate to include a fair allowance for the change of quantities. It may be argued that the fair valuation should include the loss of profit in respect of work omitted. Clause 5.6.1.2 provides for adjusting contract rates where the conditions under which the work is carried out has been changed, due to a variation. The revised rate is expressed as reflecting the change in conditions. This wording differs from the wording in the ICE conditions and is not therefore helpful to contractors wishing to argue that contract rates should be amended to take account of lost profit arising from omitted work.

9.3.4. In the case of *Wraight Ltd* v. *PH & T Holdings* (1968), a contractor's contract was wrongly determined, with the work part-completed. The determinations clause provided for the contractor to be paid 'any direct loss and/or damage caused to the contractor by the determination'. It was held by the court that this wording included loss of gross profit on the uncompleted work.

9.3.5. JCT 63, under clause 11(6), allows the contractor to recover direct loss and/or expense arising from a variation. Following the *Wraight* case, this would include loss of gross profit. Therefore, if the contractor could show that as a result of an omission profit had been lost, the loss could be recovered if the contract were worded in a like manner to JCT 63. JCT 2011, however, does not include a clause equivalent to clause 11(6) of JCT 63. Clause 4.24.1 deals with loss and expense resulting from a variation, but only applies where the regular progress of the works has been materially affected.

9.3.6. With regard to profit in a wider context, in *Bonnells Electrical Contractors* v. *London Underground* (1995), it was held that where a call-out contract was wrongly determined, the injured party was entitled to loss of profit on work which would have been carried out during the period of notice which ought to have been given.

SUMMARY

The answer to the question is not straightforward, but depends upon the wording of the contract. Standard forms in general use provide for the work to be varied, including omissions, and therefore there is no scope for claiming damages for breach of contract. Contracts which are worded along the lines of JCT 63 clause 11(6) allow the contractor to claim loss and expense where the work is varied. In *Wraight Ltd* v. *PH & T Holdings* (1968) it was held that the wording 'direct loss and/or damage' included gross profit. Therefore, it would seem that clause 11(6) of JCT 63 would also allow for loss of profit. However, this contract is rarely used, although some of its wording remains in bespoke contracts. Where wording akin to clause 11(6) is not included in the contract, the net has to be cast wider by contractors looking for friendly wording in the contract. For example, the ICE 6th and 7th Editions, by clause 52, provide for adjusting contract rates rendered inappropriate due to the varied work. Contractors may argue that in adjusting rates allowance should be made for loss of profit from work omitted. Under JCT 2011, by clause 5.6.1.2, the contractor would have an entitlement to a fair valuation where a significant omission occurred. This, it may be said, should cater for lost profit from work omitted. It is, however, quite common, where substantial work is omitted for engineers and quantity surveyors to make allowance for loss of profit.

9.4. Where work is omitted from a contract and given to another contractor to carry out, is there a liability for the employer to pay the contractor loss of profit?

9.4.1. It is within the powers of the parties to enter into a contract whose terms give the employer the right to omit work and have it carried out by others. If this be the case, the terms of the contract should then go on to indicate whether or not the contractor is entitled to claim loss of profit. The standard forms of contract in current use do not include a provision of this nature. In the absence of such a clause what remedy, if any, do contractors have where work included in their contract is omitted and the employer arranges to have it carried out by others?

9.4.2. This matter was one of the subjects of dispute in *Amec Building Ltd* v. *Cadmus Investments Co Ltd* (1996), which was referred to arbitration. The arbitrator awarded Amec sums for loss of profit in connection with a food court. The work was covered by certain provisional sums, but an architect's instruction of 9 April 1990 omitted the work from the contract. This was subsequently let to another contractor, and Amec successfully claimed loss of profit of over £12 800, plus statutory interest. There had been no agreement between Amec and Cadmus that the work should be omitted. In finding in favour of Amec on appeal, the court held:

> There is no dispute that the power is given to the architect in his sole discretion to withdraw any work from provisional sums for whatever reason if he considers it in the best interests of the contract or the employer to do so. The difficulty that arises in this case is that which arose in the Australian case [*Carr* v. *JA Berriman Pty Ltd* (1953)], namely that it would appear that the purpose was to remove it from the existing contractor and award the work to a new contractor. Without a finding that the architect was entitled to withdraw the work for reasons put forward by [counsel], and in view of the fact that the specific reasons he advances were expressly rejected by the arbitrator, it seems to me that the only conclusion I can come to is

> that the arbitrator had concluded that it was an arbitrary withdrawal of the work by the architect in order to give it to a third party other than Amec. In those circumstances, and, in particular, in view of the express finding of the arbitrator at paragraph 12.04 that the statement in *Hudson* reflects the 'generally accepted position in the industry', it seems to me that the arbitrator was perfectly correct in deciding that such an arbitrary withdrawal of work from the provisional sums and the giving of it to the third party was something for which Amec were entitled to be compensated and the compensation that he arrived at, namely the loss of the profit having accepted figures put forward to him in evidence, is one which is not open to be impugned on appeal as a matter of law. In those circumstances, therefore, albeit with some reluctance, it seems to me that I should dismiss the appeal as well.

9.4.3. The case of *Abbey Developments Ltd* v. *PP Brickwork Ltd* (2003) involved a dispute between a main contractor and its brickwork subcontractor, in relation to a contract for the construction of 69 houses. The main contractor, dissatisfied with the performance of its subcontractor, sent a letter of complaint and the following day terminated the contract and arranged to have the remainder of the work completed by another specialist brickwork company. The contractual mechanism was a 'termination at will' clause in the subcontract. Unfortunately, this letter of termination was issued only one day after the letter of complaint concerning the alleged poor performance, which did not comply with the seven-day notice required by the terms of the contract and was therefore invalid. In reviewing the situation, the judge expressed the view that a contract for the execution of work confers on the contractor not only the duty to carry out the work, but also the corresponding right to complete the work it was contracted to undertake. To take away from the subcontractor the work it had contracted to undertake and have it carried out by others was an infringement of that right. Unfortunately for Abbey, there was a clause in the subcontract which allowed them to suspend the contract and re-tender the works without vitiating the contract. However, they chose to go down the termination route, which proved to be a major mistake.

9.4.4. The Australian case of *Carr* v. *JA Berriman Pty Ltd* (1953) also dealt with the power to omit work from a contract and arrange for it to be carried out by other. The judge, in summing up, said:

> The clause is a common and useful clause, the obvious power of which – so far as it is relevant to the present case – is to enable the architect to direct additions to, or substitutions in, or omissions from, the building as planned, which may turn out, in his opinion, to be desirable in the course of the performance of the contract. The words quoted from it would authorise the architect… to direct that particular items of work included in the plans and specifications shall not be carried out. But they do not, in my opinion, authorise him to say that particular items so included shall be carried out, not by the builder with whom the contract is made, but by some other contractor.

9.4.5. The result of these judgments is that there is implied into construction contracts an obligation not to omit work and have it carried out by others. If work is omitted from a contract and arrangements made for the work to be carried out by others, this would be a breach of contract, which would carry with it an obligation to pay damages to the injured party. In the case of a contractor or subcontractor who had suffered in this way, their entitlement would be to have paid to them their loss of profit on the value of the work omitted.

SUMMARY

The parties are at liberty to enter into a contract which allows the employer to omit work included in the contract and arrange for it to be carried out by others. Such a clause should state whether or not the contractor is entitled to claim for loss of profit. The standard forms in general use do not include such a clause. It has been held in several legal cases that where work is omitted and given to others, the contractor is entitled to claim for loss of profit. This also applies to provisional sums omitted where the work is given to others to carry out.

9.5. Where, due to a variation, a contractor has to cancel an order for the supply of material, can he pass on to the employer a claim received from the supplier for loss of profit?

9.5.1. Variations issued by architects or engineers often put contractors to substantial additional expense. Standard forms of contract are not always precise as to how variations are to be valued. Often, arguments take place as to whether a particular item of expenditure may be recovered.

9.5.2. The matter was considered by the Court of Appeal when it had to make a decision on a disputed item of cost resulting from a variation in the case of *Tinghamgrange Ltd (T/A Gryphonn Concrete Ltd)* v. *Dew Group and North West Water* (1996). As a result of a variation, the contractor had to cancel an order from a supplier, who claimed for loss of profit. The contractor sought reimbursement from the employer and joined them as a third party to the action brought by the supplier.

9.5.3. Early in 1989, North West Water commissioned works at Oswestry Water Treatment Works. Dew were the main contractors and the contract incorporated the ICE 5th Edition. Part of the contract works involved new pre-cast concrete under drainage blocks, which had to be specially manufactured. North West Water inspected Gryphonn's premises and Dew placed an order with Gryphonn worth over £250,000. The reverse of Dew's order stated:

> 14. In the event of contract works in connection with which this order is given being suspended or abandoned, the contractor shall have the right to cancel any order by written notice given to the subcontractor, vendor or supplier, on any such cancellation, the subcontractor, vendor or supplier shall be entitled to payment as provided by the terms of the main contract. No allowance, however, will be made on account of loss of profit on uncompleted portions of the order.

Gryphonn had manufactured about one quarter of the blocks, most of which were either delivered to site or were about to be delivered when North West Water's resident engineer instructed Dew to cancel the order with Gryphonn, as the specification had been changed. Dew cancelled the order and wrote to North West Water stating that the change in the specification could lead to additional costs. Gryphonn claimed about £50,000 for loss of profit on the blocks which had been ordered but not delivered, plus £1,740 for the cost of the mould manufactured to satisfy Dew's order. Dew passed this claim to North West Water. North West Water eventually paid for the cost of the mould, but refused to make any payment in respect of loss of profit emanating from the cancellation of the order.

9.5.4. The matter went before the County Court in Newport, where Gryphonn claimed its loss of profit from Dew. Dew relied upon clause 14 of the contract and sought an indemnity from North West Water. Gryphonn was awarded nearly £40,000, plus interest, but the court declined to make any award by way of indemnity in the third-party proceedings. Whilst the judge expressed sympathy with Dew, as it had merely done as instructed by the engineer, there was no main contract provision which enabled the court to make any award in respect of Gryphonn's loss of profit claim.

9.5.5. Dew appealed, arguing that its claim was simply for the cost of the work it had done in accordance with the instructions received. Paying Gryphonn's loss of profit was an integral part of that cost and to exclude that part from the valuation of the work was unfair. It was held that Clause 51 of the contract...between North West Water...and the Dew Group Ltd gives the employer the right to make any variation comprised within the wide scope of clause 51(1). It is unnecessary to decide in this appeal whether a contract containing such a clause would have been legally binding – one party having the right unilaterally to vary any term of the contract – if it had not been for its inclusion of a further provision, expressed in clause 52(1), whereby a 'fair valuation' has to be made. The 'fair valuation' of the variation is intended to provide a fair compensation to the contractor for any adverse financial effect upon it, resulting from the unilateral variation. We assume, therefore,...that the validity of the head contract is not affected by clause 51, because of the provision for fair compensation in the event of loss resulting to the contractor from a unilateral variation.

The employer was, or should have been, well aware that the contractor would not itself be the supplier of the concrete blocks, but that these would be purchased under a subcontract. The employer was, or should have been, equally aware that the result of the variation which was made – the cancellation of the order for 209,874 concrete blocks – would necessarily result in a loss of profit for the subcontractor, Gryphonn, since the contractor would be contractually bound under the head contract to pass on to the subcontractor the variation consisting of the cancellation.

The court found in favour of Dew, who were entitled to recover from North West Water the sum paid to their supplier Gryphonn for loss of profit, arising from the cancellation of part of the order for supplying blocks.

SUMMARY

The engineer, when carrying out a fair valuation under the ICE conditions, in respect of a variation to omit work which includes materials received from a supplier, should provide fair compensation to the contractor for any adverse financial effect upon it resulting from the variation. This would include a legitimate loss of profit claim from the supplier. There is no reason why this principle should not apply to JCT and GC Works/1 conditions.

9.6. How are 'fair' rates defined?

9.6.1. Most forms of contract provide for the contract documents to include contract rates which are to be applied when evaluating variations. These contract rates are usually employed where work in the variation is similar to work set out in the contract. For example, JCT 2011, clause 5.6.1.1 provides for contract rates to apply to varied work, which is of a similar character to and carried out under similar conditions as, the work described in the contract documents. Work included in a variation which differs in character to work in the contract, is usually expressed as being evaluated using fair rates e.g. JCT 2011, clause 5.6.1.3.

9.6.2. Disputes often arise where fair rates are to be employed as to how the fair rate will be calculated. Courts, if asked to decide on the matter, would invariably take a subjective view and hold that the rate must be fair in all the circumstances which occurred on the project and which were relevant to the calculation of the rate. In deciding what is a fair rate as referred to in JCT 80, clause 13.5.1.3, the court in the case of *Crittall Windows Ltd* v. *Evers Ltd* (1996) considered that the constituent items which made up the rate should be provided. In this respect, Judge John Lloyd QC said:

> ...where a valuation of this nature falls to be made it is necessary for the individual items which are to be valued to be particularised and, once they have been identified, to be priced.

9.6.3. In the case of *Semco Salvage & Marine Pte* v. *Lancer Navigation Co Ltd* (1997), the House of Lords had to decide the meaning of fair rates. A dispute arose in connection with a salvaging operation at sea. The parties could not agree the amount of reimbursement to a company undertaking salvaging operations. How a 'fair rate', as referred to in the Lloyds Open Form 1990, was to be calculated formed the basis of the disagreement. Clause 1(a) of the Form states:

> The services shall be rendered and accepted as salvage services upon the principle of 'no cure – no pay' except where the property being salved is a tanker... [when] the contractor shall, nevertheless, be awarded solely against the owners of such tanker his reasonably incurred expenses and an increment not exceeding 15% of such expenses... Within the meaning of the said exception of principle of 'no cure – no pay', expenses shall in addition to actual out of pocket expenses include a fair rate for all tugs, craft, personnel and other equipment used by the contractor in the services...

The principal issue before the court was the definition of 'expenses' in article 14.3 of the International Convention on Salvage, particularly the part of it which includes in the expenses 'a fair rate for equipment and personnel actually and reasonably used in the salvage operation'.

9.6.4. Whilst the judgment was given with particular reference to the wording of the Convention and in relation to salvage, the House of Lords made some general observations as to the meaning of 'fair rate'. Lord Lloyd of Berwick had this to say:

> (1) ...'fair rate for equipment and personnel actually and reasonably used in the salvage operation' in article 14.3 means a fair rate of expenditure, and does not include any element of profit. This is clear from the context, and, in particular, from the reference to 'expenses' in article 14.1 and 2, and the definition of 'salvors' expenses' in article 14.3. No doubt expenses could have been defined so as to include an element of profit, if very clear language to that effect had been used. However this was not the case. The profit element is confined to the mark-up under article 14.2, if damage to the environment is minimised or prevented.
> (2) The first half of article 14.3 covers out-of-pocket expenses. One would expect to find that the second half of the paragraph covered overhead expense. This is what it does. If confirmation is needed, it is to be found in the reference to sub-paragraphs (h) to (j) of article 13.1...
> (3) [Counsel for the salvage operators] argued that the word 'rate' included more naturally a rate of remuneration rather than a rate of expenditure. But, as Lord Mustill points out, 'rate' is the appropriate word when attributing or apportioning general overheads to the equipment and personnel actually and reasonably used on the particular salvage operation.
> (4) [Counsel] argued that, if a fair rate means 'rate of expenditure', it would require 'a team of accountants' in every salvage arbitration where the environment has been at risk. [Opposing Counsel's] answer was that the basic rates in the present case...were agreed without difficulty between the two firms of solicitors. In any event, accountants are nowadays, as he says, part of everyday life.

9.6.5. In the *Semco Salvage* case the court clearly defined the words 'fair rate' in the context of the contract. The exclusion of profit should not, therefore, be taken as of general application when defining fair rate under different circumstances. In deciding what was a fair rate, the amount of remuneration was, in the end, preferred to the amount of expenditure, but profit was excluded.

9.6.6. By way of contrast, in the case of *Tinghamgrange Ltd* v. *Dew Group and North West Water* (1996), the Court of Appeal held that in the ICE 5th Edition a 'fair valuation' under clause 52(1) included compensation to the main contractor for a loss of profit payment to a subcontractor in respect of the cancellation of an order resulting from an engineer's variation.

9.6.7. In the case of *Banque Paribas* v. *Venaglass Ltd* (1994), the court had to decide how the 'fair and reasonable value' should be calculated in respect of a part-completed project. It was necessary to carry out the valuation as the developer had become insolvent. In the absence of any guidelines in the development agreement, the judge decided that a fair and reasonable value should be calculated on a cost or measure and value basis and not on an open-market value of the project.

9.6.8. Judge Bowsher QC, in his judgment in the case of *Laserbore Ltd* v. *Morrison Biggs Wall Ltd* (1992), had to decide the meaning of 'fair and reasonable payments for all works executed'. He had this to say regarding costs as being the correct method of fixing a fair and reasonable payment:

> I am in no doubt that the costs plus basis in the form in which it was applied by the defendant's quantum expert (though perhaps not in other forms) is wrong in principle even though in some instances it may produce the right result. One can test it by examples. If a company's directors are sufficiently canny to buy materials from stock at knockdown prices from a liquidator, must they pass on the benefit of their canniness to their customers? If a contractor provides two cranes of equal capacity and equal efficiency to do an equal amount of work, should one be charged at a lower rate than the other because one crane is only one year old but the other is three years old. If an expensive item of equipment has been depreciated to nothing in the company's accounts but by careful maintenance the company continues to use it, must the equipment be provided free of charge apart from the running expenses (fuel and labour)? On the defendant's argument the answer to those questions is 'yes'. I cannot accept that that begins to be right.

9.6.9. It can be confusing for two judges to offer differing opinions concerning a legal interpretation of commonly used wording. In the *Laserbore* case, Judge Bowsher was not in favour of applying a cost-reimbursable method in respect of 'fair and reasonable payments'. A differing approach was taken in *Weldon Plant Ltd* v. *The Commissioner for the New Towns* (2000). In this case a dispute arose in connection with the valuation of a site instruction to excavate gravel and backfill with clay in connection with the construction of the Duston Mill Reservoir. The contractor was entitled to be paid on a 'fair valuation' basis. The judge took the view that, in evaluating a fair valuation, the calculation should be based upon the reasonable costs of carrying out the work if reasonably and properly incurred. If, in the execution of the work, cost or expenditure is incurred which would not have been incurred by a reasonably competent contractor in the same or similar circumstances, then such costs would not form part of a fair valuation. Attention was drawn to what *Keating on Building Contracts* has to say on the constituent elements of a fair valuation. It states that useful evidence may include a calculation based on the net cost of labour and materials used, plus a sum for overheads and profit. The judge, influenced by what is stated in *Keating*, considered that a fair valuation should include elements for the cost of labour, the cost of plant, cost of materials, the costs of overheads and profit. In his judgement, a fair valuation has not only to include something on account of each of those elements, but also it would not be a fair valuation within the meaning of the contract if it did not do so.

9.6.10. A different argument was preferred in the case of *Serck Controls Ltd* v. *Drake and Scull Engineering Ltd* (2000). It was held that Serck was entitled to be paid a reasonable sum for the work it had completed. Drake and Scull was of the opinion that payment should be based upon normal and usual rates. This was not accepted by the judge. In assessing Serck's entitlement he commenced in many instances with Serck's costs. However, there were many allegations of inefficiency and wasted cost and the judge had to consider whether there was any precedent in English law for making adjustments to take account of such matters. Judge Hicks QC expressed the opinion that:

> The site conditions and other circumstances in which the work was carried out, including the conduct of the other party are relevant to the assessment of reasonable remuneration. The conduct of the party carrying out the work may be relevant to the assessment of reasonable remuneration. If the value is being assessed on a cost plus basis, then deduction should be

made for time spent in repairing or repeating defective work or for inefficient working. If the value is being assessed by reference to quantities, such matters are irrelevant to the basic valuation. A deduction should be made on either basis for defects remaining at completion because the work handed over at completion is thereby worth less.

Judge Hicks was influenced by the decision in *Lachhani* v. *Destination Canada* (1997). It was held in this case that where it was appropriate to calculate a fair valuation from costs actually incurred, then they would have to be reasonably, necessarily and properly incurred. If the building contractor works inefficiently and/or if the building contractor leaves defective work, then quite obviously the costs incurred must be appropriately adjusted, as the owner should not be expected to pay more than the construction work is worth.

SUMMARY

There have been a number of court cases which dealt with the meaning of fair rates and prices. In the *Crittall* v. *Evers* case, the court held that the constituent items which made up the fair rate should be provided. The House of Lords, in a salvage case, held that the words 'fair rates and expenses' mean a fair rate of remuneration, not expenditure, but does not include loss of profit. The wording could have been defined in the contract to include profit but this was not the case. The judge in the *Banque Paribas* case preferred to employ a measure and value method to establish the value of a development and not an open-market value basis, but in *Laserbore* the judge was not impressed by the argument that fair and reasonable payments should be based upon costs. In *Weldon Plant* it was held that a fair valuation should be based upon the reasonable costs of carrying out the work if reasonably and properly incurred. If the contractor works inefficiently or carries out work defectively, then in accordance with the *Serck Controls* decision, the costs should be adjusted.

These cases cannot be taken as definitions to be used universally. Courts will normally take a subjective view and hold that the rate must be fair in all the circumstances which occurred on the particular project and which were relevant to the calculation of the rate.

9.7. When do *quantum meruit* claims arise and how should they be evaluated?

9.7.1. The expression *quantum meruit* means 'the amount he deserves' or 'what the job is worth' and in most instances denotes a claim for a reasonable sum. Payment on a *quantum meruit* basis will normally arise in circumstances where a benefit has been conferred, which justice requires should result in reimbursement being made. It does not usually arise if there is an existing contract between the parties to pay an agreed sum. There may, however, be a *quantum meruit* claim in the following circumstances:

- An express agreement to pay a 'reasonable sum'.
- No price fixed. If the contractor does work under a contract, express or implied, and no price is fixed by the contract, he is entitled to be paid a reasonable sum for his labour and the materials supplied.
- A quasi-contract. This may occur where, for instance, there are failed negotiations. If work is carried out while negotiations as to the terms of the contract are proceeding, but agreement is never reached upon essential terms, the contractor is entitled to be paid a reasonable sum for the work carried out: *British Steel* v. *Cleveland Bridge* (1984), *Kitson Insulation Contractors Ltd* v. *Balfour Beatty Ltd* (1990), *Monk Construction Ltd* v. *Norwich Union* (1992).
- Work outside a contract. Where there is a contract for specified work, but the contractor does work outside the contract at the employer's request, the contractor is entitled to be paid a reasonable sum for the work outside the contract, on the basis of an implied contract. In *Parkinson* v. *Commissioners of Works* (1949), the contractor agreed under a varied contract, to carry out certain work to be ordered by the Commissioners, on a cost plus profit basis, subject to a limitation as to the total amount of profit. The Commissioners ordered work to a total value of £6,600,000, but it was held that on its true construction, the varied contract only gave the Commissioners authority to order work to the value of £5,000,000. It was held that the work that had been executed by the contractors included more than was covered, on its true construction, by the variation deed, and that the cost of the uncovenanted addition had therefore to be paid for by a *quantum meruit*.

9.7.2. In the case of *Laserbore Ltd* v. *Morrison Biggs Wall Ltd* (1992), a subcontractor, in accordance with a letter of intent, was entitled to be paid 'fair and reasonable payments for all works executed'. The contract referred to in the letter of intent was to be based upon the FCEC Blue Form, but no price or mechanism for agreeing the price was reached. It was the contractor's argument that the subcontractor's recorded costs should be the basis of payment. Judge Bowsher QC considered under the circumstances that this method was inappropriate.

9.7.3. In the decision of *Weldon Plant Ltd* v. *The Commissioner for the New Towns* (2000), it was held that a fair and reasonable payment should be based upon the reasonable costs of carrying out the work, if reasonably and properly incurred. The case of *Seck Controls Ltd* v. *Drake & Scull Engineering Ltd* (2000) again addressed the matter of the meaning of reasonable costs. It was held that, where it was appropriate to calculate a fair valuation from costs actually incurred, they would have to be reasonably, necessarily and properly incurred. If the contractor worked inefficiently or constructed work defectively, then an appropriate adjustment must be made.

9.7.4. A 'contractual *quantum meruit*' was referred to in *ACT Construction Ltd* v. *E. Clarke & Son (Coaches) Ltd* (2002), where there was no formal contract. The Court of Appeal held that, provided there is an instruction to do work and an acceptance of that instruction, then there is a contract, and the law will imply into it an obligation to pay a reasonable sum for the work undertaken. It was decided that 'reasonable remuneration' involved the sum of the contractor's costs, plus 15%. Another Court of Appeal case, that of *A.L. Barnes Ltd* v. *Time Talk UK* (2003), related to work being undertaken without a contract. In this case, account had to be taken of an overpayment in arriving at a reasonable remuneration.

9.7.5. Contractors, having entered into a contract with all the terms including the price agreed, have been known to argue that, due to changes of a critical nature and serious delays caused by the employer, the contract has been frustrated. If the argument holds water, the contract rates will no longer apply and payment should be made on a *quantum meruit* basis. This argument was the basis of the case of *McAlpine-Humberoak* v. *McDermott International* (1992), which succeeded in the lower court. The Court of Appeal did not reject the principle of payment on a *quantum meruit* basis, but found on the facts that no frustration had occurred.

9.7.6. Where part of the work has been carried out and there occurs a breach of contract by the paying party, this gives rise to a right to discharge the contract and the injured party may claim for payment on a *quantum meruit* basis for all work undertaken. Alternatively, the injured party may claim to be paid for the work carried out, at contract rates, plus loss of profit on the uncompleted work. This can be useful to a contractor or subcontractor where they have carried out part of the work on uneconomical rates and the other party then repudiates the contract, resulting in determination. Payment for the work carried out up to the repudiation may be made on a *quantum meruit* basis. The contract rates in this case are no longer applicable, though they may be used as evidence of what should be a reasonable remuneration for the work.

9.7.7. On the other hand, in the case of *Lachhani* v. *Destination Canada UK* (1997), negotiations for construction work broke down and no contract was concluded. Nonetheless, the work was carried out and payment was due on a *quantum meruit* basis. The contractor's claim for payment was well in excess of the amount of its quotation. The court held that a building contractor should not be better off as a result of a failure to conclude a contract than he would have been if his quotation been accepted. It was the court's view that the amount of the quotation must be the upper limit of his entitlement. In some cases it may be appropriate to consider the level of pricing being negotiated before work is started.

9.7.8. When evaluating a *quantum meruit* claim, the question may arise as to whether it should be based upon the value to the recipient or the cost to the party doing the work. The Australian case of *Minister of Public Works* v. *Lenard Construction* (1992) opted for the value to the recipient.

9.7.9. The following principles concerning *quantum meruit* were established in another Australian case, *Brenner* v. *Firm Artists Management* (1993):

(1) A claim for *quantum meruit* presupposes that no contract exists.

(2) The yardstick for determining a claim for *quantum meruit* is what is a fair and reasonable remuneration, or compensation for the benefit accepted, actually or constructively.

(3) If the parties have agreed a price for certain services and the services are performed, the agreed price may be given as evidence of the appropriate remuneration, but is not in itself conclusive.

(4) The appropriate method of assessing benefit in some cases may be by applying an hourly rate to the time involved in performing the services. Where it is difficult

to determine the number of hours involved, the court may make a global assessment.

SUMMARY

There is no hard and fast rule as to the basis upon which a *quantum meruit* valuation is to be based. The court will take into account all the circumstances. It may be that a contractor's costs are a fair and reasonable method of evaluating a *quantum meruit* valuation, but not by any means the only method. The court may decide that the value of the project to the recipient of the work is to be preferred to the contractor's costs as the basis for fixing a *quantum meruit* payment. Where a quotation has been submitted, but no contract concluded, the sum fixed for a *quantum meruit* payment is unlikely to be greater than the quotation.

9.8. Can the issue of a variation to the work ever have the effect of creating a separate or replacement contract?

9.8.1. Most standard forms of contract include a clause under which the employer or his representative is able to issue an instruction to the contractor to vary the works which are described in the contract. A change in shape of the scheme, the introduction of different materials, and revised timing and sequence are all usually provided for by the variations clause. It will also usually include a mechanism for evaluating the financial effect of the variation and there is normally provision for adjusting the completion date. In the absence of such a clause, the employer could be in a difficulty, should a variation to the works be required. The contractor could either refuse to carry out the work, or undertake the work and insist upon payment on a *quantum meruit*, or fair valuation basis. Calculation of the price for the extra work applying this method could involve payment well in excess of the contract rates.

9.8.2. Even where a contract includes the usual variations clause, there may be circumstances which could lead to additions or changes introduced by the employer which fall outside the variations clause. Contractors who find themselves with unattractive contract prices would find it to their advantage to argue that a change introduced by the employer fell outside the variations clause, thus leaving the way open to argue that payment for the change should be on a *quantum meruit*, or fair valuation basis.

9.8.3. In the case of *Thorn* v. *Mayor and Commonalty of London* (1876), the judge said, 'If the additional or varied work is so peculiar, so unexpected, and so different from what any person reckoned or calculated on, it may not be within the contract at all.' A mere increase in the quantities of the work will not invalidate the original contract, even though substantial. The work ordered must be totally different from that contracted for. This being the case, the contractor will be entitled to payment on a *quantum meruit* basis and not the contract rates.

9.8.4. In *Bush* v. *Whitehaven Trustees* (1888), a contractor undertook to lay a conduit pipe in the month of June and, because of delays on the part of the employer, was unable to proceed before the winter, when wages were higher and the works more difficult because of weather conditions. The court held that these changed circumstances were such that the contractor became entitled to payment on a *quantum meruit* basis.

9.8.5. In the case of *Blue Circle* v. *Holland Dredging Co* (1987), the contractor was required to remove large quantities of excavated material from the site as part of the contract. A variation was issued by the engineer, to the effect that the excavated material should be used to form a bird island, instead of removing it from the site. Again, the court held that this variation amounted to a separate contract.

9.8.6. By way of contrast, in the case of *McAlpine Humberoak Ltd* v. *McDermott International* (1992), a contract was let for the construction of four huge pallets in connection with the construction of an oil rig. The number of drawings used increased from 22 to 161, which led to many technical queries. Further, the number of pallets was reduced from four to two. The lower court was receptive to the idea of a separate contract, but this was overruled by the Court of Appeal.

9.8.7. The position in the USA is more developed. It addresses a situation where a large number of changes are instructed, which individually fall within the ambit of the variations clause, but collectively have the effect of completely changing the scope of the works. This situation is referred to as either abandonment or cardinal change and deals with the situation where the employer makes excessive changes to a project beyond what the parties reasonably could have anticipated at the time the contract was entered into. Courts will look at a number of factors in helping to decide whether the changes have been excessive. The starting points are the size, complexity and expected duration of the contract. Other factors to be considered are the number of changes, how many changes were anticipated when the project started, the magnitude of the work involved in the changes and the length of time over which such changes were made. There is no required intention on the part of the employer to abandon the contract by introducing excessive changes; this will often be implied as a result of constant interference or change. If the parties ignore the procedural provisions of the contract with regard to variations, this could help influence the court into accepting that abandonment has occurred.

SUMMARY

It is difficult to be hard and fast as to when additional or changed work will constitute a separate contract, or convert the contract into a different one. Courts in the USA seem to be more sympathetic to the contractor's case for abandonment or critical change than in most other countries. Courts and arbitrators in the UK and similar jurisdictions find themselves in the long grass when trying to decide what is due if payment is to be on a *quantum meruit* or 'fair valuation' basis. They feel more comfortable in dealing with additions or changes priced at contract rates and are inclined to play it safe in holding that the facts as presented have not resulted in a separate contract coming into being.

Employers who are perhaps starting to feel uneasy about the prospect of the USA attitudes creeping into the thinking of judges or arbitrators could give consideration to rewording the variations clause to give a wider definition of additions and changes which would provide more scope for changes to be made once the contract has been entered into.

Chapter 10
Loss and Expense/Additional Cost

10.1. Where a contractor/subcontractor is granted an extension of time, is there an automatic right to the recovery of loss and expense?

10.1.1. There is a common misconception in the construction industry that, once an extension of time has been granted, there will be an automatic entitlement to the recovery of loss and expense. Under most standard forms of contract, the entitlements to extensions of time and payment of additional cost or loss and expense are quite separate. Most of the JCT 2011 forms of contract provide for extensions of time in respect of relevant events, listed in clause 2.29, with loss and expense paid where the relevant matters listed in clause 4.24 occur.

10.1.2. Many of the reasons for awarding extensions of time also provide an entitlement to additional cost or loss and expense, an example being the late issue of drawings by the architect, contract administrator or engineer. Some grounds for extension of time, for example *force majeure* and excessively adverse weather under JCT 2011 contracts, give an entitlement to more time but no facility for financial recovery.

10.1.3. The Society of Construction Law Delay and Disruption Protocol 2002 makes the situation clear, when it states:

> 1.6.2 Entitlement to an EOT does not automatically lead to entitlement to compensation (and vice versa).

10.1.4. The Engineering and Construction Contract (NEC 3) is an exception. Within each of the alternative main contract and subcontract options, there is a compensation clause. Where, for example, one of the events listed as a compensation event in clause 60.1 occurs, the contractor is entitled to the resultant extra time and money.

SUMMARY

A contractor/subcontractor is not, under most standard forms of contract, automatically entitled to the recovery of loss and expense following the granting of an extension of time. The Engineering and Construction Contract (NEC 3) is an exception and

200 Contractual Problems and their Solutions, Third Edition. Roger Knowles.

provides for both extensions of time and the recovery of additional cost in respect of the events listed in the compensation events clause.

10.2. Where a contractor/subcontractor successfully levies a claim against an employer for late issue of drawings, can the sum paid out be recovered by the employer from a defaulting architect/engineer?

10.2.1. Employers who find themselves having to make payments to contractors as a result of the late issue of drawings by the architect or engineer usually feel aggrieved. They often contemplate sending a claim for payment to the architect or engineer, or plan to deduct such claims from fees as they fall due. For the employer to have a legal right to take such action, he must be able to show that the late issue of drawings by the architect or engineer was a breach of duty.

10.2.2. The duties owed by architects, engineers and other designers to their clients are either express or implied in their terms of engagement. The Appointment of an Architect SFA/99, published by the RIBA, provides in condition 2.1:

> The Architect shall in performing the services . . . exercise reasonable skill and care in conformity with the normal standards of the Architect's profession.

10.2.3. Late issue of drawings may result from a number of causes other than breach of duty by the architect/engineer. The employer may have delayed making a decision, or introduced a late change. There may have been a change introduced by statute or the fire officer. Information from the consulting engineer or statutory authority may have caused the delay. Late issue of drawings does not automatically mean that the architect/engineer is guilty of breach of duty.

10.2.4. Employers when appointing an architect/engineer may wish to be a little more precise in setting out their duties than the general wording provided in the RIBA 'Appointment of an Architect' form. JCT 2011 edition provides, in the fifth recital:

> the Employer has provided the Contractor with a schedule ('Information Release Schedule') which states what information the Architect/Contract Administrator will release and the time of that release.

This clause is optional, but if it is used it may be in the employer's interests to write a clause into the architect's conditions of engagement providing an obligation to issue drawings to conform with the information release schedule.

10.2.5. Where wording of a general nature is included in the conditions of appointment, such as the RIBA wording requiring the architect to exercise reasonable skill and care in the normal standards of the architect's profession, it will be necessary for an employer, if he is to be successful, to show that the drawing production by the architect fell short of what one could expect from the ordinary skilled architect.

10.2.6. In the case of *London Underground* v. *Kenchington Ford* (1998), Kenchington Ford were appointed to provide civil engineering and architectural design services in connection with the Jubilee Line station at Canning Town. They were under an express duty, as set out in their terms of engagement, to exercise all reasonable professional skill and diligence. Part of their obligations were to correct any errors, ambiguities or omissions arising, deal with questions for clarification on design matters from the works director or project director and clarify working drawings where required. The subcontractor, Cementation Bachy, was responsible for the design and construction of a diaphragm wall. The design information unfortunately included errors. The judge concluded that Kenchington Ford should have checked and discovered the errors and hence were in breach of their duty. Mowlem, the main contractor, levied a claim resulting from the incorrect design. The claim was settled and London Underground sought to recover some of the amounts paid to Mowlem from Kenchington Ford. The claim submitted by Mowlem was on a global basis and failed to provide proper details of losses alleged to have been incurred. Judge Wilcox was not impressed by this lack of detail, but he did find that Kenchington Ford were obliged to make a payment to London Underground, although in a substantially smaller amount than was claimed.

10.2.7. Employers who reach agreement with contractors to pay out in respect of claims have major hurdles to jump if they are to recover the sums paid from the architect/engineer. In the first instance, it is necessary for them to show that the late issue of design information constituted a breach of duty. Further, the employer must be able to demonstrate to the satisfaction of the court that the amount paid in respect of the claim can be linked to that breach of duty. Settlement of a contractor's claim is often the basis of a legal action for the recovery of the amount paid by the employer from the architect/engineer. To be successful in such an action, the employer must to be able to demonstrate that the amount of the settlement was reasonable: *Biggin* v. *Permanite* (1951); *P & O Developments Ltd* v. *Guy's and St Thomas' National Health Service Trust* (1998).

SUMMARY

Employers who pay out claims to contractors for late issue of design information do not automatically have a right to recover those sums from the architect/engineer. It is necessary for the employer to demonstrate a breach of duty by the architect/engineer and that the sums paid result from that breach of duty. In addition, the employer will have to show that the amounts paid out were reasonable.

10.3. Will a contractor or subcontractor substantially prejudice its case for additional payment if it fails to keep adequate accurate records?

10.3.1. Max Abrahamson, in his book *Engineering Law and The ICE Contract*, wrote, 'A party to a dispute, particularly if there is an arbitration will learn three lessons (often too late):

the importance of records, the importance of records and the importance of records'. This quotation is relevant to the case of *Attorney General for the Falkland Islands* v. *Gordon Forbes Construction (Falklands) Limited* (2003). A contract was let for the construction of the infrastructure of the East Stanley Housing Development in the Falkland Islands, using the FIDIC 4th Edition conditions. These conditions, like most standard forms, provide a procedure which the contractor is required to follow in the event of a claim being submitted. Clause 53.1 states that the contractor, if he intends to claim any additional payment, must give a written notice of his intention to the engineer within 28 days after the event giving rise to the claim. In addition, the contractor is obliged by clause 53.2 to keep such contemporary records as are necessary to support the claim. The contract goes one stage further under clause 53.3 in requiring the contractor to send to the engineer within 28 days of the written notice a detailed account of the claim. If the contractor fails to send in a written notice and detailed account there is a fall-back position in clause 53.4, which states that in the event of the contractor failing to comply with what after all are very simple obligations, his entitlement to payment will not exceed the amount the engineer or an arbitrator could consider verified by contemporary records.

10.3.2. In the case mentioned, it would seem that the contractor, in considering that he was entitled to additional payment, made his first mistake in failing to send a written notice and detailed account as required by clauses 53.1 and 53.3. The question of whether there were sufficient contemporary records to satisfy the requirements of the fall-back position became the subject of a dispute which was referred to arbitration and subsequently to the court. It would seem that the contemporary records were somewhat lacking and did not prove sufficiently convincing, as far as the engineer was concerned, to merit payment. The contractor felt this approach unjust, as he considered he was able to fill in the gaps where the contemporary records were lacking with verbal testimony from those within his organisation able to provide the necessary information. It was the engineer's view that this type of verbal evidence did not amount to contemporary records. This proved somewhat critical, as both parties were agreed that if the contractor failed to comply with these conditions his rights to payment were lost and therefore everything hung on the matter of contemporary records.

10.3.3. The judge considered that the requirements of the contract set out a clear and ordered way of dealing with any claim for an additional payment. Claims have to be notified at the time they arise, contemporary records have to be kept and regular accounts rendered. The whole contractual system is aimed at the early resolution of any queries at the time the claim arises. He then had to decide what was meant by the word 'contemporary' and felt there was little difference between contemporary and the word 'contemporaneous', i.e. occurring at the same time. Examples were given by the judge to illustrate the point. If there was a dispute as to whether labour was employed on site during a particular date it would be acceptable to produce time sheets showing the hours worked, or even a statement by the contractor indicating that this is what had taken place. It would be exceptional if the records were made more than a few weeks after the event took place. The judge posed something of a conundrum. What would the position be if the contractor wished to claim for labour costs for a four-week period? He had time sheets for weeks 1, 2 and 4 but the records for week 3 were missing. Could the court accept verbal evidence concerning week 3? The judge felt that such verbal evidence would not be acceptable.

10.3.4. The judge found in favour of the authority in deciding that, to constitute contemporary records, these must be prepared at or about the time of the events which give rise to the claim. Where there is no contemporary record, the claim will fail. Where there are contemporary records to support part of the claim, it may succeed on that part of the claim which is supported by the contemporaneous records, but not otherwise. If, however, the contemporary records are ambiguous or unclear as to their meaning, then verbal evidence will be accepted to provide clarification. The contractor, probably unaware of Max Abrahamson's words of wisdom, found out when it was too late the importance of keeping proper records.

10.3.5. Records derived from a retrospective assessment made by a witness of impeccable credentials is sometimes accepted by the courts; however, relying on such evidence can lead to uncertainty of outcome. It can also to be an extremely expensive method of producing evidence. In the case of *Bridge UK Ltd* v. *Abbey Pynford* (2007), a claim for management time, using a retrospective assessment, was submitted which was considered by the court. Mr Justice Ramsey was less than enthusiastic about the evidence, when he said:

> It must be borne in mind that such an assessment is an approximation of the hours spent and may overestimate or underestimate the actual time which would have been recorded at the time

Mr Justice Ramsey accepted the evidence, but reduced the amount awarded by 20%.

SUMMARY

Where a contract provides a procedure to be followed if the contractor, or for that matter the subcontractor, intends to claim additional payment, a failure to comply could result in a loss of the entitlement. Where records are to be contemporaneous or submitted within a specified time scale, it is crucial for them to be properly made at the time required. In the absence of proper contemporaneous records, courts may sometimes accept assessment made at a later stage. However, there is every possibility that such information will be rejected by a court.

10.4. When a contractor/subcontractor, with regard to a claim for loss and expense or additional cost, is shown to have failed to serve a proper claims notice or has not submitted details of the claim as required by the contract, can the architect/engineer legitimately reject the claim?

10.4.1. Disputes often arise between contractors or subcontractors and the employer's consultants concerning the service of a written notice in relation to a right to additional payment. Does a lack of a written notice lose the contractor or subcontractor his rights? In other words, is the procedure which has been written into the contract a condition precedent to the rights which are provided by the terms of the contract?

10.4.2. Some standard forms of contract make it clear how the lack of appropriate written notice will affect the contractor's entitlements. The ICE 6th Edition, clause 52(4)(e) and the 7th Edition, clause 53(5) deal with the problem of late notice and state:

> If the Contractor fails to comply with any of the provisions of this Clause [notice of claims/additional payments] in respect of any claim which he shall seek to make then the Contractor shall be entitled to payment in respect thereof only to the extent that the Engineer has not been prevented from or substantially prejudiced by such failure in investigating the said claim.

GC/Works/1, condition 46, states that prolongation or disruption costs will not be paid unless the contractor, immediately upon becoming aware that the regular progress of the works has been or is likely to be disrupted or prolonged, gives a notice to the project manager specifying the circumstances. The Engineering and Construction Contract (NEC 3), clause 61.3, indicates that if the contractor fails to notify the project manager of a compensation event within eight weeks of becoming aware of the event, he is not entitled to a change to the 'prices, the completion date or a key date'.

10.4.3. *Hudson's Building and Engineering Contracts*, 11th edition at paragraph 4.132, deals with the matter in general terms when it states:

> Since the purpose of such provisions is to enable the owner to consider the position and its financial consequences (by cancelling an instruction or authorising a variation, for example, he may be in a position to reduce his financial liability if the claim is justified), and since special attention to contemporary records may be essential either to refute or regulate the amount of the claim with precision, there is no doubt that in many if not most cases the courts will be ready to interpret these notice requirements as conditions precedent to a claim, so that failure to give notice within the required period may deprive the contractor of all remedy.

10.4.4. In the case of *London Borough of Merton* v. *Stanley Hugh Leach* (1985), Mr Justice Vinelott had this to say with regard to the need for a loss and expense notice under clauses 11(6) and 24(1) of JCT 63:

> The common features of subclauses 24(1) and 11(6) are first that both are 'if' provisions (if upon written application being made, etc.) that is, provisions which only operate in the event that the contractor invokes them by a written application . . .

He then went on to consider the amount of detail which must be included with the notice:

> The question of principle is whether an application under clauses 24(1) or 11(6) [of JCT 63] must contain sufficient information to enable the architect to form an opinion on the questions whether (in the case of clause 24) the regular progress of the work has been materially affected by an event within the numbered sub-paragraphs of clause 24 or (in the case of clause 11(6)) whether the variation has caused direct loss and/or expense of the kind there described and in

either case whether the loss and/or expense is such that it would not be reimbursed by payment under other provisions of the contract or in the case of 11(6) under clause 11(4).

The judge pointed out that it would not necessarily be enough simply to make what might be described as a 'bare' application, which would satisfy the requirements of clause 11(6) or clause 24(1). The application had to be framed with sufficient particularity to enable the architect to do what he was required to do. It follows that the application must therefore contain sufficient detail for the architect to be able to form an opinion as to whether or not there is any loss or expense to be ascertained. Mr Justice Vinelott also commented upon the circumstances following a contractor's application that satisfied the minimum requirements of clause 11(6) and/or clause 24(1). The architect had been able to form an opinion favourable to the contractor and was then under a duty to ascertain or instruct the quantity surveyor to ascertain the alleged loss and/or expense. He said:

> The contractor must clearly co-operate with the architect or the quantity surveyor giving such particulars of the loss or expenses claimed as the architect or quantity surveyor may require to enable him to ascertain the extent of that loss or expense; clearly the contractor cannot complain that the architect has failed to ascertain or to instruct the quantity surveyor to ascertain the amount of direct loss or expense attributable to one of the specified heads if he has failed adequately to answer a request for information which the architect requires if he or the quantity surveyor is to carry out that task.

Later, he said:

> If [the contractor] makes a claim but fails to do so with sufficient particularity to enable the architect to perform his duty or if he fails to answer a reasonable request for further information he may lose any right to recover loss or expense under [clause 11(6) or clause 24(1)] and may not be in a position to complain that the architect was in breach of his duty.

10.4.5. In the insurance case of *Kier Construction Ltd* v. *Royal Insurance (UK) Ltd* (1992) the insurance policy required the claimant to notify the insurer as soon as possible, if there was an occurrence such that in consequence a claim is to be made. The claim should have been made on 12 June 1989, but it was not submitted until 4 July 1989. The claimant lost his rights, as the notice was not served 'as soon as possible'.

10.4.6. In *Rees & Kirby Ltd* v. *Swansea City Council* (1984) the court had this to say with regard to the contractor's claim for finance charges as part of the loss and expense claim and the need for reference to be made in the notice to such charges:

> I agree with the judge's construction of clause 11(6) and 24(1) and with his conclusion that the architect can only ascertain and certify the amount of interest charges lost or expended at the date of the application. It is these charges which are the subject of the application; it is these charges which he has power to investigate, ascertain and certify. I respectfully agree with the judge that the architect would be exceeding his powers were he to take into account further financial charges or other losses accruing during these two periods, however long, and such further charges and losses would be recoverable only, if at all, under a subsequent application

or subsequent applications – although he might obtain the respondents' approval to waiving the required applications or extending the time for making them.

10.4.7. The case of *Hersent Offshore SA and Amsterdamse Ballast Beton-Waterbouw BV* v. *Burmah Oil Tankers Ltd* (1978) also deals with the question of notice, where it was held that a notice given after completion of the work could not be regarded as a notice given before work commenced, or as soon thereafter as is practicable.

10.4.8. Contractual requirements can sometimes be onerous with regard to the provision of information in support of a monetary claim and a failure to comply with the contract can prove very expensive. A good example occurred in the case of *Education 4 Ayrshire Ltd* v. *South Ayrshire Council* (2009). This case arose out of a PPP arrangement for schools in Ayrshire. All the work was subcontracted to Carillion. The contract under which the dispute arose was not one of the standard construction contracts but one designed for PPP projects. A contract was entered into for the design and construction work at six schools. The work included the removal of asbestos. A survey was carried out and the contract provided for the removal of the asbestos identified by the survey. Once work got under way, the contractor encountered more asbestos than was shown up by the survey. Removing the additional asbestos caused delay and additional cost. The contractor claimed the additional work had cause a delay of 16 weeks and additional cost amounting to £815,792, and looked to the contract for a remedy. The contract notice requirements were extremely strict and required the contractor to serve a notice within 20 business days after he became aware of the event giving rise to delay and additional cost. This notice was required to include the reasons for the delay and also a claim for any extension of time and payment of compensation. The contract made it crystal clear that, in the event of a failure to comply with this requirement, the contractor would not become entitled to more time or money. The contract stipulated that the notice must be sent to the Chief Executive of the authority by first class post, fax or by hand. A dispute arose in respect of the notice of delay and additional cost. The contractor became aware on 6 April 2007 of the additional asbestos and the likely effect on the completion date. A notice was submitted on 2 May 2007, which just met the time-scale. However, the notice reported the delay and went on to express an intention to submit a claim. This did not meet the contractual obligations, which required the actual claim to be submitted. The court had no hesitation in finding against the contractor. He had failed to comply with the contract provisions with regard to what should be included in the notice and hence lost the right to recover the sum of £815,792.

10.4.9. A failure on the part of the contractor to comply with the contract proved extremely costly in the case of *WW Gear Construction* v. *McGhee Group* (2010). The employer, McGhee Group, employed WW Gear Construction to undertake groundworks in relation to the development of the Plaza Hotel in Westminster. The contract incorporated the JCT Trade Contract Terms (TC/C) 2002, with bespoke amendments. One of these amendments required the contractor, in respect of any monetary claim, to submit in writing a fully documented and costed claim within two months of it becoming reasonably apparent that there was an entitlement. The wording of the contract made it plain that this requirement was a condition precedent to any right to additional payment. The contractor claimed a sum of £1,555,919.89, but lost the case because of a failure to submit a fully documented and costed claim, as required by the contract.

10.4.10. From the above it can be seen that a contractor or subcontractor who fails to serve a proper claims notice will in all probability lose his rights to levy a claim for loss and expense or additional cost in accordance with the terms of the contract. Contractors/subcontractors do from time to time fail to comply with the contract requirements in relation to notice or some other procedural matter and are thus prevented from levying a claim under the contract.

10.4.11. However, all may not be lost. If the event giving rise to the claim would also give an entitlement to a common law damages claim, e.g. late issue of architect's drawings, the contractor/subcontractor may be able to claim damages for breach of contract. Clause 4.26 of JCT 2011 states that the provisions of this clause are without prejudice to any other rights and remedies which the contractor may possess. Clause 24(2) of JCT 63 was similarly worded, and from the judgment of Mr Justice Vinelott in the case of *Stanley Hugh Leach* v. *London Borough of Merton* (1985) it seems clear that a claim under clause 24(2) of JCT 63 is an alternative to a claim under clause 24(1), where he observed:

> But the contractor is not bound to make an application under clause 24(1). He may prefer to wait until completion of the work and join the claim for damages for breach of obligation to provide instructions, drawings and the like in good time with other claims for damage for breach of obligations under the contract. Alternatively, he can, as I see it, make a claim under clause 24(1) in order to obtain prompt reimbursement and later claim damages for breach of contract, taking the amount awarded under clause 24(1) into account.

10.4.12. It would seem that, even without the express retention of common law rights, they would not be lost. A clause which excluded all rights except those set out in the contract would be required, if this end were to be achieved. This was the situation which arose in *Strachan and Henshaw* v. *Stein* (1997), where a contract was let using the MF/1 conditions of contract. Condition 44.4 states:

> Accordingly, except as provided for in the conditions, neither party shall be obliged or liable to the other in respect of any damages or losses suffered by the other which arise out of, under or in connection with the contract or the works

It was held by the court that a claim for breach of contract outside the provisions of the contract failed, because of the wording of condition 44.4. It would seem that a similar situation exists where the Engineering and Construction Contract (NEC 3) is employed. In clause 63.4 it states that rights (set out in the contract) to change the prices, the completion date or a key date are the only rights of the employer and contractor in respect of compensation events.

10.4.13. The failure of a contractor to submit a written notice of a monetary claim entitlement was dealt with in the case of *Maidenhead Electrical Services* v. *Johnson Controls* (1997). A key question was whether this failure affected the contractor's rights to claim damages for breach of contract. The wording of the contract stated:

> PAYMENT ... Any claims by the contractor requesting consideration for payments additional to those provided for in the subcontract or in amendments to the subcontract shall be submitted in writing to the company within 10 days of the occurrence from which the claim arises. If no notification of the claim is received by the company within 28 days of such date, then the said claim shall be automatically invalid.

The defendants argued that this clause applied to payments additional to those provided for in the subcontract and not those for extensions of time and associated monies which were claims under the subcontract. The court's considered opinion was:

> The issue as reformulated focuses on damages for breach of contract and additional monies under the contract. I do not consider that the limitation applies to claims for damages for breach of contract. I do not see that a claim for damages for breach of contract is a claim for payment additional to those provided for in the subcontract. The words are wholly insufficient to exclude liability for damages for breach of contract in the event of failure to comply with those time limits. In my view, the limitation is directed to the case where, for example, the contract rates are insufficient to cover the contractor's costs. The reference to 'amendments' supports this view. Thus, the relevant wording of condition 17 would not operate to exclude a money claim associated with an extension of time, or a disputed amendment.

10.4.14. Not every attempt to make a written notice a condition precedent succeeds. The case of *Amec Process & Energy Ltd* v. *Stork Engineers and Contractors BV* (1999) arose out of the construction of the topside of a floating oil production unit by Stork for Shell. Part of the work was subcontracted by Stork to Amec. In the subcontract conditions a clause appeared which stated that the contract price could only be changed by a variation. Amec could request a variation to the price if any of a prescribed set of circumstances arose. The clause contained several conditions regarding matters such as notice, the keeping of records and the provision of a full evaluation. Failure to comply resulted in Amec losing the right to a price adjustment. It was held that the clause was not worded with sufficient clarity to remove Amec's right to claim damages for breach. The judge explained that, if a party to a contract is to be deprived of its right to recover damages for breach, the wording should be in clear, unambiguous terms.

SUMMARY

The lack of a proper claims notice and back-up details will result in a contractor or subcontractor losing an entitlement to additional payment. Some standard contracts state that a notice is a condition precedent to a right to payment. Even where the contract fails to include this type of wording, lack of proper notice usually results in the contractor losing his rights to additional payment in accordance with the contract. Some contracts, for example the ICE 6th and 7th Editions, the Engineering and Construction Contract (NEC 3) and GC/Works/ deal expressly with the effect of lack of notice, whereas JCT 2011 does not. Loss of a right to claim under the terms of the contract may not affect a right of recovery of sums for breach of contract, unless the contract expressly excludes these rights.

10.5. With a programme shorter than the contract period, can the contractor/subcontractor claim additional payment if, because of the timing of the issue of the architect/engineer's drawings, he is prevented from completing in accordance with the shortened programme?

10.5.1. Contractors and subcontractors, when submitting tenders, are always seeking to gain a competitive edge when pricing their bid. Often, they consider that the proposed contract period is generous and they are capable of completing well within it. If the tender price is based upon a period shorter than the proposed contract period, a saving on site overheads and head office contribution will normally be achieved. Contractors and subcontractors will usually be reluctant to make the basis of their pricing known at tender stage and it is only when the programme is produced, usually after the contract has been signed, that completion is shown earlier than the contract completion date. In the event of the architect/engineer failing to issue drawings and details in sufficient time to meet the shortened programme, will this give rise to an entitlement on the part of the contractor/subcontractor to be paid any resultant additional costs?

10.5.2. In the case of *Glenlion Construction Ltd* v. *The Guinness Trust* (1987), Glenlion entered into a contract dated 10 July 1981 with The Guinness Trust. The contract incorporated JCT 63. Item 3.13.4 in the bills of quantities provided for the following:

> Progress Chart
>
> Provide within 1 week from the date of possession, a programme chart of the whole of the works, including the works of nominated subcontractors and suppliers and contractors and others employed direct including public utility companies and showing a completion date no later than the Date for Completion. The chart to be a bar chart in an approved form. Forward 2 copies to the Architect, 1 copy to the Quantity Surveyor and keep up to date. Modify or re-draft.

10.5.3. Disputes were referred to arbitration, where the arbitrator made the following decision:

- On a true construction of the contract the contractor was obliged to provide a programme showing completion of the whole of the works no later than the date for completion. Agreement or approval by the architect of the programme should not relieve the contractor of the duty to complete the whole of the works by the date for completion.
- The contractor was entitled to complete the works on a date earlier than the date for completion in the contract.

- When the contractor programmed to complete earlier than the date for completion there was no implied obligation upon the employer his servants or agents to perform the agreement so as to enable the contractor to carry out and complete the works in accordance with the programme. The reason being that the contractor was not obliged to complete early and therefore if there was an implied term it would enforce an obligation on the employer but not the contractor.

10.5.4. It follows that the contractor is entitled to complete the works earlier than the contract completion date and has a right to do so. There is no corresponding duty, however, on the part of the employer to permit him to do so, and in particular to furnish him with information or otherwise positively co-operate so as to enable him to do so. The contractor is merely free from any contractual restraint and may complete earlier. The employer must not prevent him from doing so, but this does not mean that the employer is bound to facilitate in a positive way the implementation of the contractor's privilege or liberty.

10.5.5. The *Glenlion* case, which found in favour of the employer, dealt solely with the issue of drawings necessary to enable the work to be completed. It did not address additional cost and loss and expense, in general where delays occur and in particular where caused by variations, although the judge did venture to say that it was 'unclear how the variation provisions would have applied'. Reference was made by the judge to similar authorities and, in particular, to *Keating on Building Contracts*. In the edition referred to (the supplement to the 4th edition), it is stated:

> Whilst every case must depend upon the particular express terms and circumstances, it is thought that, upon the facts set out in *Wells* v. *Army & Navy Co-operative Society* (1903) the contractor's argument is bad; and that is the case even though the contractor is required to complete 'on or before' the contract date . . . There is no authority on this point.

However, in the 5th and 6th editions of *Keating*, the author goes on to say:

> Where the programme date is earlier than the Date for Completion stated in the contract, it may be that some direct loss and/or expense may be recoverable on the grounds of disruption. However, provided that the contractor can still complete within the contract period, he cannot recover prolongation costs (*Glenlion Construction* v. *The Guinness Trust*).

The *Glenlion* decision is likely to be limited in its effect as to the late issue of drawings necessary to carry out and complete the works. Delays to an early completion programme due to variations which create additional cost may not be caught in the same way.

10.5.6. The South African case of *Ovcon (Pty) Ltd* v. *Administrator of Natal* (1991) also dealt with this matter. In like manner to *Glenlion*, the contractor showed completion in eleven months with a contract period of fifteen months. Three months' delay to the programme period was caused by the employer. The court refused to award additional prolongation costs, saying that if the contractor had taken the contemplated fifteen months these expenses would have been incurred in any event.

10.5.7. In most situations, it is not the programme which is relevant. The contractor must show that his progress was affected and that he suffered loss and/or expense thereby. For example, JCT 2011, clause 4.23, identifies circumstances giving rise to a claim for loss and expense, stating:

> If in the execution of this Contract the Contractor incurs or is likely to incur direct loss and/or expense . . . due to deferment of giving possession of the site . . . or because the regular progress of the Works or of any part of them has been or is likely to be materially affected . . . the Architect . . . shall ascertain or instruct the Quantity Surveyor to ascertain the amount of such loss and/or expense . . .

10.5.8. Programmes submitted by the contractor under clause 14 (1) of the ICE 6th and 7th Editions are for acceptance by the engineer: if the contractor submits a programme which shows early completion and this is accepted by the engineer, the situation could be different. The contractor would, one expects, apply to the engineer for drawings to be issued to meet the shortened programme. If the engineer offered no resistance to these requests, but issued the drawings later than requested, an arbitrator might consider that the information had not been issued at 'a time reasonable in all the circumstances', as required by clause 7(4)(a). This would give rise to a contractual entitlement for additional payment.

10.5.9. An exception to the general rule that there is unlikely to be an entitlement to more time or money if a contractor is prevented from completing by the date shown on an early completion programme is the Engineering and Construction Contract (NEC 3). Clause 63.3 deals with the contractor's additional time and cost entitlement in respect of compensation events, where it states:

> A delay to the Completion Date is assessed as the length of time that, due to the compensation event planned Completion is later than planned completion as shown on the Accepted Programme.

This wording brings into play the contractor's programmed completion date, when matters relating to extensions of time and additional cost are involved. In this case, if the timing of the issue of information adversely affects the programme, then the contractor will become entitled to additional time and money.

10.5.10. The Society of Construction Law Delay and Disruption Protocol 2002 expresses the opinion that a contractor will be entitled to levy a claim if unable to complete in accordance with a shortened programme, provided the employer is aware of the contractor's intention at the time the contract is entered into. The wording of the Protocol states:

> 1.12.1 If as a result of an Employer Delay, the Contractor is prevented from completing the works by the Contractor's planned completion date (being a date earlier than the contract completion date), the Contractor should in principle be entitled to be paid the costs directly caused by the Employer Delay, notwithstanding that there is no delay to the contract completion date (and therefore no entitlement to an EOT), provided also that at the time they enter

into the contract the Employer is aware of the Contractor's intention to complete the works prior to the contract completion date, and that intention is realistic and achievable.

1.12.2. Where the parties have not addressed this issue in their contract, for the Contractor to have a valid claim, the Employer must be aware at the time that contract is entered into of the Contractor's intention to complete prior to the contract completion date. It is not permissible for the Contractor after the contract has been entered into, to state that it intends to complete early, and claim additional costs for being prevented from doing so.

1.12.3 The Protocol recognises that the position it takes on this issue might be thought to conflict with the decision of Judge Fox-Andrews in the (English) Technology and Construction Court in *Glenlion Construction Ltd* v. *The Guinness Trust* (1987) 39 BLR 89, where it was held that there was no implied term of the building contract in question that the Employer in that case should so perform the contract as to enable the Contractor to complete the works in accordance with a programme that showed the works being completed before the contract completion date. Providing the Employer is aware of the Contractor's intention prior to the contract being entered into, there should be no such conflict. The Protocol considers that, as a matter of policy, contractors ought not to be discouraged from planning to achieve early completion, because of the price advantage that being able to complete early is likely to have for the Employer. But the potential for conflict reinforces why the issue should be addressed directly in every contract.

SUMMARY

The *Glenlion* case and the South African case of *Ovcon* made it clear that the architect has no implied obligation to issue drawings at such time as would enable the contractor to finish work to meet an early completion programme. There are, however, a number of angles which contractors/subcontractors may wish to consider, such as: has the delay been caused by a variation and therefore the pricing should include prolongation costs?

Would the argument stand a better chance of success if it could be shown that the timing of the issue of drawings affected the progress of the works? JCT 2011 provides for payment of additional cost resulting from a delay to the progress of the works. In the context of a contract let using the ICE 6th or 7th editions, was the contractor working to a programme accepted by the engineer? If so, the engineer might be obliged to certify additional cost on the basis that drawings had not been issued at a time 'reasonable in all the circumstances'; this being the wording used in the contract.

These matters take a different line from the *Glenlion* case, which dealt merely with the question of whether there was an implied term in a contract requiring the architect to issue drawings to meet an early completion programme. It is important to note that, with regard to the interpretation of a contract, there can never be a termed implied into the contract which overrides an expressed term.

The Society of Construction Law Delay and Disruption Protocol states that a contractor will have the right to levy a claim if it is unable to complete in accordance with a shortened programme, provided the employer is aware of the contractor's intention at the time the contract is entered into.

The Engineering and Construction Contract (NEC 3) makes it clear that an extension of time and cost recovery entitlement in respect of a compensation event will be measured against the planned completion shown on the accepted programme.

10.6. Where a contractor submits a programme which is accepted or approved, showing completion on the completion date written into the contract, must drawings be issued in good time to enable the contractor to carry out the work at the time and in the sequence indicated on the programme?

10.6.1. Most standard forms of contract provide for the contractor to produce a programme. The contractual requirements as to the detail to be provided in the programme vary from contract to contract. Timings for the issue of drawings and details by the architect or engineer are also provided in the standard forms of contract. No link, however, is expressed in these standard forms between the clauses which provide for the issue of drawings and details and the ones dealing with the contractor's programme.

10.6.2. JCT 2011 is very basic in its requirements with regard to the issue of a master programme. Clause 2.9.1.2 requires the contractor to provide the programme to the architect/contract administrator as soon as possible after the execution of the contract.

10.6.3. Clause 2.11 requires the architect/contract administrator to issue drawings and details in accordance with the information release schedule which has been produced prior to sending out tender enquiries. If the information release schedule is not to be provided, it becomes necessary to delete the fifth recital in the contract. Clause 2.12.1 applies where there is no information release schedule. This clause obliges the architect/contract administrator to issue from time to time the required information, often in the form of amplification to contract drawings and instructions, as necessary.

Clause 2.12.2 states that:

> The further drawings, details and instructions (referred to in clause 2.12.1) shall be provided or given at the time it is reasonably necessary for the Contractor to receive them, having regard to the progress of the Works, or, if in the Architect/Contract Administrator's opinion practical completion of the Works or relevant Section is likely to be achieved before the relevant Completion Date, having regard to that Completion Date.

This type of wording is extremely difficult to digest and must confuse most contractors and employer's consultants. It would seem, however, that the sub-clause requires the architect/contract administrator to have regard to the date for completion written into the contract when issuing drawings and details and not an early completion date.

10.6.4. Clause 2.12.3 deals with the contractor's obligation to request the architect/contract administrator to issue drawings, where it states:

> Where the Contractor is aware and has reasonable grounds to believe that the Architect/Contract Administrator is not aware of the time when it is necessary by which the Contractor needs to receive such further drawings or details or instructions he shall, so far as reasonably practicable notify the Architect/Contract Administrator sufficiently in advance as to enable the Architect/Contract Administrator to comply with clause 2.12.

This is another awkwardly worded sentence. It seems to be saying that, if the contractor realises that the architect/contract administrator is not aware of when the drawings are required, the contractor must notify the architect/contract administrator reasonably far in advance.

10.6.5. The ICE 6th and 7th Editions, under clause 14, require the contractor to issue a programme showing the order in which he intends to carry out the work, within 21 days after the award of the contract.

10.6.6. Clause 7(3) in the ICE 6th and 7th Editions is simple, in comparison with the JCT requirements for the issue of information. The contractor is to give 'adequate' notice in writing to the engineer of any further drawings or specifications required. By clause 7(4), in the event of the engineer failing to issue the drawings and specification at a time which is 'reasonable in all the circumstances', then any additional resultant cost will be reimbursable to the contractor under clause 60. In the case of *Neodox Ltd* v. *Swinton & Pendlebury Borough Council* (1958), Mr Justice Diplock, in deciding what was meant by 'a time reasonable in all the circumstances', said:

> What is a reasonable time does not depend solely on the convenience and financial interests of the contractor; the engineer is to have a time to provide the information which is reasonable having regard to the point of view of himself and his staff and the corporation, as well as the point of view of the contractor.

SUMMARY

It seems clear that, where the standard conditions apply, the contractor is not entitled to have information issued solely to suit his programme. The contracts seem to indicate as important:

- any information release schedule required by the contract;
- the contractor sending a request for information;
- the progress of the works;
- a reasonable time for the engineer to produce the information.

No reference is made in the standard forms to the issue of information to suit the programme, whether it has been accepted or otherwise. However, if work is progressing in accordance with the programme and proper written requests for the issue of drawings have been sent to the architect/engineer, then there may be an obligation to issue drawings in response to those requests.

10.7. Is a contractor/subcontractor entitled to recover the cost of preparing a claim?

10.7.1. When a claims situation arises, contractors are invariably put to cost in preparing a submission to go to the architect or engineer. The question often asked is, whether the cost is recoverable as part of the claim ascertainment and payment.

10.7.2. It would seem that, if the contractor in preparing and submitting the claim is merely carrying out an expressed contractual obligation to submit a claim, then there will be no entitlement to reimbursement. For example, clause 52(4)(c) of the ICE 6th Edition and clause 53(4) of the 7th Edition require the contractor to provide 'full and detailed particulars of the amount claimed'.

10.7.3. A further situation may arise where the contractor has complied with these contractual requirements, but nonetheless the architect or engineer has failed to ascertain and certify the sums due for payment. The contractor would then be entitled to claim damages for breach of contract. Vincent Powell-Smith, in an article appearing in the *Contract Journal* of 30 July 1992, had this to say with regard to claims under JCT 80, where the architect fails to ascertain loss and expense as required by the contract:

> If the contractor invokes clause 26 and does what is required, the Architect is under a duty to ascertain, or instruct the Quantity Surveyor to ascertain, whether loss or expense is being incurred and its amount. This follows from the wording of clause 26.1, which uses the word 'shall' and which thus imposes a duty on the Architect, provided that the Architect has formed a prior opinion, that the contractor has been or is likely to be involved in direct loss and/or expense as a result of the specified event(s) and which is not recoverable under any other provisions of the contract.

10.7.4. There is no doubt that the employer is liable in damages for breach by the architect of this duty, and this is so whether the architect is an employee, e.g. where the employer is a public authority, or, as is more usual, an independent consultant engaged by the employer. Where clause 4.23 of JCT 2011 says 'the Architect/Contract Administrator shall', this in effect means 'the Employer shall procure that the Architect/Contract Administrator shall'. This point is implicit in the reasoning in *London Borough of Merton* v. *Stanley Hugh Leach* (1985).

10.7.5. A contractor, when claiming damages for a breach by the employer, due to the architect or engineer not ascertaining loss and expense, or additional cost, will be governed by the rules in *Hadley* v. *Baxendale* (1854). The damages recoverable under these rules are:

- those arising naturally, i.e. according to the usual course of things, from such breach;
- such as may reasonably be supposed to have been in the contemplation of both parties at the time they made the contract.

It may be argued that both parties operating under a JCT 2011 contract would contemplate that if the architect fails to ascertain loss and expense, hence leaving the employer in breach, the parties should have contemplated that the contractor would be put to expense in preparing a fully documented claim, which should therefore be recoverable. The same type of argument would apply under ICE 6th and 7th Editions and GC/Works/1 conditions of contract, where the engineer or PM fails to certify sums due arising out of a claim.

10.7.6. There is a precedent for the payment of managerial costs resulting from a breach in the case of *Tate & Lyle Food Distribution* v. *GLC* (1982). In that case, Mr Justice Forbes said:

> I have no doubt that the expenditure of managerial time in remedying an actionable wrong done to a trading concern can properly form the subject matter of a head of special damage.

This argument may be extended to cover the cost of claims preparation following a breach, whether they be incurred in respect of the contractor's employees or outside consultants.

10.7.7. Managerial time and costs were an issue in the case of *Euro Pools* v. *Clydesdale Steel Fabrications Ltd* (2003). Euro Pools entered into a contract with the main contractor in connection with work at the swimming baths at Burntisland and East Moseley in Surrey. Filtration equipment supplied by Clydesdale to Euro Pools proved to be faulty. Part of Euro Pools' claim for breach of contract included cost in respect of time input by the managing director and an engineer, which was contested by Clydesdale. Clydesdale's case was that Euro Pools' entitlement was to recover any actual cost or loss which resulted as a direct consequence of the defective equipment. The managing director and engineers were not employed specifically in relation to the remedial works and therefore their salaries would have been paid in any event. This being the case, Euro Pools had sustained no ascertainable loss as a result of the defects. It was accepted, however, that fuel costs in relation to visits to the swimming pools were an extra cost which could be recovered. The judge was not impressed, his view being that a managing director will normally be expected to devote the whole of his time to the affairs of his employer involving company strategy, searching for new markets and the developments of new products. If he is diverted from these tasks to deal with remedying equipment obtained from a supplier, this will deny him the time which he would otherwise have spent on his normal tasks and, as such, represents a loss to his employer. The same principle applies to engineers, in that time spent on remedial work represents time which could have been profitably spent on their employer's work. The claim included the basic salaries of the managing director and engineers, plus a mark-up for overheads. The mark-up for the engineers was 350% added to the hourly charge, with a 50% increase on top of the managing director's salary costs. The percentage addition for the engineers was said to be the usual mark-up for daywork in the industry and the addition in respect of the managing director was intended to represent a reasonable amount. The judge accepted in principle that overheads are rightfully included in the claim. He felt that a contribution should be made to cover such matters as the cost of maintaining business premises, computer systems, head office and accounting staff, salesmen, estimators and general technical support.

10.7.8. The decision in *Lomond Assured Properties* v. *McGrigor Donald* (2000) and *Pegler* v. *Wang* (2000) involved claims in respect of managers who spent time investigating breaches of contract. In neither case were there any records to support time included in the claim. Oral evidence, however, was accepted by the court to support the time which formed the basis of the claim. The recovery of managerial time as damages for breach of contract seems to be an ongoing matter which is the subject of disputes. The case of *Aerospace Publishing Ltd* v. *Thames Water Utilities Ltd* (2007) attempts to lay down rules as to when managerial time is or is not recoverable. Aerospace Publishing were specialist publishers of aviation and military history. Their archive of photographic and reference material was stored in the basement of their premises in Hammersmith. Flooding occurred in the basement, for which Thames Water Utilities Ltd accepted responsibility. Aerospace Publishing Ltd included staff costs in its claim for damages. The argument put forward by Thames Water Utilities Ltd was that the staff costs would have been incurred in any event and therefore had no place in a damages claim. The Court of Appeal, in accepting the claim for managerial time, had to be satisfied that the managers involved had been diverted from other work and that if the diversion had not taken place, then they would have generated revenue in value at least to the cost of employing them.

10.7.9. The nature of the records necessary to sustain a claim for managerial time was relevant in the case of *Horace Holman Group* v. *Sherwood International Group Ltd* (2001). It was held that some basis for the quantum must be produced to support a claim for managerial time, but this does not necessarily extend to contemporaneous records. The decision in the case of *Bridge UK Com Ltd* v. *Abbey Pynford Plc* (2007) is another example of a court allowing the recovery of managerial time where there were no contemporary records. The claim for management time was based upon the memory of Mr Ruck, with regard to 120 hours of time he devoted to problems arising from the breaches of contract committed by Abbey Pynford. However, the judge considered that memory could be fallible and deducted 20% of the hours claimed in making his award.

10.7.10. Where matters are referred to arbitration, an arbitrator has a discretion to direct by whom and to whom costs shall be paid. The exercise of the arbitrator's discretion is limited with regard to costs connected with or leading up to the arbitration. Normally, the arbitrator will award costs which have been incurred after the service of the arbitration notice in favour of the successful party. However, if costs incurred before the service of the notice are in contemplation of the arbitration, then the arbitrator may include them in his award of costs. It may be argued that costs of preparing a claim document which ultimately forms part of the pleadings, but is prepared before the arbitration notice is served, falls into the category of costs in contemplation of arbitration. A note on the file before the claim is prepared to the effect that it is being prepared in contemplation of arbitration may prove helpful.

10.7.11. There is a reported case, *James Longley and Co Ltd* v. *South West Regional Health Authority* (1983), where a court had to decide whether a claims consultant's fees should be reimbursed to a successful claimant. The case arose out of a dispute concerning a successful claimant's right to recover the costs of employing a claims consultant as part of the costs of the action. An arbitration between the parties was settled after the hearing had lasted sixteen days, subject to the court's decision regarding costs. The claimant's bill of costs contained an item of £16,022 for the fees of a claims consultant. It was directed that the fees, in so far as they related to work done in preparation of the claimant's final account and to work as a general adviser to the claimants, were to be disallowed, but allowance was made for £6,452 in respect of work done in preparing the claimant's case for arbitration, namely, the preparation of three schedules annexed to the Points of Claim. This case is no authority for the proposition that costs incurred in preparing claims are always recoverable, but rather the contrary, since only the costs directly applicable to the arbitration were allowed, which were less than half the total.

10.7.12. The Society of Construction Law Delay and Disruption Protocol 2002 offers the contractor hope of recovering the cost of preparing a claim in the following terms:

> 1.20.1 Most construction contracts provide that the Contractor may only recover the cost, loss and/or expense it has actually incurred and that this be demonstrated or proved by documentary evidence. The Contractor should not be entitled to additional costs for the preparation of the information, unless it can show that it has been put to additional cost as a result of the unreasonable actions or inactions of the CA in dealing with the Contractor's claim. Similarly, unreasonable actions or inactions by the Contractor in prosecuting its claim should entitle the Employer to recover its costs. The Protocol may be used as a guide as to what is reasonable or unreasonable.

SUMMARY

It would seem that, when approaching the matter of recovery of the costs of preparing a claim, a number of questions should be addressed:

(1) It appears unlikely that, in the absence of express terms in the contract which give an entitlement to payment, the cost of producing documents in support of a claim which is a requirement of the conditions of contract such as ICE 6th and 7th editions will be recovered. In providing this information the contractor or subcontractor is merely complying with the requirements of the contract.

(2) Where the conditions of contract require the architect or contract administrator, having received notice and details from the contractor or subcontractor, to ascertain loss and expense, any failure to so ascertain will constitute a breach of contract by the employer. The cost of further preparation work regarding a claim, if it results from the breach, may well be recoverable.

(3) If it can be shown that, prior to the service of an arbitration notice, the preparation of the claim is in contemplation of such arbitration, the arbitrator may, in exercising a discretion with regard to the award of costs, include the cost of preparing the claim. The courts have now accepted as part of a claim for breach of contract the cost of managerial time spent in investigating the breaches and maintaining records.

10.8. Will the courts enforce claims for head office overheads based upon the *Hudson* or *Emden* formulae, or must the contractor be able to show an increase in expenditure on head office overheads resulting from the overrun?

10.8.1. Contractors and subcontractors when submitting claims for prolongation will normally include an item for head office costs. This can usually take one of two forms. Firstly, the argument may be that the contractor, in having resources locked into a site during the prolongation period, has lost the opportunity of using those resources on other sites where they would have earned a contribution to the costs of running the head office. For this argument to be successful, the contractor or subcontractor would have to show that work was reasonably plentiful and that, on a balance of probabilities, other work was or would have been available. This is sometimes referred to as 'unabsorbed head office overheads', in that the level of head office cost continues, but the revenue stream from the particular contract suffering the delay shows a shortfall. If the contractor is unable to demonstrate a loss of opportunity, he may be able to demonstrate that time and cost of identified resources at head office have been incurred during the overrun period. Any head office overheads recovered through extra works and the expenditure of provisional sums should be credited against sums otherwise claimed.

10.8.2. The 'loss of opportunity' method of claiming for unabsorbed overheads has been converted into three alternative formulae: *Hudson*, *Emden* and *Eichleay*.

10.8.3. *Hudson's Building and Engineering Contracts*, 11th edition at paragraph 8.182, sets out the famous 'Hudson Formula', with the head office percentage incorporated in the calculation of the contract price. The sum included in the claim for head office overheads is calculated using the Hudson Formula, comprising:

$$\frac{\text{Tender percentage} \times \text{Contract sum} \times \text{Period of delay (weeks)}}{100 \times (\text{Contract period in weeks})} = \text{unabsorbed overheads}$$

An alternative formula is produced in *Emden's Building Contracts and Practice*, 6th edition, Volume 2, page N/46:

$$\frac{\text{Head office percentage} \times \text{Contract sum} \times \text{Period of delay (weeks)}}{100 \times (\text{Contract period in weeks})} = \text{unabsorbed overheads}$$

Head office percentage – arrived at by dividing the total overhead cost and profit of the contractor's organisation as a whole by the total turnover, all extracted from the contractor's year-end accounts.

In essence, the difference between the two formulae is that *Hudson* uses the head office percentage used in the calculation of the contract sum, whereas in *Emden* the figures from which the head office percentage is calculated are extracted from the contractor's year-end accounts.

10.8.4. In *Ellis-Don Ltd* v. *The Parking Authority of Toronto* (1978), the court held that the plaintiffs had been delayed by 17.5 weeks as a result of the defendant's failure to obtain a necessary permit. As a result, the plaintiff was entitled to a weekly sum in respect of costs of overheads and loss of profits which employed as a calculating factor 3.8% of the tender sum, i.e. *Hudson*. In *Whittall Builders Co Ltd* v. *Chester-le-Street District Council* (1985), Mr Recorder Percival QC, in illustrating the *Emden* formula, said:

> Lastly, I come to overheads and profit. What has to be calculated here is the contribution to off-site overheads and profit which the contractor might reasonably have expected to earn with these resources if not deprived of them. The percentage to be taken for overheads and profits for this purpose is not therefore the percentage allowed by the contractor in compiling the price for this particular contract, which may have been larger or smaller than his usual percentage and may or may not have been realised.
>
> It is not that percentage that one has to take for this purpose but the average percentage earned by the contractor on his turnover as shown by the contractor's accounts.
>
> On that basis it is clear to me that the calculation which I have to make is as follows. I start with the figure of 14.15 agreed by the parties as the figure for overheads and profit as a percentage of turnover, and I divide that by 100, then multiply by the contract figure of £404,759 divided by 78 to give the turnover per week, and then multiply by 30; and that gives the figure of £22,028.

10.8.5. In *J.F. Finnegan Ltd* v. *Sheffield City Council* (1988), the contractor claimed *inter alia* payment in respect of additional overheads on a housing improvement contract which was subject to delay. The contract was JCT 63. Sir William Stabb QC, in finding for the contractor, said:

> It is generally accepted that, on principle, a contractor who is delayed in completing a contract due to the default of his employer, may properly have a claim for head office or off-site overheads during the period of delay, on the basis that the work-force, but for the delay, might have had the opportunity of being employed on another contract which would have had the effect of funding the overheads during the overrun period. This principle was approved in the Canadian case of *Shore & Horwitz Construction Co Ltd* v. *Franki of Canada* (1967), and was also applied by Mr Recorder Percival QC, in the unreported case of *Whittall Builders Company Limited* v. *Chester-le-Street District Council*. Furthermore, in *Hudson's Building Contracts*, at page 599 of the 10th edition, a simple formula is set out to determine the amount of the loss of funding of overheads and profit during the period of overrun.

The contractor unfortunately had not calculated his head office overhead claim using the *Hudson* formula, but one of his own invention. This led Sir William to comment:

> However, I confess that I consider the plaintiffs' method of calculation of the overheads on the basis of a notional contract valued by uplifting the value of the direct cost by the constant of 3.51 as being too speculative and I infinitely prefer the *Hudson* formula which, in my judgment, is the right one to apply in this case, that is to say, overhead and profit percentage based upon a fair annual average, multiplied by the contract sum and the period of delay in weeks, divided by the contract period.

However, Sir William obviously did not fully understand *Hudson*'s formula, which includes the head office overhead and profit percentage upon which the tender was based and not the overhead and profit percentage based upon a fair annual average, which is more in line with *Emden*'s formula. Nevertheless, judicial approval of the formula approach in appropriate circumstances is clear.

10.8.6. In North America, head office overheads are often calculated employing the *Eichleay* formula. This formula is calculated by comparing the value of work carried out in the contract period for the project with the value of work carried out by the company as a whole, for the contract period. A share of head office overheads for the company can then be allocated in the same ratio and expressed as a lump sum to the particular contract. The amount of head office overhead allocated to the particular contract is then expressed as a weekly amount, by dividing it by the contract period. The period of delay can then be multiplied by the weekly amount to give a total sum claimed. Expressed as a formula converted into English from American jargon, the formula reads (in formulaic form):

$$\frac{X}{Y} \times A = B$$

Where 'X' is the value of the contract work during the contract period; 'Y' is the *total value* of all the company's work during the contract period; 'A' is the total of all head office overhead costs expended during the contract period; and 'B' is the head office overhead costs properly allocated to the contract.

In like manner, the following calculation can be worked:

$$\frac{B}{CP} \times \text{Period of Delay} = \text{Amount Claimed}$$

Where 'B' is the Head Office overhead costs allocated to the contract; and 'CP' is the contract period.

10.8.7. The case of *Property and Land Contractors Ltd* v. *Alfred McAlpine Homes North Ltd* (1997) involved an *Eichleay* type of calculation, in which the court had to decide a contractor's entitlement in respect of the recovery of head office overheads. The dispute arose out of a contract between Alfred McAlpine Homes North Ltd and Property and Land Contractors Ltd for the construction of an estate of 22 homes for sale at Shipton. JCT 80 conditions, as amended, applied. Due to poor progress with the sale of the houses, an instruction was given to Property and Land Contractors to suspend the works, which led to a claim being submitted under clause 26 of the conditions. The claim was not resolved and the matter was referred to arbitration. One of the items in dispute related to head office overheads. A claim was submitted in the alternative. The first alternative was based upon the application of the *Emden* formula. It was the contractor's method only to undertake one major project at any one time. A second project at Tollerton was planned and it was agreed that Property and Land Contractors intended to carry out this development for its parent company after completing Shipton. It was claimed that, because of the postponement, completion of the work was delayed from 20 May 1990 until 25 November 1990 and prevented the plaintiff from carrying out the development at Tollerton. The plaintiff claimed that because of the overrun at Shipton, he lost an opportunity of carrying out the Tollerton work, which would have provided a recovery of overheads. The *Emden* formula was employed as a means of calculating the head office overheads. This argument was rejected by the arbitrator, on the facts of the case, as he was not convinced that the suspension resulted in the plaintiff being unable to work at Tollerton or anywhere else. In the alternative, the plaintiff claimed for the recovery of head office overheads actually expended. The arbitrator was persuaded that the facilities provided by head office were essential to the control and organisation of the project. He considered it irrelevant to make an attempt to show that particular head office costs were increased as a result of this delay. It was evident to him that the head office costs were related to the works for the delay period. The plaintiff's method of calculation was:

> to extract from the company's account, the overhead costs, excluding fixed costs not related specifically to progress on the site (i.e. directors' remuneration, telephone, staff salaries, general administration, private pension plan, rent, rates, light, heat and cleaning and insurance) to express such annual costs as weekly averages for both 1990 and 1991, and multiply the resulting weekly averages by the period of overrun in each year and thus to produce a figure referred to as 'C'.

The calculation gave the total overheads for the period of delay and had to be allocated between Shipton and other work being undertaken at the same time. This was achieved by a simple formula.

$$\frac{\text{Value of work at Shipton} \times \text{Total overheads(C)}}{\text{Value of work at all sites}} = \text{Amount claimed}$$

The formula, in essence, is the *Eichleay* formula. McAlpine's argument was that the arbitrator had erred in law, because he had awarded the plaintiff costs which would have been incurred in any event and could not therefore be classed as direct loss and expense. The court found in favour of the plaintiff and against McAlpine, but had the following reservations:

> All these observations like those of Lord Lloyd in *Ruxley*, of Mr Justice Forbes in *Tate & Lyle*, and of Sir Anthony May in *Keating* all suppose, either expressly or implicitly, that there may be some loss as a result of the event complained of, so that in the case of delay to the completion of a construction contract, there will be some 'under recovery' towards the cost of fixed overheads as a result of the reduced volume of work occasioned by the delay, but this state of affairs must of course be established as a matter of fact. If the contractor's overall business is not diminishing during the period of delay, so that where, for example, as a result of an increase in the volume of work on the contract in question, arising from variation etc., or for other reasons, there will be a commensurate contribution towards the overheads, which offsets any supposed loss, or if, as a result of other work, there is no reduction in overall turnover so that the cost of the fixed overheads continues to be met from other sources, there will be no loss attributable to the delay.

10.8.8. In *St Modwen Developments Ltd* v. *Bowmer & Kirkland* (1996), the arbitrator found in favour of the contractor in awarding head office overheads based upon a formula method of recovery. An appeal was lodged, not because of the use of the formula, but on the basis that no evidence had been presented to prove that the contractor was unable to use the head office resources during the period of prolongation, to generate profits on other contracts. The court seemed impressed by the learned work of Mr Duncan Wallace, in his 10th edition of *Hudson*, where he states:

> However, it is vital to appreciate that both these formulae (*Hudson* and *Eichleay*) were evolved during the 1960s at a time of high economic activity in construction. Both assume the existence of a favourable market where an adequate profit and fixed overhead percentage will be available to be earned during the delay period. Both also very importantly, assume an element of constraint – that is to say that the contractor's resources (principally of working capital and key personnel, it is suggested) will be limited or stretched, so that he will be unable to take on work elsewhere.

The court supported the arbitrator on the grounds that both expert and evidence of fact had been heard, on which the arbitrator was entitled to base his award. In any event, the arbitrator in his terms of appointment had wisely included a clause which allowed him to use his professional knowledge and experience as assistance in determining the matters in dispute. The court felt that the arbitrator would draw on his own experience to conclude from the evidence before him in the manner that he did.

10.8.9. *Amec* v. *Cadmus* (1996) is another example of a claim for the recovery of head office costs. The judge, Mr Recorder Kallipetis QC, held:

> It is for the plaintiff to demonstrate that he has suffered the loss he is seeking to recover . . . [and] this proof must include the keeping of some form of record that the time was excessive and their attention was diverted in such a way that loss was incurred.

He went on to say the plaintiff must:

> place some evidence before the court that there was other work available which, but for the delay, he would have secured . . . Thus he is able to demonstrate that he would have recouped his overheads from those other contracts and, thus, is entitled to an extra payment in respect of any delay period awarded in the instant contract.

10.8.10. In the case of *Norwest Holst Construction Ltd* v. *Co-operative Wholesale Society* (1998), the court had to decide whether the arbitrator was entitled to use a formula to ascertain direct loss and/or expense

- in relation to additional overheads; and/or
- in relation to unabsorbed overheads (in the absence of a finding by the arbitrator of a reduction of turnover directly attributable to the delay which caused the loss and expense).

A JCT contract applied. The arbitrator used the *Emden* formula, which he reduced by four-fifths when making the award. It was the court's decision that an *Emden*-style formula is sustainable when the following circumstances occur:

(1) The loss in question must be proved to have occurred.
(2) The delay in question must be shown to have caused the contractor to decline to take on other work which was available and which would have contributed to its overhead recovery. Alternatively, it must have caused a reduction in the overhead recovery in the relevant financial year or years, which would have been earned but for that delay.
(3) The delay must not have had associated with it a commensurate increase in turnover and recovery towards overheads [e.g. a variation].
(4) The overheads must not have been ones which would have been incurred in any event, without the contractor achieving turnover to pay for them.
(5) There must have been no change in the market affecting the possibility of earning profit elsewhere and an alternative market must have been available. Furthermore, there must have been no means for the contractor to deploy its resources elsewhere, despite the delay. In other words, there must not have been a constraint in recovery of overheads elsewhere.

In this claim made by CWS, it is not a situation where CWS need merely establish how it would have acted but for the delay; CWS must instead establish a number of interrelated facts:

(1) That its senior management who spent time on this contract in the period of delay, would have spent that time on other contracts on which CWS was working at the time.
(2) Had that time been spent in that alternative way, the administrative tasks that would have been undertaken on those other contracts would have caused a variety of people and a variety of contractors and suppliers with whom CWS were working to have performed more speedily, economically and efficiently, such that CWS' profit from those contracts would have improved. Only some of these individuals and companies would have been within CWS' control.
(3) As a result of CWS' additional profitability it would have earned a greater contribution to its overheads from these contracts than it actually did.

10.8.11. In the case of *Beechwood Developments Ltd* v. *Stuart Mitchell* (2001), Beechwood undertook to build houses on a site in Bearsden, which is near Glasgow, for Westpoint, a developer. Delays to the commencement of the work resulted from errors made in carrying out the site survey, which was the responsibility of Westbrook. Beechwood's claim for damages fell under three heads: direct site costs, increased cost of materials and loss of contribution to overheads and profits. With regard to the overhead claim, Beechwood argued that the delay resulted in a reduction in the turnover during the financial year in which the delay occurred. That turnover would have contributed to the cost of running the head office and the company's profit. The average overhead and profit for the two previous years, as a percentage of turnover, was 14.7%. It was claimed that by applying this percentage to the contract sum, the anticipated total sum for overheads and profit which Beechwood anticipated recovering from the project could be ascertained. The weekly amount expected to be recovered was then calculated by dividing this sum by the contract period in weeks. How much of the overrun period could be attributable to the delay at the beginning of the contract was a matter of argument. Beechwood claimed 22 weeks; the judge, on the other hand, considered that the appropriate period was only 10 weeks. The judge noted that the percentage was derived from Beechwood's accounts and did not represent the overheads built into the contract price. He did not, however, seem to think that this was too important. This, however, does raise an important point. The *Hudson* formula uses the percentage for overheads that the contractor used in building up his tender figure. *Emden*'s formula uses the overhead figures extracted from the contractor's accounts. Evidence to the effect that Beechwood's gross profit had fallen from £200,000 in the year before the problem at Bearsden, down to £9,500 in the problem year, was also convincing. In his decision, the judge awarded a percentage of 11.65% applied to the contract sum and a delay period of 10 weeks. In all, the total of the award under this head amounted to £31,007.

10.8.12. The Society of Construction Law Delay and Disruption Protocol 2002 gives qualified support to the use of the formulae in the following respect:

> 1.16.3 Unless the terms of the contract render unabsorbed overheads irrecoverable, they are generally recoverable as a foreseeable cost resulting from prolongation. The Contractor must be able to demonstrate that because of the Employer Risk Events it was prevented from taking on other overhead-earning work.
>
> 1.16.6 The three most commonly used formulae for assessing unabsorbed head office overheads are *Hudson*, *Emden* and *Eichleay*.
>
> 1.16.7 The use of the *Hudson*'s formula is not supported. This is because it is dependent on the adequacy or otherwise of the tender in question, and because the calculation is derived from a number which in itself contains an element of head office overheads and profit, so there is double counting.
>
> 1.16.8 In the limited circumstances where a head office overhead formula is to be used, the Protocol prefers the use of the *Emden* and *Eichleay* formulae. However, in relation to the *Eichleay* formula, if a significant proportion (more than, say, 10%) of the final contract valuation is made up of the value of variations, then it will be necessary to make an adjustment to the input into the formula, to take account of the fact that the variations themselves are likely to contain a contribution to head office overheads and profit.
>
> 1.16.9. The CA or, in the event of a dispute, the person deciding the dispute, should not be absolutely bound by the results of a formula calculation. It is possible that the use of a particular formula will produce an anomalous result because of a particular input into it. It is suggested that the result of the use of one formula be cross-checked using another formula.
>
> 1.16.10. The tender allowance for head office overheads may be used, if that is what the parties for convenience wish to do.

10.8.13. What the courts have not yet been asked to address is the extent to which resources are retained on site during the overrun period. The formulae would seem inappropriate if some of the site management personnel during the overrun period were in fact re-allocated to other projects, with the exception of a small essential team. If this were the case, the cost to the contractor would be reduced, but would not be reflected if the *Hudson*, *Emden* or *Eichleay* methods were employed.

10.8.14. An alternative method of presenting a claim for head office overheads was used in the case of *Tate & Lyle* v. *GLC* (1983). This is not a building case, but is of general application. The plaintiffs claimed 2.5% of the prime cost for managerial time, and the judge accepted that such a head of claim was admissible, but he did not accept the method of calculation and the application was rejected. He remarked that it was up to managers to keep time records of their activities.

> I have no doubt that the expenditure of managerial time in remedying actionable wrong can properly form the subject matter of a head of special damage. In a case such as this it would be wholly unrealistic to assume that no such additional managerial time was in fact expended. I would also accept that it must be extremely difficult to quantify. But modern office arrangements provide for the recording of time spent on particular projects. I do not believe that it would have been impossible for the plaintiffs in this case to have kept some record to show the extent to which their trading routine was disturbed by the necessity for continual dredging sessions . . . While I am satisfied that this head of damage can properly be claimed, I am not prepared to advance into an area of pure speculation when it comes to quantum. I feel bound to hold that the plaintiffs have failed to prove that any sum is due under this head.

The decision in *Babcock Energy* v. *Lodge Sturtevant* (1994) reinforced the contractor's entitlement to recover head office overheads based upon accurately recorded costs. Judge Lloyd in the *Babcock* case was influenced by the *Tate & Lyle* decision.

10.8.15. Managerial time and costs were an issue in the case of *Euro Pools* v. *Clydesdale Steel Fabrications Ltd* (2003). Euro Pools entered into a contract with the main contractor in connection with work at the swimming baths at Burntisland and East Molesey in Surrey. Filtration equipment supplied by Clydesdale to Euro Pools proved to be faulty. Part of Euro Pools' claim for breach of contract included cost in respect of time input by the managing director and an engineer, which was contested by Clydesdale. Clydesdale's case was that Euro Pools' entitlement was to recover any actual cost or loss which resulted as a direct consequence of the defective equipment. The managing director and engineers were not employed specifically in relation to the remedial works and therefore their salaries would have been paid in any event. This being the case, Euro Pools had sustained no ascertainable loss as a result of the defects. It was accepted, however, that fuel costs in relation to visits to the swimming pools were an extra cost which could be recovered. The judge was not impressed, his view being that a managing director will normally be expected to devote the whole of his time in the affairs of his employer involving company strategy, searching for new markets and the developments of new products. If he is diverted from these tasks to deal with remedying equipment obtained from a supplier, this will deny him the time which he would otherwise have spent on his normal tasks and, as such, represents a loss to his employer. The same principle applies to engineers, in that time spent on remedial work represents time which could have been profitably spent on their employer's work. The claim included the basic salaries of the managing director and engineers, plus a mark-up for overheads. The mark-up for the engineers was 350% added to the hourly charge, with a 50% increase on top of the managing director's salary costs. The percentage addition for the engineers was said to be the usual mark-up for daywork in the industry and the addition in respect of the managing director was intended to represent a reasonable amount. The judge accepted in principle that overheads are rightfully included in the claim. He felt that a contribution should be made to cover such matters as the cost of maintaining business premises, computer systems, head office and accounting staff, salesmen, estimators and general technical support.

10.8.16. The decision in *Lomond Assured Properties* v. *McGrigor Donald* (2000) and *Pegler* v. *Wang* (2000) involved claims in respect of managers who spent time investigating breaches of contract. In neither case were there any records to support time included in the claim. Oral evidence, however, was accepted by the court to support the time which formed the basis of the claim.

10.8.17. For further information, including the case of *Aerospace Publishing Ltd* v. *Thames Water Utilities Ltd* (2007), see paragraph 10.7.8. above.

10.8.18. For further information, including the cases of *Horace Holman Group* v. *Sherwood International Group Ltd* (2001) and *Bridge UK Com Ltd* v. *Abbey Pynford Plc* (2007), please see paragraph 10.7.9. above.

10.8.19. The decisions of the courts relating to managerial time, it is submitted, cast doubt on the formula methods of calculation only in so far as the contractor is unable to demonstrate a loss of opportunity to use the resources on other sites. This would often be the case during periods when the construction industry is in recession. Contractors when faced with a downturn in work often lay off resources as each job finishes, until the size of the organisation is suitable for the economic climate current at the time. It would, under these circumstances, be inappropriate to claim reimbursement of head office overheads on the basis of a lost opportunity. When applying the formula method, other matters have to be taken into consideration:

- Credit must be given for any additional overheads recovered by way of the final account.
- Allowance should be made for the situation where resources during a delay period have been substantially reduced and deployed to other sites.
- Credit should be given against the formula calculation for head office staff claimed for elsewhere, e.g. the contracts manager.

SUMMARY

There is no hard and fast rule that a contractor who has an entitlement to the recovery of additional costs due to an overrun to the completion date is entitled to be paid for additional overheads calculated on a formula basis. It will depend in the first instance upon the wording of the contract. The GC/Works/1 refers in condition 46 only to the recovery of expense, which would probably exclude any loss of contribution to overheads due to loss of opportunity to use the resources on other projects.

Whether the contractor, where appointed on other commonly used standard forms of contract would be entitled to payment based on a formula method of calculation depends upon the contractor's ability to prove that an overrun to the contract period has resulted in the loss of an opportunity to employ the resources on other work. Further, the contractor would have to show that the lost overheads had not been recovered by way of variations or other additional payment.

Where it is not appropriate to base a claim on a loss of opportunity, which would bring in a formula method, the contractor will be left to base its claim on actual head office costs incurred, such as managerial time expended on the breach of contract, or such other entitlement which forms the basis of the claim. Whichever method of calculation is employed, the courts will require supporting evidence that either opportunities have been lost or additional costs incurred.

10.9. Where a delay to completion for late issue of information has been recognised, should the loss and expense or additional cost claims in respect of extended preliminaries be evaluated, using the rates and prices in the bills of quantities?

10.9.1. Matters such as site supervision, equipment, health and safety, welfare, storage and the like normally fall within the definition of preliminaries. The time-honoured method adopted by quantity surveyors and engineers when evaluating a contractor's overrun claim is to use the preliminaries as priced in the bill of quantities. Contractors and subcontractors have in the past accepted this method with some reluctance. Nowadays, contractors and subcontractors are openly asking whether this method is correct.

10.9.2. The wording in the main claims clause in JCT 2011, clause 4.23, is: '. . . the Contractor incurs or is likely to incur direct loss and/or expense'. Other JCT contracts are similarly worded. The 8th edition of *Keating on Building Contracts*, at 19.262, page 839, has this to say concerning the meaning of direct loss and/or expense:

> This was considered by the Court of Appeal in *F G Minter* v. *W.H.T.S.O.* The court held that direct loss and/or expense is loss and expense which arises naturally and in the ordinary course of things, as comprised in the first limb in *Hadley* v. *Baxendale*. The court approved the definition of 'direct damage' in *Saint Line Ltd* v. *Richardsons* as 'that which flows naturally from the breach without other intervening cause and independently of special circumstances, whereas indirect damage does not so flow'. It follows from the decision in *Minter* that the sole question which arises in relation to any head of claim put forward by a contractor is whether such claim properly falls within the first limb in *Hadley* v. *Baxendale* so that it may be said to arise naturally and in the ordinary course of things.

10.9.3. A similar line was taken by Mr Justice Megaw in *Wraight Ltd* v. *PH & T Holdings* (1968), when he said:

> In my judgment, there are no grounds for giving to the words 'direct loss and/or damage caused to the contractor by the determination' any other meaning than that which they have, for example, in a case of breach of contract or other question of relationship of a fault to damage

> in a legal context. Therefore it follows . . . that the [contractors] are, as a matter of law entitled to recover that which they would have obtained if this contract had been fulfilled in terms of the picture visualised in advance but which they have not obtained.

10.9.4. The ICE 6th and 7th Editions define cost in clause 1(5):

> The word 'cost' when used in the Conditions of Contract means all expenditure properly incurred or to be incurred whether on or off the Site including overhead finance and other charges properly allocable thereto but does not include any allowance for profit.

GC/Works/1 defines expense in condition 46(6):

> Expense shall mean money expended by the contractor, but shall not include any sum expended, or loss incurred, by him by way of interest or finance charges however described.

MF/1 defines costs in clause 1.1j as:

> all expenses and costs incurred including overhead and financing charges properly allocable thereto with no allowance for profit.

Cost is defined in the *Concise Oxford Dictionary* as 'Price paid for thing'.

10.9.5. It would seem from the above that actual cost or loss should be the basis on which claims are based and not the preliminaries as priced in the bills of quantities.

SUMMARY

Having said that extended preliminaries should be based on actual cost, it is necessary to state that this means reasonable costs which flow from the late issue of drawings or other employer's risk items. It will be necessary to identify the periods of time which were affected and to include only those preliminary items where extra costs were incurred. From the definitions in the various contracts and textbook references, it seems clear that when evaluating loss and expense, expense or cost claims in respect of extended preliminaries, actual cost or loss should be the basis on which the evaluation is made and not the prices of the preliminaries included in the bills of quantities.

10.10. Once it is established that additional payment is due for prolongation resulting from employer delays, should the evaluation relate to the period when the effect of the delay occurs or by reference to the overrun period at the end of the contract? Will the prolongation costs incurred for the whole of the site be recoverable, or only those associated with those parts of the works which are delayed?

10.10.1. It is common practice for decisions concerning the award of an extension of time to be made before considering the matter of prolongation costs. Once an extension of time has been granted, the evaluation of the additional prolongation costs is often related to the period between the contract completion date and the extended completion date. Is this the correct method? The intention of most construction contracts is for the contractor/subcontractor to be reimbursed the additional cost which results from employer delays. This involves a comparison between the actual costs incurred and what the cost would have been had no delay occurred. Where, for example, time is lost awaiting details for external cladding, which causes a four-week delay to the critical path, evaluating the prolongation costs associated with the four weeks on site, following the contract completion date, would obviously not produce the correct answer. A more accurate evaluation would be achieved by reference to the costs incurred during the four weeks when the information was late in arriving.

10.10.2. The Society of Construction Law Delay and Disruption Protocol 2002, as at clauses 1.11.2 and 1.22.3, with regard to this matter, states:

> Arguments commonly arise as to the time when recoverable prolongation compensation is to be assessed: is it to be assessed by reference to the period when the Employer Delay occurred (when the daily or weekly amount of expenditure and therefore compensation may be high) or by reference to the extended period at the end of the contract (when the amount of compensation may be much lower)? The answer to this question is that the period to be evaluated is that in which the effect of the Employer Event Risk was felt.

10.10.3. It has been common practice, when a delay to the progress and completion of the works has been identified and agreement reached between the contractor and the employer's representatives, to evaluate the full preliminary costs for the whole site which relate to the period of delay. This method has been thrown into doubt by the decision in the case of *Costain Ltd* v. *Charles Haswell and Partners* (2009). The case arose from the design services provided by Charles Haswell and Partners to Costain, who was the main contractor, in respect of foundations at the Lostock and Rivington Treatment Works, near Bolton. It was alleged by Costain that the design work was carried out negligently and, as a result, the project was completed 12 weeks and 4 days late and additional costs incurred, which included prolongation costs. The project comprised ten structures, but only two were affected by the alleged design failure. Costains claimed payment from Haswell and Partners for all the preliminary costs which were associated with the full 12 weeks and 4 days delay to completion. The delay to completion of the works resulted from problems on only two out of the ten structures, but as they were on the critical path, the overall completion date was affected. Deputy Judge Richard Fernyhough therefore considered that Costain's entitlement to prolongation costs should be limited to those which resulted from the delays to the two structures only. In reaching his decision, Deputy Judge Richard Fernyhough said:

> There were only 10 structures to be built on the Lostock site, of which RGF and IW buildings were 2. There is no reason to suppose that as a matter of course, progress on the other 8 structures would be affected by the delays to the RGF and IW. On the face, it is hard to see why that should be the case, since there would seem to be no reason why the other structures should not be constructed independently of the RBF and IW at least for part of their construction. If

therefore, as seems likely, the other activities on the site were continuing regardless of the delays to the RGF and IW buildings, then there is no basis upon which it can be argued that Costain can recover the whole of its costs of maintaining the Lostock site, simply as a result of delays to one part of that site.

If the decision in this case is followed in many disputes regarding prolongation costs, there will be a great deal of analysis required to isolate prolongation costs to the part of the structure which is affected by delay from the remainder. It is unlikely that the recorded costs will be retained in sufficient detail to enable this calculation to be carried out accurately. In the final analysis, it may come down to making a theoretical assessment.

SUMMARY

Once it is established that additional payment is due in respect of prolongation costs resulting from employer delays, the evaluation should relate to the period when the effect of the delay occurs and not by reference to the overrun period at the end of the contract. Where the delay to completion results from a critical activity which, whilst affecting the completion date, does not impinge on all of the site, only the prolongation costs associated with the part of the work which is on the critical path will be recoverable. This would be the case if the decision in the case of *Costain* v. *Charles Haswell and Partners* were to be followed. Whether this decision will be followed in subsequent cases is a matter of conjecture.

10.11. When ascertaining contractors' claims on behalf of employers, how should consultants deal with finance charges which form part of the calculation of the claim?

10.11.1. Contractors and subcontractors when submitting claims for loss and expense, or additional cost, will invariably include sums in respect of finance charges. The argument is that they have been stood out of their money for considerable periods of time, which has involved borrowing to make up the shortfall. Interest has to be paid to the bank or, if money is taken off deposit, interest is lost.

10.11.2. There is now no doubt that the contractor is entitled to relief, by way of loss and/or expense, for the cost of financing. In his judgement in *FG Minter Ltd* v. *Welsh Health Technical Services Organisation* (1980), Lord Justice Stephenson said:

> It is further agreed that in the building and construction industry the 'cash flow' is vital to the contractor and delay in paying him for the work he does naturally results in the ordinary course of things in his being short of working capital, having to borrow capital to pay wages and hire charges and locking up in plant, labour and materials capital which he would have invested elsewhere. The loss of the interest which he has to pay on the capital he is forced to borrow and on the capital which he is not free to invest would be recoverable for the employer's breach

of contract within the first rule in *Hadley* v. *Baxendale* (1854) without resorting to the second, and would accordingly be a direct loss if an authorised variation of the works, or the regular progress of the works having been materially affected by an event specified in clause 24(1), has involved the contractor in that loss.

10.11.3. In the case of *Rees & Kirby Limited* v. *Swansea City Council* (1985) the court held, in respect of the sum of £206,629 claimed for interest as part of a claim, that the contractor was entitled both legally and morally to every penny. The Court of Appeal confirmed that financing costs were a recoverable head of loss and expense and stated that such costs should be calculated at compound interest, with periodic rests taken into account. However, the amount of interest awarded by the judge was reduced to take account of a period when negotiation took place between the employer and contractor in an attempt to settle the dispute.

10.11.4. When ascertaining the cost of financing, what should be taken into account is the appropriate rate of interest which is actually paid by the contractor, provided it is not unreasonable. In the event of the contractor paying well above or well below prevailing market rates, it seems from *Tate & Lyle* v. *GLC* (1983) that appropriate rates are those at which [contractors] in general borrow money.

The cost of finance shall be calculated on the basis that it is charged by the contractor's bank, i.e. using the same rates and compounding accrued interest at the same intervals. Where the contractor is self-financed or financed from within its corporate group, the appropriate rate of interest is that earned by the contractor (or its group) on monies it has placed on deposit. Account should be taken of actual negative cash flows by way of primary expense, i.e. expenses are incurred progressively.

10.11.5. *Amec Process & Energy Ltd* v. *Stork Engineers & Contractors No. 4* (2002) is a case where the court was asked to decide upon the matter of finance charges. Amec was Stork's subcontractor on a contract to convert a ship's hull into a floating production platform. In an earlier judgment, Amec had been awarded damages in respect of additional costs incurred due to variations in the cost and for lost hours. Amec was entitled to be paid at 'net cost plus a percentage mark up to cover all costs, overheads and profit in connection with the personnel involved'. It was held that these costs should include interest paid by Amec to enable work to be financed and for the personnel to be paid, since these finance costs are 'a percentage mark up to cover all costs, overheads . . . '. The interest awarded was compounded quarterly.

10.11.6. The principle that finance charges should be paid as part of a loss and expense claim is also recognised in Scotland, following the decision in *Ogilvie Ltd* v. *City of Glasgow District Council* (1993).

10.11.7. Some standard contract forms expressly refer to finance charges. Under the ICE 6th and 7th Editions 'cost' is defined as 'including overhead finance and other charges'. MF/1 also defines 'cost' as 'including overhead and finance charges'. GC/Works/1 deals with finance charges in condition 47(1) in a different manner, where it states:

> The Employer shall pay the contractor an amount by way of interest or finance charges (hereafter together called 'finance charges') only in the event that money is withheld from him under the contract because, either

a. the Employer, PM or QS has failed to comply with any time limit specified in the Contract or, where the parties agree at any time to vary any such time limit, that time limit as varied, or

b. the QS varies any decision of his which he has notified to the contractor.

Condition 46(6) excludes finance charges from the definition of expense.

10.11.8. The Society of Construction Law Delay and Disruption Protocol 2002 offers the following advice concerning finance charges:

Interest as damages/finance charges

1.15.4 It is the position in most areas of business that interest payable on bank borrowings (to replace the money due) or the lost opportunity to earn interest on bank deposits, is quantifiable as damages where the claimant can show:

1.15.4.1 that such loss has actually been suffered; and

1.15.4.2 that this loss was within the reasonable contemplation of the parties at the time of contracting.

1.15.5 It is recognised that, in the construction industry, it will always be in the contemplation of the parties at the time they enter into their contract that if deprived of money the Contractor will pay interest or lose the ability to earn interest. Contractors therefore need only establish that the loss was actually suffered.

Time when interest starts to run

1.15.6 There are often arguments as to the date on which interest on a Contractor's claim should start to run. Contractors will argue that it should be the date on which they incurred expenditure for which they are entitled to compensation. Employers will say that interest should run only from the date that the Contractor has provided all information needed to satisfy them that the expenditure has been incurred.

1.15.7 The appropriate starting date will not be the same in all circumstances, but generally the starting date for the payment of interest should be the earliest date on which the principal sum could have become payable, which will be the date for payment of the certificate issued immediately after the date the Contractor applied for payment of the loss and/or expense. This will be subject to any notice requirements in the contract. In contracts where there are no certificates, the Protocol recommends that interest should start to run 30 days after the date the Contractor suffered the loss and/or expense.

SUMMARY

It is clear from case law that contractors and subcontractors are entitled to be paid finance charges as part of their loss and expense or additional cost claims. This is reflected in contractual definitions included in some of the standard forms of contract. The contractor or subcontractor will, however, lose his entitlement if he fails to submit a proper notice and details required by the terms of the contract. A further restriction upon the contractor's rights may be imposed by the conditions of contract, for example GC/Works/1.

10.12. Is a contractor/subcontractor entitled to be paid loss of profit as part of his monetary claim?

10.12.1. Contractors and subcontractors usually include in their claims an amount for loss of profit. This applies whether the claim is in respect of direct loss and expense, for example where JCT standard conditions apply, or additional cost where ICE conditions have been used. Loss of profit is also normally included in claims for breaches of contract in respect of a default by the employer which is not specifically provided for by the conditions of contract.

10.12.2. *Keating on Construction Contracts*, 8th Edition, page 297, para 8.052 provided this advice concerning claims for profit:

> Contractors commonly claim a loss of profit arising out of the diminution in turnover, but it seems that to establish this claim he must show, as with overheads, that at the time of the delay he could have used the lost turnover profitably (*Sunley B and Co Ltd* v. *Cunard White Star Ltd* (1940)). A claim for loss of profit does not, it is submitted, fail merely because the contract in question was unprofitable. The question is what the contractor would have done with the money if he had received it at the proper time. Even if, at that time, the contractor's business was making a loss, a sum analogous to loss of profit is, it is submitted, reconvertable if the loss of turnover increased the loss of business.

10.12.3. A contractor is entitled to reimbursement of loss of profit if he can prove that he was prevented from earning profit elsewhere; in the normal course of his business or as a direct result of one or more of the matters referred to in the conditions of the relevant contract, as explained in the case of *Peak Construction (Liverpool) Ltd* v. *McKinney Foundation Ltd* (1970).

10.12.4. In *Ellis-Don* v. *The Parking Authority of Toronto* (1978) the judge had this to say with regard to loss of profit:

> If a contractor is entitled to damages for loss of income to cover head office overheads, why should he not also be entitled to damages for loss of income that would result in normal profit?

In *Saint Line* v. *Richardsons Westgate* (1940) a claim was submitted for defects in the main engines of a ship. In addition to an entitlement to the cost of replacement parts, the court held there was an entitlement to loss of profit whilst deprived of the use of the vessel.

10.12.5. The amount of such profit, even if proved, must not exceed the level normally to be expected. An exceptionally high profit which might have been earned on another project cannot be reimbursed, unless this fact was known to the employer when the contract was entered into: *Victoria Laundry (Windsor) Ltd* v. *Newman Industries Ltd* (1949).

10.12.6. In the case of *Inserco Ltd* v. *Honeywell Control Systems Ltd* (1996) the judge, in finding in favour of Inserco, held that in respect of overheads and profit 'weight has to be given to the industry norms, such as those found in Spons M & E as they are directly applicable to Inserco's work. Equally, the figure must be related to the cost and expectation in profit in Inserco's business.'

10.12.7. Where the contractor's employment has been determined under the contract as a result of default by the employer, the contractor is entitled to be reimbursed the amount of profit that he can prove that he would have made on that particular contract had he been allowed to complete the works, less the amount saved because of the removal of his contractual obligation: *Wraight Ltd* v. *PH & T (Holdings) Ltd* (1980).

10.12.8. However, some standard forms of contract specifically exclude profit from the definition of cost or expense. The ICE 6th and 7th Editions, in clause 1(5), state that the word cost 'does not include profit'. MF/1 and the IChemE forms are similarly worded.

SUMMARY

Contractors and subcontractors will normally be entitled to claim loss of profit where they can show that, due to the circumstances giving rise to the claim, they lost the opportunity of making a profit elsewhere. The exception lies with those standard forms of contract, such as the ICE 6th and 7th Editions, MF/1 and the IChemE forms, where profit is specifically excluded.

10.13. Is a contractor/subcontractor entitled to be paid acceleration costs as part of his monetary claim? What is meant by constructive acceleration?

10.13.1. Contracts frequently fall behind for various reasons, leaving the completion date in jeopardy. If the contractor does not voluntarily accelerate the works, it will require a separate agreement between the employer and contractor or an express term of the contract to facilitate acceleration. Clause 46(1) of the ICE 6th and 7th Editions, for example, gives the engineer power to:

> notify the Contractor in writing and the Contractor shall thereupon take such steps as are necessary and to which the Engineer may consent to expedite the progress so as substantially to complete the Works or such Section by that prescribed time or extended time.

This clause only applies where, due to any reason which does not entitle the contractor to an extension of time, the rate of progress of the works or any section is, in the opinion of the engineer, too slow to comply with the time for completion. This is a very useful provision from the employer's point of view. In the absence of such a clause, a contractor in danger of overrunning the completion date may opt to pay liquidated damages in preference to acceleration costs, on the grounds that the costs of acceleration will exceed the liquidated damages.

10.13.2. A different situation arises where the employer wants to bring forward the contract completion date. The ICE 6th and 7th Editions include in clause 46(3) for the employer or engineer to request the contractor to complete the works in a time less than the contract period or extended contract period. If the contractor concurs, special terms and conditions of payment will have to be agreed. The JCT Management Form includes detailed provisions for acceleration. In like manner to clause 46(3) of the ICE conditions, the provision can only be operated with the agreement of the contractor. Condition 38 of GC/Works/1 operates in a similar fashion. The Engineering and Construction Contract (NEC 3), under clause 36, gives the project manager the power to instruct the contractor to submit a quotation for an acceleration to achieve completion before the completion date.

10.13.3. One of the difficulties of operating an acceleration clause is proving the cost of additional resources and reduced outputs which result from acceleration measures. It will be necessary to isolate the costs of acceleration measures. To do so effectively will require the contractor to demonstrate the cost of resources which would have been employed had no acceleration measures been taken, for comparison with the actual resources which were employed.

10.13.4. Contractors and subcontractors often argue that they have been forced to accelerate the works to overcome delays caused by the architect or engineer. There may be some confusion between an acceleration claim and a loss of productivity claim. Normally, in the absence of an instruction to accelerate, the contractor is not entitled to decide unilaterally to accelerate and expect the employer to pay the costs. The contractor or subcontractor may, however, argue that he chose to accelerate faced with the architect/engineer's refusal or neglect to grant a proper extension of time. This is sometimes referred to as a constructive acceleration order. Constructive acceleration is defined by the US Corps of Engineers as:

> An act or failure to act by the Employer which does not recognise that the contractor has encountered excusable delay for which he is entitled to a time extension and which required the contractor to accelerate his programme in order to complete the contract requirements by the existing contract completion date. This situation may be brought about by the Employer's denial of a valid request for a contract time extension or by the Employer's untimely granting of a time extension.

10.13.5. In the Australian case of *Perini Pacific* v. *Commonwealth of Australia* (1969), Mr Justice Macfarlane in the Commercial Court of New South Wales indicated clearly that this type of claim could only be on the basis of some proven breach of contract by the owner – coupled, of course, with proof of damages in the form of completion to time by expenditure greater than would otherwise have been incurred. In that case, the breach consisted of a refusal or failure by the certifier to give any consideration at all to the contractor's applications. In the Canadian case of *Morrison-Knudsen* v. *BC Hydro & Power* (1978) the owner, unknown to the contractor, had secretly agreed with a government representative that no extension of time would be granted in view of the pressing need for electricity by the contract completion date. All requests for an extension of time were refused and the contractor carried on and managed to finish the project with only slight delay. The Court of Appeal of British Columbia held that the contractor could have rescinded on the basis of a fundamental breach of contract, had he known the real reason for the refusals and in that event would have been entitled, on the basis of established case law, to put his claim on a *quantum meruit* basis and so to escape from the original contract prices. The fact that the contractor had not rescinded but had completed the project limited his remedy to the usual one of damages, measured in this case in terms of the additional expense incurred in completing to time, i.e. in accelerating progress. Similarly, in *W. Stephenson (Western) Ltd* v. *Metro Canada Ltd* (1987), the Canadian Court held that, by stating that in no circumstances would the contractor receive an extension of time for completion, the employer had deliberately breached the contract and it awarded damages calculated by reference to the contractor's consequential acceleration.

10.13.6. The UK courts have been a little slow at recognising a situation where a claim for constructive acceleration would be relevant. However, the situation has been changed by the decision in *Motherwell Bridge Construction Ltd* v. *Micafil* (2002). In this case Motherwell Bridge was a subcontractor to Micafil in connection with the construction of an autoclave for BICC in Kent. Additional welding was required by Micafil, which involved a substantial amount of additional man hours. Micafil failed to grant an appropriate extension of time. In an effort to complete by the original completion date, Motherwell Bridge incurred substantial additional costs, which formed the subject of a claim. The matter was referred to court, which found in favour of Motherwell Bridge. Judge Toulman considered that Motherwell Bridge was entitled to an extension of time based upon the period of delay which would have been incurred using the original labour force and, as Micafil failed to grant an extension of time, Motherwell Bridge was entitled to be paid the acceleration costs.

10.13.7. It would seem that to recover payment in respect of constructive acceleration, a contractor or subcontractor must be able to demonstrate the following:

(1) An entitlement to an extension of time;
(2) Proper delay notice, as required by the contract, has been served;
(3) The employer or his agent has failed to grant an appropriate extension of time;
(4) The employer has requested completion by the original completion date;
(5) The contractor has incurred additional costs due to taking acceleration measures;
(6) The cost to the contractor in taking acceleration measures, when estimated in advance of the decision being made to accelerate, is less than the cost resulting from an overrun to the completion date.

SUMMARY

Whether or not a contractor or subcontractor can be required to accelerate the works will depend upon the terms of the contract. The ICE conditions give the engineer power to instruct the contractor to accelerate where, due to its own default, progress is getting behind. Where the employer requires the contractor to accelerate to overcome delays not of its making, agreement between contractor and employer is required.

If the contractor is delayed by the employer, architect or engineer and no recognition is given to the contractor's rights to an extension of time, then a claim for constructive acceleration may be successful if the contractor decides on acceleration as being less expensive than an overrun and payment of liquidated damages.

10.14. Where a written claims notice is required to be submitted within a reasonable time, or information is required to be issued by the architect or engineer within a reasonable time, how much time must elapse before the claim can be rejected as being too late, or the contractor to issue a claim for late issue of information?

10.14.1. It is common when drafting contracts where written notice is required, to include an express provision to the effect that the notice must be submitted within a reasonable time. What constitutes a reasonable time is often the subject of hot debate and usually depends upon all the circumstances of the case.

10.14.2. Mr Justice Vinelott in *London Borough of Merton* v. *Stanley Hugh Leach Ltd* (1985) had to decide a number of preliminary issues, one of which related to the written notice requirements of JCT 63 with regard to a contractor's claim for the recovery of direct loss and expense. These state that the application must be made within a reasonable time. Mr Justice Vinelott said that it must not be made so late that the architect can no longer form a competent opinion on the matters which he requires to satisfy himself that the contractor has suffered the loss and expense claimed. However, in considering whether the contractor has acted reasonably it must be borne in mind that the architect is no stranger to the project and it is always open for the architect to call for further information, either before or in the course of investigating the claim.

10.14.3. In the insurance case of *Kier Construction Ltd* v. *Royal Insurance (UK) Ltd* (1992), the insurance policy required the claimant to notify the insurer as soon as possible if there was an occurrence such that in consequence a claim is to be made. The claim should have been made on 12 June 1989, but it was not submitted until 4 July 1989. The claimant lost his rights, as the notice was not served 'as soon as possible'.

10.14.4. The case of *Hersent Offshore SA and Amsterdamse Ballast Beton-Waterbouw BV* v. *Burmah Oil Tankers Ltd* (1978) also deals with the question of notice, where it was held that a notice given after completion of the work could not be regarded as a notice given before work commenced or as soon thereafter as is practicable. Clause 52 of the contract, amongst other things, provided that:

> no increase of the contract price under sub-clause (1) of this clause or variation of rate or price under sub-clause (2) of this clause shall be made unless as soon after the date of the order as is practicable and in the case of extra or additional work before the commencement of the work or as soon thereafter as is practicable notice shall have been given in writing . . .

The dispute as to the claimants' entitlement to additional payment in respect of the variation was referred to arbitration. The arbitrator, having found the facts as set out above, determined in his award (amongst other things) that the notice of intention to claim should have been given as soon after the date of the order as was reasonably practicable, that it had not been so given and that accordingly the respondents were not liable to the claimants in respect of the variation.

10.14.5. Some of the standard forms of contract require the architect, or engineer, to issue drawings to the contractor within a reasonable time. The ICE 7th Edition, in clause 7(4), requires the engineer to issue at a time reasonable in all the circumstances, drawings, specifications or instructions requested by the contractor. In the case of *Neodox Ltd* v. *Borough of Swinton and Pendlebury* (1958), the judge, Mr Justice Diplock, had to decide whether information had been issued in accordance with this wording, or had it been issued late, giving the contractor an entitlement to a claim? He offered the following comments as to how the obligation should be addressed:

> What is a reasonable time does not depend solely upon the convenience and financial interests of the [contractors]. No doubt it is to their interest to have every detail cut and dried on the day the contract is signed, but the contract does not contemplate that. It contemplates further details and instructions being provided and the engineer is to have a time to provide them which is reasonable having regard to the point of view of him and his staff and the point of view of the corporation as well as the point of view of the contractors.

In determining what is a reasonable time as respects any particular details and instructions, factors which must obviously be borne in mind are such matters as the order in which the engineer has determined the works shall be carried out, whether requests for particular details or instructions have been made by the contractors, whether the instructions relate to a variation of the contract which the engineer is entitled to make from time to time during the execution of the contract, or whether they relate to part of the original works, and also the time, including any extension of time, within which the contractors are contractually bound to complete the works.

SUMMARY

There is no hard and fast rule as to when a notice which is required to be submitted within a reasonable time, can be rejected as being out of time. What is reasonable depends upon the individual circumstances.

10.15. What methods of evaluating disruption have been accepted by the courts and what is meant by the 'Measured Mile'?

10.15.1. One of the most difficult items to evaluate with any accuracy is disruption. The problem is usually caused by a lack of accurate records. With the spotlight on linking cause and effect having been created by the decision in *Wharf Properties* v. *Eric Cumine* (1991), claims for disruption came under greater scrutiny. It is unlikely that contractors and subcontractors will succeed where their claims for disruption are based simply upon the global overspend on labour for the whole of the contract. More detail will have to be given, which isolates the cause of the disruption and evaluates the effect.

10.15.2. Comparison of output in a period when disruption has occurred with a period where there was no disruption, often referred to as the 'measured mile', is one method of calculating disruption, which has been defined in the following terms:

> The measured mile involves a comparison of productivity during an unimpacted period of time with productivity during an impacted period of time. The productivity is calculated for both periods of time and the difference between the two is the lost productivity attributed to the impact. Ideally the measured mile is a continuous period of time during the project when the labour productivity is unimpacted. The work performed during the mile should be substantially similar in type, nature and complexity to the work affected and the composition and level of skill of the workforce should be comparable.

10.15.3. In the case of *Whittall Builders Company Ltd* v. *Chester-le-Street District Council* (1985), difficulties were experienced by the employer in giving possession of dwellings on a rehabilitation scheme. The court found that, during the period when these problems existed, the contractor was grossly hindered in the progress of the work and as a result, ordinary and economic planning and arrangement of the work was rendered impossible. However, a stage was reached in November 1974 when dwellings were handed over in an orderly fashion and no further disruption occurred. The court had to decide upon the appropriate method of evaluating disruption. Mr Recorder Percival QC, in his judgment, had this to say:

> Several different approaches were presented and argued. Most of them are highly complicated, but there was one simple one – that was to compare the value to the contractor of the work done per man in the period up to November 1974 with that from November 1974 to the completion of the contract. The figures for this comparison, agreed by the experts for both sides, were £108 per man week while the breaches continued, £161 per man week after they ceased . . .
>
> It seemed to me that the most practical way of estimating the loss of productivity, and the one most in accordance with common sense and having the best chance of producing a real answer was to take the total cost of labour and reduce it in the proportions which those actual production figures bear to one another – i.e. by taking one-third of the total as the value lost by the contractor.
>
> I asked both [Counsel] if they considered that any of the other methods met those same tests as well as that method or whether they could think of any other approach which met them better than that method. In each case the answer was 'no'. Indeed, I think that both agreed with me that that was the most realistic and accurate approach of all those discussed. But whether that be so or not, I hold that that is the best approach open to me, and find that the loss of productivity of labour, and in respect of spot bonuses, which the plaintiff suffered is to be quantified by adding the two together and taking one-third of the total.

10.15.4. The Society of Construction Law Delay and Disruption Protocol 2002 recommends 'the Measured Mile', which is described as follows:

> 1.19.7 The starting point for any disruption analysis is to understand what work was carried out, when it was carried out and what resources were used. For this reason, record keeping is just as important for disruption analysis as it is for delay analysis. The most appropriate way to establish disruption is to apply a technique known as 'the Measured Mile'. This compares the productivity achieved on an un-impacted part of the contract with that achieved on the impacted part. Such a comparison factors out issues concerning unrealistic programmes and inefficient working. The comparison can be made on the man hours expended or the units of work performed. However, care must be exercised to compare like with like. For example, it would not be correct to compare work carried out in the learning curve part of an operation with work executed after that period.

Care needs to be taken when using the 'measured mile' method to ensure that the drop in productivity hasn't been contributed to by problems for which the contractor is contractually responsible.

10.15.5. Contractors often compare output during a period of disruption with the levels of output included in the tender. It is possible to calculate a new rate by substituting the output included in the tender with the output achieved on the site. The difficulty with this method is that the contractor is often accused of including in its tender outputs which were never attainable. A more satisfactory method would be to compare actual outputs with industry standards or outputs on other projects.

10.15.6. Additional labour and plant schedules provide an alternative route. It may not be possible, because of the nature of the disrupting matters and the complexity of the project, to employ the simple approach used in the *Whittall Builders* case. Therefore it may be appropriate to attempt to isolate the additional hours of labour and plant which results from the event giving rise to disruption. This will involve recording the hours taken up by the task which was the subject of disruption and comparing them with the assessed hours which would have been involved had there been not disrupting factors. The items of work which have been disrupted can then be set out in a schedule for ease of reference, to show how the claim for disruption has been calculated.

10.15.7. The courts again have provided assistance in dealing with this problem of assessment. In the case of *Chaplin* v. *Hicks*, heard as long ago as 1911, it was held:

> Where it is clear that there has been actual loss resulting from the breach of contract, which it is difficult to estimate in money, it is for the jury to do their best to estimate; it is not necessary that there should be an absolute measure of damages in each case.

Two years later, Justice Meredith in the Canadian case of *Wood* v. *The Grand Valley Railway Co* (1913) had this to say:

> It was clearly impossible under the facts of that case to estimate with anything approaching to mathematical accuracy the damages sustained by the plaintiffs, but it seems to me to be clearly laid down there by the learned judges that such an impossibility cannot 'relieve the wrongdoer of the necessity of paying damages for his breach of contract', and that on the other hand the tribunal to estimate them whether jury or judge, must under such circumstances do 'the best

it can' and its conclusion will not be set aside even if 'the amount of the verdict is a matter of guess work'.

The Canadian case of *Penvidic Contracting Co Ltd* v. *International Nickel Co of Canada Ltd* (1975) also provides some guidance on the manner in which disruption should be evaluated where anything like an accurate evaluation is impossible. The dispute arose out of a construction agreement to lay ballast and track for a railroad. The owner was in breach in several respects of its obligation to facilitate the work. The contractor, who had agreed to do the work for a certain sum per ton of ballast, claimed by way of damages the difference between that sum and the larger sum that he would have demanded had he foreseen the adverse conditions caused by the owner's breach of contract. There was evidence that the larger sum would have been a reasonable estimate. At the hearing, damages were awarded on the basis claimed, but on appeal to the Court of Appeal this portion of the award was disallowed. On further appeal, the Supreme Court of Canada held, restoring the trial judgment, that where proof of the actual additional costs caused by the breach of contract was difficult, it was proper to award damages on the estimated basis used at trial. The difficulties of accurate assessment cannot relieve the wrongdoer of the duty of paying damages for breach of contract.

10.15.8. A method often referred to as a cost less recovery involves taking the cost of a section of work and deducting what has been allowed in the tender. This method was used by the plaintiff in the case of *Shopfitters Ltd* v. *ADI Ltd* (2003), where the judge in the case, Mr John Uff, had to consider a claim in the sum of £50,000 calculated using this system. In arriving at a decision, he concluded:

> I accept the substance of [the claimant's expert] analysis showing that the claimant incurred labour costs substantially in excess of those allowed for. The different calculations put forward by the claimant and by [the claimant's expert] all assume that the claimant was blameless and that all established additional costs should be recoverable. Having considered the evidence I am not persuaded that the whole or even the majority of the claimant's labour costs can be attributed to the matters alleged. On the contrary, there is evidence of the unproductive and wasteful use of labour. Doing the best I can on the material presented, I assess that one quarter of the minimum figure put forward on behalf of the claimant by the [claimant's expert] (£50,000) is properly attributable to additional labour costs and I allow the claimant the sum of £12,500.

SUMMARY

Disruption can often be difficult to evaluate properly. One of the most satisfactory methods is by comparison of outputs when work is disrupted with outputs when no disruption is taking place, sometimes referred to as 'the Measured Mile'. Finally, when all else fails, courts have accepted a claim for disruption based upon assessed costs.

10.16. Can a claims consultant be liable for incorrect advice?

10.16.1. A claims consultant when offering professional services will be governed by the terms which are expressed or implied in his conditions of appointment. In the absence of any express term to the contrary, a claims consultant will have an implied duty to exercise reasonable skill and care: *Bolam* v. *Friern Hospital Management Committee* (1957). If there is a failure to exercise such reasonable skill and care and as a result the client suffers loss, then the claims consultant will have a liability to the client.

10.16.2. This was illustrated in the case of *Cambridgeshire Construction Ltd* v. *Nottingham Consultants* (1996). The plaintiff contractors retained the defendant claims consultants in relation to a building contract on which they were working. They informed the consultant that they had received the final certificate, dated 13 October 1992, on 19 October 1992. Notice of arbitration had to be given within 28 days and, under JCT 80, clause 30, the certificate became final and conclusive in the absence of an arbitration notice. It was not until 11 November that the defendants advised the plaintiffs concerning the service of an arbitration notice. The plaintiffs claimed damages for professional negligence, as their right to go to arbitration had been lost. The defendants argued that the 28-day period ran from the date when the contractor received the certificate, and that, consequently, the notice was served in time.

10.16.3. It was held:

(1) The plaintiffs had been entitled to rely upon the defendants' expertise as claims consultants.
(2) Upon the true construction of clause 5.8 of the contract, the final certificate should be issued to the employer and a copy sent to the contractor. If the word 'issue' were to be taken as meaning that the certificate should reach the employer, the architect would have to wait until he knew that the employer had received it before sending the copy to the contractor. The interpretation of the clause as propounded by the defendants would, therefore, be absurd.
(3) A certificate was not issued until the architect sent it in the post. Signing it was not sufficient. On the evidence in the instant case, the certificate was posted on 13 October or, at the very latest, 14 October. The last possible date for serving notice of arbitration, therefore, was on 10 or 11 November. The notice had been given outside the 28-day period, and was thus invalid.
(4) The defendants were found liable to the plaintiffs.

SUMMARY

Claims consultants can be liable to their clients in negligence if they fail to exercise reasonable skill and care. An example which was referred to court involved claims consultants who failed to warn their contractor clients of the need to serve an arbitration notice not later than 28 days after the issue of the final certificate. The contractors lost their entitlement and it was held that they could found an action for negligence against the claims consultants due to their failure to exercise the level of skill expected of an ordinary skilled claims consultant.

10.17. If a delay in the early part of a contract caused by the architect/engineer pushes work carried out later in the contract into a bad weather period, causing further delay, can the contractor/subcontractor claim loss and expense resulting from the bad weather delay?

10.17.1. Where delays occur due to bad weather, architects and engineers are apt to consider granting extensions of time under the adverse weather clause in the contract. Obligations on the part of the contractor to pay liquidated damages will be avoided, but there will be no entitlement to the recovery of additional costs which result from the delay. An exception to this rule is the Engineering and Construction Contract (NEC 3), which provides for delay caused by weather to be regarded as a compensation event which can give rise to both an extension of time and additional cost reimbursement, if shown to occur on average less frequently than once in ten years.

10.17.2. Contractors with long experience of submitting claims will usually ask themselves whether carrying out work in the bad weather was due to the knock-on effect of an earlier delay caused by the architect/engineer. In other words, whether an earlier delay which was due to some act or omission on the part of the architect/engineer was the reason why, in a later working period, the contractor was adversely affected by weather and so, had there been no earlier delay, the weather would not have delayed the contractor. This all goes back to cause and effect. There must be a link between the cause and its effect and the important question is whether such a link can be established between the delay caused by the architect/engineer which results in the contractor being later delayed through working in bad weather conditions.

10.17.3. In the Canadian case of *Ellis-Donn* v. *The Parking Authority of Toronto* (1978), the plaintiffs were building contractors who successfully tendered to construct a new car park in Toronto. The contract period was 52 weeks, but completion was delayed for a period of 32 weeks. One of the causes of delay was a failure by the employer to obtain an excavation permit, which delayed the start of the job. The initial period of delay was some seven weeks. It was claimed by the contractor that the initial delay had a knock-on effect which further extended the delay to 17.5 weeks, made up as follows:

- Delay in obtaining the excavation permit: 7 weeks;
- Consequent delay in commencing excavation: 1.5 weeks;
- Consequent delay in obtaining crane: 6 weeks;
- Consequent delay due to extension of work into a winter period: 3 weeks.

10.17.4. The judge, in finding in favour of the contractor, was influenced by the decision in *Koufos* v. *Czarnikow* (1967). It was held in this case that a party who breaches a contract is responsible for the damages that flow from that breach, if, at the time the contract was entered into, the parties to the contract considered there was a real danger or a serious possibility that such breach would give rise to damages. In giving judgment in the *Ellis-Don* case, the judge held:

> In my view the parties to the contract, at the time the contract was entered into, contemplated, or should have contemplated, that if the defendant did not have available the necessary excavation permit until seven weeks after the plaintiff had need of it, that the plaintiff would be delayed in carrying out its work, and that such delay would throw the subcontractors off schedule, which would entail further delays and that such delays would cause the plaintiff damages of the very type, or nature it suffered.

10.17.5. Contractors and subcontractors who find themselves delayed by bad weather may be entitled to recover the additional costs, as well as time, if they are able to show that had it not been for an earlier delay caused by the architect or engineer, they would not have been affected by the weather.

SUMMARY

If the contractor is to succeed in claiming loss and expense due to inclement weather on the basis that, but for an earlier delay which was the responsibility of the employer, engineer or architect, the bad weather delay would not have arisen, he must be able to link the earlier delay with the effect on later work caused by bad weather. If there is no direct link, or the link in some way is broken, then the claim will fail, but in appropriate cases the claim for loss and expense as well as a claim for additional time will succeed.

10.18. Who is responsible for the additional costs and delay resulting from unforeseen bad ground conditions; the employer, or contractor/subcontractor?

10.18.1. If disputes are to be avoided, the contract should make it clear who is responsible for the cost in terms of time and money resulting from encountering unforeseen bad ground. In general terms, if the contract is silent on the matter and the contractor undertakes to construct the work for a lump sum, the contractor will be deemed to have taken the risk of encountering bad ground, whether it was foreseeable or not. It will be for the contractor to take a view of the risk of bad ground being encountered and make suitable provision in the price. In the case of *Workshop Tarmacadam Co Ltd* v. *Hannaby* (1995) it was submitted, on behalf of the contractor, that as the contract was a re-measurement contract, the contractor was entitled to be paid for additional work encountered due to the presence of hard rock. The court disagreed, on the basis that if the parties had intended for rock or other bad ground to be paid for as an extra, it would have been a simple matter for the contract to have made it clear. A similar situation arose in the case of *Nuttall* v. *Lynton and Barnstaple Railway Co* (1899), which related to the construction of a railway for a lump sum of £42,000. The contractor claimed extra for dealing with rock, but the Court of Appeal disagreed when it was said:

> It seems to me . . . that it is an absolute contract; whether [the contractor] found rock in a small degree, or whether [the contractor] found rock in a large degree, the agreement was to do the work for £42,000.

Some contracts make it clear that the contractor is required to bear the risk. The FIDIC 1999 Silver Book for Turnkey projects, under clause 4.12, makes it clear that the contractor is to take the risk and include in his price for 'any unforeseen difficulties or costs, except as otherwise stated in the contract'.

10.18.2. The JCT Forms are silent on the matter and, probably because most of the work will be above ground, the risk is therefore taken by the contractor. In the case of *Co-Operative Insurance Society* v. *Henry Boot* (2002), the employer provided a subsoil survey at tender stage, which proved to be inaccurate and resulted in the contractor incurring additional costs. The contract incorporated the JCT conditions and the contract documents stated:

> the contractor is deemed to have inspected and examined the site and its surroundings and to have satisfied himself as to the nature of the site including the ground and subsoil before submitting his tender.

It was held that the contractor, under this clause, took the risk of encountering bad ground.

10.18.3. The majority of the existing standard forms of engineering contract place some of the risk onto the employer. This ensures that the employer only pays for dealing with unforeseen bad ground, which is encountered and should be reflected in the contractor's price for the work. Clause 12 of the ICE 6th and 7th Editions entitles the contractor to claim for any additional time and cost which results from:

> physical conditions (other than weather conditions or conditions due to weather conditions) or artificial obstructions which conditions or obstructions could not in his opinion have been foreseen by an experienced contractor.

Condition 7(3) of GC/ Works/1 Edition includes a similar provision. Both of these contracts pass the risk of unforeseen conditions to the employer, the contractor bearing the risk of conditions which could have been foreseen.

10.18.4. The Engineering and Construction Contract (NEC 3), under clause 60.1, provides for encountering physical conditions to be a compensation event. To qualify, these have to be physical conditions which:

- are within the Site;
- are not weather conditions; and
- which an experienced contractor would have judged at the contract date to have such a small chance of occurring that it would have been unreasonable for him to have allowed for them.

10.18.5. The FIDIC 1999 Red Book, under clause 4.12, entitles the contractor to recover any additional costs incurred as a result of 'physical conditions which he considers to have been unforeseeable'. Physical conditions are defined by the conditions of contract as:

> natural physical conditions and manmade and other physical obstructions and pollutants, which the contractor encounters on the site when executing the works including hydrological conditions but excluding climatic conditions.

10.18.6. With regard to adverse conditions which were foreseeable, Max Abrahamson, in his book *Engineering Law and the ICE Contracts*, cites the case of *CJ Pearce and Co Ltd* v. *Hereford Corporation* (1968), where:

> contractors knew before tender that a sewer at least 100 years old had to be crossed in the course of laying a new sewer. What was described on the map as 'the approximate line of the . . . sewer' was shown on a map supplied to tenderers. The witnesses for both parties accepted that the word 'approximate' meant that the contractor would realise that the line of the old sewer might be 10 feet to 15 feet one side or the other of the line shown. The old sewer fractured when the contractors disturbed the surrounding soil within this area.
>
> Held: that the condition could have been 'reasonably foreseen', so that even if they had served the necessary notice they would not have been entitled to extra payment under this clause (clause 12) for renewing the old sewer, backfilling the excavation, backheading, etc.

10.18.7. The case of *Humber Oil Terminals Trustee Ltd* v. *Harbour & General Works (Stevin) Ltd* (1991) appears to have extended a contractors' facility for levying claims under clause 12 of the ICE conditions. A contract was entered into for the construction of three mooring dolphins and the reconstruction of a damaged bathing dolphin at the Immingham Oil Terminal, Humberside. For the purposes of carrying out the work the contractor used a jack-up barge equipped with a 300-tonne fixed crane. The crane was lifting and slewing a large concrete soffit in order to place it on piles which had already been prepared, when the barge lifted, became unstable and collapsed. Considerable loss and damage resulted. The contractor claimed that the collapse of the barge was due to encountering adverse physical conditions which could not reasonably have been foreseen by an experienced contractor and would therefore give an entitlement to levy a claim. It was argued on behalf of the employers that no legitimate grounds for a claim existed. They contended that clause 8(2) applied, under which the contractor was required to take full responsibility for the adequacy, stability and safety of all site operation and methods of construction. The problems arose, the employers said, due to a failure on the part of the contractor to comply with this requirement.

10.18.8. The arbitrator, in deciding whose argument was correct, had to establish the reason for the collapse of the barge. He came to the conclusion that the collapse was caused by the barge moving, due to an initial small settlement of 5 cm becoming a substantially large settlement of perhaps 20 cm or 30 cm, at which point the stability of the barge was such that further and progressive collapse occurred. He decided that the settlement was caused by adverse physical conditions which could not have been foreseen by an experienced contractor and hence gave rise to a claim under clause 12. On appeal by the employer, the court held that the arbitrator's award was correct.

10.18.9. The case of *ABP* v. *Hydro Soil Services* (2006) highlighted the requirements of clause 11 in the ICE 6th and 7th Editions, which requires the contractor to take into account information made available by the employer. Hydro Soil Services had to strengthen a quay wall as part of a design and construct contract, which would allow an adjacent berth to be strengthened. The wall proved to be weaker than the contractor expected, which deflected significantly during construction of the works. A claim under clause 12 was submitted, but the judge considered that the contractor had no entitlement, as the problem would have manifested itself to the contractor had he examined all the documents which were available to him at tender stage. A second clause 12 claim was submitted by the contractor because of the need to drill through mass concrete to form a coping beam. It was argued by the contractor that the original drawings provided by the employer at tender stage suggested that the coping rested on two inches of binding. In rejecting the claim, the judge considered that an experienced contractor should have anticipated that there were two alternative methods of constructing the coping beam and should have identified a clear risk that the original work had used mass concrete to form the coping beam, rather than gravel backfill.

SUMMARY

Which party to a contract is responsible for unforeseen bad ground should be made clear by the express terms of the contract. The ICE 6th and 7th Editions, GC/Works/1, the Engineering and Construction Contract (NEC 3) and the FIDIC Red Book are all examples of standard forms of contract which place the risk of unforeseen bad ground conditions onto the employer. If the contract is silent on the matter, the contractor will be deemed to have taken the risk.

10.19. Where one party to a contract is in breach and the injured party incurs loss, what obligations are there on the injured party to mitigate the loss?

10.19.1. Where a breach has occurred, there is a well established legal obligation placed upon the injured party, if the full extent of the loss is to be recovered, to mitigate the loss. In the case of *British Westinghouse* v. *Underground Electric Railways* (1912), Viscount Haldane LC explained the obligation in the following words:

> . . . imposes on a plaintiff the duty of taking all reasonable steps to mitigate the loss consequent on the breach, and debars him from claiming any part of the damage which is due to his neglect to take such steps.

10.19.2. The question of mitigation often arises when a breach of contract occurs resulting in loss or damage. It also can apply where a tort is committed. As can be seen from the

British Westinghouse case, the plaintiff must take all reasonable steps to mitigate the loss which results from the breach. He cannot recover any loss which he could have avoided. For example, if an employer fails to give proper access to the site at the appropriate time, the contractor would be expected to take plant off hire if it were reasonable and practicable. In other words, the plaintiff cannot recover avoidable loss.

10.19.3. A further matter to be considered is whether the additional costs incurred in taking the reasonable steps are recoverable. It was held in *Wilson* v. *United Counties* (1920) that these costs were recoverable. Using the earlier example, the contractor could recover any costs associated with taking the plant off hire. It has been argued that if a contractor is delayed by the employer, he may be obliged to accelerate the works in a reasonable attempt to mitigate the loss. This being the case, under the second rule the acceleration costs should be recoverable. This argument is flawed, as the contract normally provides a remedy for employer delays, such as an extension of time and payment of loss and expense. However, it could be a different story if the contract, such as a JCT form, provides a time-scale within which the architect must make a decision concerning an extension of time application, and the architect fails to make a timeous decision.

SUMMARY

Where a breach of contract occurs, the injured party must take all reasonable steps to keep the losses down to a minimum. The losses which could have been avoided by taking reasonable steps will not be recoverable.

10.20. What is meant in legal terms by the words 'consequential loss'?

10.20.1. The words 'consequential loss' frequently appear in construction contracts. Liability for consequential loss is often expressed as being included or excluded. It is tempting to interpret the wording in a similar way to the *Oxford English Dictionary*, being 'events that follow another'.

10.20.2. The decisions in *British Sugar* v. *NEI Power Projects* (1997) and *Simkins Partnership* v. *Reeves Lund & Co Ltd* (2003) dealt with the meaning of consequential loss. The Court of Appeal in *British Sugar* considered that the proper way to examine this question was by reference to *Hadley* v. *Baxendale* (1854). In this case, the court categorised damages under two headings commonly referred to as the first leg and second leg. The first leg deals with losses arising naturally, i.e. according to the usual course of things. By way of contrast, the second leg covers the type of losses which both parties had in mind when the contract was entered into, but which were particular to the contract, sometimes referred to as special damage. It was held that consequential loss is akin to special damage.

10.20.3. The earlier decisions in *Mondel* v. *Steel* (1841) and *Croudace* v. *Cawood's Concrete* (1978) dealt with the meaning of consequential loss in a similar manner.

10.20.4. It would seem the expression that hope dies hard in the human heart applies to some litigants and their legal advisers. In the case of *Hotel Services Ltd* v. *Hilton International*

Hotels (2000), the Court of Appeal was involved in a dispute concerning the meaning of consequential loss. The parties entered into a contract under which the defendants agreed to supply the claimants with Robobars for their hotels. A Robobar is a hotel minibar which automatically records any removal of its contents and at the same time electronically registers the item concerned on the account of the guest. Robobars were installed in the claimant's hotels on fixed 10-year agreements. They malfunctioned and, following attempts to make them work properly, which failed, they were removed. The claimant commenced an action for breach of the implied term of merchantable quality. The judge found in favour of the claimant and awarded damages for overpaid rental, cost of removal and storage and loss of profit.

10.20.5. An appeal was lodged by the defendant on the basis that the terms of the contract included a clause which excluded the liability of the defendant from indirect or consequential losses. It was argued that the judge's award included losses which fell within this category. The Court of Appeal disagreed, as a distinction should be made between normal losses and consequential losses. Normal losses are those which every claimant in a like situation will suffer, whereas consequential losses are those peculiar to the particular claimant's circumstances and not foreseeable. On this view, consequential losses are anything beyond the normal measure of damages. The losses included in the judge's award were all normal losses and fell outside the term consequential losses.

10.20.6. In the case of *McCain Foods GB Ltd* v. *Eco-Tech (Europe) Ltd* (2011), the Court of Appeal had to decide the meaning of 'consequential loss', which was excluded from the liability for breach of contract in a seller's contract relating to the installation of a system to treat waste gas. The purchaser, McCain, intended to use the waste gas arising from the system to produce electricity, which could be sold on. Because of a failure on the part of the gas treatment system, McCain sustained a loss as a result of their not being able to produce electricity. Eco-Tech claimed that this loss was consequential loss, for which they had no liability. In similar fashion to the *Hilton Hotel International* case, the Court of Appeal considered the loss fell into the first leg of *Hadley* v. *Baxendale*, and was therefore not regarded as a consequential loss.

SUMMARY

Consequential losses do not include those losses which are normal, usual and foreseeable. To be classified as a consequential loss, the expenditure must not be foreseeable, but peculiar to the particular circumstances which come to light after the contract has been entered into.

10.21. Is it possible to include in a contract a daily or weekly rate which will be paid to the contractor in respect of loss and expense or additional cost resulting from delays caused by the employer?

10.21.1. Where, due to delays on the part of the employer, the completion date is affected and the contractor incurs direct loss and expense, or additional costs, most standard forms of contract provide for the contractor to recover financially. The relevant clauses usually require the contractor to submit a written notice and supporting detail. The architect, engineer or quantity surveyor is normally empowered to ascertain the contractor's financial entitlement, which is then certified and paid.

10.21.2. This process can be time-consuming and expensive, with payment to the contractor often the subject of a long delay. It has been suggested that there should be provision in the contract for the inclusion of an all-inclusive daily or weekly sum which will be paid to the contractor, in the event of delays to completion being caused by the employer. The daily or weekly rate could be included in the contractor's tender submission or be the subject of agreement between the parties post tender submission, but prior to the contract being entered into.

10.21.3. The main disadvantage of this arrangement is that it only deals with the financial effects of a delay to the completion, leaving disruption and delays which only affect progress but not completion to be dealt with separately. Contractors will need to appreciate that the sum will also include for any legitimate claims which may be received from subcontractors. Finally, during pre-contract negotiations contractors may well find themselves under pressure to agree to an unrealistically low daily or weekly amount.

SUMMARY

It is possible to include in a contract a daily or weekly rate which will be paid to the contractor in respect of loss and expense, or additional costs, resulting from delays caused by the employer. The daily or weekly rate will include for any claims which may be received from subcontractors. Claims for disruption and delays which only affect progress but not completion will not be included and should therefore be the subject of a separate submission.

Chapter 11
Payment

11.1. Where a contract requires the contractor to give a guaranteed maximum price, does he have any grounds for increasing the price above the guaranteed maximum?

11.1.1. The term 'guaranteed maximum price', when applied to a construction contract, provides for the employer a nice feeling of financial security. However, in itself the term 'guaranteed maximum price' has no meaning. It will be necessary to read the wording of the contract carefully to ascertain which of the parties bears which risks. However, when entering into a contract of this nature, the employer is convinced that, no matter what happens, the final cost will not be above the maximum and there is a fair chance it could be lower. Any design changes which result from the specific instructions of the employer would, understandably, fall outside the guaranteed price. Guaranteed maximum price contracts have been in use for many years. IDC, a Stratford-on-Avon construction company who pioneered design and construct contracts in 1969 and the 1970s, promoted their contracts as guaranteed maximum price. It is a good selling point, which can be persuasive. With risk transfer being very much in vogue, the guaranteed maximum price contract has seen a resurgence. The intention of this procurement route is to transfer all risks to the contractor and allow for no increase in price whatsoever, other than costs which result from employer changes.

11.1.2. Construction is unlike most other commercial processes, such as the motor car industry, where a prototype is built and, when perfected, mass produced on an assembly line. Each construction project is a one-off and unique. The designer has to get it right first time and the manufacture and on-site process allows for no serious errors. The contractor who undertakes a guaranteed maximum price contract has to take these risks, together with risks such as unforeseen ground conditions, unexpected encounters with service mains, bad weather, industrial unrest, shortages of labour, plant and materials, changes in legislation, insolvency of suppliers and subcontractors, fire, storm and earthquake. Contractors are even often required to check all information received from the employer and take responsibility for its accuracy. It is hardly surprising that, when one or more of these risks becomes a reality and substantial additional costs are incurred, the contractor casts round to find good legal reasons for receiving payment above the maximum price.

200 Contractual Problems and their Solutions, Third Edition. Roger Knowles.

11.1.3. The case of *Mowlem* v. *Newton Street Limited* (2003) illustrates the difficulties which can befall a contractor who enters into a guaranteed maximum price contract. Work involved the conversion of a post office, built of reinforced concrete in 1910 in Manchester, into 104 apartments and an underground car park, with commercial units at ground level. The contract was an amended standard form. Article 10 expressly stated that the parties agreed that the contract sum was a guaranteed maximum price and that the contractor acknowledged he had taken all risks and responsibilities. Under a heading of contractor's risk, the contractor also became responsible for any incorrect or insufficient information given to him by any person, whether or not in the employment of the employer. This represented a high level of risk in view of the fact that the work involved the conversion of a building which was almost 100 years old. Difficulties arose as a result of the issue by the employer's agent of an instruction for the contractor to carry out concrete repairs to the existing structure. It was argued on behalf of Mowlem that, as there was no specific reference in the Employer's Requirements or the Contractor's Proposals to the concrete repairs, they were entitled to be paid for the work over and above the guaranteed price. The lawyers representing Newton Street Limited were of the opinion that Article 10, which placed all risks onto Mowlem, deprived them of any right to additional payment. It was the view of the judge, in finding against Mowlem, that within the scheme of the guaranteed maximum price there is nothing to displace the ordinary and unambiguous meaning of Article 10, that the risk of unforeseen defects in the existing building was the contractor's.

11.1.4. The question needs to be asked as to whether it was reasonable and sensible for a contract involving the conversion of an old post office to be let on a guaranteed maximum price basis. There was no knowing what the contractor might have encountered. The existing structures and foundations could have proved to be insufficiently robust to carry the load and to have required strengthening. Asbestos might have been discovered, which could have resulted in a substantial cost to remove it. Commercial organisations who purchase old property cannot sensibly expect to transfer all the risk of the suitability of the property for conversion onto the contractor. In attempting to price the risk in a sensible manner, the contractor is likely to end up taking a gamble. When an unpriced risk becomes a reality, the contractor, rather like Mowlem, will inevitably read the fine print to see if there is a get-out. Some types of contract more easily lend themselves to be let on a maximum price basis. For example, a project where the contractor is able properly to assess at tender stage the full extent of the work necessary to meet the employer's requirements.

11.1.5. There are no standard forms of contract which are classified as guaranteed maximum price. The one which comes the nearest is the FIDIC Silver Book, for use on international projects, and although not titled a guaranteed maximum price contract, this is its obvious intention. Employers on domestic projects who require a guaranteed maximum price are left with either amending an existing standard form or having a bespoke contract produced. Contractors who market themselves on the basis of guaranteed maximum price usually produce their own form of contract. In view of the problems encountered by Mowlem, it is hardly surprising that, despite the title, contractors will usually include somewhere in the fine print reasons for increasing the price. A contract which includes reference to a guaranteed maximum price, therefore, is not always what it appears.

SUMMARY

It is not uncommon for a contract to include in its heading 'Guaranteed Maximum Price'. The fact that this heading exists does not in itself mean that the contractor, except for employer changes, has taken on all risks. It is usually necessary to read carefully through the wording of the contract to establish that the employer has transferred all risks to the contractor. It is not unusual to find that matters such as increases in cost due to changes in legislation fall outside the guaranteed price and will be paid for as an extra.

11.2. Where a subcontract provides for 2.5% cash discount, does this mean that the discount can only be deducted if payment is made on time, or may the discount be taken even if payment is made late?

11.2.1. Many subcontractors, when submitting quotations, include for a 2.5% cash discount. Does the term 'cash discount' mean a reduction in price in return for payment on time, or a trade discount which in essence is a price reduction and applies whether or not payment is made on time?

11.2.2. Subcontractors frequently complain that many contractors pay late, but still deduct cash discount. Contractors often argue that they are entitled to deduct cash discount irrespective of whether they have paid on time or not. The dictionary definitions of cash discount make its meaning clear:

New Collins Concise Dictionary: 'Cash discount. A discount granted to a purchaser who pays before a stipulated date.'

Encyclopaedia of Real Estate Terms, 1987 edition: 'Cash discount. A reduction in price or consideration for early or prompt payment.'

11.2.3. Despite the apparent clarity of these definitions, disputes have nevertheless been referred to court concerning cash discounts. An example was *Team Services Plc* v. *Kier Management & Design* (1993). Kier were management contractors for the construction of a shopping and leisure centre at Thurrock, with Team Services appointed as a subcontractor. In a letter to Team Services dated 23 March 1989, Kier wrote:

> We hereby accept your fixed price tender sum of £14,811,000 . . . all in accordance with the attached schedule marked Appendix A inclusive of 2.5% cash discount and exclusive of VAT.

The payment terms within the subcontract dated 7 January 1991 read as follows:

> Within 7 days of the receipt of the payment due under the principal agreement pursuant to a certificate which showed a sum in respect of the subcontract works the management contractor . . . discharge such sum less:

(i) retention money at the rate specified in Appendix A
(ii) a cash discount of 2.5% on the difference between the said total value and the said retention, and
(iii) the amount previously paid.

11.2.4. It was the subcontractor's case that on a proper interpretation of the wording in the contract the management contractor was entitled to deduct 2.5% cash discount from the interim certificates only if payments had been made within the timescale provided for within the subcontract. Alternatively, the subcontractor considered there was an implied term to that effect. The management contractor's case was that, on a true construction of the terms of the subcontract, the discount could be deducted irrespective of when the interim payments were made. In any event, it was argued that it would have been a simple matter to have included appropriate wording if the intention was for discount to have been deducted only when all payments were made on time. Further, it was suggested that, as the retention could be deducted even if payment was made late, the same would apply to discount. Judge Bowsher disagreed. The word 'cash' he thought significant and it should be given due weight. Discounts, he considered, are given for a purpose; if not, it would be simpler merely to reduce the price. In finding for the subcontractor, it is interesting to note the following matters which Judge Bowsher took into account in interpreting the meaning of 'cash discount':

(1) The object is to ascertain, by the contractual words used, the mutual intentions of the parties as to the legal obligations each assumed: *Pioneer Shipping Ltd* v. *BTP Tioxide Ltd* (1982).
(2) The parties cannot give evidence themselves as to what their intentions were when drafting the contract. What should be ascertained is what the court considers a reasonable person's intentions would have been if placed in the position of the parties who drafted the contract; an objective rather than a subjective test: *Reardon-Smith Line Ltd* v. *Hansen Tangren* (1976).
(3) The words of a contract should be construed in their ordinary and grammatical sense: *Grey* v. *Pearson* (1857).
(4) Courts may consult dictionaries in order to determine the meaning of words: *Pepsi Cola* v. *Coca Cola* (1942).

The decision of Judge Bowsher was referred to the Court of Appeal, which came to a different decision. They were influenced by evidence that Team had added a sum to the tender figure to account for the discount and that the management contractor's only recovery in respect of the work undertaken by Team was the discount. Lord Justice Lord, in supporting the case put forward on behalf of the management contractor, said:

> But so far as discount is concerned, the defendants were nothing but a conduit pipe, since the prime cost payable by the employer took account of all discounts including cash discounts obtainable by the defendants. It would be an odd result of the plaintiff's construction that, if the defendant were to be a few days late in paying each instalment they should find themselves out of pocket to the extent of 2.5% of the contract price, an amount which would comfortably exceed the whole of their remuneration under the contract.

It should not be taken as the golden rule as a result of this case that cash discount in every case does not imply that it can be taken irrespective of when payment is made. The case would appear to turn on its own facts and not establish a principle of law.

11.2.5. The subject of 'cash discount' was again raised in *Wescol Structures Ltd* v. *Miller Construction Ltd* (1998), where it was held that discount could only be deducted if payments were made on time. In this case, however, the subcontract incorporated a letter sent by the subcontractor to the main contractor, which included the following wording:

> Our price is based on monthly valuations payment to be received 17 days thereafter. The 2.5% is for payment to these terms.

11.2.6. Subcontractors will always argue that discounts expressed as a cash discount can only be deducted if payment is made on time. However, subcontractors would be well advised to ensure that the terms of the contract make it clear that late payment will result in the discount being forfeited.

11.2.7. It is well worth noting that if discount can only be deducted if payment is made on time, contractors may be tempted to calculate discounts by reference to the gross amount due before the deduction of previous payments. If this method is used, the contractor need only make the final payment on time to become entitled to deduct the full discount from the gross value of work executed. The methodology would involve:

(1) Deduct discount off the gross payment before deducting previous payments.
(2) Make all interim payments late and forego the discount.
(3) Pay the final payment on time and deduct discount from the gross final account.

For example:

Final account	£100,000
Less 2.5% discount	2,500
	£ 97,500
Less previous payments	£ 90,000
Balance due	£ 7,500

The net result will be that the main contractor will have paid all interim payments late, but because the final payment was made on time discount can be deducted.

SUMMARY

The dictionary definition of 'cash discount' indicates that it is offered in return for prompt payment. However, in the light of disputes which have been referred to the courts it is advisable for subcontractors who are offering the discount to make it clear in the wording of the subcontract that it can only be deducted if payment is made on time. Unfortunately for subcontractors, the wording of many subcontracts provides for the discount deduction to be made from the gross value of the work. Hence, if interim payments are made late and discounts forfeited, the position can be rectified by timely payment of the final payment.

The advice to main contractors is that they should always make sure the final payment is on time. Subcontractors, however, should ensure that the wording of the subcontract makes it clear that the discount is to be deducted from the net payment and not the gross value.

11.3. Under what circumstances are contractors/subcontractors entitled to be paid for materials stored off site as part of an interim payment or payment on account?

11.3.1. Most standard forms of contract provide for interim payments on a regular basis, usually monthly, or by way of stage or milestone payments. It is the norm to include within those payments sums to reflect the value of work executed and also materials brought onto or adjacent to the site, but unfixed. Usually the materials are required to be adequately stored and protected.

11.3.2. Contractors and subcontractors often suffer from a heavy drain of money in respect of materials and components which are manufactured but stored off-site due to their size or value, or because of restricted storage space on site. Special provision in the contract is necessary before interim payment can be made for these items.

11.3.3. A problem from the point of view of the employer would result from payment being made for materials or components stored off-site, if the contractor or subcontractor became insolvent before they were delivered to site. Lengthy and expensive disputes could arise between employer and the receiver or liquidator, both as to the identity of the goods and components and also as to ownership.

11.3.4. JCT 2011, by clause 4.17, provides for listed items such as materials, goods or prefabricated items stored off-site to be included in interim payments. There is provision for a list which includes for specifically identified items, such as a special glazed panel, and a list where the items are not specifically identified, such as standard doors. The lists are supplied to the contractor and attached to the contract documents. To safeguard the employer in the event of insolvency of the contractor, there is provision for a bond to be supplied in respect of both the listed and uniquely identified items. The requirement for the provision of these bonds is optional.

11.3.5. JCT 2011 includes a list of requirements which the contractor must fulfil before any of the listed or uniquely listed items can be included in a certificate and paid. They include:

- Ownership of items on the lists is vested in the contractor;
- Items on the list are properly insured; and
- Clear identification at the place where items on the list are stored, of the interest of the employer and that their destination is the works.

11.3.6. ICE 6th Edition, clause 65 (1)(c) and 7th Edition, clause 60(1)(c), provide for interim payment for materials or goods stored off-site which are identified in an appendix to the form of tender.

11.3.7. GC/Works/1, condition 48C, allows for payment of 'Things Off Site', but only if stated in the abstract of particulars, and this is not applicable in Scotland. This condition requires proof of ownership and its transfer to the employer, identification, proper storage, etc. in like manner to JCT 2011.

11.3.8. The Engineering and Construction Contract (NEC 3) makes no reference to payment for goods stored off-site.

SUMMARY

Payment for materials stored off-site will only occur if the contract makes specific provision for such payment and the requisite conditions are fulfilled.

11.4. Can a contractor force an employer to set aside retention money in a separate bank account?

11.4.1. Most standard forms of construction contract provide for the employer to withhold sums of money, referred to as 'retention money', until completion. The contractor is at risk, should the employer become insolvent prior to payment of the retention. In an effort to protect contractors, clause 4.18.1 of JCT 2011 stipulates that the employer's interest in such a retention is fiduciary, as trustee for the contractor. Other JCT forms are similarly worded.

11.4.2. The protection given to contractors in respect of retention held in trust was tested in *Rayack Construction Ltd* v. *Lampeter Meat Co Ltd* (1979). The conditions of contract were JCT 63, with wording in clause 30(4) similar to JCT 2011. It was the decision of the court that the employer was obliged to set aside the retention money in a separate trust fund. JCT 2011, under clause 4.18.3, reinforces the decision in the *Rayack Construction* case in requiring the employer, unless it is a local authority, to place retention money in a separate bank account, if requested by the contractor.

11.4.3. The whole subject of withholding retention money under a JCT contract may seem simple, but a number of cases have highlighted certain difficulties.

11.4.4. In *Wates Construction (London) Ltd* v. *Franthom Properties Ltd* (1991), a contract was let for the construction of a hotel in Kent. The contract conditions were JCT 80 With Quantities and the contract sum £2,869,607. The Employer was entitled to deduct 5% as retention in respect of some of the works, with the total amount retained representing 2.5% of the total sum. Work started in June 1988 and by March 1990, when the court proceedings had commenced, the amount certified was £3,368,912 and the employer had retained £84,222. The works were certified as practically complete in August 1989. In accordance with clause 30.5.3 of the conditions of contract, the employer should have deposited the retention monies in a separate bank account. On 2 October 1989 the contractor became aware that retention monies were not being placed in a separate bank account. The contractor requested the employer to comply, but he refused. A writ was issued by the contractor claiming an injunction to compel the employer to place the retention monies in a separate account. Judge Newey, as Official Referee, decided that the contractor was entitled to the order. The employer considered the judge to be wrong and lodged an appeal. The issues considered were as follows.

(1) Under clause 30.1.1.2, the employer has a right of deduction from retention monies for such matters as liquidated damages. The employer, therefore, contended that he was a beneficiary of the trust and under no obligation to place retention monies in a separate bank account. This argument was dismissed by the Court of Appeal, as they considered the wording of clause 30.1.1.2 in the following light:

> Notwithstanding the fiduciary interest of the Employer in the Retention as stated . . . the Employer is entitled to exercise any right under this Contract of . . . deduction from monies due . . .

It was held that this suggests that the interest of the employer in the retention remains as a trustee and is not that of beneficiary. The Court of Appeal considered that, in respect of the total retention fund of £84,222, there were five beneficiaries: the main contractor and four nominated subcontractors.

(2) The employer further argued that he was only bound to appropriate a sum to the retention fund and was not bound to set it aside, in the sense of placing it in a separate identifiable account. It was the employer's contention that he could use the money in his own business as working capital. Clause 30.5.2.1 requires the architect or, if he so instructs, the quantity surveyor, to prepare a statement specifying the contractor's retention and the retention for each nominated subcontractor. The preparation of the statement, it was argued, was a sufficient appropriation of the sums involved. It was held by the Court of Appeal that the first duty of a trustee is to safeguard the fund in the interests of the beneficiaries. It would therefore be a breach of trust for any trustee to use the trust fund in his own business.

(3) The most powerful argument put forward on behalf of the employer was that clause 30.5.3, which requires him to put the retention in a separate account, had been deleted from the contract. It was said that the deletion of the clause by the parties was indicative of a common intention on their part that the employer should be under no obligation to place retention monies in a separate bank account. The Court of Appeal considered that the parties may have had differing reasons for wishing clause 30.5.3 to be deleted and therefore it was not possible to draw from the deletion of that clause a settled intention of the parties common to each of them that the general duties of the employer were other than those set out in clause 30.5.3, i.e. fiduciary, and that this could include placing the monies in a separate account if requested to do so. The Court of Appeal dismissed the appeal from the employer and upheld Judge Newey's order that the sum of £84,222 be deposited in a separate bank account.

11.4.5. The judge in *Herbert Construction* v. *Atlantic Estates* (1993) took a different view. In this case, reference in JCT 80 to retention money being placed in a separate bank account, in like fashion to the *Wates* case, was deleted. It was held, unlike the *Wates* decision, that the employer was not bound to place retention money in a separate account.

11.4.6. In *Finnegan Ltd* v. *Ford Sellar Morris Developments Ltd* (1991), a contract was let using JCT 81 With Contractor's Design. Relying on the decision in *Wates* v. *Franthom Properties*, the plaintiff sought to compel the defendant to place retention monies in a separate account. The defendant failed to do this, arguing that the plaintiff's demand of 22 April 1991 (which was 15 months after practical completion) had been made too late to comply with the provisions of JCT 80, clause 30.5. This states that the request should be made either at the commencement of the contract or before the issue of the first interim certificate. It was held that JCT 81, clause 30.5, places no restriction on the time in which a contractor should make a request for retention to be placed in a separate trust fund. Further, it was not sensible to expect the contractor to make a request each time retention is added to a monthly certificate.

11.4.7. In *MacJordan Construction* v. *Brookmount Erostin* (1991), the contractor acted too late to preserve the retention fund. The employer, who was a property developer, became insolvent and had not placed the retention in a separate bank account. MacJordan's claim to be paid retention from sums available did not take precedence over a bank floating charge. The amount of retention which the contractor was unable to recover totalled £109,000.

11.4.8. In the case of *Bodill and Sons (Contractors)* v. *Harmail Singh Mattu* (2010), Bodill was engaged by Mattu to construct new apartments and convert two warehouses in Hinchley, Leicestershire. The contract used was the JCT Standard Building Contract, Private Edition. Work was due for completion on 12 February 2007, but was not finished on time and by 12 October 2007 a sum of £3.97m had been certified, including £124,207 retention. This case illustrates the practicalities of operating a bank account under a JCT contract under which it is intended to hold retention money. The JCT contracts require the separate bank account to be designated as retention held by the employer on trust for the contractor, and for the employer to certify to the architect and the contractor that the retention money has been dealt with in this manner. Any interest which accrues in the account goes to the employer. Bodill was rather slow at taking advantage of the system for protecting the retention money and it was not until September 2007 that Mattu was asked to set up a separate bank account, into which the money was to be paid. The following month Mattu requested its bank to open a separate bank account, with the intention of transferring the money into it. The bank was rather slow at complying with the request and Bodill became impatient and applied to the court to ensure compliance with the contract requirement. It was the view of the court that, once a request had been made by the contactor to the employer for a separate bank account to be opened, it was incumbent upon the employer to get on and open one. However, neither the contract nor the law required this to be done instantaneously. Commercial realities apply and the court considered that a reasonable time for compliance would be two to three weeks. Due to inaction on the part of the bank, Mattu was held to be in default. The court also held that the name of the account was insufficiently clear to indicate that it was in fact a trust account. The name of the account was 'Harmil Singh Mattu, trading as Urban Suburban, re Bodill retention money account'. The judge considered that this wording did not make it clear enough that the bank account was a trust account.

11.4.9. Subcontractors under certain circumstances are also protected. In *PC Harrington Contractors Ltd* v. *Co-Partnership Developments Ltd* (1998), Lelliott contracted with the plaintiff as works contractor under the Works/Contract/2 subcontract under the JCT 87 Management Form. By clause 4.1 of the conditions of contract, Co-Partnership Developments, the employer under the contract and the defendant in the action, were obliged to pay to Lelliott the cost of the project, which included sums due to Harrington. Amounts paid by Co-Partnership Developments were subject to the withholding of a retention in respect of works by both the management contractor and the works contractor. Lelliott went into receivership shortly after practical completion, following the expiry of the defects liability period. As a result of the insolvency, Lelliott's employment was automatically determined. Co-Partnership Developments Ltd held the retention fund of £288,166, which was expressed as being held on trust for the management contractor and for any works contractor in clause 4.8.1 of the management contract. The dispute arose as a result of Co-Partnership Developments setting off additional costs it incurred, due to Lelliott's insolvency, from retention money held on behalf of Harrington. The court directed that the sum held by the employer under clause 4.8.1 was held in trust on behalf of the works contractor. Co-Partnership Developments were not, therefore, entitled to set off costs incurred as a result of the insolvency of Lelliott from retention held on behalf of Harrington.

11.4.10. Clause 4.19 of JCT 2011 includes in the Contract Particulars for the contractor to provide a retention bond as an alternative to the withholding of retention.

SUMMARY

Where the JCT conditions apply, employers can be forced by a court injunction to place the retention money in a separate bank account. It is not necessary for the contractor to make a request for retention to be paid into a separate bank account each time an interim certificate is due for payment. Reference to holding retention as a trustee is peculiar to JCT forms of contract. With regard to other standard forms, such as ICE 6th and 7th Editions, GC/Works/1 and the Engineering and Construction Contract (NEC 3), there is no obligation for employers to hold retention money in a separate account. Bank accounts which are opened to hold retention monies must make it clear that it is a trust account in favour of the contractor.

11.5. If an employer becomes insolvent, what liability does the contractor have for paying subcontractors who are owed money when no further sums are forthcoming from the employer?

11.5.1. When an employer becomes insolvent on construction projects, there is usually substantial money outstanding to contractors. Where the JCT contract applies, the contract provides for retention to be set aside in a separate bank account: *Rayack Construction*

Ltd v. *Lampeter Meat Co Ltd* (1979). If this has occurred, the retention money, which is held in trust, should be paid over in full by the receiver or liquidator to the contractor. Nevertheless, even where this has happened, there are usually substantial amounts due, other than the retention. Subcontractors, who are one down the line, will often be owed sums of money and disputes arise with the main contractor as to who has taken the risk of the Employer's insolvency.

11.5.2. Contractors in the past normally included in their subcontract terms and conditions, clauses which have the intention of passing down to the subcontractor the risk of non-payment or late payment by the employer for any reason, sometimes referred to as 'pay when paid' clauses. The Housing Grants, Construction and Regeneration Act 1996, section 113(1), effectively outlaws 'pay when paid' clauses, but an exception is made in respect of non-payment due to the employer's insolvency. It seems clear, therefore, that if the contractor insists on a clause in the subcontract that the risk of non-payment due to the insolvency of the employer rests with the subcontractor, then it is with the subcontractor that the loss will lie. This exception has not been amended by the Local Democracy, Economic Development and Construction Act 2008, despite aggressive lobbying by subcontracting trade associations.

11.5.3. Where the JCT Standard Building Subcontract 2010 is used, then different criteria apply. This contract does not provide for passing the risk of insolvency down to the subcontractors. Payment is usually due monthly, or at agreed stages of the subcontract works and does not relate in any way to the payment provisions under the main contract.

11.5.4. The JCT Standard Form of Subcontract provides for payment to be made by the contractor to subcontractors not later than 21 days after the date of the issue of a payment certificate under the main contract. Payment may even be due without the issue of such a certificate, if the employer becomes insolvent. In *Scobie & McIntosh Ltd* v. *Clayton Browne Ltd* (1990), Scobie & McIntosh were nominated subcontractors for the supply and installation of catering equipment, in keeping with the alternative nomination procedure under JCT 80; the subcontract conditions NSC/4 formed the basis of the subcontract. Disputes then arose concerning payment to a number of subcontractors, including Scobie & McIntosh, who had two invoices outstanding. Clayton Browne, the contractors, were faced with a battle on two fronts. They were seeking payment for their work from the employers via arbitration, whilst at the same time fending off claims for payment from their subcontractors, including Scobie & McIntosh. Their defence was that under the terms of the nominated subcontract, they were only obliged to pay a subcontractor within 17 days of the issue of an architect's certificate under the main contract. No such certificate had been issued by the architect, as there was a dispute, with the employers proceeding to arbitration. The employer determined his contract with Clayton Browne Ltd, who in turn determined the employment of Scobie & McIntosh Ltd. Scobie & McIntosh argued that the normal payment arrangement did not apply, as the subcontract had been determined. Their case was that they were in no way at fault and required payment from the main contractor. They were not content to await the outcome of the dispute under the main contract. Judge Davies QC was sympathetic to the subcontractor's position. He said the object of nomination was to permit employers to participate in the selection of subcontractors, but not to insulate main contractors against default by employers. Clayton Browne were ordered to pay on account a sum of £57,840, which was the minimum amount due. The dispute concerning the balance would be fought out in arbitration. Whilst this case involved payment under a nominated subcontract which was phased out in JCT 2005, the principles would still apply where the JCT 2005 standard form of subcontract applied, or bespoke contracts are let involving nominated subcontracts.

11.5.5. The definition of insolvency was critical to the Court of Appeal's decision in *William Hare Ltd* v. *Shepherd Construction Ltd* (2010). The dispute relates to the wording in the subcontract between William Hare and Shepherd, relating to 'pay when paid', in the event of insolvency of the employer. It became critical, as the employer on the project went into self-certifying administration and it was a matter for the court to decide whether this event was caught by the 'pay when paid' clause in the subcontract. The employer on the project was Trinity Walk, and the sum due to be paid to William Hare Ltd was £996,683.35, which wasn't paid. The wording of the 'pay when paid' clause entitled Shepherd to avoid paying William Hare if non-payment to Shepherd was the result of the employer becoming insolvent, or the making of an administration order under Part 11 of the Insolvency Act 1986. Unfortunately for Shepherd, the employer did not go into administration under the Insolvency Act 1986, but by self-certification under the Enterprise Act 2002. This appears to be a shortcoming in the wording of the clause in the subcontract, which was drawn up after the Enterprise Act 2002 came into force. It was argued, on behalf of Shepherd, that the task of the court was to give effect to the intention of the parties, and not the literal words. There was a precedent to this approach in the case of *Chartbrook Ltd* v *Persimmon Homes* (2009). This reasoning was rejected by the court, where Lord Justice Waller stated:

> if a main contractor wishes to have a pay when paid provision in a subcontract he would be bound, if it was to be effective, to identify a way in which the third party employer became insolvent as defined in the legislation. If he chose a way which was not in accordance with the legislation because he misdrafted the provision, I can see no reason why, however obvious it was that he had misdrafted the provision, the principles in (the *Persimmon* case) would come to the rescue.

The Court of Appeal held that William Hare was entitled to be paid.

SUMMARY

The contractor's obligation to pay subcontractors if the employer becomes insolvent will be dependent upon the wording of the subcontract. The JCT 2011 Standard form of Subcontract, for example, makes it clear that payment will be made irrespective of whether the employer has paid the main contractor or not. Other standard forms of subcontract are worded in a similar manner. The Housing Grants, Construction and Regeneration Act 1996, whilst outlawing 'pay when paid' clauses, makes an exception for cases where the employer becomes insolvent. Some main contractors may therefore be tempted to amend the JCT 2011 Standard Form of Subcontract, or such other standard form of subcontract which may apply, to provide for the contractor to avoid paying subcontractors where the contractor doesn't receive payment as a result of the insolvency of the employer. Bespoke subcontract conditions are likely to include such a provision, hence passing the financial risk concerning the employer's solvency status down to the subcontractor.

11.6. Can a contractor/subcontractor legitimately walk off site if payment is not made when due?

11.6.1. The law concerning a contractor/subcontractor's rights to suspend work for non-payment is incorporated in the Housing Grants, Construction and Regeneration Act 1996. The Act applies to contracts entered into after 1 May 1998 and only applies to construction contracts as defined in the Act.

11.6.2. Under the Act there is a right to suspend work for non-payment under section 112, which states:

> (1) Where a sum due under a construction contract is not paid in full by the final date for payment and no effective notice to withhold payment has been given, the person to whom the sum is due has the right (without prejudice to any other right or remedy) to suspend performance of his obligations under the contract with the party by whom payment ought to have been made (the party in default).
> (2) The right may not be exercised without first giving to the party in default at least seven days' notice of an intention to suspend performance, stating the ground or grounds on which it is intended to suspend performance.
> (3) The right to suspend performance ceases when the party in default makes payment of the amount due in full.

11.6.3. The Local Democracy, Economic Development and Construction Act 2008, which amends the Housing Grants, Construction and Regeneration Act 1996, in section 112(1) indicates that the contractor's and subcontractor's right of suspension for non-payment extends to any or all of the obligations under the contract. Section 112(3A) provides for the recovery of a reasonable amount of costs and expenses reasonably incurred as a result of exercising the right of suspension.

11.6.4. The requirements of the Construction Act, as amended by the Local Democracy Act, have been incorporated into JCT 2011. Clause 4.14 states that the contractor may suspend work if the employer fails to pay the contractor in full and the failure continues for seven days after the contractor has given the employer, with a copy to the architect/contract administrator, written notice of an intention to suspend work. The contractor is entitled to continue to suspend work until payment has been made in full. An entitlement to an extension of time as a result of the suspension of work arises under clause 2.29.6. The contractor, in addition to an entitlement to an extension of time, can recover a reasonable amount in respect of costs and expenses, as provided for in clause 4.14.2. Similar provisions exist in the JCT 2011 Standard Form of Subcontract.

11.6.5. GC/Works/1, condition 52, in like manner to JCT 2011, provides the contractor with a right to suspend work, subject to a seven-day warning notice. Condition 52(4) provides for an extension of time, but there is no reference to the payment of any costs or expenses incurred as a result of the suspension. The contractor's right, therefore, to recover a reasonable amount of costs and expenses will arise from section 112(3A) of the Local Democracy, Economic Development and Construction Act 2008.

11.6.6. There is no express provision concerning suspension of work for non-payment in the ICE 7th Edition or in the amendments to the 6th Edition which were issued following the implementation of the Construction Act, or those issued in connection with the Engineering and Construction Contract (NEC 3). The basic requirements of section 112(1) of the Construction Act provide a right to an extension of time and Section 112(3) of the Local Democracy, Economic Development and Construction Act 2008 an entitlement to recover a reasonable amount of costs and expenses as a result of the suspension.

11.6.7. The statutory right of suspension applies only to construction contracts, as defined in Section 106(1)(g) of the Construction Act. Where the provisions of the Act do not apply, what are the contractor or subcontractor's rights with regard to suspension of work, due to a failure to make proper payment as required by the terms of the contract? The decision in *D. R. Bradley (Cable Jointing) Ltd* v. *Jefco Mechanical Services* (1989) may be a pointer as to their entitlement. Jefco were main contractors for the refurbishment of Islington Town Hall. Bradley was appointed as a domestic subcontractor for the electrical installation. The subcontract between Jefco and Bradley was made verbally, following the submission of Bradley's tender. No specific payment terms of the subcontract were accepted, but it was agreed that interim payments would be made. Bradley made various requests for payment as work proceeded, but none were met in full. On 29 October 1985 Bradley made another application for an interim payment and gave notice that, in the event of non-payment they would leave site. No payment was made and on 12 December 1985, Bradley left the site and litigation commenced. It was held that the earlier underpayments by Jefco did not amount to a repudiatory breach, but the non-payment of the last application was such a breach, as it reasonably shattered Bradley's confidence of being paid. The court decided that Bradley was entitled to consider the contract at an end and for it to be terminated. From this decision it can be seen that, in the absence of an express clause in the contract, non-payment of interim certificates could still give rise to a contractor or subcontractor's entitlement to determine, if such non-payment undermined the essence of the contract. However, in the absence of an express clause giving a right to suspend work, no such right exists at common law. The options appear to be either to carry on working or, if the non-payment 'shatters confidence', to determine the contract as the only other remedy. This is still the case for contracts which fall outside the scope of the 1996 Construction Act.

11.6.8. The case of *C.J. Elvin Building Service Ltd* v. *Peter & Alexa Noble* (2003) also involved the suspension of work. A dispute arose concerning non-payment in respect of a contract to carry out extensive work at the home of Mr and Mrs Noble. The contract was not covered by the Act, as it involved undertaking work for a residential occupier. It seems, at the stage where the case came before the court, C.J. Elvin was owed in the region of £40,000. Work had been stopped due to non-payment. Mr and Mrs Noble argued that the stopping of work constituted a repudiatory breach and that non-payment by an employer is not generally a breach which will enable the contractor to treat the contract as ended. The judge held that the refusal to pay money which is due, particularly if it is a sizeable amount, or a threat not to pay any more, could amount to a repudiatory breach. It was held that Elvin leaving the site had been triggered by Mr and Mrs Noble's non-payment and could not therefore in itself be repudiatory. The court found in favour of the contractor. This case is another example of non-payment providing a right of termination of the contract, but not unless covered by the Construction Act, an entitlement to temporarily suspend the work.

SUMMARY

The Housing Grants, Construction and Regeneration Act 1996 provides for a contractor or subcontractor to suspend work for non-payment in respect of contracts classified by the Act as construction contracts. There is an entitlement to an extension of time in respect of the delay caused by the suspension. No provision is included in the Act for the recovery of loss and expense. Some contracts, such as JCT 2005, make specific provision for the recovery of loss and expense which is incurred as a result of a suspension of work. JCT 2011 provides for the recovery of reasonable costs and expenses resulting from the suspension.

The Local Democracy, Economic Development and Construction Contract Act 2009, which amends the Housing Grants, Construction and Regeneration Act 1996, in section 112(1) indicates that the contractor's and subcontractor's right of suspension for non-payment extends to any or all of the obligations under the contract. Section 112(3A) provides for the recovery of a reasonable amount of costs and expenses reasonably incurred as a result of exercising the right of suspension.

Where the provisions of the Construction Act do not apply, there is no right at common law for a contractor or subcontractor to suspend work on a temporary basis as a result of non-payment. However, it may give rise to an entitlement to terminate the contract, if it can be shown to be a repudiation of the contract.

11.7. Where a contractor undertakes work which he considers should be paid for on a daywork basis and submits daywork sheets as required under the terms of the contract to the architect/engineer, if the daywork records are not signed by the architect/engineer, how does this affect the contractor's entitlement to payment?

11.7.1. Most standard forms of contract provide, under clearly defined circumstances, for extra work to be paid for on a time and materials basis, often referred to as 'daywork'. There is usually a timescale within which the contractor is required to submit the daywork records to the architect/engineer for approval and signing. Contractors regularly complain that they go through the process of producing daywork records and submit them on time to the architect or engineer, only to have them left unsigned. The reasons for a failure on the part of the architect or engineer to sign the daywork records are not always clear. Sometimes it is due to forgetfulness, but it is often due to a disagreement with the information included in the daywork records.

11.7.2. The lack of properly authenticated and signed daywork records was the basis of dispute in the case of *JDM Accord* v. *The Secretary of State for the Environment, Food and Rural Affairs* (2004). During 2001, foot-and-mouth disease spread throughout the UK and, as a result, six million animals had to be slaughtered. A number of contractors, of whom JDM Accord was one, were appointed by the Secretary of State to slaughter the animals, prepare the burial sites for disposal of the animal carcasses and construct the necessary infrastructure to support the operation. The terms of the agreement between the parties provided for payment to be made on a time recorded basis. It was the intention for the Secretary of State to have a representative present on each of the sites to record the time taken by the operatives and plant. Time records were to be made out giving details of all the time spent, and these would then be signed each day by the representatives of both parties. Unfortunately, the Secretary of State did not have a representative on all the sites, so JDM Accord had nobody with whom they could agree the time they recorded.

11.7.3. When it came to payment, in the absence of daywork records signed by both parties, there was a refusal to make payment in full. It was considered by Judge Thornton that, as there had been no challenge to the time recorded when the timesheets were submitted, the onus was on the Secretary of State to prove that the information included on the timesheets was incorrect. Judge Thornton held that the timesheets submitted by JDM Accord should be allowed to stand without any further proof as to their accuracy, unless the Secretary of State could produce convincing evidence that the information on the timesheets was incorrect.

SUMMARY

The case of *JDM Accord* v. *The Secretary of State for the Environment, Food and Rural Affairs* (2004) makes it clear that if, having received daywork records from the contractor, the architect or engineer does not sign them, he is at risk of either having to prove at a later stage that the information included in the records is wrong, or certifying them for payment in full. If the architect or engineer challenges the information included in daywork records at the time they are submitted, it is essential for full details to be provided of the hours alleged to be inaccurate. If no agreement can be reached, then the dispute resolution procedure included in the contract may have to be implemented.

11.8. Can an architect/engineer sign a daywork sheet and then refuse to certify the sums involved for payment? Is a quantity surveyor entitled to reduce the hours included on a signed daywork sheet if he considers them unreasonable or excessive?

11.8.1. A daywork sheet is a record of time, equipment and materials employed with regard to a particular on-site operation which is described therein. A signature of the architect, engineer, clerk of works or any other employer's representative merely indicates that the hours recorded and materials listed have been employed in that operation; nothing more and nothing less. It does not indicate an intention to make payment, as the work as described on the daywork sheet may be catered for elsewhere in the contract. This applies whether or not the signature is accompanied by the words 'for record purposes only'.

11.8.2. Where, however, payment on a daywork basis is established and a daywork sheet has been signed, for example, by the architect, it is not for the quantity surveyor to reduce the hours on the grounds that he considers them excessive. If the hours are in any way incorrect, then the signature should not have been appended.

11.8.3. JCT 2011, clause 5.7, stipulates that when valuing a variation where work cannot be properly valued in accordance with the terms of the contract, then the variation shall be valued in accordance with the Definition of Prime Cost of Daywork Carried Out Under a Building Contract, issued by the RICS and Construction Federation, current at the Base Date. Clause 5.7 then goes on to state:

> Provided that in any case vouchers specifying the time daily spent upon the work, the workmen's names, the plant and the materials employed shall be delivered for verification to the Architect/Contract Administrator or his authorised representative not later than 7 Business Days after the work has been executed.

There is no reference to any input by the quantity surveyor and therefore, once a decision has been taken to pay on a daywork basis and sheets have been submitted and signed by the architect/contract administrator or his authorised representative, the quantity surveyor's role is of a very limited nature.

11.8.4. The ICE 6th and 7th Editions provide for payment on a daywork basis, if the engineer in his opinion considers it necessary or desirable. Where a quantity surveyor is involved, in like manner to a JCT contract, he has no power to alter hours which have been agreed by the engineer or his representative.

11.8.5. In the case of *Clusky* v. *Chamberlain* (1994), it was held by the Court of Appeal that the judge in the lower court was wrong to go behind the timesheets to establish an entitlement as to quantum. It was at no time suggested that the timesheets were fake. There is no entitlement to argue that the workmen did not work as expeditiously as they might.

SUMMARY

An architect or engineer may sign a daywork sheet 'for record purposes only' and then refuse to certify for payment the sum included therein. However, a quantity surveyor has no power to alter hours which he considers to be excessive on a signed daywork sheet.

11.9. Where a contractor/subcontractor includes an unrealistically low rate in the bills of quantities, can he be held to the rate if the quantities substantially increase?

11.9.1. Most quantity surveyors and engineers, if convinced that the contractor or subcontractor has included an unrealistically low rate in the bill of quantities, will insist upon the rate applying up to the quantity in the bill. Any excess in quantities over and above the bill quantity is often paid for at a fair and reasonable rate. This seems a rational and fair approach, but there are two legal cases that cast doubt upon it.

11.9.2. In the case of *Dudley Corporation* v. *Parsons & Morrin Ltd* (1959), a contract for the building of a school was let using the RIBA 1939 form, with quantities. The contract terms in issue were essentially the same as those of JCT 2011. The contractors priced an item for excavating 750 cubic yards in rock at £75, i.e. two shillings a cube. This was a gross underestimate, although it wasn't known whether or not the rock would be encountered. In carrying out the excavations shown on the drawings, the contractors excavated a total of 2,230 cubic yards of rock. The architect valued the work at two shillings a cube for 750 cubic yards, and the balance at £2 a cube. This rate was not unreasonable if no other price applied. The employer disputed this amount. The arbitrator found 'no sufficient evidence that the price of two shillings per yard cube was a mistake'. Lord Justice Pearce in the Court of Appeal, when deciding the matter related to the case, said:

> In my view, the actual financial result should not affect one's view of the construction of the words. Naturally, one sympathises with the contractor in the circumstances, but one must assume that he chose to take the risk of greatly under-pricing an item which might not arise, whereby he lowered the tender by £1425. He may well have thought it worth while to take that risk in order to increase his chances of securing the contract.

The judge did not appear to have considered the possibility that the low rate had been included in error.

11.9.3. The Court of Appeal, in *Henry Boot Construction* v. *Alstom Combined Cycle* (2000), gave a decision concerning the measurement and rating of sheet piling. Alstom was the employer for the construction of a combined cycle gas turbine power station at Connah's Quay. Boot carried out the civils work, employing an ICE 5th Edition contract. In post-tender negotiations, Boot submitted a price of £250,880 for additional and different temporary work in connection with the lowering of cold water pipework. This extra work formed part of the contract. In calculating this figure, Boot had made an error, resulting in the rate being much higher than it should have been. This error remained undetected during post-tender, pre-contract negotiations. A variation was issued to the Heat Recovery Steam Generator and Cooling Towers during the course of the work, involving additional sheet piling. Henry Boot, in pricing the sheet piling, extrapolated a rate from the price agreed for sheet piling which formed part of the £250,880 agreed in relation to lowering the cold water pipework. This rate, being higher than it should have been, produced an inflated rate for the varied work; 'windfall', to use the court's term. Alstom argued that clause 55(1)(b) of the contract provided for the use of contract rates 'as far as may be reasonable'. It was unreasonable, they said, to use a rate which was obviously included in error. Further, they argued that, as the sheet piling in the pre-contract agreed price was part of a lump sum, it wasn't reasonable to extrapolate a rate for sheet piling from this amount.

The Court of Appeal agreed with the judge in the lower court in finding in favour of Henry Boot. The reference in clause 55(1)(b) to the use of the rate being reasonable meant, not that the rate itself had to be reasonable, but that the process of applying the rate must be reasonable. The parties had agreed the rate and they were stuck with it; the mistake was irrelevant. The Court of Appeal referred the matter back to the arbitrator for him to decide whether it was reasonable to extrapolate the rate from the £250,880 lump sum, in spite of the conclusion that there was insufficient information to show how Henry Boot had calculated the sum in the first place. The process has thus gone full circle, starting with an appeal from the arbitrator's decision to the TCC, on to the Court of Appeal and back to the arbitrator.

11.9.4. Both of these cases indicate that where errors occur in contract rates, the rate will not be altered merely because of a substantial change in quantity.

11.9.5. Where a priced bill of quantities has been vetted by a quantity surveyor or engineer, any errors which are discovered should be drawn to the contractor's attention. Any failure to do so could result in the contractor becoming entitled to an adjustment of the rate.

11.9.6. A pricing mechanism issued by the contractor resulted in *Aldi Stores Ltd* v. *Galliford* (2000) coming before the courts. A contract was let for the construction of a store, employing the JCT IFC 84 conditions of contract. The price included for disposing of contaminated material to a licensed tip at a rate of £44.60 and the disposal of clean material at a rate of £8.50. During negotiations, the contractor agreed to absorb the rates for the disposal of both contaminated and clean material into the overall price for the construction of the store. The rates for the disposal of both contaminated and clean material were thus nil. When work got under way, it transpired that all the materials were contaminated and had to be disposed of at the licensed tip. The contractor claimed that this constituted a variation. No agreement was reached and the matter was referred to arbitration. The arbitrator held that there had been an error in the bill of quantities and that the contractor was entitled to an architect's variation. A rate of £36.10 was awarded, being the difference between the rate for contaminated material and the rate for clean material. The employer appealed. It was the decision of the court that the arbitrator was wrong. There had been no change in the conditions under which the work had been carried out, nor a significant change in the quantity of the work. The work had simply been moved from one category to another, i.e. from uncontaminated material to contaminated material. As the rate for both categories was nil, there was no justification in amending the price.

SUMMARY

There is legal authority for what should happen where an unrealistically low or high rate is included in a bill of quantities and the billed quantity is the subject of a substantial increase. The rate included in the bill of quantities is to apply, whether an unrealistically low or high rate applies.

11.10. Can a debtor enforce acceptance of a lesser sum in full and final settlement?

11.10.1. Pressure is often applied by a debtor trying to reduce the amount he owes by making an offer to pay a sum which is less than the amount due. The methods involved vary, from requiring the creditor to sign an agreement 'in full and final settlement', made out in a lesser amount than the amount owed, to sending a cheque for the lesser amount, stated to be 'in full and final settlement'.

11.10.2. It is common procedure for contractors and subcontractors to be expected to sign a statement to the effect that they agree to payment of the balance due on the final account and claim on a 'full and final settlement' basis. Frequently, the contractor or subcontractor is informed that payment will not be made unless the form is signed. The contractor or subcontractor often signs, as he needs the money, even though he considers a greater entitlement is due.

11.10.3. Where one party, i.e. the contractor or subcontractor, agrees to a sum which is considerably less than his due entitlement, there must be consideration for the agreement to take a lesser sum. In other words, the contractor or subcontractor must have received some benefit. For example, if the payment were to be made earlier than contractually required, then early payment would amount to consideration for payment of the lesser sum. If the early payment is accepted, the contractor or subcontractor would then have no claim to the balance. This is sometimes referred to as 'accord and satisfaction'. If an agreement is purportedly made whereby the contractor or subcontractor agrees to forego part of his entitlement in the absence of any such consideration, the agreement will not be binding.

11.10.4. It may be, however, that the final balance offered includes a sum which is disputed. The contractor considers the amount is less than he is entitled to, but the employer believes he is paying the correct amount, or more than is due. *Hudson's Building and Engineering Contracts*, 10th edition, at page 22, states:

> . . . consideration may be present in such a case as some bona fide dispute exists and a claim is given up in return for the promise to accept less.

This being the case, the agreement will be binding.

11.10.5. In the case of *D & C Builders Ltd* v. *Rees* (1965), an employer advised decorators who had undertaken work on her behalf and submitted an account in the sum of £482 13s 6d, that unless they agreed to accept a sum of £300 in full settlement she would pay them nothing at all. The decorators, who were in desperate need of the money, signed a written document agreeing to accept the reduced payment in full satisfaction of their claim. Later, they sued for the full amount. It was held by the Court of Appeal there was no true 'accord and satisfaction', as the plaintiffs had acted as a result of a threat which was without any justification and there was no consideration present.

11.10.6. In *Newton Moor Construction Ltd* v. *Charlton* (1981) the plaintiff contractor undertook some work for the defendant for a sum of £11,020. After the work was completed, the plaintiff sent in an invoice for £18,612. There had been some variations to the work, some by agreement and some not. As a consequence, completion was delayed and the defendant claimed entitlement to a set-off. He recalculated the amount he considered to be due, partly on agreed prices and partly on what he thought was a reasonable basis for the additions, deductions and delay, arriving at a figure of £8,847. A cheque for this amount was sent to the plaintiff's solicitors, with an accompanying letter saying that this was to be regarded as being in 'full and final settlement' of any outstanding sums.

The plaintiff's solicitors responded that they were only accepting it in part payment and issued a writ for the balance. The defendant maintained that the acceptance of the cheque amounted to accord and satisfaction.

It was held by the court that it didn't matter whether the sum was termed a compromise, or an accord and satisfaction; the letter, the cheque and the plaintiff's acceptance of it did not amount to accord and satisfaction without reciprocal benefit to the plaintiff.

11.10.7. In *Stour Valley Builders* v. *Stuart* (1993), it was held that a cheque sent in full and final settlement is not conclusive evidence of accord and satisfaction. The recipient of the cheque wrote indicating that he did not consider the payment as full and final, banked the cheque and successfully sued for more of the balance. The banking of a cheque sent in full and final settlement does not stop the creditor from commencing an action for a balance claimed to be due: *Auriema Ltd* v. *Haig and Ringrose* (1988).

11.10.8. The case of *Clarke* v. *Nationwide Building Society* (1998) also deals with a cheque sent in full and final settlement. The plaintiff wished to purchase a property and sought an advance from the defendant. A survey was carried out on instructions of the defendant, the cost of which formed part of the mortgage application fee. After the plaintiff moved in, damp appeared. The solicitor acting for the defendant was advised that the damp had been caused by condensation, a problem which could not have been evident when the survey had been carried out. The defendant considered there was no liability, but as a goodwill gesture wrote offering to refund the mortgage application fee. The offer was verbally rejected by the plaintiff, but nonetheless the defendant refunded the cheque for the mortgage fee 'in full and final settlement of the matter'. The cheque was presented and cleared. Nearly two years later, the plaintiff commenced an action against the defendant for negligence. The defendant's case was that the acceptance of the cheque amounted to accord and satisfaction which, in effect, put an end to the matter. It was held by the Court of Appeal that the wording of the defendant's letter did not amount to an offer capable of being accepted. Sending a cheque as a goodwill gesture in the hope of avoiding further complaint was not the same thing as offering a payment in settlement of all claims. The Court of Appeal found in favour of the plaintiff, who was allowed to continue the action.

11.10.9. Another case involving a cheque sent in full and final settlement is *Bracken* v. *Billinghurst* (2003). A dispute arose as a result of construction work carried out on the property of Mr Bracken and Ms Trickett. The dispute was referred to adjudication, which found in favour of Mr Bracken in the sum of £43,984. No payment was forthcoming and, prior to the court action, Mr Bracken rather surprisingly offered to settle for a payment of £6,000. Billinghurst, however, sent off a cheque for only £5,000 'on the strict understanding that the sum is offered to you in full and final settlement'. The letter went on to state that the settlement 'will be deemed to have been accepted by you, and therefore contractually binding if it is presented to the bank and cleared for payment'. The cheque was paid into the bank and cleared. Within a week, Mr Bracken withdrew his offer of £6,000 and demanded £38,984, being the original adjudication decision less £5,000 paid. Judge Wilcox followed the line of argument used in the *Stour Valley* case, that 'cashing a cheque is always strong evidence of acceptance (of a settlement offer) especially if it is not accompanied by immediate rejection'. The judge considered that one of the tests was, whether the words and conduct of the creditor would cause the debtor, as a reasonable person, to believe the matter was final. Taking into account all the circumstances, the court held that the payment of £5,000 was full and final.

11.10.10. In its decision in *Hurst Stores & Interiors Ltd* v. *ML Property Ltd* (2004), the Court of Appeal criticised construction managers Mace for allowing a subcontractor to enter into a 'full and final settlement' agreement with the client without the subcontractor understanding the implications. The subcontractor had not known that the settlement payment included variations, whilst Mace did. The court rectified the agreement for unilateral mistake, finding that Mace had acted unconscionably. The court also found that the project manager acting for Hurst did not possess the necessary authority to enter into a full and final settlement agreement.

SUMMARY

An agreement showing a balance due on a final account which is less than the contractor's proper entitlement will only be binding as accord and satisfaction provided the contractor receives some benefit from the agreement, such as payment earlier than otherwise would be the case. If no benefit is derived, then the agreement will not be binding. However, where a bona fide dispute exists and a delay or damages claim or part of a claim is given up by the employer in return for a promise to accept less on the part of the contractor, the agreement may be binding. When a cheque is sent in full and final settlement and is banked, it is advisable for the creditor to make it clear that the cheque is accepted as a payment on account.

11.11. How can subcontractors avoid 'pay when paid' clauses?

11.11.1. Contractors, ever ready to pass risk down the line to subcontractors when employing non-standard conditions, usually include a clause making payment to themselves a condition precedent to payment to the subcontractors. This type of clause is often referred to as 'pay when paid'.

11.11.2. The law concerning 'pay when paid' clauses has become subject to statutory control as a result of the Housing Grants, Construction and Regeneration Act 1996, which applies to contracts entered into after 1 May 1998. The Act seeks to outlaw 'pay when paid' clauses, except in respect of insolvency on the part of the original paying party. Section 113(1), which could have been drafted with phraseology of a more digestible nature, states:

> A provision purporting to make payment under a construction contract conditional on the payer receiving payment from a third person is ineffective, unless that third person, or any other person payment by whom is under the contract (directly or indirectly) a condition of payment by that third person, is insolvent.

This provision has not been amended by the Local Democracy, Economic Development and Construction Act 2008, despite aggressive lobbying by subcontracting trade associations. It was felt that subcontractors should not be expected to bear the risk of insolvency on the part of the employer, this being an exception to the general rule which outlaws 'pay when paid' clauses.

11.11.3. The standard forms of subcontract in general use provide for payment to the subcontractor in accordance with the terms of the subcontract, regardless of whether the main contractor has received payment from the employer. For example, the JCT Standard Building Subcontract provides for the main contractor to pay the subcontractor by the final date for payment, which is 21 days after the date on which the payment becomes due. The due date for payment is the date for issue of the interim certificate under the main contract, which is normally issued monthly.

11.11.4. There have been several legal cases where the courts have tried to unravel the meaning of wording in some subcontract terms which purport to be 'pay when paid' clauses. It has been the decision of the courts in some of these cases that wording included in subcontracts which contractors considered enabled them to withhold payment to subcontractors, on the grounds that they had not been paid by the employer, failed to be enforceable. These cases will still be relevant in respect of contracts which give the contractor a right to refuse payment because of the insolvency of the employer, which may occur where a contractor amends a standard form or employs bespoke subcontract conditions. The Construction Act only applies to construction contracts, as defined by the Act and, therefore, these cases are still relevant to non-construction contracts.

11.11.5. The New Zealand case of *Smith & Smith Glass Ltd* v. *Winstone Architectural Cladding Systems Ltd* (1991) throws an interesting light on the problem. The appropriate wording in the sub-subcontract was:

> We will endeavour (this is not to be considered as a guarantee) to pay these claims within 5 days after payment to Winstone Architectural Ltd of monies claimed on behalf of the subcontractor.

The court drew a distinction between an 'if' clause, i.e. if we are not paid you will not receive payment, and a 'when' clause, i.e. we will pay you when we have been paid. An 'if' clause makes it plain that payment will only be made after payment has been received. The 'when' clause was considered by the courts to indicate the time for payment only and that payment 'up the line' was not a condition of payment 'down the line'. In the case in question, the payment clause was considered by the court to be a 'when' clause and therefore non-payment of Winstone by Angus was no excuse for non-payment of Smith and Smith by Winstone. Master Towle, the judge, considered that unless the clause spells out in clear and precise terms that payment will not be made until payment is received, the clause does no more than indicate the time for payment. In arriving at his decision, Master Towle had this to say:

> While I accept that in certain cases it may be possible for persons contracting with each other in relation to a major building contract to include in their agreement clear and unambiguous conditions which have to be fulfilled before a subcontractor has the right to be paid, any such agreement would have to make it clear beyond doubt that the arrangement was to

> be conditional and not to be merely governing the time for payment. I believe that the *contra proferentem* principle would apply to such clauses and that he who seeks to rely upon such a clause to show that there was a condition precedent before liability to pay arose at all should show that the clause relied upon contains no ambiguity . . .
>
> For myself, I believe that unless the condition precedent is spelled out in clear and precise terms and accepted by both parties, then clauses such as the two particular ones identified in these proceedings do no more than identify the time at which certain things are required to be done and should not be extended into the 'if' category to prevent a subcontractor who has done work from being paid, merely because the party with whom he contracts has not been paid by some one higher up the chain.

11.11.6. In the case of *Durabella* v. *J. Jarvis and Son Ltd* (2001), a dispute arose over a 'pay when paid' clause. Jarvis was the main contractor on a project for the construction of a number of flats in Limehouse, London, for a developer. Durabella supplied and laid timber flooring as a subcontractor to Jarvis. The subcontract conditions included a 'pay when paid' clause which read:

> Our liability for payment to you is limited to such amounts as we ourselves receive from the employer in respect of your works under this order.

This is not so much a 'pay when paid' clause, but a limitation placed upon the sum which will be paid. The developer was disappointed with the performance of Jarvis and terminated their contract. As is commonplace when a termination takes place, a dispute arose over money. The dispute was not resolved and the matter was referred to court. The proceedings were settled and a compromise agreement drafted. This agreement, in terms which are commonplace, stipulated that the developer was to pay Jarvis a sum of money, in this case £550,000 in full and final settlement of all claims arising out of the development. The agreement, however, included a rather unusual term, to the effect that no value was included in the deal for works undertaken by Durabella. Disputes over payment between Jarvis and Durabella had been ongoing. Once the dispute with the developer was settled, Durabella, no doubt, thought that their dispute with Jarvis would be resolved. Jarvis, however, drew attention to the payment clause and the wording in the settlement agreement. The result was that Jarvis refused to offer any payment for the flooring work. Durabella took their grievance to court. Their first argument, that the clause was unreasonable, left the court unmoved. In principle, there is nothing to prevent a contractor passing on to subcontractors the risk of non-payment by the client, subject to the provisions of the Construction Act. The court, however, found in favour of the subcontractor. Despite the wording in the agreement, to the effect that no value had been included for work undertaken by Durabella, the court was not convinced. It was considered that the wording was a smokescreen, included at the behest of Jarvis, to avoid paying Durabella. The court considered that Jarvis had every opportunity to ascertain how the £550,000 was made up and failed to do so. Jarvis, therefore, had no valid reason for refusing to make payment to Durabella.

11.11.7. The exception to the status of 'pay when paid' clauses under the Act, which applies when the employer becomes insolvent, was the subject matter of the decision in *Hills Electrical and Mechanical plc* v. *Dawn Construction Ltd* (2003). Hills, a subcontractor, entered into a contract with Dawn which included a 'pay when paid' clause. Part way through the subcontract work, the employer became insolvent. Prior to the insolvency the 'pay when paid' clause was inoperable. However, in accordance with the Act, contractors can avoid paying subcontractors by operating a 'pay when paid' clause where the employer has become insolvent. The payment clause in the subcontract provided a final date for payment of 28 days after the due payment date. Had the Scheme for Construction Contracts applied, this period would have been 17 days and Hills would have been paid. Unfortunately for Hills, the insolvency took place between the 17 days and 28 days. They argued that the subcontract failed to provide dates by which the subcontractor should make application for payment. This being the case, the whole of the payment provision would cease to apply and therefore the Scheme would become operative. The court did not agree, as it considered that the Scheme could be used in isolated cases of non-compliance by the subcontract conditions to fill the gaps, without having to import major sections.

11.11.8. In the case of *William Hare Ltd* v *Shepherd Construction Ltd* (2010), the wording of the 'pay when paid' clause entitled Shepherd to avoid paying William Hare if non-payment to Shepherd was the result of the employer becoming insolvent on the making of an administration order under Part 11 of the Insolvency Act 1986. Unfortunately for Shepherd, the employer did not go into administration by reference to the Insolvency Act 1986, but by self-certification under the Enterprise Act 2002. This appears to be a shortcoming in the wording of the clause in the subcontract, which was drawn up after the Enterprise Act 2002 came into force. It was argued on behalf of Shepherd that the task of the court was to give effect to the intention of the parties and not the literal words. The court disagreed and found in favour of William Hare.

SUMMARY

Following the implementation of the 1996 Act, contractors' 'pay when paid' clauses became limited in their effect to situations where the employer has become insolvent. The standard forms of contract have abandoned entirely 'pay when paid'. Some contractors may decide to amend the standard forms or to use bespoke conditions that make it clear that 'pay when paid' still applies when the employer is insolvent. In New Zealand the courts have differentiated between 'pay when paid' clauses referred to as 'if' clauses and those termed 'when' clauses. A 'when' clause indicates merely the time for payment whereas an 'if' clause makes it plain that payment will only be made after payment has been received. The *William Hare* case highlighted a shortcoming in the wording of a 'pay when paid' clause. From a contractor's point of view, if 'pay when paid' is to be used relating to the insolvency of the employer, very clear wording is required. The Construction Act only applies to construction contracts as defined by the Act, and hence 'pay when paid' clauses are still likely to appear in non-construction contracts.

11.12. Once the value of a contractor/subcontractor's work has been certified and paid, can it be devalued in a later certificate?

11.12.1. Contractors and subcontractors rely upon interim certificates and payments to keep their businesses going. Major difficulties can arise where work certified and paid for in the early part of a contract is later revalued at a lower price and an adjustment made in a subsequent certificate. The contractor who has already paid a subcontractor based upon the earlier certificate may have difficulty in recovering the overpayment. Legal textbooks and decisions of the courts make it clear that interim certificates and payments are in fact payments on account of the final sum which is due. Amounts which have been certified and paid may be the subject of a reduced valuation in a later certificate, which in essence amounts to a clawback of sums already paid.

11.12.2. *Hudson's Building and Engineering Contracts*, 11th edition, at paragraph 6.187, says:

> As a rule, the payments contemplated by such provisions only represent the approximate value (or a proportion of it) of the work done, and possibly also of materials delivered to the site, at the date of payment, and, in the absence of express provision, they are not conclusive or binding on the Employer as an expression of satisfaction with the quality of the work or materials. It makes no difference that they are frequently expressed to represent the value of work properly done, since such a qualification is an obvious one in any provision for payment on account.

It also has this to say, at paragraph 5.007:

> Even though a building owner may have accepted the work so that a liability to pay the price of it arises, that will not (in the absence of a provision in the contract making the acceptance binding on the Employer) prevent the building owner from showing that the work is incomplete or badly done; he may either counterclaim or set off damages in an action by the builder, or he may pay or suffer judgment to be obtained against him for the full price and later bring a separate action for his damages . . .

and again, at paragraph 8.116:

> Moreover, it should not be forgotten that waiver of a breach, or a renunciation of the right to damages, or a liability to pay for the work, will not, in general, and in the absence of express provision, be implied from acceptance of the work by the building owner or his Architect, even in the case of patently defective work.

11.12.3. In *Fairclough Building* v. *Rhuddlan Borough Council* (1985), a nominated subcontractor, Gunite, had its employment determined. The architect issued a renomination, requiring the work to be finished off by Mulcaster. Gunite had been paid for work which was subsequently shown to be defective. In a subsequent certificate, the amount certified in favour of Gunite was reduced, by which time Gunite was insolvent. It was held by the court that the employer was entitled to be credited with the reduced value of work carried out by Gunite. Unfortunately, the main contractor had paid Gunite before it became insolvent. The loss therefore fell upon the main contractor's shoulders.

11.12.4. An exception to this rule applies under the ICE 6th and 7th Editions, where payment has been certified and paid to a nominated subcontractor. Clause 60(8)(a) and (b) states that the engineer will not delete or reduce any sum previously certified and paid to a nominated subcontractor. A further exception seems to occur in clause 2.3.3 of JCT 2011, which states:

> In respect of any materials, goods and workmanship, as comprised in executed work, which are to be to the reasonable satisfaction of the Architect in accordance with clause 2.1, the Architect shall express dissatisfaction within a reasonable time from the execution of the unsatisfactory work.

Where this clause applies, architect and employer may have some difficulty if work is certified and paid for and, much later, they try to reduce the valuation of the work certified on the grounds that it does not meet the architect's satisfaction. Whether it will be possible to reduce the valuation of work once certified will depend upon the circumstances, but this will put the architect under pressure to reject substandard work early, otherwise the contractor could argue that the architect can only reject work if the contract requires it to be to his satisfaction, if he does so within a reasonable time.

11.12.5. The Engineering and Construction Contract (NEC 3), under clause 50.5, provides for the project manager to correct any wrongly assessed amount in a later certificate.

SUMMARY

Payment of an interim certificate represents a payment on account of the final sum due. It is always open to the architect or engineer to certify a sum and later reduce the amount certified in respect of work executed. There are exceptions; for example, sums certified and paid in respect of nominated subcontractors' work under ICE 6th and 7th Editions cannot be reduced in a later certificate. There must, however, be specific wording in the contract which prevents the reduction in later certificates of sums which have been certified in earlier certificates.

11.13. Can a contractor deduct claims for overpayments levied on one contract from monies due on another in respect of a subcontractor's work?

11.13.1. It is quite common for a contractor who has an indisputable debt due to a subcontractor to refuse to make payment on the grounds that legitimate claims have been levied, or overpayments have been made, on another contract or contracts which are greater than the sum due. Hence no payment is made.

11.13.2. Where money is owed to a company which is insolvent, claims against the insolvent company can be used to reduce or eliminate the amount of money owed in compliance with the provisions of the Insolvency Act 1986. This is sometimes referred to as statutory set-off or mutual dealings. In the case of *Bouygues (UK) Ltd* v *Dahl-Jensen* (2000), an application was made to the court for the enforcement of an adjudicator's decision. It was refused, on the grounds that accounting between the main contractor and subcontractor should be undertaken pursuant to the Insolvency Rules 1986.

11.13.3. If both the creditor and debtor are solvent, money owed to the creditor can be reduced in respect of debts which are due and payable, applying what is known as legal set-off. Legal set-off is not applicable if the amount which is the subject of a set-off is in dispute.

11.13.4. Where the creditor is solvent and there is no facility for a legal set off, then the debtor, in an attempt to avoid settlement altogether, or at least reduce the sum to be paid, may consider equitable set-off. This situation arose in the Court of Appeal case of *B. Hargreaves Ltd* v. *Action 2000* (1992). In this case, 12 subcontracts were entered into between the same main contractor and subcontractor. The subcontractor was due to be paid £104,160 in accordance with three interim certificates. No sum was paid, as the contractor argued that on some of the contracts overpayments had been made which exceeded the amount of the certificates. Therefore, it said, it was entitled in equity to set off. The court held that, for the contractor to set off the overpayments, the other contracts must be closely and inseparably connected to the one where the certificates were due. In giving judgment, Judge Fox-Andrews QC stated:

> The law was restated by the Court of Appeal in *Dole Dried Fruit & Nut Co Ltd* v. *Trustin Kerwood* (1990). In his judgment Lord Justice Lloyd, with which Lord Justice Beldam agreed, said:
>
>> But for all ordinary purposes, the modern law of equitable set-off is to be taken as accurately stated by the Court of Appeal in *Hanak* v *Green* (1958). It is not enough that the counter-claim is 'in some way related to the transaction which gives rise to the claim'. It must be 'so closely connected with the plaintiff's demand that it would be manifestly unjust to allow him to enforce payment without taking into account the cross claim'.
>
> The other two subcontracts were made between the same parties, on the same day with the same building owner as the Irlam subcontract. Each of these three subcontracts related to the construction of a petrol station. But Action was under no contractual obligation to Petrofina to have entered into each of these three subcontracts with Hargreaves. Taking all circumstances into account the equitable set-off plea fails.

11.13.5. A second argument used by the main contractor was that, as the overpayments amounted to debts, they had a legal right to set-off. The court held that a claim of mutual debts was only available when the claims on both sides were in respect of liquidated debts or money demands which were readily and without difficulty ascertained. The figures sought to be set off were assessed by Action's surveyor and it was not suggested that Hargreaves accepted the figure. Without a full hearing the amount due, if any, could not be ascertained.

11.13.6. The Court of Appeal, in the case of *Geldof* v *Simon Carves Ltd* (2009), had to decide whether equitable set-off applied to two contracts, one for the supply of pressure vessels and the other for the installation of storage tanks. The Court, in overturning the lower court's decision, followed the ruling in *Hanak* v *Green* (1958). It was held that equitable set-off is allowed where the cross-claims are so closely connected with the claimants' demands for payment that it would be manifestly unjust to allow it to enforce payment without taking into account the cross-claim. The appeal was allowed, as a result of Geldof bringing the two contracts into an intimate relationship with each other. This occurred because Geldof made payment under the supply contract a condition precedent for completing the installation on the other contract. Both contracts related to the same bioethanol plant; the pressure vessels were of no use unless the storage tanks had been properly installed.

11.13.7. Contractors often overcome the difficulties of setting off overpayments and claims arising on one contract from monies due on another by including an express right in the terms of the subcontract. The main contract GC/Works/1 gives the employer such rights of set-off where, in condition 51, it states, in rather long winded wording:

> Without prejudice and in addition to any other rights and remedies of the Employer, whenever under or in respect of the Contract, or under or in respect of any other Contract between the Contractor or any other member of the Contractor's Group and the Employer or any other member of the Employer's Group, any sum of money shall be recoverable from or payable by the contractor or any other member of the Contractor's Group by or to the Employer or any other member of the Employer's Group, it may be deducted by the Employer from any sum or sums then due or which at any time thereafter may become due to the Contractor or any other member of the Contractor's Group under or in respect of the Contract, or under or in respect of any other contract between the contractor or any other member of the Contractor's Group and the Employer or any other member of the Employer's Group. Without prejudice and in addition to any other rights and remedies of the Contractor, each member of the Contractor's Group shall have rights reciprocal to those of each member of the Employer's Group under this Condition.

SUMMARY

It would seem that, for a contractor to be able to set off from monies due on one contract overpayments or claims on another, he must be able to demonstrate one or more of the following:

- An equitable right, on the basis that the contracts in question are closely and inseparably connected (for example, separate phases of a development which have been let as separate contracts).
- The insolvency of the party to whom money is owed.
- A legal right to set off mutual debts, which must be liquidated and not contested (an unlikely contingency in the field of contested claims).
- A clause in the subcontract which gives the contractor the necessary powers of set-off.

11.14. When a contractor completes significantly early, may the architect/engineer legitimately delay certification to match the employer's ability to pay from available cashflow?

11.14.1. Most standard forms provide for the contractor completing early.

- JCT 2011, by clause 2.4, states that the contractor 'shall begin the construction of the Works or Section and regularly and diligently proceed with the completion of the same on or before the relevant Completion Date'.

- ICE 6th and 7th Editions, by clause 43, require the contractor to complete the work 'within the time as stated in the appendix'.
- The Engineering and Construction Contract (NEC 3), by clause 30.1, states that 'completion is on or before the completion date'.
- GC/Works/1, in condition 34(1), refers to completion 'by the Date or Dates for Completion'.

It is clear from all these clauses that the contractor is entitled to finish early.

11.14.2. With regard to the issue of the completion certificate, the architect or engineer has no option but to issue one once the contractor has finished.

- ICE 6th and 7th Editions, by clause 48, require the engineer to issue a certificate of substantial completion when the works are substantially completed.
- JCT 2011, in clause 2.30, provides for the architect or contract administrator to issue a certificate of practical completion when, in his opinion, practical completion for the purposes of the contract has taken place.
- GC/Works/1, condition 39, states: 'The PM shall certify the date when the Works or any Section . . . are completed'.
- The Engineering and Construction Contract (NEC 3), by clause 30.2, states 'The PM certifies completion within one week of completion'.

11.14.3. With regard to payment, the standard forms in regular use normally provide for certification and payment of the value of the works executed on a monthly basis. Certificates and payments must therefore be made on this basis, whether or not the contractor is ahead or behind programme and even if he is likely to finish early or has indeed already completed. For example:

- JCT 2011, clause 4.9.2;
- ICE 6th and 7th Editions, clause 48;
- GC/Works/1, condition 47;
- Engineering and Construction Contract (NEC 3), clause 50.

All provide for certification and payment of work properly carried out at the date of the certificate or other method of stating the value of work. Some standard forms, such as GC/Works/1 and JCT 2011 Design and Build, provide for milestone or stage payments. None of these provisions relate in any way to the employer's ability to pay. The advantage of provisions for milestone or stage payments is that the employer is able to manage his cashflow in a more structured manner. FIDIC 4th Edition, in clause 14.3, attempts to address this problem by requiring the contractor to produce a cashflow forecast. This allows the employer to plan his cashflow in advance, on the assumption that a programme is produced by the contractor to complement the stage payment schedule and that the contractor adheres to his programme.

SUMMARY

Most standard forms of contract provide for completion of work on or before the completion date written into the contract. The payment provisions require certification and payment of work properly executed. There is no facility for the architect or engineer to reduce the value of amounts to be certified or delay certification to suit the employer's ability to pay.

11.15. Where an architect/engineer undercertifies, is the contractor/subcontractor entitled to claim interest?

11.15.1. Contractors frequently allege that engineers and architects deliberately undervalue work in interim certificates. The tendency has been to ensure that contractors are not overpaid where there is a recession in the construction industry, which may result in many contractors becoming insolvent. When undercertification occurs, there will often be a catching-up of payments after practical completion, with contractors being paid substantial sums long after the work has been completed. The amounts involved can be much greater where genuine disputes occur which are ultimately resolved. Contractors often claim interest on these late certifications, but rarely receive payment.

11.15.2. A great deal of the law as it applies in this country is judge-made. When judges fail to agree, confusion and additional cost is inevitably the result. Take, for example, the alternative ways in which the courts have interpreted clause 60(6) of the ICE 5th Edition. This clause states:

> In the event of failure by the Engineer to certify or the Employer to make payment in accordance with sub-clauses (2), (3) and (5) of this clause the Employer shall pay to the contractor interest on any payment overdue.

11.15.3. Different judges in recent times have placed more than one interpretation as to what is meant by 'failure by the Engineer to certify'. In the Scottish case of *Nash Dredging* v. *Kestrel Maritime* (1987), Lord Ross stated:

> Accordingly if it appeared at the end of the day that the sum certified by the engineer was less than ought to have been certified in my opinion the engineer could not be said to have failed to have certified, provided that it had been his honest opinion that the sum certified by him was the amount then due.

In other words, there would be no failure to certify if the engineer, for example, issued a certificate concerning, say, a claim under clause 12 and, having given the matter further consideration, later increased the amount certified, provided he acted in good faith. This decision was clear and free from ambiguity.

11.15.4. The second case to deal with an interpretation of clause 60(6) is *Hall and Tawse* v. *Strathclyde Regional Council* (1990), another Scottish decision. In this case the judge followed *Nash Dredging*, using the following wording:

> I agree with Lord Ross that there would not be a failure on the part of the engineer to certify merely because the sum certified turned out to be less than the sum which the court or arbiter thought was due.

The judge went on to suggest a further situation which may call for interpretation under clause 60(6), but declined to express a view:

> It is not necessary for present purposes to consider whether there would be a failure on his [the engineer's] part if he had proceeded on an interpretation of the contract or some other point of law relating to matters on which his opinion is required in order to provide a certificate which is later found to be erroneous.

11.15.5. A contrasting view was taken in England, when an arbitration award was the subject of an appeal in *Morgan Grenfell Ltd and Sunderland Borough Council* v. *Seven Seas Dredging Ltd* (1990). The matter was referred to Judge Newey. His view was:

> If the engineer certifies an amount which is less than it should have been, the contractor is deprived of money on which he could have earned money . . . If the arbitrator revises his [the Engineer's] certificate so as to increase the amount, it follows that the engineer has failed to certify the right amount.

Judge Newey upheld the decision of the arbitrator in holding that interest would be payable under clause 60(6) if the engineer, acting in a *bona fide* manner, undercertified. This is in stark contrast to the Scottish decisions, which deprived the contractor of a right to interest if the engineer acts honestly. There was no appeal against Judge Newey's decision, which must have sent shockwaves through establishments employing civil engineering contractors in England and Wales.

11.15.6. The Commercial Court differed from Judge Newey's decision in *The Secretary of State for Transport* v. *Birse Farr Joint Venture* (1992). The case arose out of a contract to construct part of the M25, which involved the use of the ICE 5th Edition. The arbitrator found in favour of the contractor with regard to interest, which was to be computed from a date three months after each valuation date. An appeal was lodged by the Secretary of State for Transport. Mr Justice Hobhouse in the Commercial Court, impressed by the views expressed by Mr Justice Buckley in the case of *Farr* v. *Ministry of Transport* (1960), commented as follows:

> A distinction clearly emerges from this case between the issue of a certificate which *bona fide* assesses the value of the work done at a lower figure than that claimed by the contractor and a certificate which, because it adopts some mistaken principle or some errors of law, presumably in relation to the correct understanding of the contract between the parties, produces an under-certification.

Mr Justice Hobhouse, in finding in favour of the Secretary of State for Transport, said that interest under clause 60(6) would only be due where the engineer under-certifies due to some mistaken principle, or some error of law. A certificate which *bona fide* assesses the value of the work done at a lower figure than is due to the contractor and which does not involve a contractual error or misconduct of the engineer will not rank for interest under clause 60(6). This shows a marked difference from what was said by Judge Newey in the *Morgan Grenfell* case, when allowing interest on undercertification, even where the engineer acted honestly without making any contractual errors.

11.15.7. Another case on the subject is *Kingston upon Thames* v. *Amec Civil Engineering Ltd* (1993) where the court, on hearing an appeal from an arbitration, followed the line of the decision in *Birse* v *Farr*.

11.15.8. In *BP Chemicals* v. *Kingdom Engineering* (1994), the ICE 5th Edition applied, with clause 60(6) deleted. It was held that the arbitrator could only award interest from the date of his award.

11.15.9. The ICE 6th and 7th Editions, in clause 60(7), have expanded the wording of clause 60(6) of the 5th Edition, making it clear that interest is payable on undercertification and thus following the *Morgan Grenfell* decision. The clause 60(7) wording states:

> In the event of
>
> (a) failure by the Engineer to certify or the Employer to make payment in accordance with sub-clauses (2) (4) or (6) of this Clause or
> (b) any finding of an arbitrator to such effect
>
> the Employer shall pay to the Contractor interest compounded monthly for each day on which any payment is overdue, or which should have been certified and paid at a rate equivalent to 2% per annum above the base lending rate of the bank specified in the Appendix to the Form of Tender.
>
> If, in an arbitration pursuant to Clause 66, the arbitrator holds that any sum or additional sum should have been certified by a particular date in accordance with the aforementioned sub-clauses but was not so certified, this shall be regarded for the purposes of this sub-clause as a failure to certify such sum or additional sum.

11.15.10. In the case of *Amec Building* v. *Cadmus Investment Co Ltd* (1996), a contractor's claim arising out of a JCT form of contract included interest for undercertification. The arbitrator didn't follow the decision in *BP Chemicals* v. *Kingdom Engineering* (1994). He awarded interest from the date of undercertification and the employer appealed. In finding for the contractor, the judge said:

> I respectfully concur in the reasoning of the authors of the *Building Law Reports* at page 116E, that the party seeking to review a certificate had a cause of action at the date of the undercertification and not, as Judge Harvey held, only upon the publication of the award of the arbitrator. I accept the argument that if monies previously unpaid to a contractor are subsequently found to be due by reason of the determination in the arbitration, that an award of interest should be made to compensate the contractor for the period during which such monies have been withheld from him. It would also remove the benefit, unjustly obtained as

a result of the arbitrator's award, which would accrue to the employer by withholding sums which were properly due to the contractor. For all these reasons it seems to me that the arbitrator's award of simple interest in this case was perfectly proper and I therefore dismiss the head of appeal.

11.15.11. Interest claimed in respect of money set off during a period of insolvency, was the subject matter in the case of *Charles Brand Ltd* v. *Orkney Islands Council* (2001). The contract, under which the ICE 5th Edition conditions applied, was for various works relating to new ferry terminals at Loth and Rapness in the Orkney Isles. A sum of £280,880 had been certified by the engineer in three certificates, but not paid by the employer. It was the employer's case that it was owed money by the contractor in respect of a contract to supply stone and other services. The employer argued that it was entitled to a set-off as the contractor was insolvent. Subsequently, the contractor received an injection of working capital from its parent company. It was agreed by all concerned that, as the contractor was no longer insolvent, the money which had been set off should be, and was in fact, paid. The contractor lodged a claim for interest. A right of set-off existed whilst the contractor was insolvent, but this came to an end following the injection of capital. Clause 60(6) of the contract obliges the employer to pay interest to the contractor should payment not be made in accordance with the terms of the contract. It was the contractor's case that the employer had failed to make payment as required by the contract and, therefore, payment of interest was due. Understandably, the employer argued that, due to the contractor's insolvency, a right of set-off arose and therefore there was no question of a failure to make payment. Rather surprisingly, there seem to be few cases which give much guidance as to the meaning of the wording 'failure to make payment'. The court took the view that the employer was entitled to withhold payment of the principal sum to provide it with security for a debt under a separate contract. It was not a matter of withholding money due to a breach of contract. The interest would, under the circumstances, continue to accrue and as the employer had the use of the money which was withheld, the interest should be paid to the contractor.

SUMMARY

The situation with regard to interest on undercertification had been extremely unclear, because of conflicting legal decisions. Where the ICE 5th Edition and similarly worded conditions apply, it would seem that only in the event of a failure to certify at all, or undercertification due to influence by the employer, or a misunderstanding of the contract by the engineer, gives an entitlement to interest. Interest will, however, be recoverable under ICE 6th and 7th Editions, even where undercertification occurs in good faith.

Where the JCT conditions apply, courts seem to take the view that an arbitrator has power to award interest from the date the certificate should have been issued. The contract does not make specific provision, but in some legal cases, the court has taken the view that it is equitable to allow the contractor compensation for not having use of the money during the whole period.

11.16. Can an architect/engineer refuse to include an amount of money in a certificate in respect of materials stored on site if the contractor or subcontractor cannot prove he has good title to the materials?

11.16.1. Suppliers in the recent past have been at risk, after having delivered materials to site, of not receiving payment. From time to time, a contractor may become insolvent without having paid the supplier for his materials. The supplier will often lose out, as the liquidator or receiver will take the benefit of the materials and pay out to the supplier only a small fraction of their invoiced price for the materials and often nothing at all. Many suppliers set out to protect themselves from this type of risk by including in their terms of trading what has become known as a 'retention of title clause'. The purpose of this clause is to enable the supplier to retain ownership until he has been paid. If the supplier has delivered the materials to site and the contractor becomes insolvent before paying him, the goods can be reclaimed or payment in full demanded from the receiver, employer or whoever wishes to make use of them. This is subject to the retention of title clause being effective. The court held in *Hendy Lennox (Industrial Engineers) Ltd* v. *Grahame Puttick Ltd* (1983), that the following represented an effective retention of title clause:

> Unless the company shall otherwise specify in writing, all goods sold by the company to the purchaser shall be and remain the property of the company, until the full purchase price thereof shall be paid to the company.

11.16.2. Employers are often concerned that they may pay for materials stored on site and, subsequently, the contractor becomes insolvent before paying subcontractors and suppliers. If the unpaid subcontractors or suppliers have effective retention of title clauses in their contracts with the contractor, they can demand payment from the employer if he intends to use the materials. Under these circumstances, the employer may finish up paying twice for the materials, as happened in *Dawber Williamson Roofing Ltd* v. *Humberside County Council* (1979).

11.16.3. Architects, engineers and quantity surveyors often consider they have a duty to protect the employer from the risk of having to pay twice for materials delivered to site. Some therefore insist upon the contractor producing proof of title before agreeing to include materials delivered to site but unfixed in an interim certificate.

11.16.4. The standard forms in common use expressly provide for payment in respect of materials delivered to site but unfixed: ICE 6th and 7th Editions (clause 60(1)(b)), JCT 2011 (clause 4.16.1.2) and GC/Works/1 (condition 48). None of these clauses make any reference to certifying amounts only in respect of materials for which the contractor is able to demonstrate that he holds good title. It would be necessary for a special clause to be included which limits the contractor's entitlement to payment for materials on site, to those which he actually owns. Such a clause would, however, be almost impossible to apply. The contractor, for example, could pay his supplier, only to discover later that a sub-supplier (who had not been paid by the supplier) had an effective retention of title clause in his terms of trading. Under these circumstances, the employer might not be protected.

11.16.5. It should be noted that, once materials have been built into the structure of the building, the retention of title clause will no longer be effective.

11.16.6. The Engineering and Construction Contract (NEC 3) refers to payment being made in respect of 'the Price of the Work Done', with no specific reference to payment for materials stored on site.

SUMMARY

An employer who pays for materials delivered to site, but unfixed, may find that he finishes up paying twice where the main contractor becomes insolvent, if the supplier has an effective retention of title clause in his terms of trading. Attempts can be made to avoid this, but it seems clear that the standard forms of contract place this type of risk onto the employer. The contracts in regular use, e.g. ICE 6th and 7th Editions, JCT 2011, GC/Works/1 and the like, all include for payment in interim certificates of unfixed materials on site. No mention is made in any of these contracts to proof of title being demonstrated by the contractor as a condition precedent to payment. Quantity surveyors, architects and engineers may decide to warn employers of the possible shortcomings of this type of payment clause, but are not entitled to exclude unfixed materials on site from interim certificates solely on the grounds that good title cannot be proved. It would require a specially drafted clause in the contract fully to protect the employer, and even this may be difficult to implement.

11.17. Can an employer refuse to honour an architect/engineer's certificate on the grounds that he considers the sum certified is incorrect, or is he legally obliged to pay the contractor the sum certified by the architect/engineer?

11.17.1. It had been traditional for employers to honour the architect/engineer's payment certificate without question. The first chink in the armour appeared in the case of *CM Pillings & Co Ltd* v. *Kent Investments* (1985), where an employer refused to honour an architect's certificate and he convinced the court that there was a *bona fide* dispute as to its accuracy. The court ordered a stay of the application for summary judgment for the sum certified, so that the matter could be referred to arbitration.

11.17.2. Another decision dealing with the same principle is *John Mowlem & Co Plc* v. *Carlton Gate Development Ltd* (1990). John Mowlem was the contractor and Carlton Gate Development the employer under the terms of the contract, which incorporated a non-standard form. For contractual and administrative reasons, the development was divided into sections. The architect issued two interim certificates, which totalled about £2.25m. On the last date on which payment was due, the architect sent to John Mowlem letters which purported to be notices under the conditions of contract authorising deductions from certified interim payments in the case of delays in completion of any one section of the work. In addition, Carlton Gate raised various counterclaims and, relying on the architect's purported notices and their alleged rights of set-off in respect of the counterclaims, they deducted half the sum certified by the architect before making payment. John Mowlem contested Carlton Gate's right of deduction on the following grounds:

- The contents of the letters and the surrounding circumstances showed that they were not written in good faith.
- The counterclaims arose after the date when money due on the certificates should have been paid, and for that reason, as a matter of law could not be set off against the liability to pay on the interim certificates.
- The letters written by the architect did not satisfy the requirements of the contract.

John Mowlem applied for summary judgment under Order 14 on the grounds that, as there was no dispute as to the debt, the court should therefore order that payment be made. The employer contested the application, requesting a stay of the court proceedings for the matter to be referred to arbitration.

11.17.3. Judge Bowsher decided in favour of the employer. He referred to the case of *Home & Overseas Insurance* v. *Mentor Insurance* (1989), where it was said that the purpose of Order 14 is to enable a plaintiff to obtain a quick judgment where there is plainly no defence to the claim. Judge Bowsher, when giving his decision, had in mind the interests of other litigants who might be delayed by a lengthy hearing and the extreme pressure on the court's time, referring to *British Holdings Plc* v. *Quadrex* (1989), where the judge considered applications for summary judgment inappropriate with facts so complex that days were required for the hearing and a huge weight of evidence was necessary in order to understand the issues. However, Judge Bowsher did comment that, where construction contracts provide for interim payment, it is usually for the very good reason that the contractors need money to continue with the project. The sums involved, he thought, were so large that even the bigger construction companies would feel the pinch when payment is withheld. It was his view that, in appropriate cases, the court should not shrink from dealing with an Order 14 summons even if the evidence is bulky. However, if the inappropriate cases can be discouraged, those cases in which it is appropriate to give summary relief will receive earlier attention. In considering the three points put forward by John Mowlem, the judge held as follows:

- It was inappropriate at a hearing for an Order 14 application to consider whether an architect had acted in bad faith.
- The contention that the employer's counterclaims, because of their timing, were invalid was a fundamental point of law, which should be decided only after due deliberation and would therefore not be a matter for an Order 14 summons.
- The claim that the architect's letter concerning deduction from certificates did not comply with the contract could not stand on its own.

The judge refused to grant John Mowlem's application for summary judgment and granted a stay for the matter to be referred to arbitration. This case is not authority for stating that Carlton Gate were correct in the actions they took, but that disputes of this nature have no place in Summary Judgment applications.

11.17.4. In the case of *RM Douglas Ltd* v. *Bass Leisure Ltd* (1991) a dispute arose out of a contract which incorporated the standard JCT Management Contract 1987. The contractor applied for summary judgment in the sum of £1.3m due on interim certificates. Bass Leisure applied for the action to be stayed and the matter referred to arbitration. The defendant's reasons for not honouring the certificates were twofold:

- It was contended that there were grounds for questioning whether the sums certified were in fact due under the interim certificates.
- It was argued that there was an entitlement to set off damages in respect of delay and other alleged breaches of contract.

Judge Bowsher listened to arguments as to what constitutes a dispute to be referred to arbitration. He was influenced by the decision of Judge Savill in *Hayter* v. *Nelson and Home Insurance* (1990), where Judge Savill interpreted the words 'there is not in fact any dispute' as meaning the same as 'there is not in fact anything disputable'. Judge Bowsher went even further to conclude that the defendant, to defeat an application for summary judgment, need only 'in good faith and on reasonable material raise arguable contentions' as to whether sums certified are due. It can readily be seen that if a defendant to an application has to demonstrate that there is a *bona fide* dispute, this can be a far more taxing matter than merely showing that a proposition is disputable or merely raises contentions. Judge Savill had summed up the situation in the following terms:

> Only in the simplest and clearest cases, that is where it is readily and immediately demonstrable that the respondent has no good grounds at all for disputing the claim, should that party be deprived of his contractual rights to arbitrate.

In the main, Judge Bowsher found in favour of the employer. He considered there were reasonable grounds for challenging the certificate, with the exception of £236,253.20, which he ordered to be paid to the contractor, the balance being referred to arbitration.

11.17.5. These two decisions run contrary to the way in which the industry had operated in times past, best illustrated when Lord Denning said that an architect's certificate was like a bill of exchange, i.e. money in the hand and should be honoured. In *Enco Civil Engineering* v. *Zeus International* (1991) an employer refused to honour a certificate issued by the engineer, where the ICE conditions applied. The court refused to award summary judgment and referred the matter to arbitration. In the *Bank of East Asia* v. *Scottish Enterprise and Stanley Miller* (1996), it was held that the employer was entitled to deduct the costs of remedying defective work from sums certified.

11.17.6. In the light of the Arbitration Act 1996, courts have no discretion to hear disputes where an arbitration clause exists in the contract. Moreover, the decision in *Halki Shipping* v. *Sopex Oils* (1997) leads to the conclusion that if a reason for non-payment is raised, the court will be unlikely to investigate whether the reason is *bona fide* or not; the proceedings will be stayed whilst the matter is referred to arbitration. All an unscrupulous employer needs to do is to raise a well-orchestrated smoke screen to delay payment.

11.17.7. It would seem that contractors who are not paid sums certified, as a first port of call applied to the court for summary judgment. It is clear that, where the employer alleges that there is a dispute relating to the sums certified, the courts will refuse summary judgement. Rather than go down the long route of arbitration or litigation, contractors are now more inclined to refer the dispute to adjudication under the Housing Grants, Construction and Regeneration Act 1996.

SUMMARY

In the light of case law it would seem that employers may, if a *bona fide* dispute arises as to the accuracy of a payment certificate, withhold payment whilst the dispute is referred to arbitration. Moreover, the court may be unwilling to investigate whether a reason for non-payment is *bona fide* or not, as explained in *Halki Shipping* v. *Sopex Oils* (1997). The Arbitration Act 1996 gives comfort to an employer who resists paying a sum certified. Under this Act a court has no discretion to hear a dispute where the contract includes an arbitration clause. The matter must be stayed and referred to arbitration. Contractors, in view of the reluctance of the courts to order summary judgment when a certificate isn't honoured, often prefer to refer the dispute to adjudication under the Housing Grants, Regeneration and Construction Act 1996.

11.18. Where a cheque is issued in respect of construction work undertaken, can it be stopped before it is honoured if it subsequently becomes obvious that the work is defective?

11.18.1. The case of *Isovel Contracts Ltd* v. *ABB* (2001) involved a dispute concerning payment by cheque for construction work. The work comprised the construction of a leisure centre. ABB was the subcontractor who carried out the mechanical and electrical work, some of which they subcontracted to Isovel. An application for payment was made by Isovel and a cheque in the sum of £70,000 issued by ABB in payment for the work. It appears that under the terms of the subcontract, Isovel was entitled to payment of sums paid to ABB under the main contract by Miller Construction. After ABB had issued the cheque to Isovel, it became clear that Miller had reduced the value of the mechanical and electrical work, because of non-compliance. ABB then stopped the cheque made out to Isovel, on the basis that the value of the work done by Isovel merited a nil valuation. Isovel commenced proceedings for the payment of the cheque. The judge in the case explained that a cheque is like a bill of exchange and is as good as cash in hand. The party who issued the cheque is not entitled to argue that the work had no value and therefore there is no consideration for the payment, neither can a counterclaim be used to avoid payment. ABB was therefore required to make payment of the sum of £70,000 to Isovel.

11.18.2. It is clear from the *Isovel* case that cheques, once made out and issued, cannot be stopped by the sender.

SUMMARY

Where a cheque is issued, it cannot subsequently be stopped if it transpires the work is defective. In *Isovel Contracts Ltd* v. *ABB* (2001), the court considered that a cheque is like a bill of exchange and is as good as cash in the hand.

11.19. Where agreement is reached whereby one party to a construction contract agrees to pay the other a sum of money, can the paying party refuse to make the payment on the grounds that he was financially coerced into the agreement? What is meant by economic duress?

11.19.1. If an agreement is reached under duress in the form of economic pressure, it is possible to have the agreement set aside. This could occur where a small contractor has completed the work and is promised payment of a final balance only if he accepts a sum which is less than his due entitlement. As the contractor is short of money, he accepts the offer. Such an agreement is unlikely to be enforceable, leaving the contractor to demand payment of his due entitlement, which was the situation in *D & C Builders Ltd* v. *Rees* (1966).

11.19.2. The case of economic pressure could also be used to secure agreement to a payment in excess of the due entitlement. A subcontractor who has a fixed-price contract may coerce the main contractor into agreeing to make additional payments where it was impossible to secure a replacement subcontractor. An agreement reached under these circumstances is unlikely to be enforceable.

11.19.3. The courts will not always hold that an agreement which is secured under financial pressure is unenforceable. In the case of *Williams* v. *Roffey Bros & Nicholls (Contractors) Ltd* (1990), a subcontractor underpriced the carpentry work in connection with the refurbishment of a block of flats. Part way through the work, the subcontractor got into financial difficulties and threatened to stop work unless the contractor agreed to increase the contract prices. The contractor considered that, if the subcontractor stopped work, it would cost considerably more to engage a replacement and therefore agreed to an extra payment. The court held that the agreement was binding, as the contractor derived the benefit of retaining the subcontractor's services and thus avoided paying more to have the work completed. This seems a strange decision, but the judge, who is the only person whose opinion is relevant, considered the facts did not demonstrate economic duress.

11.19.4. It is often a case of the large organisation taking financial advantage of a smaller one, but this is not always the situation. The case of *Carillion Construction Ltd* v. *Felix (UK)* (2000) tells a familiar tale. Carillion was the main contractor for the construction of an office building for Hammerson UK Properties. The design, manufacture and supply of the cladding was subcontracted to Felix UK Ltd. Work was commenced by Felix in September 1999, before the subcontract had been entered into. A subcontract became binding upon the parties on 14 January 2000. Complaints were being made by Carillion concerning delays caused by Felix, due to late delivery of cladding panels. Disputes were occurring at about this time concerning the likely final value of the work undertaken by Felix. In particular, the parties were experiencing difficulties in reaching agreement on what constituted variations and their value. Much of the work which Felix claimed to be the subject of variations, Carillion considered to be a part of the original works. Felix, anxious to secure the agreement of Carillion, resorted to what may be considered strong-arm tactics. They refused to make deliveries until agreement was reached. Carillion became concerned that there could be a delay to the completion of the main contract works, which could involve payment of liquidated damages to the employer. A significant difference as to the final sum payable to Felix had arisen by the end of March. Carillion considered that a total sum of £2.75m would become payable on completion of the subcontract works. Felix stated that their entitlement was £3.11m. Carillion felt under great pressure to agree a final account.

The parties reached a settlement agreement in the sum of £3.2m, which was expressed as full and final. Carillion indicated their displeasure at being forced into the agreement and reverted to their original figures once deliveries had been made. Felix brought an action to try and enforce the agreement. The court, however, considered that the pressure applied by Felix was illegitimate and without justification and refused to enforce the terms of the agreement.

11.19.5. Cases of this kind depend upon the facts; however, courts have been consistent in stating that for an agreement to be set aside, it must be shown that it was obtained by illegitimate pressure and where the victim had no realistic practical alternative but to submit to the pressure. This was the situation in the case of *Adam Opel GMBH and Renault S.A.* v. *Mitras Automotive (UK)* (2007), where the claimant, who was a car manufacturer, sought the recovery of monies paid out pursuant to a compromise agreement entered into with the respondent, a supplier of component parts. Mitras was the sole supplier of the front bumper of a particular type of van. Opel had sought to change the design of the van, which included the use of a different bumper from what was being supplied by Mitras. Opel gave Mitras six months' notice that they intended to use a different supplier with regard to the new type of bumper. Mitras demanded a substantial increase in the price of the bumpers they were supplying and would continue to supply until the new van came on stream. Negotiations took place between the parties, but a stage was reached, because of the 'just in time' system which operated, where Opel had only 24 hours' supply of bumpers which were supplied by Mitras in stock. In order to maintain a continuity of supply, Opel agreed to the revised price Mitras were seeking, which, in total, over the period of continued supply, amounted to a total sum of £450,000.

11.19.6. At a later date, Opel sought the repayment of the excess sum and the matter was referred to the courts, who found in favour of Opel. It was clear from the correspondence and negotiations that the respondent had threatened to stop supplies unless its demands for a price increase were met. The pressure created by the threats was both illegitimate and caused Opel to agree to the demands. With only 24 hours' supply of bumpers in stock,

Opel had no realistic practical alternative but to agree with the demands being made by Mitras.

11.19.7. A similar situation occurred in the case of *Kolmar Group AG* v. *Traxpo Enterprises PVT* (2010), where Traxpo refused to supply methol at a price which was included in the contract unless the price was increased. With no alternative readily available, Kolmar Group agreed to the price increase. The court held that Kolmar Group were entitled to recover the excess represented by the price increase.

11.19.8. It is not uncommon, where recessionary forces are at work, for contractors to put pressure on their subcontractors or suppliers to reduce prices in the name of pain-sharing, or the issue of threats to seek alternative subcontractors or suppliers for future contracts. Where subcontractors and suppliers submit to such threats, they would be fully entitled to seek recompense at a later stage, on the basis that they had no alternative to submitting to the financial pressure being applied. It is not necessary to protest at the time the threats are made and it doesn't mean that the payments made as a result of the coercion were made voluntarily.

SUMMARY

Courts will not enforce an agreement which has been reached by coercion using financial pressure. However, financial pressure in itself is not sufficient to have the agreement set aside. It is recognised that parties are constantly bargaining and haggling, which is all part of the hurly-burly of commercial negotiating. The financial pressure must be illegitimate, by being accompanied by a threat to breach the contract, for example not finishing off the work or refusing to continue supplying goods. The injured party must also be able to demonstrate that there was no practicable alternative but to submit to the financial pressure.

11.20. Are there any circumstances when a standard form of construction contract is applicable, whereby an employer will be obliged to pay in full the amount included in a contractor's application for payment, even though the sum has not been certified and is overvalued?

11.20.1. JCT With Contractor's Design, 1998 Edition, when first published, was unusual as a standard form of construction contract in allowing, under certain circumstances, a contractor to submit an application for payment and be entitled to be paid the sum claimed, even though the amount included in the application may be overvalued.

11.20.2. JCT With Contractor's Design 1998 requires the contractor to submit an application for payment. If Alternative A applies, the contractor is entitled to stage payments and the application will include work in respect of completed stages. Where Alternative B applies, the contractor's application should include the value of work properly executed, together with any changes to the Employer's Requirements and materials delivered to site.

11.20.3. Clause 30.3.3 requires the employer, not later than five days from receiving the contractor's application for payment, to produce a written statement indicating the sum it is proposed will be paid and to make payment by the final date for payment. A problem may arise for the employer if the written statement required by clause 30.3.3 is not produced. Should the employer fail to send a written statement as required by this condition, clause 30.3.5 obliges the employer to pay the contractor the amount included in the contractor's application.The wording of the clause 30.3.3 states:

> Where the Employer does not give any written notice pursuant to clause 30.3.3 and/or 30.3.4 (withholding notice) the Employer shall pay the Contractor the amount stated in the Application for Payment.

11.20.4. Bearing in mind that the employer has only five days within which to respond to the contractor's application, it is hardly surprising that payment is often due in accordance with the contractor's application. It is not uncommon for the employer's agent to dispute the amount included in the contractor's application and hence considers an overpayment is made. Under normal circumstances this does not cause a major problem, as the valuation can be corrected in the next payment and only cashflow is affected. The situation, however, may be different if the work is almost complete, or the contractor becomes insolvent after payment has been made, but prior to the final date for payment of the next sum due. Where either of these circumstances occurs, the employer could sustain a significant loss.

11.20.5. When drafting JCT 2005, the JCT obviously had second thoughts concerning this matter. Clause 4.10.5 stipulates that if the employer fails to produce the written notice in response to the contractor's application indicating the amount it intends to pay, then the contractor will be entitled to a sum calculated in accordance with the terms of the contract. The wording of clause 4.10.5 states:

> Subject to any notice given under clause 4.10.4 (withholding notice) the Employer shall no later than the final date for payment, pay the Contractor the amount specified in the notice given under clause 4.10.3 (Employer's notice) or in the absence of a notice under clause 4.10.3 the amount due to the Contractor as determined in accordance with clause 4.8.

11.20.6. JCT 2011 Design and Build, which was published to bring the contract into line with the provisions of the Local Democracy, Economic Development and Construction Act 2009, reverts back to the provisions of JCT 1998 With Contractor's Design. The contractor, in line with clause 4.8, is entitled to submit an interim payment application. The due date is the specified date for the interim application or the date the employer receives the interim application, whichever is the later. Not later than five days after the due date, the employer must issue a payment notice, setting out the sum it considers due and the basis of the calculation. If, however, the payment notice isn't produced on time, or is deficient in not providing details of the sum due, clause 4.9.3 states that the contractor is entitled to be paid the sum included in the interim application. The employer has, however, a fall-back position in sending a 'pay less' notice under clause 4.9.4, indicating the sum to be paid and the reason why it is less that the sum included by the contractor in his payment application. However, the 'pay less' notice must be given not later than five days before the final date for payment.

11.20.7. A similar situation arises with regard to the final payment. Clause 30.5.5 of JCT 1998 and 4.12.1 of JCT 2005 requires the contractor, within three months of Practical Completion, to submit a Final Account and Final Statement for agreement by the employer. Clause 30.5.5 of JCT 1998 goes on to state that within one month from

- The end of the Defect Liability Period;
- The date named in the Notice of Completion of Making Good Defects; or
- The date of submission of the Final Account and Final Statement,

whichever is the latest, the Final Account and Final Statement:

> will be conclusive as to the balance due between the Parties in accordance with the Final Account and Final Statement except that the Employer disputes anything in that Final Account and Final Statement before the date on which, but for the disputed matters, the balance would be conclusive.

The wording of clause 4.12.6 in JCT 2005 and JCT 2011 is similar. It is clear from this requirement that the employer's agent needs to act fairly quickly, once the contractor's Final Account and Final Statement has been received, to respond to those documents; otherwise the employer will be obliged to pay the contractor the amount included in the Final Account and Final Statement.

11.20.8. JCT 98, 2005 and 2011 are alike with regard to the risks borne by the employer in not responding within the stipulated timescale to the contractor's Final Statement. All three contracts make it clear that, if the employer does not respond to the contractor's Final Statement within the stipulated timescale, the amount claimed by the contractor becomes conclusive. However, only JCT 98 and 2011 put the employer financially at risk in not responding to the contractor's application for payment on time.

11.20.9. The JCT 2011 contracts which provide for employer-designed projects incorporate amendments to cater for the Local Democracy, Economic Development and Construction Act 2009, which increases the financial risk to the employer if the architect fails to issue a payment certificate and the employer or his agent doesn't respond on time to the contractor's application for payment. The Act provides, in section 143(3), a revised mechanism for payment, which has been incorporated into the terms of JCT 2011. These terms include a requirement for the contractor to submit an interim payment notice if the architect fails to issue an interim certificate. The interim payment notice must state the sum the contractor considers to be due and the basis on which it has been calculated. Unless the employer, or the architect or quantity surveyor on his behalf, serves on the contractor a pay less notice not later than five days before the final date for payment, then the employer is bound to pay the contractor the sum included in the contractor's interim payment notice. The pay less notice must state the sum considered to be due and the basis on which it has been calculated. It should be stressed that these provisions only apply if the architect fails to issue an interim certificate, which would be rather unusual.

The process for certification and payment comprises:

- The contractor may make an application to the quantity surveyor for payment not less than seven days before the due date, stating the sum he considers is due and the basis on which it has been calculated. This is referred to as an 'interim application notice'.
- Interim valuations are to be made by the quantity surveyor for the purpose of ascertaining how much is due to be included in the interim certificate.
- The architect is required to issue an interim certificate not later than five days after the due date and is required to state the sum due and the basis on which it has been calculated.
- If the architect fails to issue an interim certificate, the contractor's interim application becomes a interim payment notice.
- The employer will be obliged to pay the contractor the amount included in the contractor's interim payment notice unless he serves on the contractor a 'pay less notice', which states the sum the employer considers is the contractor's due entitlement. This pay less notice must state the sum which he considers is due to the contractor and the basis on which it is calculated and may be sent on his behalf by the architect or quantity surveyor. It must be sent not later than five days before the final date for payment.

SUMMARY

JCT 1998 With Contractor's Design requires the Contractor to submit an application for payment. If the employer fails to respond within five days, then the sum included in the contractor's application has to be paid. This provision has not been carried forward to JCT 2005. Under the provisions of this contract, if the employer fails to respond to the contractor's application for payment within five days, the contractor is only entitled to be paid a sum calculated in accordance with the contract. JCT 2011 Design and Build, in line with the Local Democracy, Economic Development and Construction Act 2009, reverts back to the provisions of JCT 1998 With Contractor's Design, whereby, if the employer fails to produce a payment notice on time, the contractor becomes entitled to payment in accordance with his application.

JCT 1998, 2005 and 2011, where the work has been designed by the contractor, require the contractor to submit a Final Statement within three months of Practical Completion. A period of one month from the issue of the Final Statement, or the end of the Defects Liability/Correction Period, or the date of the Notice of Making Good Defects, whichever is the latest, is allowed for disputes to be raised by the employer. If the employer fails to dispute any matters, then the Final Statement becomes conclusive.

The Local Democracy, Economic Development and Construction Act 2009, section 143(3), has resulted in a revised payment mechanism, which has been incorporated into JCT 2011 contracts for employer-designed projects. These new provisions include a requirement for the contractor to send in an interim payment notice if the architect fails to issue an interim certificate. The interim payment notice must state the sum the contractor considers to be due and the basis on which it has been calculated. Unless the employer, or the architect or quantity surveyor on his behalf, serves on the contractor a pay less notice not later than five days before the final date for payment, then the employer is bound to pay the contractor the sum included in the contractor's interim payment notice. The pay less notice must state the sum considered to be due and the basis on which it has been calculated. It should be stressed that these provisions only apply if the architect fails to issue an interim certificate, which would be rather unusual.

11.21. What is a project bank account and how do the advantages compare with the disadvantages?

11.21.1. The main purpose of a project bank account is to ensure payment security and shorten the time required for payment to find its way down the supply chain. A mechanism is required whereby money is passed direct to major subcontractors without it passing through the hands of the main contractor. This is achieved by opening a project bank account, into which the employer pays sums certified and which in turn pays out direct to subcontractors. It has been estimated by the Office of Government Commerce that a saving of 2.5% can be achieved if project bank accounts become the norm. Payment through the agency of a project bank account will mean that the subcontractors will receive payment at the same time as the main contractor and, as this is likely to be less than 21 days, it therefore represents a major step forward from the 40, 60 and sometimes 80 days which are commonplace in the construction industry. The Office of Government Commerce's estimate of a saving of 2.5% may be kite-flying, but whatever the correct percentage happens to be, the sums involved would be very substantial.

11.21.2. Where a project bank account is used, the normal methods of ascertaining and certifying sums which are due to the main contractor and subcontractors, as required by the standard contracts in common use, will remain unchanged. Once the bank account has been set up, it will normally require signatories on behalf of the employer and main contractor to enable money to be moved out of the account. There is also a need for the account to be transparent for all subcontractors who are to be paid from the account, in order for trust to be developed. They will be informed when a payment is due to arrive into the account and the time when payments to subcontractors will be made. The account will also be subject to audit.

11.21.3. The use of a project bank account will not overcome matters which lead to disputes down the supply chain, such as the value of the work and rights of set-off. Following the certification process, a sum will be paid into the project account and the signatories to the account, normally the employer and main contractor, will inform the bank as to the amounts to be paid out of the account to the main contractor and subcontractors.

11.21.4. Difficulties could arise in respect of a project bank account if the main contractor were to become insolvent. A problem occurred in the case of *British Eagle v Air France* (1975) resulting from money being held in a central fund from which airlines were paid, but this was held to be incompatible with the Insolvency Rules. These Rules state that all unsecured creditors rank equally. Once the main contractor has become insolvent, any money in the account or due to be paid into the account earmarked for subcontractors, would be frozen. What money remained following a sale of assets, collection of money due and payment of secured creditors and expenses would be available for payment to the subcontractors. However, there is usually little money left for unsecured creditors, which would leave each subcontractor with only a small fraction of its original entitlement. In an attempt to overcome this problem the account is given Trust Status, in that the money is held in trust on behalf of all those who are involved in the process. It is anticipated that the Trust Status of the money held in the project bank account will take precedence over the Insolvency Rules.

11.21.5. It is unlikely that all subcontractors will be involved in the arrangement. Those included need to be on monthly payment terms and the project should be reasonably large to justify the time and cost involved in setting up and administering the account.

11.21.6. The JCT produced a Standard Bank Account Agreement in 2009, with the following features:

- The parties are the employer, contractor and a number of subcontractors;
- At the times indicated in the main contract and subcontracts, the account holders will identify payments to be made under the building contract and subcontracts;
- The employer is required to pay into the account the monies due under the building contract;
- The account holders will promptly instruct the bank to make payments to the contractor and subcontractors from the account, as soon as the payment by the employer is cleared. They will then inform each subcontractor of the amount which is to be paid to them from the account.

11.21.7. Like any new idea, there will always be doubters and it will take a number of years before it is adopted on a large scale. The process is intended to provide a system for ensuring that all members of the supply chain receive payment promptly. It will not remove the disputes concerning the value of work executed, nor will it curb those who are swift to apply set-off. A test case will be needed to ensure that the Trust Status of the money held in the project bank account takes precedence over the Insolvency Rules. However, it seems a positive move for the industry and should be welcomed.

SUMMARY

The purpose of the project bank account is to offer to subcontractors a secure and speedy method of payment. Where a project bank account is set up, money due under the main contract is paid into the account by the employer. Once the money is cleared into the account, subcontractors receive their entitlement, which is paid promptly from the account. No changes to the certification process under the contract are required. Where a main contractor becomes insolvent, money paid into the account would normally be frozen and the Insolvency Rules applied; that is, all unsecured creditors rank equal. This means that the subcontractors would receive little of the money they were owed. In an attempt to overcome this problem, the money in the account is held in trust for the subcontractors. It will, however, require a court decision to establish that the Trust Status takes precedence over the Insolvency Rules. Finally, the JCT published a Project Bank Account Agreement in 2009, which is suitable for use on JCT main contracts. Like any new idea, there will always be doubters; however, it seems a positive move for the industry and should be welcomed.

11.22. What is the difference between set-off and abatement?

11.22.1. Where claims for money are made, those on the receiving end may often be looking for a way of reducing or eliminating the sum claimed. There are two main methods by which this can be achieved. A reduction or elimination of the sum claimed may be achieved by way of set-off; or the use of a technique known as 'abatement' may be applied.

11.22.2. Where the party who receives the claim has a counterclaim, he may reduce or eliminate the amount claimed by way of set-off. To qualify, the counterclaim must be for a liquidated debt or money claims which can easily and without difficulty be ascertained, which may be appropriate even if disputed, as explained in the case of *Axel Johnson Petroleum AB* v *M G Mineral Group AG* (1992). If the amount of the counterclaim is not for a liquidated debt, then to qualify for set-off the sum must be so related to the claim that it would be manifestly unjust to allow the claim without taking into account the counterclaim, referred to as 'equitable set-off'. In the case of *Hanak* v *Green* (1958), the employer claimed damages for failure to complete properly. The contractor was held to be entitled to an equitable set-off in respect of a claim for extra work done outside the contract, loss caused by the employer's refusal to admit the contractor's workmen and damages for trespass to tools. Liquidated damages for delay and claims for correcting defects on the same contract will usually be regarded as grounds for equitable set-off. In *Gilbert Ash (Northern)* v *Modern Engineering* (1974), a main contractor was held entitled to set off its claims for defective work and delays from sums certified by the architect as due to a subcontractor.

11.22.3. Contracts often include provisions for set-off. For example, the JCT forms of contract allow the employer to set off from sums due to the contractor in respect of liquidated damages. An employer would not, however, be entitled to set off a claim in respect of an unconnected contract, unless given the right by the terms of the contract under which the contractor is making the claim. The parties are entitled to restrict or eliminate the right of set-off by means of a term in the contract.

11.22.4. Abatement arises where there is a difference of opinion in respect of the value of the work. For example, a subcontractor in its application for payment may include a value for the work completed at a sum which is much higher than that placed upon it by the main contractor. Before making payment, the main contractor will usually substitute his own valuation of the work; the value of work is thus abated. In the case of *Multiplex Construction (UK) Ltd* v *Cleveland Bridge UK Ltd* (2006), which related to the construction of Wembley Stadium, a dispute arose out of the value of the steelwork. It was alleged by Multiplex that there were a great many defects in the work and it applied a set-off before making payment. The court held that in respect of defective work it was appropriate to abate the sum claimed by reference to the cost of the remedial works necessary as a consequence of the defects.

11.22.5. The judge in the *Multiplex* case drew attention to a number of cases where abatement was at issue. In the case of *Mondel* v *Steel* (1841), it was held that a shipowner was entitled to abate the price of the ship as it had not been built in accordance with the specification.

11.22.6. The judge in the *Multiplex* case summed up the law relating to abatement as follows:

- In a contract for the provision of labour and materials where performance has been defective, the employer is entitled, under common law, to maintain a defence of abatement.
- The measure of abatement is the amount by which the product of the contractor's endeavours has been diminished in value as a result of that defective performance.
- The method of assessing diminution in value will depend upon the facts of the case.
- In some cases, diminution in value may be determined by comparing the current market value of that which has been constructed with the market value it would have had if properly constructed. In other cases, diminution in value may be determined by reference to the cost of remedial works. In the latter situation, however, the cost of remedial works does not become the measure of abatement. It is merely a factor that may be used, either in isolation or in conjunction with other factors, while determining diminution in value.
- Claims for delay, disruption or damage caused to anything other than that which the contractor has constructed cannot feature in a defence of abatement.

11.22.7. The difference between set-off and abatement became relevant in respect of the requirement for a withholding notice under section 111 of the Housing Grants, Construction and Regeneration Act 1996, if payment was not to be made in respect of sums which were otherwise due. It was held in the case of *KNS Industrial Services (Birmingham) Ltd* v *Sindall Ltd* (2001) that whilst a withholding notice was required in respect of a set-off, it was not applicable where the dispute relates to abatement. The judge, in arriving at a decision, said 'one cannot withhold what is not due'. The wording of Section 111 has been amended by the Local Democracy, Economic Development and Construction Act 2011 and it will be a matter of debate as to whether a withholding notice will, under the new wording, be applicable in respect of abatement.

SUMMARY

Where claims for money are made, those on the receiving end may often be looking for a way of reducing or eliminating the sum claimed. There are two main methods by which this can be achieved. A reduction or elimination of the sum claimed may be sought by way of set-off, or by the use of a technique known as abatement.

Where the party who receives the claim has a counterclaim, he may, subject to certain rules, reduce or eliminate the amount claimed; this is referred to as set-off. To qualify, the counterclaim must be for a liquidated debt or for money claims which can easily and without difficulty be ascertained and may be appropriate, even if disputed. If the amount of the counterclaim is not for a liquidated debt, then to qualify for set-off the amount of the set-off must be so related to the claim that it would be manifestly unjust to allow the claim without taking into account the counterclaim; this is referred to as equitable set-off.

Abatement arises where there is a difference of opinion in respect of the value of the work. For example, a subcontractor, in its application for payment, may include the value of work completed at a sum which is much higher than that placed upon it by the main contractor. Before making payment, the main contractor will usually substitute his own valuation of the work; the value of the work is thus abated.

11.23. Where a contract requires the contractor to provide information for use by the employer, for example heath and safety documents, manuals and built drawings, is the contractor legally entitled to refuse to supply them on the grounds that money is owed by the employer? What other remedies are available as a result of a failure on the part of the employer to make proper payment?

11.23.1. Contractors who are owed money often use whatever leverage there is to hand in an effort to secure payments which they feel are due. Threats to withhold 'as built' drawings, QA and Health and Safety manuals are from time to time used in this manner.

11.23.2. Where a party to a contract fails to comply with one of its obligations, often referred to as a breach of contract, the innocent party is not necessarily discharged from further performance of the works. However, if an employer fails to make payment as required by the contract, this would entitle the contractor to suspend work, which would include any obligation to supply 'as built' drawings, QA and Health and Safety manuals.

11.23.3. The Housing Grants, Construction and Regeneration Act 1996, section 112, provides a contractor with the right to suspend work if money owed isn't paid in full by the final date for payment. A seven-day notice of an intention to suspend work must be given, prior to suspension taking place. The right to suspend performance ends when payment is received. The standard forms of contract in general use, such as JCT and GC/Works, have introduced clauses into their contracts which reflect the statutory right of suspension. In the case of JCT 2011, the contractor is not required to suspend all the work but may choose to suspend a part only: for example, supplying documents. The ICE and Engineering and Construction Contract (NEC 3) do not have suspension clauses, which leaves the contractor to operate the suspension provisions set out in the Act.

11.23.4. A breach of contract, such as a failure to make payment by the final date for payment, will also entitle the innocent party to sue for damages. The damages normally recoverable are interest. Where the breach is a serious one which goes to the root of the contract, the remedy may be to terminate the contract and hence provide a discharge from any further performance of the contract. In addition there would be a right to recover damages which resulted from the breach. This would be a severe remedy, for use only as a last resort.

11.23.5. Standard forms of contract, such as the JCT and the Engineering and Construction Contract (NEC 3), provide specific remedies for certain breaches of contract, such as non-payment. Where the JCT and the Engineering and Construction Contract (NEC 3) apply, a failure on the part of the employer to make payment as required by the contract gives a contractor the right to terminate. In the absence of any provision in the contract, the rights of the innocent party to terminate can be dealt with by way of common law remedies. Unless the remedy is specifically excluded, a contractor would be entitled to terminate in accordance with either the requirements of the contract or at common law.

11.23.6. Where a termination occurs for failure on the part of the employer to make payment in accordance with the terms of the contract, it is important for the contractor to ensure compliance with any procedure provided for by the terms of the contract. In the case of the JCT contract, a 14-day warning notice is required before the right to terminate will arise. The termination applies only in respect of the contractor's employment. Other existing contractual rights will continue after the termination has taken place. The Engineering and Construction Contract (NEC 3) provides for the contractor to terminate its obligations if the employer has failed to make payment of a sum certified by the project manager within 13 weeks of the date of the certificate. There is no provision for a warning notice.

11.23.7. Where there is no express provision in the contract for termination because of a failure on the part of the employer to make payment, the contractor must ensure that the failure to make payment constitutes a repudiation of the contract on the part of the employer. A repudiation may consist of an express or implied refusal to comply with the contract. A single failure to make a payment may not be regarded as a repudiation. It could, for example, result from an oversight or a temporary cashflow problem. A contractor should avoid being too hasty in terminating a contract on the grounds of a failure of the employer to make payment as required by the terms of the contract.

11.23.8. The case of *Alan Auld Associates Ltd* v *Rick Pollard Associates and Another* (2008) is a good example of consistent late payment constituting a repudiation. Dr Pollard provided consultancy services as a sub-consultant to Dr Auld in connection with work at Dounreay Power Station, for which Dr Auld made a profit of £20 per hour. Dr Pollard submitted 19 invoices to Dr Auld between December 2004 and April 2006, all of which were paid late. By the end of May 2006 Dr Pollard was owed £21,000. As from June 2006, Dr Pollard provided his services direct to the client at Dounreay Power Station. Dr Auld commenced an action against Dr Pollard for the loss of profit on his services. The court held that the late payments were substantial, persistent and cynical, with every likelihood that future payments would be made late; as such, they constituted a repudiatory breach. The decision was the subject of an appeal, but the Court of Appeal upheld the decision of the lower court.

11.23.9. Difficulties can be encountered if the contractor commences these remedies where a dispute regarding payments exists. The contractor will be in breach of contract if, in any legal proceedings, it transpires that the employer is proved to be correct.

11.23.10. A further difficulty for contractors who threaten to withhold key documents is that it may be met with the employer making an application to the court for an injunction to prevent the documents being withheld. In the case of *Amec Group Ltd* v *Universal Steels*

(Scotland) (2009), the defendant, due to a dispute over payment, refused to hand over essential QA documentation required by Amec. The importance of the QA documentation was that they were required by the MOD, the employer, who was responsible for a new berthing facility at its naval dockyard. In the absence of the QA documentation, installation could not progress to meet the next window of opportunity involving tides, which would not occur again for a period of five months. The court held that an injunction should be granted. In view of the circumstances, it was held that damages would be an adequate remedy for any consequences of the withholding of the documents.

SUMMARY

A contractor would not be entitled to withhold documents as of right, because of a failure on the part of the employer to make payment as required by the terms of the contract. It would involve suspending work in accordance with the Construction Act, requiring a seven-day notice. JCT 2011 provides for suspending part only of the work if payment has not been received when due, which would include submitting documents. A right of termination due to a failure to pay on time may arise, which is usually set out in the terms of the contract; this is also a remedy which is available. Alternatively, a failure to make payment may be regarded as a repudiation, giving the contractor the right of termination at common law. Threats to withhold essential documentation may be met by an application to the court for an injunction to prevent documents from being withheld. However, it would only be in exceptional circumstances that a court would issue an injunction.

11.24. Where a party to a construction contract is due to make a payment to the other party, any intention to reduce the sum due under the contract by way of set-off, for example in respect of delay, will require a withholding notice to be served. What information must be included in the withholding notice to ensure that it is valid?

11.24.1. Section 111 of the Housing Grants, Construction and Regeneration Act 1996 includes the following provisions regarding withholding notices:

> A party to a construction contract may not withhold payment under the final notice of payment of a sum due under the contract unless he has given an effective notice of intention to withhold payment. The notice mentioned in section 110(2) may suffice as a notice of intention to withhold payment if it complies with the requirements of this section.
>
> To be effective the notice must specify:
>
> - The amount proposed to be withheld and the grounds for withholding payment or
> - If there is more than one ground, each ground and the amount attributable to it, and must be given not later than the prescribed the prescribed period before the final date for payment.

The 'prescribed period' provided in the Scheme for Construction Contracts is seven days before the final date for payment.

11.24.2. The courts have given some guidelines regarding withholding notices, which include the following:

- To be effective, a withholding notice must be in writing: *Strathmore Building Services* v *Colin Greig* (2001);
- To be effective, a withholding notice must be issued at the requisite time before the final date for payment: *VHE* v *RBSTB* (2001);
- The courts take a practical view of the contents of a withholding notice and will not accept any contrived or artificial arguments concerning the form of the notice, aimed at trying to establish that they are invalid.

11.24.3. A dispute arose in the case of *Windglass Windows Ltd* v *Capital Skyline Construction Ltd, London and City Group Holdings Ltd* (2009) relating to the validity of a withholding notice. Under a contract date 10 November 2008, Windglass undertook to supply and install glazed windows, doors and screens at a site known as City Link Court. Two applications for payment were made by Windglass, one in the sum of £45,000 and the other for £121,000, but were not paid by Capital. It was argued by Capital that they had no obligation to make payment as Windglass had not submitted the applications for payment in sufficient detail and in the form which the parties had agreed. Windglass argued that they had not agreed to make applications in the form suggested. The case for Capital was that they had set out the reasons for not having paid the money in a letter to Windglass in a letter dated 13 March 2009 in respect of the first application and a second letter, dated 13 April 2009 in respect of the later application.

11.24.4. The wording in the letter was:

> Our financial director has returned this application and is not willing to process this amount due to insufficient supporting information. Please note that our company policy is such that each sub-contractor valuation must be presented in a standard format, copy attached and authorised by the appropriate site manager before your application can be processed. Could you kindly represent your application with the correct supporting information and our office will process it immediately.

11.24.5. The matter was referred to an adjudication, where it was decided that the letter sent by Capital did not meet the requirements of the Act as it did not provide the grounds for withholding money. It was also the adjudicator's finding that there was no binding obligation that the applications for payment be made in any particular form. The adjudicator concluded that Capital had not produced an effective withholding notice, as they did not specify the amount to be withheld or the grounds for the money being withheld.

11.24.6. Capital also alleged that Windglass had caused delay and that some of the work was defective, but, in the absence of an effective withholding notice the adjudicator considered the set-off to be invalid.

11.24.7. The adjudicator, in his decision, ordered Capital to pay Windglass a sum of £185,033. Capital refused to make payment, on the grounds that the adjudicator had exceeded his jurisdiction when he concluded that the withholding notice was invalid due to there being no stated grounds for withholding. The court rejected this contention and ordered Capital to pay Windglass the sum included in the adjudicator's decision.

SUMMARY

The Housing Grants, Construction and Regeneration Act 1996 makes it clear that there is no entitlement to set-off unless a valid withholding notice has been served in the correct timescale. The notice must state the amount to be withheld and the grounds for it being withheld. The courts will not allow any deviation from these requirements. In the absence of proper compliance, it will not be possible to resist making a payment of a sum due under the contract, despite there being valid entitlement to cross claims for matters such as delay.

11.25. Where payment is made late, is there a legal entitlement to claim interest?

11.25.1. Late payment in the construction industry creates a regular problem for contractors and subcontractors. This situation affects cashflow and, quite often, an ability to trade in the normal manner. For many years, the law in England refused to recognise an entitlement to compensation for late payment. The case which was frequently quoted in support of a refusal to pay compensation is *Chatham and Dover Railway Co* v *South Eastern Railway Co* (1893). In this case the House of Lords decided that the common law does not permit the award of interest by way of general damages for delay in the payment of a debt when it was contractually due.

11.25.2. The law progressed slowly on this matter; however, in the Court of Appeal case of *Wadsworth* v *Lydall* (1981), it was recognised that under certain circumstances damages for breach of contract in the form of interest could be recovered. In this case, the defendant had failed to pay an agreed sum which the plaintiff required to finance the purchase of property. The plaintiff raised the money by way of a mortgage and claimed interest as special damage, which was accepted by the court. This decision was given the seal of approval by the House of Lords in *President of India* v *La Pintada* (1985).

11.25.3. The House of Lords brought the law into line with commercial practice in the case of *Sempra Metals Ltd (formerly Metallgesellschaft Ltd)* v *Inland Revenue* (2007). The case related to an overpayment of Advanced Corporation Tax, where it was held that interest should be paid in respect of the money reclaimed. The court considered that the entitlement should reflect the loss of use of the money. If money had to be borrowed to make up the shortfall, the loss would comprise either the cost of borrowing the money or the loss of interest which would have been gained, depending upon the claimant's financial position. Alternatively, the claim could be framed in restitution, which would reflect the benefit gained by the defendant in not paying the money when it was due.

11.25.4. The House of Lords in this case also recognised an entitlement for the claimant to be paid interest on a compound basis, when it was said:

> . . . the loss on the late payment of a debt may include an element of compound interest, but the claimant must claim and prove his actual interest losses if he wishes to recover compound interest, as is the case where the claim is for a sum which includes interest charges . . .

11.25.5. Interest is payable on certain debts under a term implied into a contract by the Late Payment of Commercial Debts (Interest) Act 1998. The act applies to contracts for the supply of goods or services where both parties are acting in the course of business. It is stipulated that the interest is to be statutory interest, in other words simple interest. The fixed interest is 8% above the Bank of England rate and is intentionally penal, as a means of encouraging prompt payment in the business world. Interest begins to run from the last date for payment, which may be stated in a contract. If none is stated, then interest is to run 30 days from the date of the performance of the obligation to which the debt relates.

11.25.6. Following the Late Payment of Commercial Debts (Interest) Act 1998 becoming law, most standard forms of contract were amended to allow for interest to be claimed on payments due under the terms of the contract which were made late. The Act does not apply if the contract contains a substantial remedy for late payment. The JCT family of contracts provides for a rate of 5% above the rate fixed by the Bank of England.

11.25.7. The law with regard to interest is different in Scotland, where it is governed by the Interest (Scotland) Act 2007. Prior to the Act coming into force, interest on debt claims ran from the date on which the court action is raised. This can be contrasted with claims for damages, where interest claims start at a date which is left for the court at its discretion to decide. The Act combines the two and allows for interest to run from the date when the payment becomes due. A rate of 1.5% above the Bank of England rate has been fixed, which, unlike the law as it applies in England, is not intended to be penal. Parties to a contract, however, may provide for different arrangements with regard to interest from the provisions of the Interest (Scotland) Act 2007.

SUMMARY

There are three possible ways is which interest may be claimed in the event of late payment.

1. It may be argued that the late payment constitutes a breach of contract. This being the case, there would be an entitlement to claim interest on a compound basis. To succeed with this type of claim it would be essential to demonstrate that the interest claimed equated to an actual loss sustained by the claimant.
2. The Late Payment of Commercial Debts (Interest) Act 1998 provides for simple interest to be claimed from the date payment was agreed to be made. If there was no such agreement, then from a date 30 days after the date of the performance of the obligation provided for by the contract. The interest rate has been set at a penal

level of 8% above Bank of England rate, to encourage prompt payment. In Scotland, an interest rate of 1.5% above Bank of England rate, for interest on both late payment and damages, has been fixed by the Interest (Scotland) Act 2007.

3. This Late Payment of Commercial Debts (Interest) Act 1998 does not apply where the contract includes a substantial remedy for late payment. Most standard forms of contract now include a clause requiring interest to be paid in the event of late payment. The JCT family of contracts has an interest clause which provides an interest rate of 5% above the Bank of England rate.

11.26. Where, under a JCT contract, an interim payment is due to be made to the contractor, but the final date for payment has passed without a payment having been made by the employer and no withholding notice has been served, and subsequently the contractor becomes insolvent, can the employer use insolvency as a legitimate reason for not making the payment?

11.26.1. Insolvency is a common occurrence in the construction industry, often at a time when the contractor is owed money by the employer. There is then often a dispute between the representatives of the insolvent contractor and employer concerning the unpaid sums. This was the situation which resulted in the House of Lords making its first decision relating to the Housing Grants, Construction and Regeneration Act 1996 in the case of *Melville Dundas Ltd (in receivership) and others* v *George Wimpey (UK) Ltd and others*.

11.26.2. Wimpey engaged Melville Dundas in relation to the construction of a housing development in Scotland for the sum of £7,088,270. The contract incorporated the terms and conditions of the JCT Standard Form of Building Contract With Contractor's Design 1998 Edition. On 2 May 2003, Melville Dundas applied for an interim payment in the sum of £396,630. The final date for payment was 16 May 2003, but Wimpey had failed to make payment by this date. Administrative Receivers were appointed for Melville Dundas on 22 May 2003. This was followed, on 30 May 2003, with Wimpey terminating the employment of Melville Dundas, as they were entitled to do, by clause 27.3.4 of the conditions of the contract. Wimpey did not pay the sum of £396,630 or any sum in respect of that interim application, relying on the wording of clause 27.6.5.1, which stated that with regard to clause 27.3.4 of the conditions of contract relating to payment after termination:

> the provisions of this contract which require any further payment or release of or further release of retention to the contractor shall not apply.

The wording of this clause means that, once the contractor's employment has been terminated, the employer is under no obligation to make any further payments to the contractor.

11.26.3. The provisions of clause 27.6.5.1 seem in conflict with the requirement of section 111 of the Housing Grants, Construction and Regeneration Act 1996, which states:

> A party to a construction contract may not withhold payment after the final date for payment of a sum due under the contract unless he has given an effective notice of intention to withhold payment.

11.26.4. It had generally been considered that the provisions of the Act would always take priority over any term in a construction contract. This was the view taken by the Inner House of the Court of Sessions in Scotland, which favoured the case argued on behalf of Melville Dundas.

11.26.5. The House of Lords struggled with the problem, but by a majority of three to two, found in favour of Wimpey. Lord Hoffman, in supporting the position taken by Wimpey, said:

> It is not suggested that the JCT conditions failed to provide for payment by instalments or for the matters mentioned in s110(1) (the requirement for a mechanism for determining what payments are due and the final date for payment). The question is whether they could in addition provide that in the circumstances specified in cause 27.6.5.1 an instalment payment which had previously been payable should cease to be payable. Apart from the requirements of ss 109(1) and 110(1) the Act does not purport to interfere with the freedom of the parties to make their own terms about interim payment.

11.26.6. A further reason for allowing Wimpey to refuse any further payment to Melville Dundas related to the timing of the withholding notice, as required by the JCT contract, and the date of the appointment of the receivers. The JCT contract requires a withholding notice to be served not later than five days before the final date for payment. In this case the final date for payment was 16 May 2003 and, thus, five days prior would be 11 May 2003. The appointment of the receivers did not occur until 22 May 2003, by which time it was too late to serve the withholding notice.

11.26.7. S111(10) of the Local Democracy, Economic Development and Construction Act 2009 mirrors the Melville Dundas decision.

SUMMARY

It seems clear from the decision in the case of *Melville Dundas Ltd* v *George Wimpey Ltd*, that where a JCT form of contract applies, once a contractor has become insolvent, the employer can rely on the wording of the contract to terminate the contractor's employment and then refuse to pay any sums where the final date for payment has passed by the time the contractor's employment had been determined. This is now the situation as provided for by the Local Democracy, Economic Development and Construction Act 2009.

11.27. Architects, engineers and quantity surveyors are often required to provide services to commercial organisations at risk. In the absence of a specific undertaking to work indefinitely without fee, is a stage reached when a right to payment for services arises?

11.27.1. The majority of architects, engineers and quantity surveyors engaged by commercial organizations have at one time or another undertaken to provide services on a risk basis. If the project goes ahead, then payment is forthcoming; if not, it is a matter of leaving empty-handed. When the arrangement is entered into, some thought is usually given as to how much resource will be used and the likely sums of money which are at risk. It is common for the process to drag on with a much greater input being involved than anticipated, leaving substantial losses to be borne in the absence of the project proceeding. Employers rarely respond in a positive way to requests for some form of payment, as they usually draw attention to the original arrangement that the work was provided on a risk basis.

11.27.2. Occasionally, it is argued that the arrangement for work to be undertaken on a risk basis came to an end at some stage, and thereafter there was an undertaking to make payment. This was the argument used by an architect in the case of *Dinka Letchin Associates* v *General Mediterranean Holkins SA* (2003). Mr Latchin, the architect, produced designs in connection with four projects in Tunisia. The first involved producing sketches and design studies for refurbishing an existing hotel, Villa France, to bring it up to a four-star standard. A further project involved upgrading the design to enable the hotel to be given a five-star classification. The development of the Old Tennis Club into apartments represented another project. A final scheme involved the production of drawings for a private villa.

11.27.3. It was accepted by both parties that Mr Latchin had agreed with Mr Auchin on behalf of General Mediterranean, the defendants, for the original design work to be undertaken on a risk basis. The involvement of Mr Latchin on the four projects commenced in April 1994 and continued until August 1996. For a part of the work Mr Latchin employed an assistant, a Mr Ciric, to whom he paid a sum of between $15,000 and $20,000. Several meetings took place between Mr Latchin and Mr Auchi; unfortunately, neither bothered to take notes and there was disagreement as to whether some of the meetings took place at all. In the cases where they both accepted that a meeting had taken place, there was a divergence of opinion as to what was discussed at the meeting. Mr Latchin, for example, argued that a meeting took place in May 1994, when Mr Auchin gave instructions for the hotel project to commence. It was Mr Auchin's recollection that no such meeting took place.

11.27.4. The schemes as designed by Mr Latchin did not proceed and in March 1999 he submitted a fee account in the sum of £240,000. The fee account was rejected by Mr Auchin in principle and also as to the amount. A substantial amount claimed related to trips made by Mr Latchin to Tangiers and it was suggested by Mr Auchin that the real purpose of these trips was to allow Mr Latchin the opportunity of meeting a lady who lived in Tangiers.

11.27.5. Recorder Mr John Uff held that the parties did in fact enter into a binding contract on 1 September 1994 and therefore Mr Latchin was entitled to receive payment. Any work undertaken prior to this date was held to be on a risk basis. The decision was the subject of an appeal. It was argued on behalf of the defendant that Recorder Uff, in deciding that work undertaken after 1 September 1994 should be paid for, had merely plucked a date out of the air. The appeal was unsuccessful, as the Court of Appeal held there were no grounds for finding fault with Recorder Uff's judgment.

SUMMARY

As most architects, engineers and quantity surveyors who provide services in the private sector are well aware, it is part of the culture in the construction industry for work to be undertaken on a risk basis. Without this type of service, many schemes would probably never get under way. There is often no firm agreement as to the extent of the risk services to be provided. There is no hard and fast rule as to when the risk services come to an end and an entitlement to payment begins. For the benefit of both parties to such an agreement, it is essential, if disappointment and sometimes expensive legal actions are to be avoided, for the extent of the risk work to be properly defined and set out in writing. If a time arrives when the at risk agreement comes to an end and further work is required which will be the subject of payment, it is important, to ensure there is no misunderstanding, for a letter to be sent to the employer explaining the situation. It is also essential for the basis of the charges to be properly set out in writing and agreed.

11.28. Where money remains unpaid, is the service of a statutory demand and a petition to the court for a winding-up petition an effective method of debt collection?

11.28.1. Unpaid invoices are one of the main reasons why companies go out of business; it is the lack of ready cash rather than lack of profits which spells the end. Pressure, by one means or another, is usually applied to the debtor to try and secure payment. The service of a statutory demand for the money is considered a legitimate method, which may get results.

11.28.2. A statutory demand is a written warning, giving a debtor 21 days to pay a debt. It can be served as soon as the debt is due, but only applies if the debt is more than £750. Should the debt remain unpaid at the end of the 21-day period, a petition can be made to the court for a winding-up order to be made. If the debtor is unable to pay the amount due, the service of a statutory demand may result in the debtor agreeing to a payment plan involving instalments, or to an offer of an asset as security.

11.28.3. Once a winding-up order has been granted by the court, it will be advertised in the *London Gazette*. At this point, the process of closing down the business has begun. Once the winding-up petition has been granted, the bank account will be frozen. The company is then not allowed to sell or transfer any of its assets without permission of the court.

11.28.4. The directors of the company will have the opportunity to challenge a winding-up petition at a court hearing. If the challenge is rejected, the court will appoint a liquidator and the business will be closed.

11.28.5. If the debtor disputes the claim, he or she can apply to have the statutory demand set aside. The proceedings will be halted if there is any *bona fide* dispute in connection with the sum outstanding. It is relatively easy for a debtor to have a statutory demand set aside on the grounds of a *bona fide* dispute and the process can result in an order for costs being made against the creditor. An application to have a statutory demand set aside must be made within 18 days of the statutory demand being served.

11.28.6. There have been a number of examples where a successful party to adjudication proceedings tried unsuccessfully to enforce payment by the use of a statutory demand. A good example is the case of *Shaw* v *MFP Foundations and Piling* (2010), where the judge, in setting aside the statutory demand because of a genuine cross-claim, said:

> Where a statutory demand is founded on an adjudicator's decision, if the debtor can show that he has a substantial cross-claim, the insolvency regime does not contemplate that he should be shut out from raising those matters in opposition to bankruptcy proceedings simply because he could have, or even unsuccessfully did, also raise those matters before the adjudicator.

The judge, when arriving at a decision, took into account the fact that statutory adjudications are a 'pay now, argue later' approach to the dispute resolution process.

11.28.7. The service of a statutory demand can be a useful tool to secure payment, if the debtor is solvent but reluctant to pay the outstanding account. This can be due to any number of reasons, such as a temporary cashflow problem which is affecting the debtor's ability to pay. The receipt of a statutory demand may result in the debtor deciding to make payment a priority, when otherwise payment would have to await an improvement in the cashflow. If the service of a statutory demand has not resulted in a payment being made, a successful petition to the court to wind up the company will probably mean that the debt will never be paid.

11.28.8. Courts, however, have been very robust in their adverse criticism of using a statutory demand as a means of debt collection, as it is regarded as an abuse of the court process.

SUMMARY

A statutory demand is a written warning to a debtor to pay a debt which contains the threat that, in the event of a failure to pay, the court will be petitioned to liquidate the debtor. The debtor may have failed to pay the debt as a result of a *bona fide* dispute. This being the case, the court will set aside the statutory demand. The threat of having a company liquidated can have the effect of securing payment when a debtor is suffering a temporary cashflow problem. However, courts have been very critical of the use of a statutory demand as a means of collecting debts, as it is regarded as an abuse of the court process.

11.29. Where a claim is made by a subcontractor against a main contractor for matters such as delays which have been caused by the employer or its agents, can the subcontractor be forced to accept payment based upon a settlement made between the employer and main contractor in respect of these matters?

11.29.1. Subcontractors who incur additional expenditure due to delays are often advised by the main contractor that the cause of the delay lies with the employer or its agent, such as the architect or engineer. The main contractor in turn will probably have been the subject of delay, for which a claim has been made against the employer, including the subcontractor's claim. As a result of negotiations, the main contractor reaches a settlement with the employer which includes an amount in respect of the claim received from the subcontractor. In any legal proceedings between the main contractor and subcontractor, can the main contractor legitimately resist the claim from the subcontractor for the payment of more money than included in the settlement with the employer?

11.29.2. The subcontractor may argue the settlement made between the employer and the main contractor, which included a sum in respect of the subcontract work, was well below the subcontractor's entitlement. It may be considered that, as it was the intention to pass down to the subcontractor entirely the element of the settlement which related to the subcontract work, the main contractor did not press that element of the claim as forcefully as it ought to have done.

11.29.3. The English legal system encourages settlement in preference to legal proceedings. In pursuit of this objective, the courts have enforced settlements between disputing parties which bind third parties who are not parties to the action. In the case of *Biggin* v *Permanite* (1951), Biggin sold adhesive to the Dutch government which was defective and resulted in a claim being made. A settlement was reached and Biggin sought to recover the amount of the settlement from Permanite, who supplied the adhesive to Biggin. The Court of Appeal held that, provided the settlement was reasonable and even at the upper end of what was reasonable, then the amount of the settlement would have to be paid by Permanite.

11.29.4. The principle was more recently considered in the case of *John F Hunt Demolition* v *ASME Engineering* (2007). Kier Whitehall entered into a contract with Kier Build to construct new commercial premises. Kier Build in turn engaged Hunt Demolition to undertake demolition work. Hunt Demolition appointed ASME to provide a steel frame to support the building facade. ASME, whilst cutting steelwork on the site, started a fire which caused substantial damage. A claim was made by Kier Whitehall and Kier Build, in the sum of £248,145, against Hunt Demolition. The claim was passed down the line by Hunt Demolition to ASME, who together commissioned a quantity surveyor independently to value the sum claimed. A sum of £151,545 was included in the quantity surveyor's valuation and, as a result, £152,500 was offered by Hunt Demolition to Kier Build in settlement of the claim, which was accepted.

11.29.5. Hunt Demolition then sought to recover the amount they had agreed with Kier Build from ASME, who resisted the claim on the grounds that there was no liability under the contract between Kier Whitehall and Kier Build for the payment of this amount.

However, whilst accepting this as correct, Hunt Demolition argued that there would have been a liability under the law of tort. This was denied by ASME, who obviously looked to escape without making any payment.

11.29.6. The court concluded that the losses sustained by the fire were £108,987 by Kier Whitehall and £43,512 by Kier Build. Further, the court held that Kier Build had no liability in tort to Kier Whitehall, because of the wording of the main contract. The maximum liability for Hunt Demolition would have related to the loss of £43,512 sustained by Kier Build. The question to be answered was: whether the sum of £152,500 paid by Hunt Demolition to settle the dispute in respect of a claim which could not exceed £43,512, was reasonable. It would appear on face value that the settlement was unreasonable, being three times greater than the maximum liability. Judge Coulson did not agree, as, having examined all the authorities, he concluded that in deciding whether a sum paid by way of a settlement was reasonable did not depend upon it being proved that there was a liability to pay the amount of the settlement amount. He concluded that the reasonableness of a settlement is almost exclusively a matter of fact, so that the question whether the settlement between Kier Build and Hunt Demolition was reasonable was a matter of fact. This was not the end of the dispute, as Judge Coulson was required to deal with preliminary issues only. However, in the light of his observations, it is likely that the parties reached an out-of-court settlement.

11.29.7. With these decisions in mind, it is clear that a subcontractor can be bound by an agreement made between employer and main contractor relating to the subcontractor's work, in which the subcontractor has had no say in the matter. This is subject, however, to the settlement and the allocation of the sum agreed in respect of work subcontracted, based upon the facts, being reasonable. Main contractors are advised, when reaching a settlement with employers, to ensure they have a comprehensive build-up of how the settlement sum has been calculated, where the sum includes amounts which reflect work which is subcontracted. This will be essential if subsequently they are seeking to pay, or recover, amounts included in the settlement from subcontractors. Often contractors enter into a lump sum settlement figure with the employer without there being a build-up. This would restrict the options available to the contractor when attempting to reach a settlement with a subcontractor whose dispute with the main contractor formed a part of the deal struck with the employer.

SUMMARY

There is a principle in English law, where disputes arise, whereby negotiated settlements are encouraged. This principle, when applied, can involve third parties being bound by agreements which affect their financial entitlements, even though they have no involvement in the negotiations. For the settlement to be binding upon third parties based upon the facts, they must be reasonable.

Where a claim is made by a subcontractor against a main contractor for matters such as delays which have been caused by the employer or its agents, the subcontractor can be forced to accept a settlement made between the employer and main contractor in respect of these matters, provided that, based upon the facts, the settlement is reasonable.

11.30. Can a contractor or subcontractor refuse to commence work until satisfactory bank and trade references are provided?

11.30.1. One of the principal fears when providing goods or services is the inability on the part of the purchaser to make payment when due. Some comfort can be achieved if bank and trade references are produced by the purchaser. Should this information be slow in arriving, there may be a reluctance on the part of the contractor or subcontractor to commence work. A refusal to commence work would be understandable, but will the contractor or subcontractor be at fault in doing so?

11.30.2. A case came before the court in Scotland, *Miller Fabrications Ltd* v *JD Pierce (Contracts) Ltd* (2010), which had to decide this issue. Pierce was a subcontractor to Muir Construction in relation to a project in Edinburgh. Pierce subcontracted the supply and fix of a mezzanine floor to Miller. Miller fabricated the floor, but refused to deliver it to site unless Pierce provided bank and trade references. The trade references arrived, but none were provided by the bank, even though requested to do so. Muir attempted to intervene by asking Miller if they would work direct for them. Miller agreed, but Pierce objected to this arrangement. In the meantime, Miller did not commence work. Pierce, in frustration, notified Miller that the contract was terminated and made arrangements to undertake the work themselves. Miller's responded by letter dated 23 June 2004, saying they would complete the work if a banker's reference was provided. Pierce responded on 25 June, indicating that they had completed the work. The matter was then referred to court.

11.30.3. The Sheriff decided that a written quotation and verbal acceptance formed the basis of the contract between the parties, but did not include a term to the effect that bank and trade references would be provided. It was held that Pierce was entitled to terminate the contract, as Miller was not entitled to insist on bank and trade references in the absence of a requirement in the contract. Miller was ordered to pay Pierce the sum of £12,835.88.

11.30.4. Miller appealed to the Sheriff Principal, who supported the decision of the Sheriff. He concluded that Miller was not entitled to be provided with bank and trade references. The refusal to deliver the mezzanine floor was a material breach of contract, giving Pierce the right to terminate. If the refusal to perform is made before performance of the contract is due, this will be regarded as an anticipatory breach, giving the other party the right of termination.

SUMMARY

It makes good commercial sense to require the provision of bank and trade references, before committing resources to a contract. However, there must be a provision included in the contract that these references will be provided. In the absence of a provision in the contract, there will be no right to insist upon them being produced. Any attempt to apply pressure to have them produced by refusing to perform the contract will be considered a breach of contract. If the threat is made before the date required for performance, it may be regarded as an anticipatory breach of contract. In either case, a right of termination will arise.

Chapter 12
Practical Completion and Defects

12.1. How are practical completion and substantial completion under the JCT and ICE conditions defined?

12.1.1. A point of debate on building contracts frequently relates to practical completion. The issue of a certificate of practical completion by the architect will usually lead to a sigh of relief from the contractor. Any liability for liquidated damages ceases; the employer becomes obliged to insure; retention is released and the defects period begins to run. It is, therefore, not surprising that architects and contractors frequently argue as to whether or not practical completion has been achieved.

12.1.2. Where a dispute occurs, a first port of call is usually the conditions of contract. JCT 2011 does not, unfortunately, define practical completion. Clause 2.30, however, provides for the following:

> When in the Architect/Contract Administrator's opinion practical completion of the Works is achieved . . . he shall forthwith issue a certificate of practical completion.

12.1.3. The question of practical completion under a JCT form of contract has been the subject of referral to the courts on more than one occasion. Unfortunately, there is still no precise definition.

12.1.4. In *J. Jarvis & Sons* v. *Westminster Corporation* (1978) the House of Lords had to decide whether the main contractor was entitled to an extension of time under JCT 63, where delays occurred due to remedial works undertaken by the nominated piling subcontractor after the piling work was completed. Practical completion of the subcontract works became relevant, as the court held that no extension of time was due in respect of delays caused by remedial works to the piling, after the piling had been completed. Lord Justice Salmon started the ball rolling with this version of what is meant by practical completion:

> I take these words [practical completion] to mean completion for all practical purposes, that is to say for the purposes of allowing the employer to take possession of the works and use them as intended. If completion in clause 21 meant completion down to the last detail, however trivial and unimportant, then clause 22 would be a penalty clause and as such unenforceable.

200 Contractual Problems and their Solutions, Third Edition. Roger Knowles.

Contractors no doubt were more than pleased with this definition, but unfortunately the waters were muddied with the definition given in the same case by Lord Dilhorne, who said, in a dissenting judgment:

> The contract does not define what is meant by practical completion. One would normally say that a task was practically completed when it was almost but not entirely finished, but practical completion suggests that that is not the intended meaning and what is meant is the completion of all the construction that has to be done.

12.1.5. The courts made a further attempt at a definition in *H.W. Nevill (Sunblest) Ltd* v. *William Press & Son Ltd* (1981). In this case defects occurred after practical completion of a preliminary works contract, which delayed a follow-on contract. Again, what constitutes practical completion was relevant. It would appear that the judge in this case sided with the views of Lord Dilhorne, in saying:

> I think the word 'practically' in clause 15(1) gave the architect a discretion to certify that William Press had fulfilled its obligation under clause 21(1), where very minor *de minimis* work had not been carried out, but that if there were any patent defects in what William Press had done, the architect could not have given a certificate of practical completion.

12.1.6. Practical completion was at issue in *Emson Eastern Ltd* v. *EME Developments Ltd* (1991). Emson were the contractors and EME developers for the erection of business units. JCT 80 formed the basis of the contract. Judge John Newey QC, in arriving at the meaning of completion of the works, took account of what happens on building sites. He considered that he should keep in mind that building construction is not like the manufacture of goods in a factory. The size of the project, site conditions, use of many materials and employment of various types of operatives make it virtually impossible to achieve the same degree of perfection as can a manufacturer. His view was that it must be rare for a new building to have every screw and every brush of paint correct. Further, a building can seldom be built precisely as required by the drawings and specification. Judge Newey, in considering the meaning of practical completion, thought he stood somewhere between Lord Salmon and Viscount Dilhorne in the *Jarvis* case.

12.1.7. The Court of Appeal of Hong Kong in *Big Island Contracting (HK) Ltd* v. *Skink Ltd* (1990) had to decide whether practical completion had been achieved. The plaintiffs were contractors for work on the 12th and 13th floors of the defendant's building under a contract which provided payment of 25% of the price upon practical completion. The defendants went into occupation of the building. The plaintiffs contended that the works had been practically completed and issued proceedings seeking 25% of the agreed price, i.e. nearly HK$100,000. District Judge Yam found that practical completion had not been achieved, since there were defects which would have cost between HK$40,000 and HK$60,000 to rectify and that the plaintiffs had failed to execute an important part of the modification of the sprinkler system on the 13th floor, for which a provisional sum of HK$20,000 had been allowed. The judge found that this defect affected the safety of the system and would take between two and ten days to correct. He gave judgment for the defendants. The plaintiffs' appeal against the decision was dismissed. Occupation by the employer does not in itself, therefore, constitute practical completion.

12.1.8. The judge in another Hong Kong case, *Mariner Hotels Ltd* v. *Atlas Ltd* (2006), had to decide whether the fact that occupation of the hotel constituted practical completion. It was claimed on behalf of the hotel that practical completion would not be achieved until 'a state of affairs in which the hotel has been completed free from any patent defects other than those ignored as trifling' was reached. The contractor argued that, once the hotel had been handed over and was capable of being operated, practical completion had been achieved, even though certain works remained incomplete. The judge preferred the case presented on behalf of the hotelier.

12.1.9. The case of *Memolly Investments* v. *Cerap* (2009) involved the purchase of a development. The agreement for sale required the buildings to be practically completed before they were to be taken over by the purchaser. The case revolved around whether or not the buildings had achieved practical completion. Following an examination of the judgement in the case of *J Jarvis and Sons* v. *Westminster Corporation* (1978), the judge concluded that:

> Practical completion is perhaps easier to recognise than to define. No clear answer emerges from the authorities as to the meaning of the term.

12.1.10. A comprehensive definition of 'substantial completion' as it applies to the ICE conditions appears in *Engineering Law and the ICE Contracts*, 4th edition, by Max W. Abrahamson, at page 160, as follows:

> The *Concise Oxford Dictionary* equates 'substantial' with 'virtual' which is defined as 'that is such for practical purposes though not in name or according to strict definition'. It is at least clear on the one hand that the fact that the works are or are capable of being used by the Employer does not automatically mean that they are substantially complete ('any substantial part of the Works which has both been completed . . . and occupied or used') and on the other hand that the Engineer may not postpone his certificate under this clause until the works are absolutely completed and free of all defects. The many reported cases on the question of 'substantial' completion in relation to payment under an entire contract, a different legal problem, are of doubtful relevance. Obviously both the nature and extent of the uncompleted work or defects are relevant, and to say that substantial completion allows for minor deficiencies that can be readily remedied and which do not impair the structure as a whole is probably an accurate summary of what is a question of fact in each case.

12.1.11. The difference between JCT 2011 and the ICE 6th and 7th Editions is that under the latter the engineer may issue a certificate of substantial completion with outstanding work still to be done, provided the contractor gives an undertaking to complete the outstanding work in the maintenance period.

12.1.12. The Engineering and Construction Contract (NEC 3) takes a different approach from the JCT and ICE contracts by providing a definition of 'Completion', which occurs when the contractor has:

- done all the work which the Works Information states he is to do by the Completion Date;
- corrected notified Defects which would have prevented the Employer from using the works and others from doing their work.

12.1.13. The RIAI (2002) Edition, used in the Republic of Ireland, under clause 31 states that practical completion means 'the works have been carried out to such a stage that they can be taken over and used by the Employer for their intended purpose . . .'

SUMMARY

Disputes concerning whether or not a building has achieved practical completion are common, as difficulties often arise in deciding at what point has it been achieved. The judge in the *Menolly* case considered that practical completion is easier to recognise than define. From the various authorities available on this subject, Judge Newey in *Emson Eastern* v. *EME* seems the most sensible. He advised architects that, when issuing a certificate of practical completion, they should bear in mind that construction work is not like manufacturing goods in a factory. His view was that it must be rare for a new building to have every screw and every brush of paint correct. NEC 3 and the RIAI contract define completion, which is very useful.

12.2. Where an employer takes possession of a building or engineering facility before all the work has been completed, can the contractor rightfully claim that practical completion or substantial completion has been achieved in relation to the part taken over?

12.2.1. Some contracts specifically state that, where a building or facility is completed to the stage where it is fit for occupation and by the employer, then practical completion has been achieved. The RIAI (2002) Edition used in the Republic of Ireland, under clause 31, states that practical completion means 'the works have been carried out to such a stage that they can be taken over and used by the Employer for their intended purpose. . . . Other contracts, such as those published by the JCT, merely make reference to the architect being obliged to issue a certificate of practical completion when the works have reached the stage of being practically completed.

12.2.2. Practical completion of the works is a defining moment on most construction contracts. Retention becomes due for release, liability for damage passes to the employer and, of course, any obligation on the contractor's part to pay liquidated damages ceases. Most standard forms of contract include both a clause for practical completion of the whole of the works and also for the employer taking possession of part of the works. The wording normally states that if the employer, during a period of overrun, takes possession of part of the works before practical completion of the whole of the works, then the contractor's obligation to pay liquidated damages will be at a reduced level.

12.2.3. The decision in the case of *Skanska Corporation* v. *Anglo-Amsterdam Corporation* (2002) shows that complications can arise where part possession by the employer occurs. Skanska Corporation undertook to construct an office facility in Edinburgh for the Anglo-Amsterdam Corporation, employing a JCT 81 With Contractor's Design standard form of contract. Clause 16 of the contract deals with practical completion and the process requires the employer's agent to provide a written statement to indicate when it has been achieved. The standard wording had been amended to read that the statement will only be issued when the employer's agent is satisfied that any unfinished work is very minimal and of a minor nature. A dispute arose between the parties as to the date by which the contractor had achieved practical completion. The matter was referred to an arbitrator, who had to decide whether practical completion took place on 12 February 1996 or 25 April 1996.

12.2.4. It seems that Anglo-Amsterdam had leased the premises to ICL, who were anxious to gain access to enable the fit-out to commence. By the time 12 February 1996 had arrived, work was still incomplete. The main problem was the air-conditioning, which was not fully functioning, and Skanska's failure to produce operating and maintenance manuals. The seasoned observer may consider that this is not unusual. The situation, however, should not hold up the fit-out and, therefore, ICL moved in. Skanska did not finish off the incomplete items until 25 April 1996, and as a result Anglo-Amsterdam levied liquidated damages for a failure to complete on time. Skanska argued that, as ICL had moved in on 12 February 1996, liquidated damages could not be deducted for the period after that date. This called for a close examination of the wording in the contract. Clause 16 was stringent, bearing in mind that there was no entitlement to an employer's agent's statement that practical completion had occurred if work, except for that of a very minimal and minor nature, was still outstanding. No help here for Skanska, as the air-conditioning was not working by 12 February 1996. The wording of clause 17 did not seem to apply either, as it only dealt with a situation where the employer takes possession of a part of the works, in other words partial possession. What appears to have occurred is that ICL took partial possession of the whole of the works. They did not take full possession of the whole of the works, as Skanska were engaged in finishing off the work, while ICL were undertaking the fit-out.

12.2.5. Judge Thornton QC took a very mature and practical view of the situation in finding in favour of Skanska. Clause 17, which deals with partial possession, states that if the employer takes over a part of the building, then as far as that part of the works is concerned the contractor is deemed to have achieved practical completion. Responsibility for the part taken over rests squarely with the employer for such matters as damage and health and safety. This would apply, even if the clause 16 definition of practical completion had not been achieved. Whilst the contract did not deal with the situation where the employer takes possession of the whole of the works before practical completion, the same principle should apply as where the employer takes possession of part of the works. In other words, practical completion is deemed to have taken place. The word 'deemed' is an interesting one, in that it means that for all legal purposes an event has happened, even if as a matter of fact it has not. In other words, if practical completion is deemed to have occurred it would not matter if the building had no roof on. In this case, it was only the air-conditioning which was not completed.

12.2.6. It is not in all cases of employers gaining access that practical completion occurs, which became clear in the case of *Impresa Castelli SpA* v. *Cola Holdings Ltd* (2002). Clause 23.3.2 of the JCT contracts gives the employer an entitlement to gain access to part or all of the works. This presence on the site does not affect his entitlement to deduct liquidated damages. Cola, the employer, engaged Impresa Castelli for the construction of Kingsley Hall Hotel in Great Queen Street, London. Work was running late, an obligation to pay liquidated damages arose and Cola levied a claim of £1.2m. Impresa Castelli argued that Cola had taken partial possession of the whole of the works under clause 17.1 and in keeping with the *Skanska* decision no liquidated damages were payable. Judge Thornton, who heard the *Skanska* case, did not agree. Based upon the facts, he considered that occupation by Cola did not amount to taking partial possession as envisaged by clause 17.1. The situation led the judge to believe that Cola had merely gained access to the site, as envisaged by clause 23.3.2, and therefore liquidated damages were payable. There is obviously a significant difference between an employer taking partial possession of the building and one who merely is allowed by the contractor to have access.

12.2.7. The ICE 7th Edition, under clause 48(4), provides for the engineer to issue a certificate of substantial completion in respect of parts of the work. Access to the site is provided for under clause 37. Whilst the wording in the ICE conditions is not as detailed as the JCT, it is considered that if the ICE 7th Edition conditions had applied in the *Skanska* and *Impresa Castelli* cases, the decisions of the court would have been no different. The Engineering and Construction Contract (NEC 3) is silent on the matter of partial possession and access.

SUMMARY

It would make the situation clear if contracts state that, should the employer at any time before practical completion of the works take possession of a part of or the whole of the works, then practical completion of the part or the whole of the works taken into possession is deemed to have taken place. In the absence of such wording, but in the light of the decision in *Skanska Corporation* v. *Anglo-Amsterdam Corporation* (2002), it is likely that, if an employer takes possession of a building or facility, practical completion will be deemed to have occurred. This, however, may be confused with access being granted to the employer where practical completion is not deemed to have occurred. There is always likely to be scope for argument as to whether an employer has been granted occupation, when practical completion will be deemed to have occurred, to be contrasted with the employer gaining access only, which does not involve practical completion.

12.3. When does practical completion occur under a JCT Standard Form of Building Subcontract?

12.3.1. It is important for contractors or subcontractors to establish with some precision the date when practical completion of the subcontract has been achieved. Uncertainty may affect the contractor's right to claim for late completion against the subcontractor. Liability for damage to the subcontract works and release of retention are also affected.

12.3.2. In the case of *Vascroft (Contractors) Ltd* v. *Seeboard plc* (1996) a dispute arose as to when practical completion of a subcontract works occurred. The conditions of the subcontract were DOM/2 and the dispute was referred to arbitration in accordance with the provision of the contract. Under DOM/2, clause 14.1, the subcontractor is required to notify the contractor, in writing, of the date when, in his opinion, the subcontract works were practically completed. Clause 14.2 states that if the contractor dissents in writing within 14 days of receipt of the notice, practical completion occurs on such date as is agreed and, in the absence of agreement, practical completion is deemed to have occurred on the date of practical completion of the main contract works, as certified by the architect. If the contractor fails to respond within 14 days of receipt of the subcontractor's notice, practical completion of the subcontract works will be deemed to have occurred on the date indicated in the notice. The Arbitrator found in this case, where the subcontractor failed to provide a notice as required by the terms of the subcontract, practical completion would still not be deemed to have occurred until the Architect had certified completion of the main contract. Vascroft appealed against this finding. The judge said:

> In my judgment, where, as here, the parties have entered into a standard form of contract (with some amendments to bring it into line with the nature of the work) it is necessary to look at the whole of the contract. It is quite clear that there are many provisions of this contract which depend for their effective operation on a firm date for practical completion being established under clause 14, either by operation of the mechanism there provided or by agreement.
>
> The machinery of clause 14 starts with a notice being given by the subcontractor. In my judgment, the words 'The subcontractor shall notify the contractor in writing of the date when, in his opinion, the subcontract works are practically completed' impose an obligation on the subcontractor to do so. The contractor will undoubtedly have his own view as to whether or not the subcontract works are complete, but this contract does not require him to express it unless and until he receives a notice from the subcontractor. If the subcontractor had given a notice, but the contractor duly dissented from it and no agreement could be reached, then the deeming provisions would establish a firm date, i.e. the date of practical completion of the main contract works. I do not consider it right that, if a subcontractor is in breach of his obligation to serve written notice, the date for practical completion fails to be established as a matter of fact. If this interpretation were correct, it could, in many instances, suit a subcontractor not to give any notice at all rather than a notice which he knew would not be accepted.

The judge concluded that in the absence of a written notice from the subcontractor, the situation with regard to practical completion would be the same as would occur if the subcontractor served notice which was dissented from by the contractor and no agreement reached. In other words, in the absence of notice, practical completion of the subcontract will occur on the date of practical completion of the main contract as certified by the architect. This type of requirement puts pressure on the subcontractor to ensure the notice is submitted and if the contractor dissents, to agree a practical completion date with the main contractor. Any failure to do so results in practical completion of the subcontract being delayed until the main contract works have been certified as complete. The wording in clauses 2.20.1 and 2.20.2 of the Standard Building Subcontract 2011 is almost identical to that in clauses 14.1 and 14.2 of DOM/2, except that where the main contractor dissents, he is obliged to give reasons.

SUMMARY

Clause 14.1 of DOM/1 and DOM/2 requires the subcontractor to serve notice on the main contractor when in his opinion the subcontract works are practically completed. If the main contractor dissents in writing, practical completion occurs on such date as is agreed. In the absence of agreement, practical completion is deemed to occur on the date of practical completion certified by the architect under the main contract.

If the subcontractor fails to serve notice, then again practical completion of the subcontract work will be deemed to occur on the same day as practical completion certified by the architect under the main contract. The wording in clauses 2.20.1 and 2.20.2 of the Standard Building Subcontract 2011 is almost identical, except that where the main contractor dissents he is required to provide reasons. Subcontractors operating under DOM/1, DOM/2 or the Standard Building Subcontract 2011, or subcontracts containing similar wording are advised to ensure that written notice is sent to the main contractor as soon as the subcontractor considers his work has been completed. If the main contractor dissents, the subcontractor needs to agree a completion date to avoid the completion date becoming the same as that under the main contract.

12.4. Where at the end of the defects liability/rectification period/maintenance period the architect/engineer draws up a defects list but, due to an oversight, omits certain defects and a second list is prepared after the defects on the first list have been completed, will the contractor/subcontractor be obliged to make them good?

12.4.1. To prevent stale claims being levied for breach of contract, the Limitation Acts lay down periods of time within which actions must be commenced. These periods of time seem generous. The Limitation Act 1980 states, by section 7:

> An action to enforce an award, where the submission is not by an instrument under seal, shall not be brought after the expiration of six years from the date on which the cause of action accrued.

and by section 8:

> An action upon a speciality shall not be brought after the expiry of twelve years from the date on which the cause of action accrued.

This means, in reality, that if the contract is a verbal or written contract, an action will be time barred if not commenced within six years of practical completion. Where the contract is a speciality, in other words under seal or expressed as a deed, the period is 12 years. Where a construction contract includes a defects or maintenance period which is less than the timescales laid down in the Limitation Acts, do the provisions of the defect or maintenance clauses shorten the effect of the Limitation Acts?

12.4.2. JCT 2011 and other JCT forms, ICE 6th and 7th Editions and GC/Works 1/, 1998 Edition and the Engineering and Construction Contract (NEC 3) all include a defects liability or maintenance period. The purpose of these defects or maintenance periods is to allow the contractor an opportunity of making good his own defects. Whilst it may not be obvious to many contractors and subcontractors, these clauses actually bestow a benefit upon them. In this context *Keating on Building Contracts*, 5th edition, at page 247, states:

> The contractor's liability in damages is not removed by the existence of a defects clause, except by clear words, so that in the absence of such clear words, the clause confers an additional right and does not operate to exclude the contractor's liability for breach of contract . . . But it is thought that most defects liability clauses will be construed to give the contractor the right as well as to impose the obligation to remedy defects which come within this clause.

In other words, defects in construction work amount to a breach of contract entitling the employer to claim damages: *HW Nevill (Sunblest)* v. *William Press and Son* (1981). In the absence of a defects clause and where defects in the work appear after practical completion, the employer would be within his rights to employ others to make good the defects and to charge the contractor with the cost. The presence of a defects clause gives the contractor the right to remedy his own defects, the cost of which should be less than would be the case if others carried out the work.

Hudson's Building and Engineering Contracts, 11th edition, at paragraph 5.050 says:

> Since such work can be carried out much more cheaply and possibly more efficiently by the original contractor than by some outside contractor brought in by the building owner, defects clauses in practice confer substantial advantages on both parties to the contract.

12.4.3. JCT 2011, under clause 2.38, requires the architect to prepare a schedule of defects which is to be issued not later than 14 days after the end of the rectification period. No specific reference is made in either the ICE 6th or 7th Edition, GC/Works/1 or the Engineering and Construction Contract (NEC 3) to a defects list, but it is nonetheless common practice.

12.4.4. A question often raised is whether a contractor is obliged to make good defects if, under a JCT 2011 contract, the architect produces the defects schedule outside the 14-day period or, alternatively, if, having issued a schedule with which the contractor has complied, he produces a second schedule which lists more defects. When answering the question, the comments made earlier should be borne in mind. Defects in the contractor's works amount to breaches of contract for which the employer is entitled to damages and this is not excluded by the defects clause. All the defects clause does is to give the contractor the right to make good defects. The contractor would be well advised to make good defects on this and subsequent lists, if these situations were to occur. Should the contractor refuse to make good defects on a list which was issued outside the 14-day period, the employer would be entitled to deal with the matter like any other breach of contract. He could have the correction work undertaken by someone of his choosing and send the bill to the contractor.

12.4.5. The case of *Pearce and High Ltd* v *Baxter* (1999) well illustrates this point. It was held that an employer could recover damages for defective work the presence of which was known and which should have been included on the defects list but wasn't.

SUMMARY

It would seem that a failure by the architect/engineer to issue a defects schedule on time or the issue of a second one would not amount to a waiver of the employer's rights. The contractor may be able to make out a case that a payment should be made for the costs of making two visits as a result of the architect's failure to include all defects on the original list. It is, however, advisable for him to take the cheaper option of exercising his right to remedy his own defects, rather than have the employer call in an outsider and sue for the costs. The defects or maintenance clauses do not overrule the provisions of the Limitation Acts.

12.5. Is a contractor/subcontractor absolved from any liability if the employer refuses him access to make good defects because he chooses to make them good himself?

12.5.1. The standard forms of contract provide for contractors to make good defects following practical or substantial completion during the defects period. Clause 2.38 of JCT 2005 provides for the architect to give instructions to the contractor that certain or all defects are not to be made good. This being the case, the clause goes on to provide that an appropriate deduction is to be made from the contract sum. No assistance is given in the clause as to how the amount of deduction is to be calculated. Most of the other standard forms make no provision for making good defects to be omitted.

12.5.2. Problems can arise where employers make their own arrangements to make good defects. The decision in *William Tomkinson & Sons Ltd* v. *The Parochial Church Council of St Michael and Others* (1990) aptly deals with this point. William Tomkinson, the contractor, was employed by the church council to carry out restoration works at a parish church in Liverpool. The contract was let using the JCT Minor Works 80. A dispute arose concerning certain defects and damage to the works, including damage to plasterwork caused by hammering, damage to woodblock flooring, plasterwork, roof-lights and their flashings, plus many more. The contractor's defence was that either the damage was not of his making or, alternatively, he was protected by the provisions of clause 2.5 of the conditions of contract. This clause makes the contractor liable for any defects, excessive shrinkage or other faults which appear within three months of the date of practical completion because materials or workmanship are not in accordance with the contract. The defects and damage which formed the basis of the church council's case had been discovered and remedied by other contractors, on instruction of the church council, prior to practical completion. Both parties agreed that for clause 2.5 to be effective, notice of the defects must be given to the contractor. It was also common ground that some, but not all items of defect or damage, resulted from defective workmanship by the contractor. No warning was given to the contractor prior to the church council instructing other contractors to carry out the work. It was argued on behalf of the contractor that the church council, in arranging for having defective work corrected by others, prevented the contractor from exercising his rights to correct defects, and therefore the contractor had no liability.

12.5.3. The court, in arriving at a decision, was influenced by the wording in *Hudson's Building and Engineering Contracts*, 10th edition, at pages 394–7:

> It is important to understand the precise nature of the maintenance or defects obligations. It is quite different from the Employer's right to damages for defective work, under which he will be able to recover the financial cost of putting right work either by himself or another contractor. Since maintenance (defects) can usually be carried out much more cheaply by the original contractor than some outside contractor . . . so the contractor not only has the obligation but also in most cases it is submitted the right to make good at its own cost any defects.

The court went on to consider that workmanship which falls short of the standard required by the contract, and which the employer remedies, prior to practical completion, still constitutes a breach of contract. It was held, in finding in favour of the church council, that their entitlement was to recover damages from the contractor, subject to proof that these were attributable to workmanship or materials which fell below the contractual standard. The amount of damages which the church council would be entitled to recover was not, however, their outlay in remedying the damage, but the cost which the contractors would have incurred in remedying them if they had been required to do so; a sum anticipated to be much less than the actual remedial costs. The answer to the question, therefore, is that a contractor is not absolved from liability if he is refused access, but damages recoverable by the employer are limited to the amount it would have cost the contractor had he been given an opportunity to make good the defects.

SUMMARY

JCT 2011 gives the architect power to instruct the contractor not to make good defects, in which case an appropriate deduction will be made from the contract sum. Most of the other standard contracts make no such provision. If defects are not made good by the contractor, but the employer arranges for the work to be carried out by others, the employer will only be entitled to recover from the contractor what it would have cost the contractor, had he in fact made good the defects. Agreement as to what those costs might have been could be difficult to achieve.

12.6. Most subcontracts provide for the release of the final balance of retention only when the period included in the contract for correcting defects has expired and all defects under the main contract have been made good. If the main contractor or other subcontractors are dilatory in making good defects, is there any mechanism to enable the subcontractor to secure an early release of retention?

12.6.1. Subcontractors regularly complain of the huge amounts of money tied up in retention. The sums usually exceed a year or two's profits. Where a JCT form of contract is employed, subcontractors usually have to wait until the architect has issued the certificate of making good defects under the main contract before they receive the final balance of retention. This will only occur when all defects have been made good to the architect's satisfaction. Other standard forms of contract, such as GC/Works/1 and ICE, are similarly worded. Subcontractors are often disadvantaged when they have corrected their own defects but the certificate of making good defects has not been issued, due to a failure on the part of the main contractor or other subcontractors.

12.6.2. In the case of *Pitchmastic* v. *Birse Construction* (2000) a dispute arose between the main contractor and a subcontractor concerning the release of retention. The main contract, for the construction of a new distribution centre for Tesco in Milton Keynes, employed a JCT With Contractor's Design contract. Pitchmastic were subcontractors for the roofing works, the subcontract being an amended DOM/2. The subcontract works were completed, according to Pitchmastic, on 26 February 1998, followed by practical completion of the main contract works on 22 May 1998. A defects liability period of 12 months was included in the main contract. As at 21 May 1999, Pitchmastic's work was defect-free, but defects remained to be corrected under the main contract. The final half of retention, in the sum of £33,651, was held in respect of Pitchmastic's work. Pitchmastic commenced an action for the recovery of balances it claimed to be due under the final account, including release of the retention.

12.6.3. Birse Construction argued that release of retention was not due until the certificate of making good defects had been issued under the main contract. In finding in favour of the main contractor, Judge Dyson said:

> It is not sufficient for Pitchmastic merely to show that the architect withheld the certificate because there were outstanding defects in the work that had been carried out by Birse, or other subcontractors. Provided that Birse and its subcontractors were proceeding with reasonable diligence, to make good the defects, Birse was not preventing the issue of the certificate. Some defects take longer to put right than others.

SUMMARY

A situation may arise where the period for correcting defects has expired. One of the subcontractors has completed making good defects in its work, but there is still defects correction required on the part of the main contractor and other subcontractors. It was held in *Pitchmastic* v. *Birse Construction* (2000) that, provided the correction of outstanding defects is proceeding with regular diligence, the contractor will not be obliged to release the outstanding retention to the subcontractor.

12.7. Can an employer recover from the contractor the costs involved where it became necessary to employ an external expert to demonstrate that work was defective?

12.7.1. Telling a contractor that the defects liability period provides him or her with a benefit is likely to be an invitation for a rude response. Most contractors consider that an obligation to return to site some six, nine or twelve months after the architect or engineer has issued a certificate signifying that satisfactory completion has been achieved provides no discernable advantage.

12.7.2. The decision in the Scottish case of *Johnston* v. *W.H. Brown Construction (Dundee) Ltd* (2000) provides otherwise. A contract was let to construct an office block employing the Scottish Building Contract With Contractor's Design. Following practical completion, defects appeared in the external envelope. The employer appointed a firm of architects, HRP, to investigate. Their report identified a number of defects resulting from a failure to follow the Building Regulations. These defects were remedied by the contractor, without admitting liability. The employer commenced an action against the contractor to recover the fees paid to the architect, together with solicitor's fees in connection with the provision of contractual advice and the employer's management costs. In all, a sum of around £80,000 was claimed.

12.7.3. The court found against the employer, in that clause 16 of the contract provides its own remedy where defects are discovered during the defects period. It is the duty of the employer's representative to produce a defects schedule at the end of the defects period and the contractor's obligation to put them right. If the contractor carries out the remedial work, the employer has no other remedy.

12.7.4. There is no provision in the contract which allows the employer to recover the costs of identifying the defects, taking advice, instructing the contractor and subsequently inspecting the remedial works. In the absence of such a clause, defects appearing after completion would be treated as a breach of contract, leaving the employer to recover the expenditure as common law damages.

SUMMARY

Where the standard forms of contract are used, an employer is unable to recover the cost of employing external consultants to demonstrate that work is defective. The situation may be different where standard forms are not used, when this type of cost may be recovered as damages for breach of contract.

12.8. Where a dispute arises between employer and contractor which includes defective work carried out by a subcontractor and is the subject of legal proceedings which are settled by the contractor making a payment to the employer, is the subcontractor obliged to reimburse the contractor for the cost of remedying defective work, even though he considers the subcontract works contain no defects?

12.8.1. Where a dispute arises which involves employer, contractor and subcontractor, difficulties can arise in getting all three to agree to a compromise arrangement. It is often the case that legal proceedings are commenced involving all three. Pressure is applied for the parties to settle the dispute and it is not uncommon for the contractor and employer to reach a settlement, leaving the contractor to arrive at an accommodation with the subcontractor. The contractor will usually attempt to impose upon the subcontractor that part of the settlement with the employer which relates to the subcontractor's work.

12.8.2. In the case of *Bovis Lend Lease Ltd* v. *RD Fire Protection Ltd* (2003), a dispute arose concerning a claim from the management contractor, Bovis, against RD Fire Protection, a subcontractor who undertook the dry lining and fire protection. Bovis had been appointed as the management contractor on the Braehead Shopping Centre in Glasgow. A major row broke out between Braehead, the developer, and Bovis, with claims flying in both directions. Bovis alleged that they were owed £37m as the balance due on the final account and damages for breach of contract, whilst Braehead counterclaimed in the sum of £65.8m in respect of alleged mismanagement, defective work and overvaluation of the works. Proceedings were commenced and as part of the process Bovis joined RD Fire Protection as a joint defendant on the basis that part of the counterclaim related to the quality of work undertaken by RD Fire Protection. Bovis and Braehead entered into a compromise arrangement, whereby Braehead paid Bovis a sum of £15m in full and final settlement of all matters in dispute, which included the alleged defective work carried out by RD Fire Protection.

12.8.3. The settlement reached between Braehead and Bovis was on a global basis, in that there was no build-up which showed how the amount had been calculated. The cost of correcting the defective work formed part of the counterclaim, but it was not clear whether or not the agreed figure included any provision in respect of the alleged defective work. Bovis, however, claimed from RD Fire Protection the full cost of the remedying the RD work as pleaded by Braehead. They sought to bypass the settlement entirely and proceed as if it had never taken place. This was obviously wrong, as Bovis were only entitled to recover whatever loss they could prove resulted from the defective work. It was possible for there to be some provision in the global settlement figure for the defective work; the more likely explanation, however, was that as the settlement figure was on a global basis, the sum could not be identified. It was open therefore for Bovis to request the court to ascertain the sum involved, which could then be passed on to RD Fire Protection. Bovis did not use this approach, but instead passed on the total sum originally claimed by Braehead. The court rejected the case brought by Bovis and found for RD Fire Protection.

12.8.4. It is possible to pass down to a subcontractor a part of a settlement agreement reached with an employer, if it can be shown that the sum to be passed down is reasonable. This situation was explored in *Biggin* v. *Permanite* (1951), *P&O Developments* v. *The Guys and St Thomas's National Health Trust* (1998) and *Sainsbury* v. *Boardway Malyan Earnest Green* (1998). It was decided in these cases that the question is not whether the plaintiff who entered into the compromise and seeks a contribution acted reasonably, but whether the settlement was a reasonable one. This being the case, that part of the reasonable settlement which is due to fault on the part of the third party is recoverable from the third party.

SUMMARY

When a dispute between an employer and contractor involving defective work carried out by a subcontractor is settled, the contractor cannot as of right recover the amount included in the settlement in respect of the subcontractor's work. It is open to the contractor, if the dispute is the subject of legal proceedings, to request the court or arbitrator to identify the sum in a settlement agreement. In the absence of the amount being identified by the court or arbitrator, then the amount which the subcontractor is obliged to pay must be a reasonable sum.

12.9. What is the difference between patent defects and latent defects?

12.9.1. Defects, which may be latent or patent, can arise on construction projects due to design, materials or workmanship which fail to comply with the requirements of the contract. Patent defects are those which, on reasonable inspection, are plain, evident or conspicuous and can normally be seen with the naked eye. By way of contrast, Mr Justice David Steel, in the case of *Baxall Securities Ltd* v *Sheard Walshaw Partnership* (2002), explained that a defect was only latent when:

> it would not be discovered following the nature of inspection that (one) might reasonably anticipate the article (i.e. the building) to be subjected to.

The standard forms of contract in regular use and legal principles deal with patent and latent defects in different ways.

Patent Defects

12.9.2. When a contractor has completed the work included in the contract, most standard forms of contract provide for the employer's representative to issue a document to signify that the work has been completed in accordance with the requirements of the contract. The JCT contracts provide for the architect or contract administrator to issue a certificate; in the case of the Design and Build contract, the employer issues a statement to the effect that the works have reached practical completion. The certificate or statement will not be issued if there are any patent defects which still require putting right. Some contracts, such as the ICE 7th Edition and FIDIC, allow for undertaking small parts of the works after completion.

12.9.3. Most standard forms of contract provide for a period of time to be identified following practical completion, sometimes referred to as a maintenance period, defects correction period, or rectification period. This period normally extends to 12 months following practical completion. A 12-month period is usually chosen to provide for the works to be occupied and used whilst exposed to the weather conditions during the four seasons of the year. Once this period has elapsed, there is usually a process which involves an inspection of the works to identify any faults which may have appeared. In the case of the JCT contracts, a list of 'defects, shrinkages or other faults in the works' is prepared by the architect or contract administrator, the employer in the case of the Design and Build contract. The list must be presented to the contractor not more than 14 days after the end of the rectification period.

12.9.4. Some forms of contract, the JCT being a good example, provide for a document to be issued to the contractor which signifies that the defects notified at the end of the aforementioned period have been remedied. Where an architect or contract administrator is involved, the document is referred to as a certificate of making good; where a Design and Build form of contract is employed, the document is referred to as a notice of completion of making good.

Latent Defects

12.9.5. Defects which come to light after the patent defects have been made good and the appropriate documentation issued are normally referred to as latent defects. The contractor is not excused in any way for latent defects merely because of the issue of documentation which signifies that the employer or his representative is satisfied at the time the documentation was issued. However, there is a limited period of time within which the employer can commence an action with regard to these latent defects to avoid falling foul of the Limitation Acts and being regarded as out of time.

12.9.6. Where latent defects come to light, the employer had a choice of remedies. He could have opted to claim that the latent defect represents a breach of contract, in that some element of the works has not been carried out as required by the contract. Alternatively, he was able to claim that the defective work results from negligence on the part of the contractor and commence an action under the law of tort. These rights, however, have been restricted by the decision in *Robinson* v *Jones* (2011), in which it was held by the Court of Appeal that there could be no concurrent rights in both contract and tort in respect of work undertaken by a builder. The reason being that it was held by the court that the builder does not take on an assumption of responsibility to his customer outside of the terms of the contract. The court, however, held that the decision does not apply in the case of the provision of professional services, as the professional does take on the assumption of responsibility independent of the contract between the parties.

12.9.7. The Limitation Act 1980 applies to a breach of contract action, which in the case of a simple contract must be commenced within six years of the date the action accrued. If the contract is a deed, the period is 12 years from the date the action accrued. The action accrues on the date of the breach. If the defect is discovered outside the limitation period, then the employer will lose any rights to bring an action for breach of contract against the contractor. There is, however, some consolation for employers in that the date of the breach is often not the time of the act of constructing the defective work, but on practical completion of the whole of the work: *Thameside Metropolitan B.C.* v. *Barlow Securities Group* (2001).

12.9.8. The limitation period may be extended where any fact which is relevant to the claimant's right of action has been concealed from them by the defendant. This is well illustrated by the case of *King* v *Victor Parsons* (1973), where in 1962 a developer sold to the plaintiff a plot of land on which the foundations and concrete floor slab had already been constructed, together with two courses of brickwork. The developer undertook to complete the work to the plaintiff's reasonable satisfaction. The plaintiff went into occupation in 1962. Large cracks appeared in 1968, because the house had been built on a rubbish tip. The developer had ignored advice as to the type of foundations required. As this information had not been conveyed to the plaintiff, it amounted to deliberate concealment and, hence, the cause of action accrued in 1968.

12.9.9. Due to defects often coming to light after the limitation period has expired, employers could, if the action was founded on breach of contract, find themselves out of time. However, an action in the tort of negligence may present itself, in which case the limitation period is different from that relating to breach of contract. In the case of *Pirelli* v *Oscar Faber* (1983), cracks appeared near the top of a chimney which had been designed by the defendant. Unfortunately, it was decided by the House of Lords that the right of action accrued when the damage occurred and not when, with reasonable diligence, it could have been discovered. The damage appeared in 1977, but occurred in 1970, and hence the action was held to be out of time. This right of action in the law of tort, however, only applies in relation to an action against a professional and is not available in respect of a dispute between the employer and contractor: *Robinson* v *Jones* (2011).

12.9.10. As a result of the *Oscar Faber* case and others, The Latent Damage Act 1986 was passed. This statute changes the limitation period for actions brought in the tort of negligence by providing an added option to the general rule, as applied in *Oscar Faber*. An action for damages in respect of negligence can now be commenced within three years of the date when the claimant had both the knowledge required to bring an action together with the right to bring an action for damages. This may be some years after the date when the damage occurred. There is, however, a longstop with regard to actions in negligence, except in respect of personal injury. The action must be commenced not later than 15 years from the act of negligence by which the damage is alleged to have been caused. The Latent Damage Act 1986 does not apply to action for breach of contract.

Final Certificate

12.9.11. Some of the JCT contracts, for example JCT 2011 With Quantities, provide for the architect or contract administrator to issue a final certificate when all defects have been made good and the final account agreed. The wording in clause 1.9.1.1 has a limiting effect on an employer's rights of action once a 28-day period has elapsed following the issue of the final certificate. Where the contract documents call for particular goods or materials to be to the satisfaction of the architect or contract administrator, the Final Certificate is conclusive evidence that the satisfaction of the architect or contract administrator has been secured. No action with regard to the unsatisfactory nature of those goods or materials can be commenced beyond the 28-day period. None of the other standard forms in general use, such as ICE 6th and 7th editions, GC/Works/1 or the Engineering and Construction Contract (NEC 3), contains such a restriction.

SUMMARY

Most standard forms of contract have a procedure for correcting patent defects. Prior to practical completion, the contractor will be expected to put right all patent defects except those of a trivial nature, before practical completion will be recognised by the employer or his representatives. Once a set period of time stipulated in the contract, which is usually 12 months, has elapsed, following practical completion, often referred to as the maintenance, defects correction or rectification period, the employer or his representative will carry out an inspection and instruct the contractor to correct any latent defects which have come to light in that period. Once corrected, the employer or his representatives will issue documentation signifying satisfaction with the correction work.

Once the documentation has been issued, defects which have hitherto been latent may come to light. However, there is a statutory limitation period in which the employer must commence an action to prevent his rights being out of time. These periods are provided in the Limitation Act 1980. In the case of a breach of contract action in respect of defective work, the action must be commenced within six years of the right of action accruing. Normally, the period starts on the date of practical completion. If the contract is a deed, the period is 12 years. Actions in the tort of negligence in respect of defective work have a different limitation period from actions for breach of contract. The basic rule is that the action must be commenced within six years from the date the damage which resulted from the negligent act occurred. However, there is an option under the Latent Damage Act 1986 which allows an action to be commenced within three years of the damage being discovered. There is, however, under this Act a longstop, which prevents any action in the tort of negligence from being commenced more than 15 years after the date the negligent act occurred. In the light of the Court of Appeal decision in *Robinson* v *Jones* (2011), a right of action in tort is not available for an employer against a contractor. The same restriction does not apply with regard to rights of action which an employer may have against his professional adviser.

Some of the JCT contracts, for example JCT 2011 With Quantities, provide for the architect or contract administrator to issue a final certificate when all defects have been made good and the final account agreed. The wording, in clause 1.9.1.1, has a limiting effect on an employer's rights of action, once a 28-day period has elapsed following the issue of the final certificate. Where the contract documents call for particular goods or materials to be to the satisfaction of the architect or contract administrator, the Final Certificate is conclusive evidence that the satisfaction of the architect or contract administrator has been secured. No action with regard to the unsatisfactory nature of those goods or materials can be commenced beyond the 28-day period. None of the other standard forms in general use, such as ICE 6th and 7th editions, GC/Works/1 or the Engineering and Construction Contract (NEC 3), contains such a restriction.

12.10. Are there any circumstances where a quantity surveyor could be liable for defective work, payment for which has been made in accordance with the quantity surveyor's interim valuation?

12.10.1. A recurring nightmare scenario for quantity surveyors occurs when a contractor becomes insolvent and it is discovered that the work has been overvalued, certified and paid. This may be due to an error on the part of the quantity surveyor when valuing the work, or the result of some of the work included in the valuation being defective. Where this occurs could the quantity surveyor have a liability? The decision in *Dhamija v Sunningdale Joineries, Lewandowski Willcox and McBains Cooper Ltd* (2010) provides a clue regarding the answer to this question.

12.10.2. Mr Dhamija appointed an architect, Lewandowski Willcox Ltd, to design a house for him, located in Virginia Water. The architect recommended McBains Cooper Ltd to act as quantity surveyors, which was accepted and an appointment duly made, although no formal contract was ever drawn up. Sunningdale Joineries Ltd entered into a JCT contract with Mr Dhamija to construct the house. When the house was completed, Mr Dhamija was not satisfied, as he considered that it contained a large number of defects. An action was commenced by Mr Dhamija against the contractor, architect and quantity surveyor seeking damages in respect of the alleged defects.

12.10.3. The case against the quantity surveyor was that there was an implied term in its conditions of appointment to:

> only value work that had been properly executed by the contractor and was not obviously defective.

12.10.4. In the absence of a formal contract, the court had to decide what duties a quantity surveyor was legally obliged to exercise. Mr Justice Coulson was impressed by the views expressed in *Professional Negligence and Liability*, by Mark Simpson, where he states:

> It is submitted that an independent assessment of the work carried out must be made by the quantity surveyor each month, in order to arrive at a proper valuation. It is clear however, that

whilst the quantity surveyor must check the quantities of work carried out he is not obliged to investigate whether or not the work is defective.

12.10.5. Attention was drawn to the decision in *Sutcliffe* v *Chippendale and Edmundson* (1971), where it was made clear that the architect, if aware of any defects in the work, should notify the quantity surveyor. This would give the quantity surveyor an early warning which would be reflected in the next interim valuation. This was echoed in *Jackson and Powell on Professional Liability*, where it is stated:

> . . . Where a quantity surveyor is also engaged by the employer, the architect should keep him continually informed of any defective or improperly executed work observed so as to give him the opportunity of excluding it from interim valuation.

12.10.6. In arriving at a decision, Mr Justice Coulson considered that there was an implied term in McBain's contract in order to give it business effectiveness, in that:

> McBains would act with the reasonable skill and care of quantity surveyors of ordinary competence and experience when valuing the works properly executed for the purpose of interim valuations.

12.10.7. Mr Justice Coulson sought to draw a distinction between what he considered was the implied obligation of McBains and that suggested on behalf of Mr Dhamija. The implied term argued on behalf of Mr Dhamija sought to impose an absolute obligation, in a similar fashion to a guarantee. The implied term, if accepted, would impose a positive duty on McBains to inspect the works to enable them to advise the architect of work which they considered to be obviously defective. Mr Justice Coulson decided that there was no implied term in the appointment of McBains as suggested by those representing Mr Dhamija. The case being brought by Mr Dhamija against the contractor and architect, no doubt, will continue, but Mr Coulson undermined the case being brought against McBains.

12.10.8. One of McBains' standard procedures provided assistance to its case. The interim valuations they prepared made it plain on the face that all issues of defective work were for the architect to address. This type of wording, if not already standard, should appear on all quantity surveyors' interim valuation forms.

SUMMARY

What duties a quantity surveyor undertakes will be governed by the conditions of engagement. In the absence of any specific obligation, it is clear from the authorities that it is not part of a quantity surveyor's implied obligations to inspect the work to decide whether or not it is defective. It is clear, however, that the architect has an obligation to notify the quantity surveyor of any defects in the work of which he is aware.

12.11. Work has been completed and defects in the contractor's work identified. The architect instructs the contractor to make good the defects, but the contractor either refuses or neglects to undertake the work. It is left to the employer to make his own arrangements to appoint another contractor and make a charge in respect of the costs incurred against the defaulting contractor. The work is quite extensive and the employer is legally obliged to make a payment to a tenant who was forced to vacate the premises for a period whilst the work is carried out. Is the employer entitled to recover the payment it was obliged to make to the tenant from the defaulting contractor?

12.11.1. The standard forms of contract include a period of time within which defects in the contractor's work must be put right by the contractor. The JCT contracts provide for the Defects Rectification Period to be stated in the Contract Particulars, or if none is stated, the period is six months, commencing at the date work was completed and a Certificate of Practical Completion issued. The contractor is obliged to make good any defects which appear in the Defects Rectification Period. Similar provisions are included in the other well-used standard forms of contract.

12.11.2. There is little in the way of detail of what the employer may recover from a contractor who has refused or neglected to make good the defects. The employer may incur direct costs involved in employing another contractor; there may also be supervision costs and, possibly, design costs. In addition, there is the possibility of incurring other costs, such as the cost of finding alternative premises, if the remedial work makes the premises or part of them unusable. Should the premises be let, there may be compensation payable to tenants who have had to find temporary accommodation.

12.11.3. *Hudson's Building and Engineering Contracts*, 13th Edition, at 8.157, has this to say regarding claims against the employer from third parties affected by the contractor when they are making good defects:

> . . . liabilities of the owner to third parties arising from a contractor's defective work may be recoverable by the owner as damages resulting directly from the contractor's breach of contract, subject to the rules of remoteness of damage.

12.11.4. The matter of an employer's loss arising from correcting defects was an issue in the case of *H W Neville (Sunblest) Ltd* v *William Press & Son Ltd* (1981). The defendants contracted to construct the foundations and drainage work in connection with the construction of a new bakery. The superstructure was constructed by another contractor. Following practical completion, defects came to light, which were corrected by the contractor; the final account was agreed and the final certificate issued. Unfortunately, the making good of the defects had the effect of delaying the contractor undertaking the construction of the superstructure and other work related to the construction of the bakery. Additional payments became due as compensation for these delays, which the employer sought to recover from the contractor. It was argued on behalf of the contractor that it was protected by the Final Certificate. The court disagreed, as it considered that the certificate of the architect related only to workmanship and materials and did not in any way deal with subsequent costs.

12.11.5. The *H W Neville* case established clearly that these subsequent costs were recoverable from the defaulting contractor. The same would apply in respect of any additional design and supervision costs, as they are foreseeable costs which may occur as a result of the contractor's failure to make good defects. However, as a result of the decision in *Johnston* v *WH Brown Construction (Dundee) Ltd* (2000), the employer would not be entitled to recover the costs incurred in paying a consultant to demonstrate that the work was defective.

SUMMARY

It was established in the case of *H W Neville (Sunblest) Ltd* v *William Press & Son Ltd* (1981) that an employer is entitled to claim from a contractor who neglects or refuses to make good defects in the work the costs arising from having others undertake the work. These costs would include subsequent costs which were foreseeable, such as payment to tenants who were obliged to vacate premises whilst the work was undertaken. There would, however, be no entitlement to recover the cost involved in paying a consultant to demonstrate that the work was defective.

Chapter 13
Rights and Remedies

13.1. Where a contract requires the parties to act in good faith, is it enforceable?

13.1.1. The UK government and many others who regularly commission construction work have abandoned the appointment of contractors on the basis of lowest price and adopted best value as a replacement. It is now quite common to see clauses written into contracts which require the parties to act toward each other in good faith. The 1998 version of GC/Works/1, for example, includes the following wording:

> The Employer and the Contractor shall deal fairly in good faith and in mutual co-operation with one another and the Contractor shall deal fairly in good faith and in mutual co-operation with all his subcontractors.

JCT Constructing Excellence, published in 2006, is another example, which in clause 2.1 requires the parties 'to work together in a co-operative and collaborative manner in good faith'.

13.1.2. The difficulty with an obligation such as acting in good faith is that it is extremely vague and lacking in definition as to its meaning. We can all imagine actions which we consider display a lack of good faith and no doubt, in any cross section of people, a consensus as to the meaning of 'good faith' may prove elusive. It may be argued that where contracts contain an obligation to act in a co-operative and collaborative manner and in good faith, disputes are less likely to arise. This has not, however, prevented cases coming before the courts, leaving the judge to decide whether the action of one of the parties lacked good faith. Judges have attempted to provide a definition, where contracts have included a good faith obligation and the parties have disagreed as to its meaning. A common factor is that the circumstances in each case have in the main been ad hoc and unlikely to be repeated. In the case of *Interfoto Picture Library Ltd* v *Stilleto Visual Picture Programmes* (1989), the judge said good faith 'does not simply mean that the parties should not deceive each other ... its effect is perhaps more aptly conveyed by such metaphorical colloquialisms as playing fair, coming clean or putting one's cards on the table'.

13.1.3. There have been other examples of good faith requirements being included in contracts in the UK. A dispute arising from good faith obligations expressed in a contract arose

200 Contractual Problems and their Solutions, Third Edition. Roger Knowles.

and was referred to the Court in the case of *Gold Group Properties* v *BDW Trading Ltd* (2010). Gold Group, the freeholder of a parcel of land, entered into a Development Agreement with Barratt for the latter to construct 100 houses and flats for sale and 16 affordable flats for a housing association. Dates for commencement and completion were agreed, together with the minimum prices for the properties, and the parties arranged a share of the revenue. Due to the financial climate, the work was never started and hence Barratt was in breach of the Agreement. Gold considered that this amounted to repudiation and determined the contract.

13.1.4. It was contended by Barratt that Gold should have been prepared to accept a lower minimum price, in view of the financial conditions which prevailed. There was an express obligation in the contract for the parties to act in good faith. The minimum price clause, it was alleged by Barratt, was inserted for the sole benefit of Gold and their failure to accept a reduction amounted to a breach of the good faith obligation.

13.1.5. Barratt's failure to develop the land in accordance with the Development Agreement was clearly an act of repudiation. The application to the Court had been made by Barratt, who wished to defend the action on the basis of a breach of the good faith obligation. In making a decision the court, owing to the lack of any precedents in the English Courts, cast its eyes to the Australian case of *Overlook* v *Foxtel* (2002), and was impressed by the following passage:

> It must be accepted that the party subject to the obligation [to act in good faith] is not required to subordinate the party's own interests, so long as pursuit of those interests does not entail unreasonable interference with the enjoyment of a benefit conferred by the express contractual terms so that the enjoyment becomes [or could become]... nugatory, worthless or perhaps seriously undermined. The duty is not a duty to prefer the interests of the other contracting party. It is rather a duty to recognise and to have due regard to the legitimate interest of both parties in the enjoyment of the fruits of the contract as delineated by its terms.

13.1.6. Judge Furst considered that good faith required the parties to act in a way that would allow them to enjoy the anticipated benefits of the contract. It does not require either party to give up a freely negotiated financial advantage clearly embodied in the contract. Barratt claimed that Gold had failed to act in good faith in failing to accept a reduction of £2.8 m payable to it under the terms of the contract. This offer, if accepted by Gold, would have the effect of transferring a loss of £2.8 m on the contract form Barratt to Gold. Judge Furst did not consider that Gold in refusing the offer was acting in bad faith.

13.1.7. The case of *Cable & Wireless plc* v. *IBM United Kingdom Ltd* (2002) arose out of a dispute concerning a contract in which there was a clause obliging the parties to act in good faith. The parties entered into a global framework agreement for the supply of information technology services. Clause 41 set out the following obligations:

> (1) The parties shall attempt in good faith to resolve any dispute or claim arising out of or in relation to this agreement . . . promptly through negotiations between the respective senior executives of the parties who have authority to settle the dispute pursuant to clause 40.

(2) If the matter is not resolved through negotiation, the parties shall attempt in good faith to resolve the dispute or claim through an alternative dispute resolution (ADR) procedure . . .

The claimant argued that this procedure was only an agreement to negotiate and not binding. It was held, however, that the reference to ADR was analogous to an agreement to arbitrate and represented a free-standing agreement which was ancillary to the main contract. As such, it was capable of being enforced by a stay of proceedings or by injunction. The parties were therefore required to seek a settlement of the dispute by way of negotiations and if that failed, to undertake an ADR procedure.

13.1.8. In the absence of an express clause which requires the parties to act in good faith, the courts may in the final analysis decide that there is a requirement to act in good faith implied in the contract. In the case of *Balfour Beatty* v. *Docklands Light Railway* (1996), the Court of Appeal held that the employer was required to act honestly, fairly and reasonably. The implied requirement for a party to draw attention to the onerous nature of clauses in its contract was recognised by the court in the case of *Thornton* v. *Shoe Lane Parking Ltd* (1971) and is an example of a requirement to act in good faith. In the case of *Interfoto Picture Library Ltd* v. *Stilleto Visual Programmes Ltd* (1989), the judge said of good faith:

> that it does not simply mean [the parties] should not deceive each other . . . its effect is perhaps most aptly conveyed by such metaphorical colloquialisms as playing fair, coming clean or putting one's cards on the table. It is in essence a principle of fair and open dealing.

There are other examples of legal requirements for the parties to act in good faith. In the case of *Blackpool & Fylde Aero Club* v. *Blackpool Borough Council* (1990) the court held that a tenderer, if he complies with the requirements of the tender enquiry, is, as a matter of contractual right, entitled to have his tender considered along with the others which have been submitted.

13.1.9. A failure to act in good faith, whether the requirement is an express or implied term in the contract, will entitle the injured party to recover the costs incurred as a result of the failure. In some cases the loss arising from the breach can be fairly established in principle, for example failing to consider a tender which has been properly submitted; but the sum involved may be a little more difficult to ascertain with any accuracy. The same may be said of a party who regularly refuses or neglects to attend a team meeting to try and identify cost savings. The courts, however, have never been slow at being inventive when it comes to ascertaining loss which in principle is recognised but at first glance is difficult to establish with accuracy. If it is impossible to demonstrate loss due to a failure to act in good faith, then there will be no case to answer.

SUMMARY

There is a growing tendency to include in contracts a requirement that the parties are to act in good faith. Where there is no express provision, the courts may decide that the requirement of good faith is implied. There is a need to be able to define what is meant by good faith and it is helpful if the wording in the contract is expressive of the meaning. Where a breach of good faith occurs, the injured party will be entitled to recover the costs which have been incurred as a result of the breach. It may often be the case that, whilst the principle of loss can be easily established, its assessment with anything like accuracy becomes a difficult matter. The courts, however, are inventive in their approach to deciding on an amount where this occurs.

13.2. What obligation does a contractor, subcontractor or supplier have to draw attention to onerous conditions in his conditions of sale?

13.2.1. There is extensive legislation protecting consumers from unfair and ill-considered contracts, but these rules and many of the provisions of the Unfair Contract Terms Act 1977, only apply to 'consumers', i.e. not to companies, or even persons acting in the course of their business. The general rule of *caveat emptor* will normally apply to a commercial contract, which means 'let the buyer beware'. Courts constantly inform litigants that they do not have any power or obligation to correct contracts which are one-sided. Like all rules, however, there are exceptions. A court will from time to time intervene if it considers that a term in a contract is so onerous that its presence should have been drawn to the attention of the other party to the contract at the time the contract was entered into. This will only apply in exceptional circumstances where one of the parties is using its own standard conditions.

13.2.2. It was held in the case of *Interfoto Picture Library Ltd* v. *Stiletto Visual Programmes* (1987) that the party seeking to enforce a particularly onerous condition in a contract has to demonstrate that the other party was sufficiently aware of the condition in question. If this could not be demonstrated, then the condition would not be incorporated into the contract. The defendants in this case failed to return some transparencies to the plaintiff agency and incurred a charge of £5 per day plus VAT per transparency. The total sum in contention was £3,783.50. It was held by the court that, when the contract was entered into, the plaintiff should have specifically drawn the defendant's attention to the charge for the late return of the negatives. Because of this failure, the court reduced the charge from £5 per day per transparency to a more reasonable charge of £3.50 per transparency per week.

13.2.3. The case of *Worksop Tarmacadam Co Ltd* v. *Hannaby and Others* (1995), heard in the Court of Appeal, dealt with an onerous term in a civil engineering contract. The plaintiffs were a small firm of civil engineering contractors, the defendants were architects and engineers and other parties concerned in a housing development. Roads and sewer works had to be undertaken for the development and the architects and engineers sought tenders for the same. The contractor submitted a tender in the sum of £24,268, which included the following relevant matters:

- At the foot of the tender, it stated 'N.B. THIS IS NOT A FIXED PRICE'.
- On the reverse were a number of conditions, clause 4(e) of which stated: 'At any time before completion of the contract, the company shall be entitled to vary the price and take into account the following factors . . .'
- Clause 15 provided that all work would be measured upon completion.

13.2.4. The plaintiffs entered into a contract and completed the works. During construction, they encountered unforeseen ground conditions which necessitated extra work. On re-measurement of the completed work, the value of excavating the unforeseen ground was ascertained at a sum of £3,007. The court found against the contractor, holding that there was no entitlement to the amount claimed for unforeseen ground conditions. An appeal was lodged against the decision.

13.2.5. The Court of Appeal commented on the conditions of contract as follows:

> (1) [The appellants'] submission is that clause 15, read literally and in context, is sufficiently wide to permit [them] to charge for the additional work that they encountered because of hard rock. I disagree. Had the plaintiffs wished to make such a provision in the event of unforeseen ground conditions being encountered, it would have been the easiest thing in the world for them so to have provided in specific terms. They did not do so.
>
> Thus, although perhaps by a somewhat different route, I reach the same conclusion as did Judge Bullimore; namely that the plaintiffs are not entitled to the additional £3,007 because that cost is not caught by any of the conditions to be found on the reverse of the tender.
>
> (2) From the learned County Court judge's point of view, this was not the end of the matter because, in a very extensive judgment, prepared with skill and care, the judge went on to make some specific findings in relation to whether clause 15 ever formed part of the contract between the parties. He came to the conclusion that the plaintiffs were unable to rely upon it, not only on the grounds already indicated in this judgment, but because it had not been sufficiently brought to the defendants' attention. I would prefer not to express any definitive view on that aspect of the case but to proceed on the basis that the clause was brought to the attention, certainly of Mr Hannaby, who must be taken to have experience in this form of transaction.
>
> That said, however, although it is unnecessary for the reasons I have given in this judgment to make any express finding upon it, I take the view that clause 15, if it is not corollary to clause 4 and 8 providing machinery for ascertaining the true cost to the plaintiffs, was in any event so vague and in some ways so onerous a term that, without more specific attention being directed to its terms and those terms being brought specifically to the attention of Mr Hannaby, it did not form part of this contract.
>
> (3) For all these reasons, and in agreement with the learned judge, although perhaps adopting a slightly different approach to the problems as he did, I have come to the conclusion that his ruling was correct that the plaintiffs were not, and are not, entitled to the additional £3,007.

13.2.6. Onerous terms were again an issue in the case of *P4 Ltd* v *Unite Integrated Solutions* (2006); the terms of the contract in issue related to retention of title. P4 supplied a particular type of emergency light, which was used by Unite in the accommodation units it constructed for workers and students. Unite did not contract direct with P4, but the light fittings were supplied to Tudor, an electrical subcontractor, who fixed them. Tudor became insolvent, owing P4 a substantial amount of money. P4 alleged there was a retention of title clause in its terms of trading, of which Unite was aware and therefore they sought payment for the light fittings from Unite. P4 had as part of its terms and conditions a retention of title clause which appeared on the reverse side of its quotation.

The quotation was faxed by Tudor to Unite, which of course only showed the face of the quotation and not the retention of title clause on the reverse side. The court held that the quotation from P4 did not form part of the contract, and therefore Unite had no liability for the cost of the light fittings.

13.2.7. To succeed, P4 would have had to prove that the retention of title clause had been incorporated into the contract with Tudor and that Unite had knowledge of the clause. If a retention of title or any other onerous clause is to apply, specific attention must be drawn to it. Onerous terms which appear in conditions of contract on the reverse side of quotations, to be valid, must be referred to on the face of the quotation. In this case, even if P4's conditions of contract had been included in the contract, the retention of title clause wouldn't have applied, as there was no reference to it on the face of the quotation. To make matters worse, the quotation was sent by fax, which would not show the conditions of contract.

SUMMARY

The consumer protection legislation does not normally apply to commercial contracts. However, where contractual terms are contained in one party's own standard terms and are considered by the court onerous, they must be brought to the other party's attention at the time the contract is entered into, otherwise they may be unenforceable. In the case of *Worksop Tarmacadam* v. *Hannaby* (1995), conditions on the reverse side of the quotation referred to all work being the subject of re-measurement. The conditions were not standard form conditions and on the face of the quotation, it appeared to offer the work for a lump sum. The judge held that such a clause was unenforceable, as it was not specifically drawn to the purchaser's attention.

13.3. Is there an unfettered power to reject work given to an architect/engineer, where the specification calls for the work to be carried out to the architect's/engineer's satisfaction?

13.3.1. The standard forms in general use require the contractor to carry out the work to the satisfaction or reasonable satisfaction of the architect, engineer or supervising officer. For example, JCT 2005 clause 2.3.3 states:

> Where and to the extent that approval of the quality of materials or goods or the standard of workmanship is a matter for the Architect/Contract Administrator's opinion such quality and standards shall be to his reasonable satisfaction.

The ICE 7th Edition, in clause 13(1), requires the contractor

> in so far as is legally and physically possible carry out and complete the works in strict accordance with the contract to the satisfaction of the Engineer.

13.3.2. Vincent Powell-Smith, in *The Malaysian Standard Form of Building Contract*, had this to say on the matter:

> There is no definition of what is meant by reasonable satisfaction anywhere in the contract . . . 'Reasonable satisfaction' might appear to suggest that the test is an objective one, but in truth the test is the subjective standards of the particular Architect, and there is a strong element of personal judgment in that opinion. It is reviewable in arbitration under clause 34 and the expression of satisfaction or otherwise by the Architect can be challenged by both the Employer and the contractor, provided a written request to concur in the appointment of an arbitrator is given by either party before the issue of the final certificate or by the contractor within 14 days of its issue: see clause 30(7).

13.3.3. The manner in which an architect exercises his duties was examined in the case of *Sutcliffe* v. *Thackrah* (1974). In this case, it was stated that the employer and the contractor enter into their contract on the understanding that in all matters where the architect has to apply his professional skill he will act in a fair and unbiased manner. It would seem that, under JCT 2011, where the term 'reasonable satisfaction of the Architect' is used, the architect when exercising his powers cannot demand standards which exceed those specifically referred to in the specification. Further, he must act in a fair and unbiased manner. Should the contractor or subcontractor be dissatisfied with the architect's decision, then the remedy lies in a reference to adjudication in accordance with the Construction Act 1996 or, if the contract so permits, to an arbitrator who has powers to 'open up, review and revise any certificate, opinion, decision . . . [of the architect/contract administrator].

13.3.4. The case of *Cotton* v *Wallis* (1955) is unusual in its interpretation of work required to be to the reasonable satisfaction of the architect. The contractor who was appointed-produced work which contained a substantial number of defects. In the words of the trial judge, the work had been 'scamped'. The contractor had submitted a low price for carrying out the work and the judge considered that this was a determining factor as to whether the architect had been negligent in not having many of the defects corrected. Under the circumstances, the architect was entitled to apply lower standards.

13.3.5. The ICE 7th Edition includes slightly different words in clause 13(1), which states that:

> . . . the Contractor shall construct and complete the Works in strict accordance with the Contract to the satisfaction of the Engineer . . .

There is no reference to 'reasonable' satisfaction. Does this mean that the engineer has a greater discretion under the ICE conditions than that of an architect under JCT conditions? In all probability, a court would hold that the contract contained an implied clause that the engineer in exercising his powers must act reasonably. A previous reference has been made to the decision in *Sutcliffe* v. *Thackrah*, when it was held that an architect is obliged to act in a fair and unbiased manner. The same would apply to an engineer. It is submitted that, in practical terms, there is little difference between a contractor or subcontractor's obligation to carry out work to the 'satisfaction' or 'reasonable satisfaction' of the architect or engineer.

SUMMARY

There is no definition as to what is meant by 'satisfaction' or 'reasonable satisfaction'. It might appear to suggest that the test is an objective one, but in truth the test is the subjective standards of the particular architect or engineer. If the contractor is not satisfied, his recourse is to refer the matter to adjudication or arbitration. The architect and engineer, however, are duty bound to act in a fair and unbiased manner, when performing their duties of ensuring that the contractor carries out the work in accordance with the requirements of the contract.

13.4. If an estimate prepared by an engineer or quantity surveyor proves to be incorrect, can the employer claim recompense?

13.4.1. For an employer to recover financially from an engineer, or quantity surveyor, as a result of an incorrect estimate it will be necessary to prove breach of an obligation and a resultant financial loss. The consultancy agreement between the employer and engineer or quantity surveyor can take many forms. It may be that a specially drafted agreement is used or a standard ICE or RICS agreement is employed. A simple exchange of letters or even a verbal agreement may be the basis of the contract. The terms of the contract may be express or implied. It is normal for a contract such as this to indicate that the engineer or quantity surveyor will have an obligation to carry out his duties with reasonable skill and care. If there is no express term to this effect, then one will normally be implied by law. In the case of *Bolam* v. *Friern Hospital Management Committee* (1957), the House of Lords had to define 'reasonable skill and care'. Mr Justice McNair had held that it was not necessary to achieve the highest possible professional standards, when he said:

> A man need not possess the highest expert skill at the risk of being found negligent. It is well established law that it is sufficient if he exercised the ordinary skill of an ordinary competent man exercising that particular art.

13.4.2. The employer may, as an alternative, when seeking to recover costs resulting from an incorrect estimate, seek to demonstrate that the error constituted a breach of warranty. In the case of *Copthorne Hotel (Newcastle) Ltd* v. *Arup Associates* (1997), an action was brought by an employer against one of its consultants as a result of an incorrect estimate, on the basis of breach of warranty. In 1987, the plaintiff decided to construct a hotel in Newcastle and, after an initial abortive design which had been prepared by a local architect, the plaintiff engaged Arup as engineers, architects and quantity surveyors for the project. Arup produced a number of budgets or estimates. Rush & Tompkins were engaged as the contractor, but went into receivership and the hotel was completed by Bovis Construction. Practical completion was achieved in February 1991 for a total sum of £15,205,000. The plaintiff withheld fees from Arup as they were dissatisfied with their work and, consequently, Arup terminated the contract in July 1991. In 1994, the plaintiff commenced court action, claiming that Arup had given inaccurate cost estimates and had failed to design within the construction costs estimated. Further, they had failed to control costs, co-ordinate the work or supply prompt information, and there were defects in the works. Arup counterclaimed, for fees allegedly owed.

In 1995 the Official Referee directed that the trial should comprise specified questions in relation to certain paragraphs in the statement of claim and the defence and counterclaim. These included, with regard to liability:

- the existence and scope of duty;
- whether there were any breaches, and if so, what breaches.

Paragraph 11 of the statement of claim read:

> At a meeting held in mid-April 1988, attended by Norman Cooke and Kenneth Hunt of the plaintiff and Peter Foggo, Michael Lowe and Dick Lee of the defendant, the defendant confirmed that the hotel to be designed by the defendant could be completed for the sums proposed by the plaintiff with a margin of plus or minus 5% namely, an overall cost of £12 m, of which sum the construction costs would be £8 m approximately.

The overall cost of the hotel proved to be £21.2 m, of which building costs accounted for £15,205,000.

13.4.3. In finding in favour of Arup, Judge Hicks QC said:

> Paragraph 11 uses the word 'confirmed', but there is no suggestion of any prior statement susceptible of confirmation, so it should presumably be understood as meaning 'stated'. However that may be, there is certainly no use of any technical term such as 'represented', 'advised' or 'warranted' to indicate what is alleged as to the legal status of the words used, or as to the nature of the relationship between the parties. In paragraphs 22, 23 and 24, however, what is alleged in paragraph 11 is referred to as 'advice'. Although I have no doubt, as I have indicated . . . above, that Arup was offered and accepted within a few days of 12 April 1988 the appointment which became much later the subject of a contract in writing, I am equally clear that there was no contractual relationship between the parties at the meeting referred to in paragraph 11 of the re-amended statement of claim. Whatever was said then was not, therefore, professional advice given pursuant to a contract. The most natural analysis, in my view, is that, if anything of the kind alleged was said, it was a representation intended to induce the plaintiff to engage Arup. That is not expressly pleaded.

In other words, there may possibly have been a misrepresentation inducing the employer to enter into the contract, but there was not any warranty which could be said to have been breached by the cost overrun, as there was at that time no contract. The consultant may have been a little fortunate, as it seems clear that an estimate was produced which proved to be inaccurate. However the court concluded that in providing the figure, Arup did not provide a warranty as to its accuracy.

13.4.4. In the Singapore case of *Paul Tsakok & Associates* v. *Engineer & Marine Services (Pte) Ltd* (1991), an architect brought an action against a client for unpaid fees. The architect had given an estimate for a project to the client. Eventually, the client decided not to call any tenders, because he considered that tenders would exceed the estimate, but instead ordered the architect to revise all the drawings so that the tenders would come within the estimate given. The architect declined and demanded his fees. The client refused to pay. The court held that it was never the parties' intention that the estimate was guaranteed. The architect could not be held responsible in this case and was entitled to be paid the proper full fees, pursuant to the SIA terms of engagement. This case shows that consultants are not expected to be perfect and that clients cannot dismiss them as and when they so desire. The architect, quantity surveyor or engineer does not normally guarantee an estimate he gives to a client. However, although the court found in favour of the architect, the judge considered that both parties had behaved unreasonably and the matter could have been resolved without going to court. Because of this, he only awarded the architect one-half of his legal costs.

13.4.5. The decision in the case of *Nye Saunders* v *Alan E Bristow* (1987) paints a different picture. The claimant architects were appointed to submit a planning application in respect of the renovation of a substantial house. A budget figure of £250,000 for the work was provided by the employer for whom the work was to be undertaken. The architect appointed a quantity surveyor, who produced a budget estimate in the sum of £238,000 for carrying out the work. Unfortunately, the quantity surveyor had omitted to include in his estimate for inflation or contingencies and failed to draw attention to these omissions. Following receipt of planning permission and the development of the design, it became clear that the estimate produced by the quantity surveyor was too low. A revised estimate of £440,000 was produced, which brought about a sharp intake of breath. As a result of the substantial increase in the estimate, the project was cancelled. At the time, inflation was running at a high level and it was the usual practice either to stipulate that a budget estimate included or excluded inflation. As a substantial amount of the increase in the estimated price resulted from inflation, the architect was held to have been negligent in failing to make proper provision in the first estimate or to draw attention to its omission. Whilst it was the quantity surveyor who made the error, it was the architect who was liable to the client for its accuracy.

SUMMARY

An estimate which proves to be incorrect, in itself will not provide the employer with a right of redress. To be successful, the employer must demonstrate that the engineer or quantity surveyor warranted the accuracy of the estimate, or show that it was exceeded because of a lack of reasonable skill and care on the professional's part. It is possible for the estimate to be inaccurate for reasons outside the control of the engineer or quantity surveyor; for example, a change in market conditions. Where an estimate excludes an item of cost which is likely to happen but isn't certain, such as inflation, the estimate must make it clear that no provision has been made for such an item of likely cost. If it can be shown that, had the estimate been correct, the employer would in any event have proceeded with the project, the recoverable loss will be nominal, even though an action is otherwise successful. This is because the employer, in continuing with the scheme, will be unable to demonstrate loss, as the incorrect estimate did not affect his decision. If, on the other hand, once tenders are received above the estimate, the scheme is abandoned on financial grounds, the employer should be able to demonstrate loss.

13.5. When defects come to light after the architect/engineer has issued the final certificate, does the contractor/subcontractor still have a liability, or can he argue that once the certificate has been issued the employer loses his rights?

13.5.1. The Limitation Act 1980 provides periods of time within which actions must be commenced, after which they are 'statute barred', providing the defendant with an unassailable defence. Section 7 states:

> An action to enforce an award, where the submission is not by an instrument under seal, shall not be brought after the expiration of six years from the date on which the cause of action accrued.

Section 8 states:

> An action upon a speciality [i.e. a contract under seal or expressed as a deed] shall not be brought after the expiration of twelve years from the date on which the cause of action accrued.

With construction contracts where a breach has occurred, for example defective work, those periods will usually begin to run from the date of practical completion. Claimants therefore have six years or, in the case the contract is under seal or expressed as a deed, twelve years after practical completion within which to commence a court action.

13.5.2. Questions are regularly asked as to how these periods are affected by the issue of a final certificate. Could the issue of a final certificate shorten the periods of limitation set down by the Limitation Act?

13.5.3. With regard to JCT contracts, JCT 80, prior to the issue of amendment 15 in July 1995, makes reference in clause 30.9.1 to the final certificate being conclusive evidence as to the architect's satisfaction, where it states:

> . . . the Final Certificate shall have effect in any proceedings arising out of or in connection with this Contract (whether by arbitration under Article 5 or otherwise) as . . . conclusive evidence that where and to the extent that the quality of materials or the standard of workmanship is to be of the reasonable satisfaction of the Architect the same is to such satisfaction.

An exception would occur if legal proceedings were commenced either before the issue of the final certificate or within a period up to 28 days after its issue. The Court of Appeal gave a decision on the effect of that clause in the case of *Crown Estate Commissioners* v. *John Mowlem and Co Ltd* (1994). It was held that the final certificate would safeguard a contractor from any claims made after its issue. Sir John Megaw had this to say in giving judgment:

> Where the final certificate thus becomes conclusive evidence, the effect is that any claim in an arbitration which seeks to support some provision of the final certificate is bound to succeed, and any claim which seeks to challenge a provision of the final certificate is bound to fail, without any hearing on the merits.

13.5.4. The effect of the *Crown Estate* case was that, once the final certificate had been issued and a further 28 days had elapsed, either party could commence an action but could not produce evidence to support its case, which would make it almost impossible for an action to succeed. Employers, in particular, were placed in a less advantaged position in using JCT contracts than would have been the case if other forms of contract had been employed. The drafters of the JCT forms therefore issued amendment 15 in July 1995, to negate the effect of the *Crown Estate* case, where it states:

> . . . but such Certificate (Final Certificate) shall not be conclusive evidence that such or any other materials or goods or workmanship comply or complies with any other requirement or term of this Contract . . .

The final certificate remains conclusive in so far as it relates to qualities of materials or goods or any standard workmanship expressly described in the contract documents to be to the approval of the architect. As a result of the *Crown Estates* case and amendment 15, specifications are now carefully worded to ensure they do not make reference to specific materials or goods being to the satisfaction of the architect. Clause 1.9.1 in JCT 2011 is worded in a similar manner to amendment 15. The Scottish courts, in *Belcher Food Products Ltd* v. *Miller & Black and Others* (1998), gave a different view as to the meaning of clause 30.9.1 of JCT 80. It was the opinion of the Scottish Outer House in respect of a defective floor screed that the decision in *Crown Estates* could prove unfair to the employer. It was the court's decision that the final certificate is conclusive only in respect of those matters expressly reserved by the contract to be to the architect's satisfaction.

13.5.5. In the case of *London Borough of Barking & Dagenham* v. *Terrapin Construction Ltd* (2000), the Court of Appeal was asked to address the matter in respect of a contract let under a JCT 81 With Contractor's Design form. Clause 30.8.1.1 of the With Contractor's Design is similar to clause 30.8.1 of JCT 80, in that it states 'as conclusive evidence that where it is stated in the Employer's Requirements that the quality of material or the standard of workmanship are to be to the reasonable satisfaction of the Employer the same are to such satisfaction'. It was argued on behalf of the employer that in the particular case, as there was no specific document referred to as embodying the Employer's Requirements, the clause was inoperable. Further, unlike the JCT 80 contract, where there was an architect who had to be satisfied, there was no like person under the With Contractor's Design form. The Court of Appeal was unimpressed. It was held that clause 1 of the agreement required the execution, completion and maintenance of the works to be in all respects to the satisfaction of the employer. Further, the employer may choose to perform the functions assigned to him personally or through a qualified agent. The conclusiveness applied to all defects, whether latent or patent. The wording of clause 30.8.1.1 refers to defective goods or materials, but does not lock out claims resulting from defective design. It was also held that the wording of clause 30.8.1.1 does not exclude claims based upon any failure to meet any statutory requirements which, under the provisions of the contract, are the responsibility of the contractor.

13.5.6. The Model Form MF/1 also refers to the final certificate as being conclusive, with exceptions. Clause 39 states:

> A final certificate of payment shall be conclusive evidence:
>
> that the works or section to which such certificate relates is in accordance with the contract;
>
> that the contractor has performed all his obligations under the contract in respect thereof; and
>
> of the value of the works or section.

Payment of the amount certified in a final certificate of payment shall be conclusive evidence that the purchaser has performed all his obligations under the contract in relation to the works or section thereof to which the certificate relates. A final certificate of payment shall not be conclusive as to any matter dealt with in the certificate, in the case of fraud or dishonesty relating to or affecting any such matter.

> A final certificate of payment shall not be conclusive if any proceedings arising out of the contract whether under clause 52 [Disputes and Arbitration] or otherwise shall have been commenced by either party in relation to the works or section to which the certificate relates,
>
> before the final certificate of payment has been issued, or
>
> within three months thereafter.

13.5.7. The Institution of Chemical Engineers' Model Conditions of Process Plant 1981 Edition and the Red Book, at clause 38.5 includes the wording:

> The issue of the final certificate for the plant as a whole or, where for any reason more than one final certificate is issued in accordance with this clause, the issue of the last final certificate in respect of the works, shall constitute conclusive evidence for all purposes and in any proceedings whatsoever between the purchaser and the contractor that the contractor has completed the works and made good all defects therein in all respects in accordance with his obligations under the contract

The effect of this was subject to dispute in *Matthew Hall Ortech Ltd* v. *Tarmac Roadstone Ltd* (1997) in respect of a contract for the design, erection and commissioning of a mixed processing plant. Matthew Hall carried out the work and it was alleged by Tarmac Roadstone, the other party to the contract, that 22 steel bunkers were suffering from what appeared to be structural damage as a result of design and construction deficiencies emanating from Matthew Hall's breaches of contract. The responsibility for issuing the final certificate lay with Tarmac Roadstone, but no such certificate was ever issued. Matthew Hall argued that a final certificate should have been issued, and, if it had been, it would have acted as a bar to any claim from Tarmac Roadstone. It was Tarmac's argument that, even if they had issued a final certificate, they would not be prevented from bringing a claim against Matthew Hall. The matter was referred to arbitration. It was held by the arbitrator that a final certificate would bar contractual claims against Matthew Hall for defects which it was accused of not correcting, but would not prevent either a contractual claim for defects allegedly put right, but later discovered to have been done badly, or a claim for latent defects. The court did not fully agree with the arbitrator. It was considered that there would appear to be commercial justification for the contract to provide a definite cut-off point, once plant had been constructed, tested, provided and made good in all respects in conformity with the contract. The court considered that Matthew Hall were correct in their contention that a final certificate is conclusive evidence that all work has been completed in accordance with the requirements of the contract. It was also held that the absence of a final certificate when one ought to have been issued would not be allowed to defeat the object of clause 38.5.

13.5.8. The ICE 6th and 7th Editions and GC/Works/1 1998 conditions do not make the final certificate in any way conclusive; therefore, the periods laid down in the Limitation Act within which actions must be commenced apply. The ICE Conditions, at clause 61(2), put it very clearly:

> The issue of the Defects Correction Certificate shall not be taken as relieving either the Contractor or the Employer from any liability the one towards the other arising out of or in any way connected with the performance of their respective obligations under the Contract.

Clause 39 of the GC/Works/1 is not so emphatic, but it makes no reference to the certificate being final.

13.5.9. In the decisions in *Gray* v. *TP Bennett & Son* (1987) and *William Hill Organisation Ltd* v. *Bernard Sunley & Sons Ltd* (1982), it was held that a final certificate could not be conclusive where, due to fraudulent concealment, the defects could not have been detected following reasonable inspection.

13.5.10. There is no reference to a Final Certificate in the Engineering and Construction Contract (NEC 3).

SUMMARY

JCT 80 (in England but not Scotland), prior to amendment No. 15 issued in July 1995, MF/1 and the IChemE Forms make a final certificate conclusive evidence that work has been satisfactorily carried out. Actions under the JCT wording must be commenced within 28 days of the issue of the final certificate and, in the case of MF/1, within three months to be effective. Other commonly used standard forms of contract give the final certificate no such effect and the Limitation Act periods apply. JCT amendment No. 15 is now incorporated in JCT 2011 and indicates that the final certificate is conclusive only in respect of particular qualities of materials or goods, or any standard of workmanship expressly described in the contract documents as being to the reasonable satisfaction of the architect. A final certificate is not conclusive if, because of fraudulent concealment, the defect could not have been detected following reasonable inspection.

13.6. Who is responsible if damage is caused to a subcontractor's work by a person or persons unknown, the subcontractor, contractor or employer?

13.6.1. Contractors normally like to pass down to subcontractors the risk of damage to the subcontract works. Non-standard subcontracts are often worded in such a manner that the subcontractor is expressly required to protect the subcontract works to prevent damage. Courts will be obliged, where a dispute arises, to place an interpretation on such wording. In the case of *WS Harvey (Decorators) Ltd* v. *HL Smith Construction Ltd* (1997) the terms of the subcontract required the subcontractor to provide 'all necessary and proper protection'. The court held that 'all necessary protection' means such protection as is necessary to prevent damage to the works from whatever cause. Further, the clause stated that the subcontractor:

> will be held responsible for the adequacy of the protection afforded and shall make good or re-execute any damaged work at his own expense.

The judge, in finding for the contractor, said that the wording imposed the obligation of protecting the works firmly and squarely upon the subcontractor.

13.6.2. Damage is categorised under three headings, where the standard subcontract forms for use with JCT 2011 apply.

- That caused by the specified perils, e.g. fire, storm, tempest etc. is an insurance risk, for which the contractor or employer will be liable by clause 6.7.1.1 of the Standard Building Subcontract;
- That caused by any negligence, omission or default of the contractor, his servants or agents or any other subcontractor will be the responsibility of the main contractor, under clause 6.7.1.2 of the Standard Building Subcontract;
- Where materials or goods have been fully, finally and properly incorporated into the works, but before practical completion of the subcontract works, the contractor will be responsible under clause 6.7.2 of the Standard Form of Building Subcontract.

Some difficulties have been experienced in deciding when the main contractor becomes liable under the third category above. In particular, an interpretation of the wording 'fully, finally and properly incorporated into the works' is required.

13.6.3. It has been argued by some main contractors that this stage cannot be achieved until all the subcontract works have been completed and accepted on behalf of the employer. This cannot be correct, as clause 6.7.2 refers to materials or goods having been fully, finally and properly incorporated into the works before the Terminal Date, which is defined as practical completion of the subcontract works. The wording obviously contemplates the stage being reached before practical completion of the subcontract and hence this argument does not hold good. The essence, therefore, is the wording 'fully, finally and properly incorporated into the works'. In the *Concise Oxford Dictionary*, the words are defined as:

- fully – completely, without deficiency;
- finally – coming last;
- properly – suitably, rightly.

13.6.4. A reasonable interpretation would therefore be that liability lies with the main contractor for damage to the subcontractor's materials or goods once they have been put without deficiency (i.e. all in place and nothing missing) in their final position and are suitable in respect of the contract requirements. With regard to materials manufactured off site, for example ceiling tiles or wall tiles, when they are fixed in position with nothing further to be done to them and comply with the requirements of the contract, then they are 'fully, finally and properly' incorporated into the works. Where wet trades are involved, such as plaster, paint or asphalt, once the wet material has been applied or laid and dried off, then the materials or goods, if defect free, are fully, finally and properly incorporated into the works. The purpose behind the wording would seem to be that as the subcontract works progress and parts of the work are completed, the subcontractor will move on, leaving the completed parts behind. These completed parts become the responsibility of the main contractor, as he and his following trades will by then be working in those completed areas.

13.6.5. There has been an attempt made to avoid disputes relating to this matter by introducing into the 2011 subcontract item 14 in the Subcontract Agreement. This requires words to be inserted relating to specific Elements of the Sub-Contract Works, to indicate the extent to which each is to be carried out so as to be regarded as fully, finally and properly incorporated in the works. If the main contractor, who normally completes the Sub-Contract Agreement, before sending it to the Subcontractor for signature, considers what wording to include, he may be tempted to state that fully, finally and properly incorporated into the Main Contract works only occurs when practical completion of all subcontract works has been achieved.

13.6.6. The CECA Blue Form for use with the ICE 6th and 7th Editions is worded along different lines. Clause 14 provides for insurance to be taken out in accordance with the requirements of the Fifth Schedule for the risks set out therein. The wording of clause 14(2) states:

It will be necessary for the subcontractor to make sure that Part II of the Fifth Schedule fully covers damage from all causes to the subcontractor's materials and equipment. In the absence of adequate wording in the Fifth Schedule, clause 14(2) places the risk on the subcontractor's shoulders in the following terms:

> Save as aforesaid the subcontract works shall be at the risk of the subcontractor until the main works have been substantially completed under the main contract, or if the main works are to be completed by sections, until the last of the sections in which the subcontract works are

comprised has been substantially completed, and the subcontractor shall make good all loss of or damage occurring to the subcontract works prior thereto at his own expense.

SUMMARY

Main contractors, where non-standard forms are used, like to include terms which place the risk of damage to the subcontract works onto the subcontractor. Where the Standard Building Subcontract conditions for use with the JCT main conditions apply, the subcontractor is liable for damage to goods until such time as they are 'fully, finally and properly incorporated into the Works', unless the damage has been caused by the specified perils (fire, storm, tempest, etc.) or is due to negligence by the main contractor or other subcontractors. In the case of the CECA Blue Form of subcontract, the subcontractor is at risk until the main contract works have been substantially completed, unless Part II of the Fifth Schedule states the contrary.

13.7. How is the term 'regularly and diligently', as used in the standard forms of contract, to be defined?

13.7.1. The employer who is operating under most standard forms of contract is entitled to determine the contractor's employment if he fails to proceed 'regularly and diligently' with the works. A definition of the term 'regularly and diligently' has been provided in the case of *West Faulkner Associates* v. *The London Borough of Newham* (1992).

13.7.2. An action was commenced by the architects for the recovery of fees and damages for wrongful repudiation of their contract. By way of defence and counterclaim the council alleged, among other things, default by the architects in not serving a default notice to Moss, the main contractor, under clause 25(1)(b) concerning a failure to proceed 'regularly and diligently' with the works. If such a notice had been served, it would have given rise to the council's entitlement to determine. The failure by the architects to serve the notice left the local authority powerless to effect a determination. Instead, they were obliged to pay a substantial sum to the contractors for them to leave the site. The court had to decide the meaning of the words 'regularly and diligently' and whether, in fact, the contractor had failed to proceed in that manner. Judge John Newey QC, having listened to expert evidence and had his attention called to various precedents, decided:

> In the light of the judgments, textbooks and expert evidence I conclude that 'regularly and diligently' should be construed together and that in essence they mean simply that contractors must go about their work in such a way as to achieve their contractual obligations. This requires them to plan their work, to lead and to manage their workforce, to provide sufficient and proper materials and to employ competent tradesmen, so that the works are fully carried out to an acceptable standard and that at all times, sequence and other provisions of the contract are fulfilled.

13.7.3. Judge Newey concluded that Moss did not plan their work properly, did not provide efficient leadership or management and some, at least, of their tradespeople were not reasonably competent and, therefore, they had failed to proceed regularly and diligently with work. It was Judge Newey's opinion that Moss' failures were very extreme and the architects should have realised that Moss were not proceeding regularly and diligently and therefore served the notice. As a direct result of the architects' failure to serve Moss with the notice, the council were prevented from terminating the contractor's employment. The council suffered loss by having to pay the new contractors more than they would have had to pay Moss. They also incurred additional costs in respect of site supervision, additional quantity surveyors' fees, payments to tenants and lost rent. All these losses Judge Newey considered flowed from the breach by the architects.

SUMMARY

Contractors who are required to carry out work 'regularly and diligently' must go about their work in such a way as to achieve their contractual obligations. This requires them to plan their work, to lead and manage their workforce, to provide sufficient and proper materials and to employ competent tradesmen, so that the works are fully carried out to an acceptable standard and that all contractual time, sequence and other provisions are fulfilled.

13.8. Are there any circumstances under which a contractor/subcontractor could bring an action for the recovery of damages against an architect/engineer for negligence?

13.8.1. A failure on the part of an architect/engineer to exercise reasonable skill when issuing payment certificates or performing other functions under the contract could prove expensive for the contractor. Does this leave the contractor with a right to recover his losses from the negligent architect/engineer?

13.8.2. The matter was considered in the case of *Arenson* v. *Arenson* (1977), when Lord Salmon said:

> The Architect owed a duty to his client, the building owner, arising out of the contract between them to use reasonable care in issuing his certificates. He also, however, owed a similar duty of care to the contractor arising out of their proximity: see *Hedley Byrne & Co Ltd* v. *Heller & Partners Ltd* (1964), *Sutcliffe* v. *Thackrah* (1974).

In *Michael Salliss & Co Ltd* v. *Calil* (1987), Judge Fox-Andrews also held that an architect had a duty to the contractor to act fairly when certifying:

> It is self-evident that a contractor who is party to a JCT contract looks to the architect or supervising officer to act fairly as between him and the building employer in matters such as certificates and extensions of time. Without a confident belief that that reliance will be justified,

in an industry where cash flow is so important to the contractor, contracting would be a hazardous operation. If the architect unfairly promotes the building employer's interest by low certification or merely fails properly to exercise reasonable care and skill in his certification it is reasonable that the contractor should not only have the right as against the owner to have the certificate reviewed in arbitration but also should have the right to recover damages against the unfair architect.

13.8.3. In *Pacific Associates* v. *Baxter and Halcrow* (1988), the Court of Appeal took a different view. Pacific entered into a contract with the Ruler of Dubai for the dredging of a lagoon in the Persian Gulf. Halcrow was appointed as the engineer. The contract incorporated the FIDIC conditions (2nd Edition, 1969). Condition 86 of the contract provided as follows:

> Neither any member of the Employer's staff nor the Engineer nor any of his staff, nor the Engineer's representative shall be in any way personally liable for the acts or obligations under the contract, or answerable for any default or omission on the part of the Employer in the observance or performance of any of the acts matters or things which are herein contained.

The work was delayed because of the presence of hard materials. Pacific made claims for extensions of time and additional expenses which were rejected by Halcrow. Pacific then made a formal submission in accordance with condition 67 for the decision of Halcrow. When this too was rejected, Pacific referred its claims to the ICC for arbitration in accordance with condition 67. The proceedings were compromised when the Ruler of Dubai agreed to pay Pacific some £10 m in full and final settlement of its claims against him. In March 1986 Pacific issued a writ claiming £45 m from Halcrow, being the unrecovered balance (including interest) of its claim against the Ruler of Dubai. The Court of Appeal held:

(1) In considering whether a duty of care existed it was relevant to look at all the circumstances, and these included the contract between the Ruler of Dubai and Halcrow.
(2) There had been no 'voluntary assumption of responsibility' by Halcrow relied upon by Pacific, sufficient to give rise to a liability to Pacific for economic loss in circumstances in which there was an arbitration clause. The position might well have been otherwise if the arbitration clause or some provision for arbitration had not been included in the contract.

13.8.4. *Keating on Construction Contracts*, Eighth Edition, considers that the decision in *Pacific Associates* v. *Baxter and Halcrow* represents the law on the matter, when at 13-080 it states:

> The court reviewed dicta in the House of Lords and at first instance (including Michael Sallis) suggesting that architects or engineers might owe contractors a duty of care of the kind alleged, but concluded that there was no basis on which a duty of care to prevent economic loss could be imposed on an architect or engineer in favour of a contractor.

SUMMARY

It would seem from the decision in *Michael Salliss & Co Ltd* v. *Calil* (1987) that architects and engineers could have a liability to a contractor if they fail to act fairly. However, the Court of Appeal decision in *Pacific Associates* v. *Baxter and Halcrow* (1988) went against the contractor, as it was held that there was no evidence that the engineer had assumed any responsibility towards the contractor. Leaving architects and engineers open to claims from contractors could prove to be an unsatisfactory manner of carrying out construction works. Courts may in future cases continue to play the lack of 'voluntary assumption of liability' card to resist contractors' claims.

13.9. What is a contractor's liability to the employer for failing to follow the specification, where it is impractical to take down the offending work?

13.9.1. Most of the standard forms of contract provide the architect or engineer with power to instruct the contractor to take down and remove work which does not comply with the contract. These powers extend beyond practical completion into the defects period. A situation may arise where work is defective, but it is not practical for it to be corrected. JCT 2011, clause 2.38, provides for an appropriate deduction to be made in respect of defects, shrinkages or other faults which are not required to be made good.

13.9.2. The question of what level of recompense an employer would be able to recover from a contractor whose work was defective arose in the House of Lords case of *Ruxley Electronics and Construction Ltd* v. *Forsyth Laddingford Enclosures* (1995). The dispute concerned the construction of a swimming pool; the maximum depth to which the pool was constructed being 6 feet 9 inches, which differed from the 7 feet 3 inches depth specified. The respondent sought damages for breach of contract for the cost of demolishing the existing pool and rebuilding it to the required depth. The trial judge had found that the pool, as constructed, was safe to dive into and that the deficiency had not decreased the value of the pool. Further, he was not satisfied that the respondent intended to rebuild the pool and that the cost of rebuilding would be wholly disproportionate to the disadvantage of having a pool which was less than a foot too shallow. Only damages for loss of amenity of £2,500 were awarded. The Court of Appeal allowed an appeal, finding that the only way the employer could have achieved the object of the contract was to reconstruct the pool at a cost of £21,560 and that this was reasonable. The contractors appealed. It was held by the House of Lords that the award of damages was designed to compensate for an established loss and not to provide a gratuitous benefit to the aggrieved party. Therefore, it followed that the reasonableness of the award was linked directly to the loss sustained. It was unreasonable to award damages for the cost of reinstatement if the loss sustained did not extend to the need to reinstate. A failure to achieve a contractual objective does not necessarily mean that there is a total failure. In the instant case, the employer had a perfectly serviceable pool, even if it was not as deep as it should have been. His loss was not the lack of a usable pool and there was no need to construct a new one. Reinstatement was not the correct measure of damages in this case. The claimant was therefore entitled to £2,500, for loss of amenity only.

Lord Jauncey offered an *obiter* comment to the effect that, in the normal case, the court has no concern with the use to which a plaintiff puts an award of damages for a loss which has been established. Intention, or lack of it, to reinstate can have relevance only to reasonableness and, hence, to the extent of the loss which has been sustained. Once that loss has been established, intention as to the subsequent use of the damages ceases to be relevant.

13.9.3. A case which deals with a similar matter is *RJ Young* v. *Thames Properties* (1999). The work provided for the construction of a car park. 100 mm of limestone should have been laid but, instead, the contractor provided only 30 mm. It was held by the court that if the contractor undertakes work which departs from what is required by the contract, there is nevertheless an entitlement to payment for what work was actually carried out, unless the work was of no benefit, entirely different or incomplete. The measure of damage was based upon the value of the car park as laid and not the cost of correcting the work.

13.9.4. The Scottish case of *McLaren Murdock and Hamilton* v. *The Abercromby Motor Group* (2002) had to deal with a defective heating system and the amount of money which should be paid by the offending party. The judge applied the principle that the innocent party is generally entitled to be placed in as good a position financially as it would have been, had the breach not occurred. The innocent party is entitled to recover the cost of remedial work, unless these costs are disproportionate to the benefit to be gained from the remedial works. It was held that the cost of replacing the heating system was not disproportionate to the benefit to be obtained.

13.9.5. Disputes relating to defective work may relate to whether the employer wishes to claim for the costs of demolition and rebuilding, by way of contrast to the contractor, who considers repair is adequate. The leading case that dealt with this problem is *The Board of Governors of the Hospitals for Sick Children and Another* v. *McLaughlin and Harvey* (1987), often referred to as the *Great Ormond Street* case, followed by a number of other cases, including *McGlinn* v. *Waltham Contractors Ltd and Others* (2007) and *Linklaters Business Services* v. *Robert McAlpine* (2010). In all three cases, the court in arriving at a decision concerning whether damages for breach of contract resulting in defective work should be based upon repair or replacement, had to decide the most reasonable solution in all the circumstances based upon the evidence presented. The court, in arriving at a decision, would in all probability take into account the advice provided by an expert. Judge Newey in the *Great Ormond Street* case, summed up the matter when he said:

> The plaintiff who carries out either repair or reinstatement of his property, must act reasonably. He can only recover as damages, the cost which the defendant ought reasonably to have foreseen that he would incur and the defendant would not have foreseen unreasonable expenditure. Reasonable costs do not, however mean the minimum amount which with hindsight, it could be held would have sufficed. When the nature of the repairs is such that the plaintiff can only make them with the assistance of expert advice, the defendant should have foreseen that he would take such advice and be influenced by it.

SUMMARY

The House of Lords decision gives authority to the view that the employer will be entitled to recover the cost of rectification if work carried out by a contractor is defective. However, if rectification was not a reasonable solution, because of the high cost compared with minimal benefit, rectification costs would not be awarded. An award based upon loss of amenity may be more appropriate.

13.10. Do retention of title clauses still protect a supplier or subcontractor where a main contractor becomes insolvent, or have there been cases which throw doubt on their effectiveness?

13.10.1. Where a company or organisation becomes insolvent, there is usually insufficient money available for all to whom payment is due. In an effort to protect themselves, some suppliers and subcontractors include in their terms of trading what has become known as a 'retention of title' clause. In essence, the clause states that the goods supplied to the purchaser remain in the ownership of the supplier until payment has been made in full. In the event of a failure to make full payment, the clause usually provides for their return. If, therefore, a purchaser becomes insolvent before paying for the goods, the liquidator or receiver should either make payment or allow the goods to be returned to the seller.

13.10.2. The basic principle is contained in the maxim *nemo dat quod non habet,* and this is embodied in the Sale of Goods Act 1979, section 21(1). It means that a person cannot transfer ownership or title in something which he does not himself own; he cannot pass on good title if he does not have it himself. There are exceptions to the *nemo dat* rule. Where the contract is one of sale of goods or materials, the most significant exception is that created by the Sale of Goods Act 1979, section 25, which provides for a buyer in possession to transfer possession with the consent of the owner following an agreement to sell, even though the buyer does not own the goods. In such circumstances, provided that the third party purchasing from the buyer has no notice of the absence of the title in the first buyer, he, the second buyer, will obtain a good title even against the true owner. In other words, if a contractor is paid for materials by an employer who is ignorant of the retention clause in the supplier's contract with the contractor, the employer can acquire a good title even though the supplier has not been paid. However, to ensure the employer receives a good title over that of the supplier, where a contractor is in possession of goods acquired from a supplier with a retention of title clause in the contract of sale, then the transfer of ownership to the employer from the contractor must be 'under any sale, pledge, or other disposition thereof'. It was held in the case of *P4 Ltd* v. *Unite Integrated Solutions plc* (2006) that, once materials are delivered to site and the contract, such as JCT 2011, prevents them from being removed without the employer's consent, then there has been a disposition.

13.10.3. In the case *Aluminium Industrie Vaassen bv* v. *Romalpa Aluminium* (1976), it was established that a seller who supplies goods under retention of title and authorises his buyer to sell them on condition that he accounts for the proceeds of sale has an equitable right to trace those proceeds and to recover them from the buyer. This was followed by the Court of Appeal decision in *Clough Mill* v. *Martin* (1985), where a retention of title clause protected a seller of yarn which was to be made into fabric. The contract provided for the yarn to be used in the manufacturing process, with the supplier of the yarn becoming the owner of any produce manufactured from the yarn until paid. In the absence of payment, the seller was held to be entitled to claim back the goods incorporating the yarn which had been sold. *Border (UK)* v. *Scottish Timber Products* (1981) deals with a retention of title clause related to resin used in the manufacture of chipboard. It was held that, once the resin had been incorporated into the chipboard, the retention of title clause was no longer effective. This decision can be contrasted with *Hendy Lennox* v. *Grahame Puttick* (1984), where a retention of title clause was held to be valid with regard to the sale of engines to be attached to generator units after the engines had been fixed. To take out the engines from the generators took several hours, but still the court held the retention of title clause to be valid.

13.10.4. An example of an effective reservation of title clause in the construction industry is to be found in the case of *W. Hanson (Harrow) Ltd* v. *Rapid Civil Engineering Ltd and Usborne Developments Ltd* (1987). In this case, Hanson were suppliers of timber and timber products, Usborne was a development company engaged in developing some residential sites in London and the building contractor was Rapid. Hanson brought a claim against Usborne alleging the wrongful use of building materials supplied by Hanson to Rapid. Hanson had been suppliers to Rapid since 1979 and their terms of business had always been the same. They were set out on all of their documentation, i.e. consignment notes, delivery notes and invoices. One of those terms, namely condition 10, dealt with the retention of title in the following form:

> 10 Transfer of Property
>
> a. The property in the goods shall not pass to you until payment in full of the price to us;
>
> b. The above condition may be waived at our discretion where goods or any part of them have been incorporated in building or constructional works.

Rapid and Hanson carried on business on a running account. Rapid were apparently slow payers. In order to assist Rapid with their cash flow, Usborne agreed to make more frequent payments against the contract price, by including valuations of goods on site as well as for work done. On 16 August 1984 Hanson made a delivery to one of the residential sites. Shortly afterwards, they found out that Rapid had gone into receivership on 15 August. They demanded payment or return of the goods, but both were refused by the receivers, although the receivers permitted them to enter the site on 17 August to make inventories of goods for which payment had not been made. Hanson marked the goods so as to identify them. Hanson reminded the receivers of their contract with Rapid and, in particular, the retention of title provision and enclosed lists of the goods, stating their intention to collect them. Hanson also notified Usborne of their claim to retention of title to the goods and reserved their right to claim damages against Usborne for conversion if they used the goods. They failed to get an assurance from Usborne that Usborne would not use the goods and, accordingly, proceedings were commenced. The question to be decided by the court, therefore, was whether the title to the goods remained in Hanson (the supplier); or was Usborne (the developer) protected by section 25 of the Sale of Goods Act 1979? The judge held, firstly, that there was no delivery or transfer of the goods on site by Rapid to Usborne by way of any sale or other disposition under section 25. The contract between Rapid and Usborne provided for monthly payments of 97% of the value of the work executed, including the value of all materials on site. The contract further provided that the property in the goods supplied should not pass to Usborne until payment of the instalment in which the supply was contained. The judge referred to section 2 of the Sale of Goods Act 1979, which provides, *inter alia*, as follows:

> A contract of sale of goods is one by which the seller transfers or agrees to transfer property in the goods;
>
> Where under such a contract of sale the property in the goods is transferred from the seller to the buyer, the contract is called a sale;
>
> Where under such a contract of sale the transfer of the property in the goods is to take place at a future time or subject to some condition to be fulfilled, then the contract is called an agreement to sell.

The judge made the point that section 25 only applies to delivery or transfer of goods under a 'sale or otherwise disposition' and that an agreement to sell only becomes a sale when any conditions are fulfilled, subject to which the property in the goods is to be transferred. As between Rapid and Usborne, therefore, the building contract between them only operated to create a sale within the meaning of section 25 when any conditions subject to which the property in the goods was to be transferred were fulfilled. In this case, it meant that payment of the valuation in which the supply of materials was contained was required before any sale took place. Until then, there was only an agreement to sell but not a sale. Accordingly, Usborne did not obtain any title to the goods and Hanson's claim to title was not defeated by the operation of section 25.

Hanson had effectively retained title to the goods, as permitted by section 19 of the Sale of Goods Act 1979, and the title in those goods did not pass to Usborne. In so far as Usborne used them, they had wrongly converted them to their own use. If, in this case, Usborne had paid Rapid for the goods delivered to site, then the agreement to sell would have been converted into a sale. Provided Usborne were unaware of the retention of title clause between Rapid and Hanson, Usborne would have obtained a good title to the goods under section 25.

13.10.5. Another case, also involving a judicial interpretation of section 25 of the Sale of Goods Act 1979, is *Archivent Sales & Development Ltd* v. *Strathclyde Regional Council* (1984), where Archivent were sellers and agreed to sell ventilators to the contractors for incorporation into a primary school. The contractors had a contract with Strathclyde based on JCT 63. The *Archivent* conditions of supply provided as follows:

> Until payment of the price in full is received by the company, the property in the goods supplied by the company shall not pass to the customer.

Ventilators were delivered to site and their value included in an interim certificate under the main contract. The employer paid the contractor, but the contractor failed to pay Archivent before going into receivership. Archivent sued the employer for the value of the ventilators, arguing that title to the property remained in them. The employer contended that section 25 operated to give them an unimpeachable title. Archivent challenged this, saying that its operation depended upon a transfer of the goods from the contractor to the employer. The goods in this case were at all times under the control of the employer (not the contractor), by virtue of the provisions in clause 14(1) of JCT 63, requiring the architect's consent to any removal after they were delivered to site by the subcontractor. Archivent contended, therefore, that there could be no transfer by the contractor to the employer so as to qualify for the protection afforded by section 25. It was held that section 25 conferred good title upon the employer. The ventilators were in the contractor's possession in law even if not under their direct control. They were transferred to the employer's possession upon the employer's surveyor making provision for payment for them in a payment certificate.

13.10.6. If the clause does not fully reserve legal title, it will be merely a charge, which will be ineffective and void as against a liquidator etc., unless registered under the Companies Act 1985. If the goods lose their identity by being completely submerged with other goods so as to produce a new material, the chances are that the retention of title clause will cease to operate.

13.10.7. Where the goods or materials are simply supplied by a subcontractor as part of their work there will be no sale of goods, either to the main contractor or to the employer. The goods will pass either to the main contractor or to the employer under a contract for work and materials. This being so, section 25 of the Sale of Goods Act 1979 will not apply, unless the contractual machinery is such that the goods or materials are separately sold apart from the work element. Here, a straightforward reservation of title clause, i.e. the subcontractor retaining title in the goods or materials until he is paid, could well be effective until the goods become incorporated into the building. However, where standard forms of JCT contract and subcontract are used, this is no longer possible. JCT 80 was amended, as was NSC/4 and DOM/1, as a result of the decision in *Dawber Williamson Roofing Ltd* v. *Humberside County Council* (1979), where the main contract was based on JCT 63. These amendments are now included in JCT 2011 and the Standard Form of Building Subcontract. The domestic subcontractor's contract was based on the standard form of domestic subcontract, referred to as the Blue Form. By clause 14 of the main contract it was stated that any unfixed materials or goods delivered to and placed on or into the works should not be removed without consent, and that when the value of those goods had been included in a certificate under which the contractor had received payment, such materials or goods should become the employer's property. By clause 1 of the domestic subcontract, the subcontractor was deemed to have notice of all the provisions of the main contract, apart from prices. It should be noted that there was no express provision in the subcontract concerning when and if the property in the subcontractor's materials or goods was to pass to the main contractor. The subcontractor delivered 16 tons of roof slates to the site and submitted invoices to the main contractor. An interim certificate which included the value of the slates was issued under the main contract. The employer paid the appropriate sum to the main contractor. Therefore, ownership in the slates would vest in the employer, according to the main contract, as the amount had been certified and paid. The main contractor did not pay the domestic subcontractor and went into liquidation. The subcontractor claimed that he was still the owner of the slates and therefore was entitled to their possession. The court held that the slates were still owned by the subcontractor. They were never at any time owned by the main contractor, and therefore he could not pass title in them to the employer. The employer had to pay for them again.

13.10.8. Once goods or materials become incorporated into the works, the effectiveness of a retention of title clause will be lost. In the case of *Peoples Park Chinatown Development Pte Ltd* v. *Schindler Lifts (Singapore) Pte Ltd* (1993), Schindler was appointed as a nominated subcontractor to the main contractor. The developer was Peoples Park Chinatown, who became insolvent before the project was completed. Before Schindler was nominated, it entered into an agreement with Peoples Park Chinatown to accept deferred payment terms. Schindler supplied and installed ten escalators, but did not test or commission them, nor supply and fix the finishings, such as balustrades. The liquidator of Peoples Park Chinatown sold the building, together with the ten escalators, to a third party. Schindler claimed against the liquidator for payment for the ten escalators. The liquidator refused, on the grounds that the escalators had become attached to the building (i.e. the land) and therefore belonged to Peoples Park Chinatown. He claimed a right to sell them, leaving Schindler as an unsecured creditor. It was held:

(1) The escalators had been fixed to the building in such a way as to become a permanent feature of the building. They became part of the land, notwithstanding they were not commissioned. Therefore they became the property of Peoples Park.
(2) There was no direct agreement between Schindler and Peoples Park. The escalators had been supplied as part of the main contract between Peoples Park and the main contractor. There was no provision in the main contract that would give Schindler the right to recover the escalators from the building after they were fixed. However, they could have recovered any unfixed materials (e.g. the balustrades).

When this case was heard at first instance, the judge found in favour of Schindler, which seemed just in the light of the terms of payment agreed between Schindler and Peoples Park. However, the Court of Appeal reverted to the classical position in English land law. When items become permanently fixed to a building they become part of the land and they belong to the owner of the land. The court followed the principal English authority on this point: *Seath & Co* v. *Moore* (1886).

13.10.9. As time goes by, retention of title clauses continue to be referred to the courts, whereby different aspects of these clauses are examined. In the case of *P4 Ltd* v. *Unite Integrated Solution plc* (2007), the court examined a retention of title clause relating to the supply of electrical and mechanical products by P4. Interim payment had been made in accordance with the DOM/2 standard conditions. However, in respect of materials delivered to site, a lump sum had been allowed in the payment. It was argued that the payment included a lump sum for materials on site and hence, in accordance with the terms of the contract, ownership was transferred at the same time. The court did not agree and considered that, as payment had been made on a lump sum basis, ownership of materials did not automatically pass. The judge, having examined all the authorities, said:

it is necessary to identify with particularity the materials and goods that are the subject of payment. A general lump sum interim valuation is insufficient

When acting for employers, the quantity surveyor or whoever is charged with recommending or certifying payment must ensure that, where materials delivered to site are included, full details of the materials included in the payment have been provided. With materials being delivered on a regular basis and fairly rapidly built into the works, good records need to be retained. P4, in any event, lost the case as their terms and conditions, which included the retention of title clause, were held not to form part of the contract of sale.

13.10.10. The Insolvency Act 1986 imposes certain restrictions on retention of title clauses. Once a petition has been presented to the court for the appointment of an administrator to a company, section 10(1)(b) of the 1986 Act provides that no steps may be taken to repossess goods in the company's possession under any retention of title agreement, except with the leave of the court and subject to such terms as the court may impose. Once an administrator has been appointed, the embargo on repossession continues, though an administrator, as well as the court, may consent to the recovery of goods by the supplier. Moreover, an administrator is empowered to sell goods which have been supplied subject to a retention of title agreement, if he can persuade the court that disposal would be likely to promote one or more of the purposes specified in the administration order.

SUMMARY

Reservation of title clauses, if properly drafted, will normally provide protection to a supplier who delivers goods to a purchaser who becomes insolvent before payment is made. However, no protection exists if the goods have been incorporated into the works. A further difficulty can arise for the seller under section 25 of the Sale of Goods Act 1979. Where this section applies, a purchaser acting in good faith without notice of the reservation of title may acquire a good title, despite the provisions of the clause. However, to qualify under section 25, the buyer, such as an employer on a construction project, must have acquired the goods, which may be part of a materials and workmanship contract, under a sale, pledge or other disposition; for example, by way of JCT 2011.

The Insolvency Act 1986 provides that, where a petition has been presented to the court for the appointment of an administrator, goods cannot be repossessed under a reservation of title clause, except with the leave of the court.

13.11. Can the signing of time sheets which make reference to standard conditions of contract form the basis of a contract?

13.11.1. The courts seem ever willing to infer that a contract has come into being, in preference to the uncertainty where goods are bought and sold, or work carried out in the absence of a contract. Agreement between the parties is one of the main ingredients in deciding whether there is a contract. In trying to establish whether agreement has taken place, courts are prepared to examine the business dealings between the parties.

13.11.2. This matter was the subject of the decision in *Grogan* v. *Robin Meredith Plant Hire and Triact Civil Engineering* (1996). In 1992 Triact, a civil engineering contractor, was laying pipes on a site. The principal of Meredith, a plant hire company, approached the site agent and it was agreed that Triact would hire a driver and machine from Meredith at an all-in rate of £14.50 per hour from 27 January 1992. No formal agreement was mentioned. At the end of the first and second weeks, Meredith's driver presented the site agent with time sheets for checking and signature. At the bottom of the sheet it stated: 'All hire undertaken under CPA conditions. Copies available on request.' During the third week there was an accident and the plaintiff was injured. He issued proceedings against both defendants and they consented to judgment in the sum of £82,798.17, Meredith paying one-third and Triact two-thirds. Meredith claimed that Triact was liable to indemnify it, in accordance with the CPA conditions referred to on the time sheets. The issue was whether the presentation of the time sheets by the driver amounted to a variation of the plant hire contract sufficient to incorporate the CPA conditions by reference? The central question was whether the time sheet had a contractual effect. It was held that normally a document such as a time sheet, invoice or statement of account does not have a contractual effect, in the sense of making or varying a contract. Time sheets do not normally contain evidence as to the terms of a contract, and in this case they were intended merely as a record of a party's performance of an existing obligation. The signed time sheets did not have, nor purport to have, contractual effect. Therefore, Triact was not liable to indemnify Meredith.

SUMMARY

If time sheets which make reference to standard conditions of contract are signed, it doesn't follow that those conditions will form part of the contract or become substituted for existing conditions of contract.

13.12. Can suppliers rely upon exclusion clauses in their terms of trading to avoid claims for supplying defective goods, or claims based on late supply?

13.12.1. Suppliers to the construction industry, when drafting their conditions of trading, will usually include a clause the effect of which will be to reduce, or even in some cases eliminate, their liability to reimburse a purchaser where goods are delivered late or are in some way defective. In an effort to regulate this type of limitation or exclusion clause, the Unfair Contract Terms Act 1977 was enacted. The statute is unfortunately worded, as it deals only with clauses which seek to limit or exclude liability and not unfair terms in general.

13.12.2. Section 3 of the Act states that it applies to contracts made after 1 February 1978 and relates to all clauses excluding or restricting liability which are contained in 'written standard terms of business'. A contracting party who wishes to rely on any such term must demonstrate to the satisfaction of the judge (or arbitrator) that the exclusion or limitation is something which it was 'fair and reasonable' to have included in the contract, 'having regard to the circumstances which were, or ought reasonably to have been, known to or in the contemplation of the parties when the contract was made'. There have been a number of cases related to the construction industry where courts have been called upon to decide whether an exclusion or limitation clause is valid.

13.12.3. In the case of *Rees Hough Ltd* v. *Redland Reinforced Plastics Ltd* (1984), the court had to decide whether the following limitations clause which formed part of Redland's standard terms of trading was reasonable and therefore enforceable.

> The company warrants that the goods shall be of sound workmanship and materials and in the event of a defect in any goods being notified to the company in writing immediately upon the discovery thereof which is the result of unsound workmanship or materials, the company will, at its own cost at its option, either repair or replace the same, provided always that the company shall be liable only in respect of defects notified within three months of delivery of the goods concerned. Save as aforesaid, the company undertakes no liability, contractual or tortious, in respect of loss or damage suffered by the customer as a result of any defect in the goods (even if attributable to unsound workmanship or materials) or as a result of any warranty, representation conduct or negligence of the company, its directors, employees or agents, and all terms of any nature, express or implied, statutory or otherwise, as to correspondence with any particular description or sample, fitness for purpose or merchantability are hereby excluded.

In deciding that the limitation clause was unfair and therefore unenforceable, the court took the following matters into account:

- The strength of the bargaining positions of the two parties;
- Whether the customer received an inducement to agree to the term or had an opportunity of entering into a similar contract with others without such a term;
- Whether the customer knew of the term;
- Where the contract excluded or restricted liability for breach of condition, whether it was reasonable to expect compliance with it; and
- Whether the goods were manufactured to the special order of the customer.

Section 11(5) of the Act provides that: 'It is for those claiming that a contract term . . . satisfies the requirement of reasonableness to show that it does.'

13.12.4. In *Chester Grosvenor Hotel Co Ltd* v. *Alfred McAlpine Management Ltd* (1991) a dispute arose concerning whether an exclusion clause in a management contract was reasonable. The wording stated that McAlpine would undertake to take all practical steps to enforce the construction contractors' contracts, to secure performance of the obligations under those contracts and to recover damages. Such action was to be in Grosvenor's name and at their expense. The clause also said that Grosvenor were not entitled to recover from McAlpine, by set-off or other action, any sums greater than those which McAlpine recovered with Grosvenor's consent from the construction contractors. Further,

Grosvenor could not recover any such sums from McAlpine before McAlpine themselves had recovered them from the construction contractors. Grosvenor argued that the exclusion clause was subject to and invalidated by section 3 of the Unfair Contract Terms Act 1977. The judge, in finding the clause was reasonable, identified the following factors in favour of McAlpine:

- The equal bargaining power of the parties;
- The fact that the clause was designed to act as an agreed division of risk between commercial entities dealing at arm's length;
- The fact that this allocation of risk was reflected in McAlpine's remuneration;
- The relatively slight risk to which Grosvenor were exposed under the clause (viz., that of McAlpine's insolvency, it being plain that Grosvenor still had a right to proceed directly against the construction contractors under their direct contracts);
- Grosvenor's right to control that risk by vetoing the selection of construction contractors under clause 1(c) and 19 of the management contract and by making inquiries about those contractors' financial status;
- The fact that it was within the parties' contemplation that, if the risk to Grosvenor eventuated, the results would probably not be disastrous;
- The fact that McAlpine's contracts had been presented to Grosvenor as open to negotiation and not on a 'take it or leave it' basis;
- The fact that Grosvenor could have contracted on different terms with other management contractors;
- The substantial time and general opportunities available to Grosvenor for consideration of the terms;
- The availability to Grosvenor of independent advice;
- The intelligibility of the exclusion clause; and
- The fact either party could have insured against the relevant risk, but that in either event the cost would have fallen on Grosvenor.

13.12.5. Other cases involving limitation or exclusion clauses relating to construction work include the following:

- *Charlotte Thirty and Bison* v. *Croker* (1990);
  In this case it was held that an exclusion clause in a contract to supply batching plant was unreasonable and therefore unenforceable.
- *Barnard* v. *Marston* (1991);
  In this case a subcontractor's exclusion clause was held to be valid.
- *Barnard Pipelines Technology* v. *Marston Construction* (1992);
  An exclusion clause in this case was held to be reasonable and therefore valid. The fact that the purchaser was aware of the exclusion clause when the contract was entered into influenced the judge.
- *Stewart Gill* v. *Horatio Myer* (1992);
  In this case a right of set-off was excluded by the terms of trading. The court held that this was unreasonable and therefore invalid.

13.12.6. In the case of *British Fermentation Products Ltd* v. *Compair Reavall Ltd* (1999), the court had to check whether terms and conditions which applied to a contract were one party's 'standard terms of business', as referred to in the Act. It was held that for the wording 'standard terms' to apply, the terms must 'invariably or at least usually be used by the party in question'.

13.12.7. It is not uncommon for a supplier's terms and conditions to include a clause which requires the purchaser to inspect the goods on delivery. The clause usually goes on to either exclude or limit liability for defects which could have been discovered at the inspection stage. In the case of *Expo Fabrics (UK) Ltd* v. *Naughty Clothing Co Ltd* (2003), the Court of Appeal had to decide whether an exclusion clause in a contract to supply fabrics passed the reasonableness test. Clause 6A in the terms stipulated that any claim based upon a defect in the quality or condition of the goods should be notified in writing within 20 days of delivery. The clause went on to say that outside this period, Expo would have no liability for faulty goods. Clause 6B dealt with the situation where faults were discovered in the goods. When this occurred, Expo was obliged to replace the goods free of charge or refund the price. Lord Justice Waller in the Court of Appeal considered the clause reasonable, as the 20-day period provided plenty of time to check the goods.

13.12.8. Disputes as to whether exclusion or limitation clauses are reasonable are a constant source of litigation, as can be seen from the following cases:

- *The Salvage Association* v. *CAP Financial Services* (1992);
- *Trolex Products* v. *Merrol Protection Engineering* (1991);
- *Fillite Runcorn* v. *APV Plastics* (1993);
- *Edmund Murray* v. *BSP International Foundations* (1992);
- *St Albans City & District Council* v. *International Computer Ltd* (1996);
- *Omega Trust* v. *Wright Son and Pepper* (1996);
- *Governor & Company of the Bank of Scotland* v. *Fuller Peiser* (2001);
- *Mostcash plc (Formerly UK Paper), Fletcher Challenge Forest Industries Ltd and Metsa-Serla (Holdings) Ltd* v. *Fluor Ltd* (2002);
- *John Moodie & Co and Others* v. *Coastal Marine (Boatbuilders) Ltd* (2002);
- *Bacardi-Martini Beverages Ltd* v. *Thomas Hardy Packaging* (2002);
- *Britvic Soft Drinks Ltd and Others* v. *Messer UK Ltd and Another* (2002);
- *Amiri Flight Authority* v. *BAE Systems plc and Crossair Limited Company for Regional European Air Transport* (2002);
- *SAM Business Systems Ltd* v. *Hedley & Co (A Firm)* (2002);
- *Rolls Royce Engineering plc and Another* v. *Ricardo Consulting Engineers Ltd* (2003);
- *Farrans Construction Ltd* v. *RMC Ready Mixed Concrete (Scotland) Ltd* (2003); and
- *University of Keele* v. *Price Waterhouse* (2004).

13.12.9. The temptation of drafters of conditions of contract to eliminating liability for certain breaches of contract needs to be carefully considered. The case of *Regus* v. *Epcot* (2007) involved failed air-conditioning in serviced premises let by Regus to Epcot. Under the terms of the service accommodation provided by Regus for Epcot, Regus was in breach of the agreement with regard to the failed air-conditioning. Whilst there was no doubt that Regus had a liability, it relied upon an exclusion clause when it came to the payment of damages. The standard form, which was regularly used by Regus, excluded any liability for loss of business, loss of profits, loss of anticipated savings, loss or damage to data, third-party claims, or any consequential loss. The clause was held by the court to be unreasonable and therefore unenforceable. The exclusion clause was construed as meaning that Regus would not in any circumstances be liable for any damages Epcot incurred, even where a given loss was the only loss it was likely to incur. The judge therefore held the clause to be unreasonable and hence unenforceable. The reason given by Mr Justice Ramsey in the case of *Lobster Group Ltd* v. *Heidelberg Graphic Equipment Ltd* (2009) were similar to those in the *Regus* v *Epcot* decision. It was held by Mr Justice Ramsey that an exclusion clause which sought to exclude liability for immediate loss or increased costs or expenses or for direct damage was unreasonable. He considered that if the clause were imposed, it would leave the claimant without a meaningful remedy.

13.12.10. A very useful decision, from the point of view of interpreting the enforceability of clauses which seek to limit liability, is *Shepherd Homes* v. *Encia Reclamation Ltd and Green Piling Ltd* (2007). There were several matters in dispute between the parties, not least a potential claim in the sum of £10 m from Shepherd Homes against Encia as a result of defective piling. Encia intended to pass the claim down to its piling subcontractor, Green Piling. However, Green Piling in its terms and conditions, which formed part of the subcontract between Encia and Green, included a clause which limited Green's liability for breach of contract to the amount of the contract sum. It was held by the court that this clause was reasonable and therefore enforceable. It may be useful to compare this decision with that in the case of *George Mitchell* v. *Finney Lock Seeds* (1983). In this case the supplier of cabbage seeds supplied the wrong type. Liability was limited to the cost of supplying the seed, which totalled £192. The loss involved, however, was £61,000. It is perhaps not surprising that the court held the limitation clause to be unreasonable and therefore unenforceable. The court will obviously take into account the amount of the contract price when compared with the damages suffered in deciding whether such a limitation clause is reasonable.

13.12.11. The unfortunate aspect of this statute is that the courts, whilst trying to be consistent in the way in which they have decided when limitation or exclusion clauses are unreasonable and therefore unenforceable, have left suppliers with a dilemma. When entering into contracts to supply their products, they do not know whether the exclusion or limitation clause will be valid, should they have need to make use of it. This being the situation, how do they price the risk?

SUMMARY

Suppliers are entitled to include in their terms of trading clauses which exclude or limit their liability if goods they supply prove to be defective or are delivered late. However, under the Unfair Contract Terms Act 1977, if such clauses are to be enforceable the supplier is required to demonstrate that they are reasonable. It has been held that an attempt to eliminate all liability for loss or damage resulting from a breach of contract is likely to fail as being unreasonable. Liability which is limited to the contract price has been upheld in one well-reported case whilst being held to be unreasonable in another.

In making a decision on this type of limitation of liability the court will no doubt take into consideration the amount of the contract price when compared with the damages suffered.

13.13. What level of supervision must an architect provide on site?

13.13.1. Where work proves to be defective and the contractor becomes insolvent and unable to correct the defects, questions are often asked as to whether the architect has any liability to the employer for the cost of correcting the defects. This raises the matter as to the extent to which the architect is obliged to supervise the work of the contractor. A typical clause used by architects when drafting their own terms is:

> At intervals appropriate to the stage of construction visit the works to inspect the progress and quality of the works and determine that they are being executed generally in accordance with the contract document.

When deciding the extent of this type of obligation, it is generally accepted that the Architect is not required personally to check every detail. He should, however, check important matters, such as the formation below a concrete floor slab before it is laid.

13.13.2. The matter of the obligation of the architect with regard to supervision has been the subject of a number of cases. In *Alexander Corfield* v. *David Grant* (1992), the plaintiff architect sought unpaid fees of £23,750.95 plus interest for work undertaken at a listed building whose use the defendants wished to change to a hotel. The defendants counterclaimed for damages for breach of contract. The defendants had a set time scale for the conversion work, in order to get the hotel into the 1990 guide books, which were to go to press in May 1989. In a letter, the plaintiff set out a proposed timetable and recommended that the RIBA fee scale be used. The defendant wrote back accepting this and agreeing that plans must be submitted from time to time to meet the planning requirements and stressed that this should be done in time for the District Council planning meeting. The defendant also accepted the plaintiff's recommendation as to the builder to be employed. The defendant submitted that the plaintiff made so many mistakes that he was in repudiatory breach of the contract. The court accepted that the plaintiff continually did or omitted things to such an extent that his conduct showed an intention not to perform the contract to a reasonable standard and that the defendant reasonably lost confidence in him, so as to entitle them to accept the breaches as repudiation of the contract and to dismiss him.

The court, in arriving at a decision, advised on the level of supervision on site which an architect is required to provide. What was adequate by way of supervision and other work was not in the end to be determined by the number of hours worked, but by asking whether it was enough for the job. In this case the plaintiff needed, but did not have, at least one skilled and experienced assistant, with the result that the project was an inadequately controlled muddle and the plaintiff was in continuous breach of contract. Judge Stabb held:

> I think that the degree of supervision required of an architect must be governed to some extent by his confidence in the contractor. If and when something occurs which should indicate to him a lack of competence in the contractor, then, in the interest of his employer, the standard of his supervision should be higher. No one suggests that the architect is required to tell a contractor how his work is to be done, nor is the architect responsible for the manner in which the contractor does the work. What his supervisory duty does require of him is to follow the progress of the work and to take steps to see that those works comply with the general requirements of the contract in specification and quality. If he should fail to exercise his professional care and skill in this respect he would be liable to his employer for any damage attributable to that failure.

13.13.3. Judge Stabb reflected what Lord Upjohn said about the nature of an architect's duty in circumstances where he knew about the incompetence of the contractor. This was in *East Ham Corporation* v. *Bernard Sunley & Sons Ltd* (1964), where he said, with regard to supervision by an architect:

> As is well known, the architect is not permanently on the site but appears at intervals, it may be a week or a fortnight, and he has, of course, to inspect the progress of the work. When he arrives on the site there may be very many important matters with which he has to deal: the work may be getting behind-hand through labour troubles; some of the suppliers of materials or the subcontractors may be lagging; there may be physical trouble on the site itself, such as finding an unexpected amount of underground water. All these are matters which may call for important decisions by the architect. He may in such circumstances think that he knows the builder sufficiently well and can rely upon him to carry out a good job; that it is more important that he should deal with urgent matters on the site than that he should make a minute inspection on the site to see that the builder is complying with the specifications laid down by him . . . It by no means follows that, in failing to discover a defect which a reasonable examination would have disclosed, in fact the architect was necessarily thereby in breach of his duty to the building owner so as to be liable in an action for negligence. It may well be that the omission of the architect to find the defect was due to no more than error of judgment, or was a deliberately calculated risk which, in all the circumstances of the case, was reasonable and proper.

13.13.4. In *Sim & Associates* v. *Alfred Tan* (1997), a Singapore case, Alfred Tan bought a piece of land on which to build a two-storey bungalow and engaged Sim & Associates, a firm of architects, planners and engineers, to provide the various services in connection with the construction of the bungalow. Sim & Associates, on behalf of Alfred Tan, awarded the contract for the building project to Hok Mee Construction. Alfred Tan sued Sim & Associates, alleging that they had breached their duties as architect by certifying defective works and not calling upon the main contractor to rectify the defective works and complete the uncompleted works. Sim & Associates counterclaimed for the balance of their professional fees and disbursements. The trial judge, although he rejected both grounds of claim (partly for lack of evidence and partly because the contractors refused to return because of their disagreements with Tan over various issues), nevertheless went on to hold that the building defects were attributable to a lack of proper supervision and a failure to require the contractor to make good defective works. Sim & Associates appealed. It was held that an architect is merely required to give the building works reasonable supervision.

(1) The defective works were ultimately the result of the failure of Tan and the main contractor to arrive at an acceptable compromise and not in any way attributable to any failure by Sim & Associates to perform their duties properly as architects.
(2) Tan had not succeeded in showing that he had suffered any real loss. The total cost of the rectification works carried out did not exceed the final sum due to the architects.

The mere existence of defective works does not, of itself, translate into a finding of lack of supervision against the architect in a building contract.

13.13.5. The case of *McGinn* v *Waltham Contractors and Others* (2007) involved a claim made against the contractor, the architect and the rest of the design team in respect of the construction of a residential property in Jersey, which, because of extremely poor workmanship, was demolished. The contractor became insolvent and the employer was left to seek his entitlement from the architect and the rest of the design team. The court had to decide the extent to which the architect was responsible for the contractor's defective work. Judge Coulson set out the following six legal principles which governed the work of the architect with regard to the contractor's workmanship:

1. When deciding the number of visits to be made to the site to inspect the work, the architect should be guided by the nature of the work being undertaken by the contractor.
2. The architect should instruct the contractor not to cover up work prior to inspection, dependent upon the importance or stage of the work.
3. The fact that defective work was carried out when the architect was absent from the site does not absolve him from being found negligent.
4. If an element of the work is particularly important, the architect must ensure that he visits the site at the outset of the work.
5. Reasonable inspection of the work does not require the architect to go into every matter of detail.
6. The mere fact that the contractor's work is defective does not in itself mean that the architect has been negligent.

SUMMARY

The architect's duties with regard to what level of supervision he must provide are rarely given in any detail in the architect's conditions of appointment. In the final analysis, the court will make a subjective judgment based upon the facts of each case. However, as a general rule, an architect is merely required to provide a reasonable level of supervision. The fact that work is certified which turns out to be defective does not in itself mean that the architect has been negligent.

13.14. Where a specification includes a named supplier 'or other approved', can the architect/engineer refuse without good reason to approve an alternative supplier proposed by the contractor/subcontractor?

13.14.1. The question of the architect's right to refuse to give approval to an alternative supplier proposed by the contractor where the contract documents named a supplier 'or other approved' was the subject of a dispute in *Leedsford Ltd* v. *The City of Bradford* (1956). The defendant council wished to have a new infants' school built. Contract documents including bills of quantities were prepared and submitted to contractors for pricing. Provision was made in the bills for artificial stone, which was to be used in the following terms:

> Artificial Stone . . . The following to be obtained from the Empire Stone Company Limited, 326 Deansgate, Manchester or other approved firm.

13.14.2. The successful contractor proposed that the artificial stone should be supplied by HK White (Precast Concrete Works) Ltd and sought the architect's approval. The cost of stone from Empire Stone was £1,250, compared with £500 from the alternative supplier. The architect refused to approve the alternative supplier, insisting the stone be obtained from Empire Stone.

13.14.3. The matter was referred to the court by the contractor, who claimed the difference in the cost of stone. The contractor contended that those words had been inserted for its benefit. It was argued that they had the right to put forward and obtain approval for any firm who would supply stone of the proper quality at a price less than the price which the Empire Stone Company Ltd would charge. It was further argued that the architect had breached the contract when he insisted that the stone was obtained only from the Empire Stone Company Ltd.

13.14.4. It was held in the Court of Appeal that the words 'or other approved firm' did not give the contractor an option to submit any firm of their choice for the architect's approval. The words 'to be obtained from the Empire Stone Company Limited . . . or other approved firm' should be analysed in the following way: The builder agrees to supply artificial stone. The stone is to be Empire Stone unless the parties both agree some other stone, and no other stone can be substituted except by mutual agreement. The builder fulfils his contract if he provides Empire Stone, whether the Bradford Corporation want it or not; and the architect can say that he will approve of no other stone except the Empire Stone.

Accordingly, the position was that there was an absolute obligation on the contractor to supply Empire Stone unless the architect should give his approval to some other stone. Moreover, the architect was not bound to give any reasons for withholding approval of any other firm. The most that was required of the architect was that he should act in good faith, and no allegation against his good faith was made.

13.14.5. This decision would not now apply on contracts let in the public sector, where there is an EU restriction on the use of a single specified supplier.

SUMMARY

Where a contract calls for goods to be supplied by a named supplier, 'or other approved,' the architect may refuse to approve an alternative supplier proposed by the contractor without giving reasons. The architect is only required to act in good faith. This does not apply to public contracts, where, under EU rules, there is a restriction on the naming of a sole supplier in a specification.

13.15. If a subcontractor is falling behind programme and in danger of completing late because of his own inefficiencies, can the contractor bring other labour onto the site to supplement the subcontractor's efforts, to ensure completion on time?

13.15.1. Late completion of a project will usually result in substantial extra costs being incurred by the contractor. Where delays result from poor progress on the part of a particular subcontractor, the main contractor will usually apply pressure on the subcontractor to increase productivity to ensure that completion on time is achieved. Frustration, on the part of the contractor, often sets in if there is no obvious improvement in progress. Contractors' minds may then turn to the idea of bringing onto the site more operatives to supplement those of the defaulting subcontractor, in an effort to make up lost time. Subcontractors are understandably resistant to operatives who are working for either the contractor or other subcontractors being employed on their project, and often a stalemate occurs, or the contractor takes matters into his own hands and goes ahead and recruits more operatives. Is the contractor entitled to act in this fashion, or is he subject to some form of legal restriction?

13.15.2. This matter was the subject of a dispute which came before the courts in the case of *Sweatfield* v. *Hathaway Roofing* (1997). Sweatfield, who were formally 'IT Design and Build', subcontracted the roofing and cladding of a leisure centre to Hathaway Roofing. The relationship between the parties was strained and was not improved because Hathaway Roofing's work fell behind schedule. A payment which was due to be made to Hathaway Roofing was withheld by IT Design and Build and, as a result, Hathaway Roofing withdrew its labour from site. Negotiations between the parties resulted in Hathaway Roofing returning to the site. Progress did not improve very much, resulting in a letter being sent by IT Design and Build to Hathaway Roofing accusing them of failing to proceed regularly and diligently with the works. Hathaway Roofing's response was to allege that their progress was being hampered by the poor workmanship of IT Design and Build. Hathaway Roofing was then instructed by IT Design and Build to increase the numbers of on-site operatives. Hathaway Roofing refused, on the grounds that it was impractical to accommodate any more operatives. IT Design and Build decided to take matters into their own hands and brought onto the site a gang of sheeters from another subcontracting company. Hathaway Roofing's reaction was to withdraw its labour from the site, claiming that, in bringing another subcontractor onto the site to carry out some of its work, the contractor was in breach of contract. The court decided that the action of IT Design and Build was of great gravity and concluded that it was in breach of contract.

13.15.3. It is possible to write into a contract a clause which gives the contractor a right to bring onto site operatives to supplement those employed by the subcontractor.

SUMMARY

A contractor who brings operatives onto the site or introduces another subcontractor to supplement the workforce employed by one of its subcontractors, risks a claim being levied for breach of contract. A clause, however, can be drafted into the subcontract which gives the contractor express power to bring on extra labour. If a subcontractor is failing to proceed regularly and diligently with the subcontract work, there is usually an entitlement under the terms of the subcontract for the contractor to terminate the subcontract.

13.16. Can parties to a dispute be forced to submit the matter to mediation?

13.16.1. The perceived wisdom concerning commercial disputes is that a resolution process involving ADR is preferable to the more formal court or arbitration procedures. Mediation has come to the forefront as a leading method of ADR and has, in many instances, been very successful. Where disputes are referred to court or arbitration, significant costs are usually incurred by both parties. Under normal circumstances, the judge or arbitrator has a discretion to issue an order to the effect that the losing party will be obliged to pay the winning party's costs. Whilst in most cases ADR is less expensive than litigation or arbitration, nevertheless the process can often involve the parties in substantial costs. Parties in the construction industry are being steered into mediation where a dispute has arisen. JCT 2011 provides for mediation as the first stage on the road to the dispute being resolved. If matters are referred to court where the contract does not provide for mediation, the parties are often required by the judge to try mediation as a way of avoiding the full court process.

13.16.2. Mediation is a consensual method of resolving disputes. It would seem, therefore, if either of the parties have no desire to refer a dispute to mediation, forcing their hand would appear to be an expensive and time-consuming process, with little or no chance of success. Despite the rather obvious drawbacks, courts, on occasion, seem insensitive to the prospect of wasting time and cost, as can be seen from the examples of parties to contracts being obliged to take part in mediation. In the case of *Cable and Wireless plc* v. *IBM United Kingdom Ltd* (2002) the court had to decide whether, under the terms of a contract to supply information technology, the parties could be forced to refer the matter for resolution by ADR in accordance with clause 41. The court held that the parties were obliged to submit to ADR. In *Balfour Beatty Construction Northern Ltd* v. *Modest Corvest* (2008), Judge Coulson gave support to the decision in the *Cable and Wireless* case that, where ADR provisions were included in the contract, they must be used as the forum for settling disputes and the court will enforce such provisions. However, in the *Balfour Beatty* dispute, Judge Coulson considered that the mediation provisions in the contract were merely an agreement to agree, which wasn't enforceable.

13.16.3. Under the Civil Procedures Rules (1998), High Court judges have power to recommend parties who are involved in litigation to submit their dispute to mediation.

13.16.4. Courts are even prepared to penalise parties who refuse mediation without justifiable grounds. In the case of *Dunnett* v. *Railtrack* (2002), a dispute was referred to the Court of Appeal, where Railtrack were successful. The normal rule that the losing party pays the winner's costs was expected to apply. Despite Railtrack's success, the Court of Appeal refused to order that Dunnett pay Railtrack's costs, as Railtrack had refused the offer of mediation.

13.16.5. Courts, however, do on occasion realise that mediation would be a waste of time and act accordingly. The matter of a costs order was again the subject of a dispute in the case of *Hurst* v. *Leeming* (2002). Hurst was a partner in a firm of solicitors and fell out with his partners. The matter was referred to court and Hurst instructed Ian Leeming QC to represent him. The case failed in the Lower Court, the Court of Appeal and in the House of Lords. Hurst considered that Leeming had negligently handled his case and commenced an action in the High Court. Following advice from the judge, the case was discontinued. Normally, Hurst would have been left to pay Leeming's costs, but after the commencement of proceedings Hurst had invited Leeming to partake in mediation, but this offer was refused. In view of the decision in the Railtrack dispute, Hurst sought to have Leeming's refusal of mediation to be taken into account. The judge, however, was persuaded that, as Hurst had no real prospect of success, this was justification for Leeming refusing to take part in mediation.

13.16.6. In *Malkins Nominees Ltd* v. *Société Finance* (2002), the successful party lost 25% of its overall cost recovery after refusing an offer of mediation.

13.16.7. Rather surprisingly, in the case of *Shirayama Shokusan Co Ltd and Others* v. *Danova Ltd* (2003), the court decided that it had the power to force the parties to mediate, even though neither wished to take this route. Mediation is very much a process of settlement when both parties are prepared to enter into meaningful negotiations. If one or both parties are looking for an outright victory, mediation could become a waste of time and cost.

13.16.8. The Court of Appeal in *Halsey* v. *Milton Keynes General Hospital Trust* (2004) in being less strident in its attitude to forced mediation, laid down some guidelines as to when it would be inappropriate for mediation to be used as a method of attempting to resolve a court action:

(1) If the whole object of the litigation is to determine an issue of law or lay down a legal principle;
(2) Where the claimant's position is unreasonably weak and is attempting to use mediation as tactical play in a nuisance value action;
(3) Where a defendant has made a reasonable offer to settle, which has been rejected;
(4) The cost of mediation would be disproportionately high;
(5) If the prospect of mediation would delay the date for trial; and

(6) Where mediation has little chance of success because of an intransigent attitude of one or both parties.

13.16.9. A question which has been raised on a number of occasions is whether, when a dispute has been referred to mediation which does not achieve a settlement, and litigation or arbitration takes place, can one of the parties can be ordered to pay the other's costs incurred in relation to the mediation? This matter was the subject of a court decision in the case of *Lobster Group Ltd* v. *Heidelberg Graphic Equipment Ltd* (2008). The dispute in this case arose out of the purchase of a printing press which proved to be defective. An action was commenced by the purchaser to recover the alleged losses. Prior to the commencement of the proceedings, mediation took place in January 2005 using the CEDR Model Form of Procedure, but unfortunately no agreement was reached. The claimant was placed in administration in July 2005 and went into liquidation in July 2006. Proceedings commenced in May 2007, following which the defendant raised the question of security for their costs. This is a procedure whereby a defendant faced with an action from a claimant who is insolvent can seek from the court an order for security of its costs. The intention is to ensure that if, in the final analysis, the defendant is successful and the court orders that the claimant is to pay the defendant's costs, there will be funds available from which to make the payment. The defendant's application for security of costs included the pre-action costs, together with those incurred in the mediation process. The judge decided that the legal costs incurred in participating in mediation convened after proceedings commenced are, in principle, recoverable. However, legal costs incurred prior to the commencement of proceedings are not in principle recoverable. A further factor which the judge took into account in deciding that the costs of the mediation would not be recoverable was the provision in the CEDR Model Form which states that each party will bear its own costs.

SUMMARY

A party to a dispute which arises under a contract where the disputes resolution procedures provide for mediation can be forced to have the dispute referred to mediation. In the absence of provision within the contract for mediation, should the matter be the subject of litigation, the judge has power to recommend that the parties refer the matter to mediation. If one of the parties refuses to submit to mediation, it may affect its entitlement to the recovery of costs in the event of it receiving a favourable court award.

13.17. Do contractors, subcontractors, architects or engineers who have been involved in the construction of a dwelling house, in the absence of any contractual link, have any legal liability to subsequent owners if, due to faulty design or construction, the dwelling is not fit for habitation?

13.17.1. The Defective Premises Act 1972 imposes a duty on all persons who take on work in connection with the construction of a dwellinghouse to ensure that the work is done in a workmanlike or professional manner with proper materials, so that the dwelling will be fit for habitation. This duty is owed to those who ordered the dwelling to be constructed and those with a legal or equitable interest in the dwelling. Purchasers of dwellings, from contractors or developers and subsequent purchasers, benefit from this legislation. A dwelling will include a house or flat, whether new, a conversion or an enlargement.

13.17.2. A person owing a duty under the statute may be released from this obligation if they are acting on the claimant's instructions, except where there exists a duty owed to the person to warn of any defects in the instruction. This may take the form of an exclusion of liability in the conditions of contract. The obligations under the Act extend to the contractor who constructed the dwelling, together with the subcontractors and also the architect and consulting engineers.

13.17.3. When can a dwelling be considered not fit for habitation? One may be forgiven for thinking that to qualify, the dwelling, due to its condition, could not be occupied. Whilst this may be a reasonable interpretation, it would be wrong. In the case of *Bole and De Haak* v *Huntsbuild and Richard Money Associates* (2009), the claimants continued to live in the house, but were nonetheless successful in an action under the statute. The reason is that the court considered the Act relates to defects of quality as well as dangerous defects. Defects in a part of the dwelling, rendering that part unsuitable for habitation, would suffice. For example, if, due to defects, one bedroom was damp and could not be occupied, leaving the remainder of the house fit for habitation.

13.17.4. An action was commenced by the claimant against the contractor and the consulting engineer engaged by the contractor. The basis of the case was that cracking had occurred because of heave resulting from the roots of a tree which had been felled shortly before the commencement of work. If the foundations had been constructed to a greater depth there would not have been a problem. It was the consulting engineer's case that his drawings showed foundations to go at least 500 mm below the last evidence of roots or desiccation of the clay. This explanation did not impress the court, which considered that it was not appropriate to leave the contractor to decide how deep to construct the foundations. The court held that the contractor was liable for breach of contract and the consulting engineer liable under the Defective Premises Act. The consulting engineer had failed to have proper regard to the NHBC and BRE Guidance on building near trees. They were jointly required to reimburse the claimant for the cost of the construction of a piled raft foundation.

13.17.5. The act stipulates that any action for a remedy for the purposes of the Limitation Act 1939 and the Law Reform (Limitation of Actions etc.) Act 1954 must be commenced within six years of the completion of the construction of the dwelling. The limitation period was an issue in the case of *Alderson* v *Beetham Organisation Ltd* (2003). In this case the claimant bought two basement flats from Beetham in a development involving the conversion of an existing property in Liverpool. The purchase occurred in January 1995, following which the claimant and her daughter took occupation. In April 1995 black mould and fungus was noticed in the bedroom walls of both flats which had been caused by damp. Beetham undertook some remedial work, in the form of re-laying flagstones outside the building and laying extra drainage pipes. The work did not solve the problem as, following heavy rain in September 1995, the basement flats flooded.

13.17.6. An independent chartered surveyor was engaged by the claimant, who reported that the problem resulted from a lack of proper tanking. An action under the Defective Premises Act was commenced in January 2001. It was argued by Beetham that the action was time barred, as it should have been commenced not later than six years after the date the original work was completed. The action, in accordance with the argument made by Beetham, should therefore have commenced not later than May 2000.

13.17.7. The claimant's case was that the limitation period commenced when the unsuccessful remedial work was undertaken. If this was correct, the action would not be time barred. Beetham countered by saying that, as the work they undertook to overcome the damp problem was unsuccessful, it should not be considered when calculating the limitation period. The lower court agreed with Beetham, but the Court of Appeal had other views. They said that, if Beetham's argument was correct, the contractor who neglected to put right a problem would be in a better position than one who had made an attempt, albeit futile. The Court of Appeal considered this could not be correct and in finding in favour of the claimants, overturned the decision of the lower court.

SUMMARY

In accordance with the Defective Premises Act 1972, a contractor or subcontractor who constructs a new house or flat, or is involved in the conversion or enlargement of one, has a duty to ensure that the work undertaken is done in a workmanlike manner with proper materials so that the dwelling will be fit for habitation when completed. An obligation is placed upon architects and engineers to provide services in a professional manner in order that the completed dwelling is fit for habitation when completed. This obligation extends to the person for whom the dwelling was constructed and any subsequent purchasers. There have been disputes as to the meaning of fitness for habitation. The courts have held that it applies to parts of a dwelling, such as a bedroom which due to damp cannot be occupied. It is not essential for all the dwelling to be uninhabitable.

13.18. Parties to a dispute often make offers to settle which are stated to be 'Without Prejudice'. The intention of an offer being 'Without Prejudice' is that, in subsequent litigation, evidence of the offer cannot be used to support the offeree's case, as the offer was made 'Without Prejudice' and as such, is privileged. Are there any circumstances where the 'Without Prejudice' safety net could fail and does it apply to other dispute resolution processes, such as statutory adjudication?

13.18.1. The purpose of allowing offers to be made without prejudice is founded on public policy and is intended to encourage parties who are in dispute to negotiate freely, without fear that details of the offer will be given in evidence in any subsequent litigation. This rule applies irrespective of whether the offer is given orally or in writing. The application of the rule does not depend upon the use of the phrase 'without prejudice'. If it is clear from the surrounding circumstances that the parties were attempting to reach a settlement of the dispute details of the negotiations will not be allowed in evidence: *Chocoladefabriken Lindt and Sprungli AG* v *Nestle Co Ltd* (1978).

13.18.2. If, following the without prejudice negotiations, a settlement is achieved, if there is a dispute as to whether or not a settlement in fact took place, the protection of without prejudice will not prevail in attempting to resolve that issue: *Rush and Tompkins* v *Greater London Council* (1989). This might well include objective facts communicated by one party to the other during negotiations. The process of interpretation should in principle be the same, whether negotiations were without prejudice or not, if it were necessary to establish whether the dispute had been resolved. It was held in *Oceanbulk Shipping and Trading SA* v *T MT Asia Ltd and Others* (2010) that, as without prejudice exchanges are admissible once a settlement has been achieved, they should also be admitted to show whether an agreement had occurred and as to what was the basis of the agreement.

13.18.3. Evidence of without prejudice correspondence may be admissible if, at the time of the exchange, there was no dispute. In *Buckinghamshire CC* v *Moran* (1990), it was held that a without prejudice letter which set out one of the parties' rights but not amounting to an offer to negotiate, was held not to be privileged and was therefore admissible as evidence. It was held, however, in the case of *South Shropshire District Council* v *Amos* (1986), that it was not necessary that the correspondence resulted in an offer being made; all matters disclosed or discussed in the without prejudice communications are protected.

13.18.4. In the case of *Framlington* v *Barnetson* (2004) it was held that without prejudice exchanges concerning a variation were not protected as they arose before any litigation commenced or was contemplated.

13.18.5. Adjudication which arises under the Construction Act 1996 is still in its infancy and the application of without prejudice protection to what would otherwise be admissible evidence is still unclear. In the case of *Specialist Ceilings Services Northern Ltd* v *ZVI Construction (UK) Ltd* (2004) it was accepted by both parties that the adjudicator had been sent without prejudice documents. The adjudicator was requested by the responding party to withdraw, but he refused. His reason was that any knowledge of a without prejudice offer would not affect his impartiality. The court held the adjudicator's decision to be enforceable, on the grounds that he had not seen or been aware that an offer had been made and, although aware that negotiations had taken place, he had not been influenced by the without prejudice documents which he had seen. It may be concluded that an adjudicator may have access to without prejudice correspondence and hence be aware that negotiations took place, without it affecting his position. It may be another story, however, if the adjudicator becomes aware that an offer had been made.

13.18.6. In the case of *Glencot Developments and Design Co Ltd* v *Ben Barratt and Son (Contractors) Ltd* (2001), an adjudicator had, partway through the adjudication, taken on the role of mediator. When the mediation failed to achieve a settlement, the adjudication resumed. The court held that as the adjudicator, when acting as mediator, had access to without prejudice communications, his decision was not enforced.

SUMMARY

Without prejudice offers to settle disputes cannot be produced as evidence in any subsequent court proceedings, on the basis that they are privileged. This is due to public policy, which is to encourage parties who are in dispute to reach a settlement. The without prejudice offer need not be in writing, as oral offers are also protected, to ensure that the parties are free to speak openly to secure a settlement. Once a settlement has been achieved, the without prejudice protection falls away. If there is a dispute as to whether or not a settlement has been reached, the without prejudice protection ceases to apply, for the purposes of making a decision on the matter. For the without prejudice protection to apply, there must be a dispute which is in the process of litigation or where litigation is contemplated.

Adjudication under the Construction Act is governed by the same rules, but as it is in its infancy the relevant case law is only developing. It would seem, however, that as the adjudication process is very short, with adjudicators who are well-versed in the ways of the construction industry, a little more license may be given to adjudicators who become aware of without prejudice exchanges.

13.19. Where a contractor takes over work which is part completed, does it have any responsibility for correcting work which was incorrectly undertaken by the contractor it replaced?

13.19.1. It is common practice, where a contractor becomes insolvent and ceases trading, for work to be left incomplete on several projects. Either the employer or the person who is appointed to manage the contractor's affairs, such as a receiver, will normally engage another contractor or contractors to finish off the work. Following the recommencement of work by the replacement contractor, it is not unknown for work which had been undertaken by the original contractor to contain defects. Does the replacement contractor have any liability for the cost of correcting the defective work?

13.19.2. The ideal arrangement would be for it to be made unambiguously clear when the replacement contractor is appointed where responsibility lay for non-compliant work which had been carried out by the original contractor. Unfortunately, this is not always the case; as often as not the contract documents are silent on the matter. Such was the situation in the case of *CJ Pearce Developments Ltd* v *Oakbridge Street Mellion Building Ltd* (2002).

13.19.3. This case involved a development of 20 detached holiday homes and access roads. Different contractors were appointed to construct the roads and homes. CJ Pearce undertook the construction of the homes, where a JCT Standard Form of Building Contract With Contractor's Design 1981 was used. The contractor who was contracted to construct the roads became insolvent before completing the work and CJ Pearce was appointed to finish off the work, employing the same JCT form of contract employed on the contract for the homes.

13.19.4. Following commencement of the task in finishing off the roadworks, CJ Pearce encountered defects in the work undertaken by its predecessor, which they corrected and for which they sought payment. A dispute arose, as Oakbridge, the developer, which was the other party to the CJ Pearce contract, argued that there were two related but discrete obligations with regard to the completion of the roadworks. There was an obligation to complete the road, but also there was an acceptance of responsibility for the workmanship of the original contractor. CJ Pearce contended that their obligation was limited to completing the work, but with no responsibility for the defective work on the part of the original contractor.

13.19.5. There was no clear answer and the judge was left to carefully examine the contract documents to arrive at a conclusion. The detailed drawings for the access roads had been prepared by Oakbridge; part of the work had been completed by the original contractor and CJ Pearce was contracted to finish off the work. During the completion of the roadworks, Oakbridge instructed CJ Pearce to correct certain work undertaken by the original contractor, the cost of which they were seeking to recover.

13.19.6. The Contractor's Proposal included the following wording:

> The access road being unfinished by others will be taken on and completed by ourselves.

The judge had to decide whether the words 'taken on' limited CJ Pearce's obligation merely to finishing off the roadworks, or whether it went further and included taking on responsibility for the work undertaken by the original contractor. Part of the roadworks had been completed by the original contractor and it wasn't clear from the contract documents whether CJ Pearce was required to take over all of the road, including the part fully completed, or merely the incomplete part.

13.19.7. The judge concluded that none of the contract documents made it clear whether the scope of the works included the assumption of responsibility by CJ Pearce for the work undertaken by the original contractor. In the circumstances, he gave the wording 'taken on' a narrow meaning, in confining it to work undertaken by CJ Pearce. It did not extend to responsibility for work completed by the original contractor.

13.19.8. It is clear from this decision that express wording is required if a contractor is to take on responsibility for correcting defects in work undertaken by others. In the absence of wording of this nature, a replacement contractor is unlikely to be found liable for defects in such work.

SUMMARY

There is no basic ruling which applies, where a contractor takes over and finishes off work commenced by others, as to whether the replacement contractor is responsible for the work undertaken by the others. It would appear from the case of *CJ Pearce Developments Ltd* v *Oakbridge Street Mellion Building Ltd* (2002) that, in the absence of clear wording to the effect that responsibility lies with the contractor appointed to finish off the work, no liability will lie with him. The contract documents for the finishing off work should make it clear whether the intention is for the contractor who is appointed to finish off the work is also to be responsible for correcting defects in the original contractor's work.

13.20. Where a contract requires a notification to be in writing, sent by post, or actual delivery, will an email or fax suffice?

13.20.1. Some contracts provide for a wide variety of methods by which one of the parties is required to communicate with the other. JCT 2011, for example, provides the following methods:

- All notices and communication expressly required by the contract between the employer or his agents, e.g. the architect and the contractor, must be in writing.
- The written notice or communication must be transmitted, whether electronically or otherwise, in a manner agreed from time to time by the parties.
- The notice or communication may be served by any effective means, including delivery by hand, or sent by pre-paid post.

13.20.2. The Engineering and Construction Contract (NEC 3) is less prescriptive with regard to 'each instruction, certificate, submission, proposal, record, acceptance reply and other communication which this contract requires', in that they are required to be sent 'in a form which can be read, copied or recorded.'

13.20.3. The effectiveness of notices required by contracts have been the subject of several court cases. In the Singapore case of *Central Provident Fund* v *Ho Bok Kee* (1981), a written notice was required by the contract to be sent from the employer to the contractor, by registered post or recorded delivery, concerning any decision to terminate the contract. The employer in this case sought to determine the contract by delivering the termination notice by hand. It was held by the court that the requirements of a termination clause must be strictly complied with. A wrongful forfeiture, as in this case, is treated as a repudiation.

13.20.4. A notice and its validity were the subject of the decision in *Construction Partnership Ltd* v *Leek Developments Ltd* (2006). Leek Developments engaged Construction Partnership to carry out refurbishment in Macclesfield, employing a JCT Intermediate Form of Contract, 1998 Edition. Leek Developments failed to pay two certificates issued by the contract administrator and Construction Partnership sought to determine its employment in accordance with the terms of the contract. The contract conditions required a warning notice specifying the default, to be followed by a notice of determination if the fault was not rectified. A failure to make payment in accordance with the terms of the contract was a fault which could give rise to a determination. The clause which caused the dispute was clause 7.1, which states:

> Any notice which includes a notice of determination shall be in writing and given by actual delivery or by special delivery or recorded delivery. If sent by special delivery or recorded delivery, the notice or further notice shall, subject to proof to the contrary, be deemed to have been received 48 hours after the date of posting, excluding Saturday and Sunday and public holidays.

13.20.5. On 23 December 2005, Construction Partnership sent a warning letter to Leek Developments concerning a failure to make payment in accordance with the terms of the contract. The notice was sent by both fax and post. In the absence of any payment being received, a letter of determination was sent by Construction Partnership to Leek Developments on 17 January 2006. It was argued on behalf of Leek Developments that the warning notice sent on 23 December 2005 did not comply with the contract, in that it was sent by fax, which is not 'actual delivery' as required by the contract. It was argued on behalf of Construction Partnership, that for actual delivery to occur, somebody must go along to the reception and hand the letter over. This argument was rejected by the court as being unrealistic. The judge stated:

> It is commonplace in modern commercial practice for documents to be sent by post or even more commonplace for documents to be sent by fax these days. A fax, it seems to me, clearly is in writing; it produces, when it is printed out on the recipient's machine, a document, and that seems to me is clearly a notice in writing. The question is, is that actual delivery? It seems to me, if it has actually been received, it has been delivered. Delivery simply means transmission by an appropriate means so that it is received, and the evidence in this case is that the fax has actually been received.

The directors of Leek Developments submitted that they had not read the fax until they returned to work after the Christmas break. The judge considered that this was not relevant, as the fax arrived at the offices of Leek Developments at 8.46 am on 23 December 2005 and the office did not close down for the Christmas break until noon on that day.

13.20.6. The editors of *Construction Industry Law Letter* dated July/August 2006 state that, in the light of the amount of business which is now conducted by email, email would be considered an appropriate means of transmission for the purposes of actual delivery.

13.20.7. The sending of an email to commence arbitration proceedings was the subject of the decision in the case of *Bernuth Lines Ltd* v *High Seas Shipping Ltd* (2005). In this case, the matter in dispute related to whether an email is a valid method of commencing arbitration proceedings. The court also answered the question that if email is appropriate for commencing arbitration proceedings, does this still apply if the email is sent to an address which had not been communicated to the party commencing the proceedings? It transpired that the email was sent to an email address shown in the Lloyds Maritime Directory and on the defendant's website, neither of which had been communicated to the referring party. The court held that Section 76 of the Arbitration Act 1996 had been drawn purposely wide and contemplated that any means of service would suffice, provided that it would be effective in delivering the document to the party to whom it was sent. The judge explained that arbitration proceedings tend to be used by businessmen with access to lawyers and their habitual means of communication is by email. He did not consider the email address to which the arbitration notice had been sent as particularly relevant, as the Arbitration Act was not specific as to which email address communications were to be sent. This is by way of contrast with the Crown Prosecution Rules, which are prescriptive as to when communications by email are acceptable.

SUMMARY

It is essential for those involved in sending notices required by a contract to ensure that the method employed accords with the requirements of the contract relating to communications. It would seem that a communication sent by email or fax will be regarded as being in writing. An email or fax will be regarded as delivered once received, irrespective of whether or not it has been read. However, it is submitted that the communication should arrive during business hours to comply with the requirement of being delivered.

13.21. Where an engineer is employed by the employer, is he legally obliged to warn of dangers associated with the temporary works?

13.21.1. Engineers who are engaged by the employer do not usually see themselves as guardians of the temporary work, as this is a matter for the main contractor. Does this encompass temporary works which may cause damage to the permanent works, if inadequate? This matter was examined in the case of *Hart Investments Ltd* v *Fidler and Larchpark* (2007).

13.21.2. The claimant was the owner of a building in Muswell Hill in London which was being redeveloped. As part of the redevelopment, the front and side facades had to be retained. A deep basement was constructed, requiring a substantial amount of propping to the existing retaining wall. Because of a lack of supports, part of the facade collapsed on 5 and 6 February 2004. The main contractor went into liquidation and the employer looked to the engineer, Mr Fidler, for the recovery of his losses sustained as a result of the collapse.

13.21.3. Mr Fidler was engaged by Larchpark in March 2003 to provide advice in relation to the temporary works. Hart Investments engaged Mr Fidler as engineer in June 2003 with regard to the permanent works, but no formal contract was ever entered into. It was therefore implied that Mr Fidler would undertake the normal duties to be expected of a structural engineer. Mr Fidler therefore had a dual role, namely that of advising Larchpark in respect of the temporary works and advising Hart Installations for the permanent works.

13.21.4. Mr Fidler provided Larchpark with various versions of the temporary works before the collapse occurred. He stated that one of his designs provided propping in the form of raking props to the retaining walls for when the excavation took place. Mr Fidler visited the site regularly, including the morning of 3 February 2004, when he inspected reinforcing bars at the rear of the site, which must have involved going past the area of the site where the propping of the retaining wall should have been. He would at that time have seen whether or not propping had been installed. Mr Fidler, in his evidence, stated that the excavation had not taken place on his visit and that Larchpark had suddenly, without notice to him, undertaken the excavation work contrary to his drawings. This evidence was rejected by the judge on the basis that, according to the programme, the excavation work was to be undertaken in early January 2004 and had progressed to the point of being dangerous by the time Mr Fidler visited the site on 3 February 2004.

13.21.5. The case against Mr Fidler was that he had a liability in both contract and tort to the building owner. It was alleged that Mr Fidler had failed to design any, or any appropriate, scheme for the temporary support of the underpinning which surrounded the deep basement excavation. In addition, he should have ensured that the contractor had taken precautions so as to support the basement wall, when he saw that it was unsupported. This was denied by Mr Fidler on the basis that he had no contractual liability to the claimant for the state of the temporary works, including the underpinning. He argued that he had designed an entirely appropriate scheme and, finally, he did not observe anything to suggest that the contractor had not carried out the scheme which he had designed.

13.21.6. The judge considered that Mr Fidler owed a duty, when commissioned to design the permanent works, to ensure the safe execution of these works by ensuring that the temporary works were constructed safely. It was the opinion of the judge that:

> In my judgment if an engineer, employed by an owner in respect of permanent works, observes a state of temporary works which is dangerous and causing immediate peril to the permanent works in respect of which he is employed, he is obliged to take such steps as are open to him to obviate that danger.

Had Mr Fidler performed this obligation properly, the collapse would not have taken place.

13.21.7. By way of conclusion, the judge considered that Mr Fidler:

- Knew that temporary propping was required;
- Knew it was his duty to inspect the works to ensure it was carried out properly; and
- Failed to notice that the basement excavation was being carried out without the propping.

13.21.8. The judge found that Mr Fidler had a liability for breach of an implied term in his contract with Hart Investments. He also found that Mr Fidler had a liability in tort. Whilst an action in tort based upon economic loss will normally fail, a liability existed based upon the special facts in the case as explained in *Henderson* v *Merrett Syndicates Ltd* (1995).

SUMMARY

Engineers who are engaged by the employer with regard to the design and inspection of the permanent works need to be very vigilant with regard to the construction of the temporary works. It will be an implied term of the engineer's engagement that if he notices the state of the temporary works are likely to cause a peril in respect of the permanent works, he must take action to obviate the danger. This obligation will extend beyond taking action only when the engineer notices that there is a problem, and will include making regular inspections to ensure that no dangers arise. Any shortcoming in respect of the temporary works will in the first instance be addressed to the main contractor, but if no longer in business, the spotlight could well fall on the engineer.

13.22. What is a repudiatory breach?

13.22.1. A breach of contract occurs when a party to a contract fails to perform one of its obligations under the contract. For example, a contractor in its provision of brickwork does not adhere to the specification but uses a brick of a different quality. Breach of contract may result in one of two alternative remedies. There will always be a right to claim financial damages where there is a failure to comply with an obligation under the terms of the contract. However, if the obligation which is breached is sufficiently serious, it gives the innocent party an option to bring the contract to an end, on the grounds that it has been repudiated. The word 'repudiated' is ambiguous in that it has more than one meaning. However, if one party so acts, or so expresses itself, as to show that it does not mean to accept the obligations of the contract any further, this constitutes repudiation and, if accepted by the innocent party, both parties are released from further performance of the contract.

13.22.2. Every breach of contract entitles the innocent party to damages as compensation for the loss sustained as a consequence of the breach. However, for the right on the part of the innocent party to terminate the contract, there must be either:

- A right to terminate provided by the terms of the contract; or
- The effect of the breach is to deprive the other party of substantially the whole benefit which it was the intention of the parties should have been gained from the contract.

When a repudiatory breach occurs, the innocent party has the option of accepting the breach, thus bringing the contract to an end and claiming damages, or alternatively treating the contract as subsisting and claiming damages.

13.22.3. An acceptance of repudiation need not be conveyed in any particular manner. In the case of *Heymans* v *Darwins* (1942), the judge said with regard to acceptance:

> An act of acceptance of repudiation requires no particular form: a communication does not have to be couched in the language of acceptance. It is sufficient that the communication or conduct clearly and unequivocally conveys to the repudiating party that the aggrieved party is treating the contract as at an end.

13.22.4. An example of an unusual form of acceptance of repudiation occurred in the case of *Angus Joinery Ltd* v *James and Valerie McKay* (2010). A contract was let to Angus Joinery by the McKays for the supply and installation of six specially made windows. It was intended to be a special 40th wedding anniversary present from Mr McKay to his wife. Unfortunately, the work was done badly, the workmen were messy in the way they went about their duties and a badly stained carpet was attributable to negligence on the part of the workmen. It was alleged that Mr McKay made threats of a physical nature to employees of Angus Joinery, which was regarded by Angus as a repudiatory breach. They refused to return to site to correct the defects and commenced an action for payment of money which they alleged was due. The court found it laughable that the employees of Angus Joinery would be afraid of Mr McKay, who was short in stature and well beyond 60 years of age and, in so doing, awarded in favour of Mr and Mrs McKay.

13.22.5. It is not unknown for both parties to be committing a fundamental breach, which occurred in the case of *Alan Auld Associates Ltd* v *Rick Pollard Associates and Another* (2008). Dr Auld and Dr Pollard were old friends. Dr Alan Auld was the principal of Alan Auld Associates, who successfully bid to provide consultancy advice relating to the removal of radioactive waste at the Dounreay Power Station. The rate to be paid with regard to the consultancy work was £70 per hour, plus expenses. It was agreed that most of the work in relation to the commission would be undertaken by Dr Pollard, who was the principal of Rick Pollard Associates, at a rate of £50 per hour, plus expenses, to be invoiced at the end of each month, indicating the number of hours worked. This information would allow Dr Auld to invoice the authority. The authority was a prompt payer, usually taking two to five weeks from receipt of invoice to make payment. It was an implied term of the consultancy arrangement between Dr Pollard and Dr Auld that Dr Pollard would be paid immediately payment was received by Dr Auld from the authority.

13.22.6. Dr Pollard submitted 19 invoices between December 2004 and April 2006 which were all paid late. Payment periods varied between one and nine months, the average being four months. By the end of May 2006, Dr Pollard was owed £21,000. Dr Auld's excuse for not paying this sum was that he was owed considerable sums of money by clients other than the authority. Dr Pollard did no work in May 2006 and subsequently provided the advice direct to the authority. An action was commenced by Dr Auld claiming his lost £20 per hour which had resulted from Dr Pollard providing his services direct to the authority. It was argued by Dr Auld that in providing his services direct to the authority, Dr Pollard had breached an implied term of mutual trust and loyalty. His claim was for the recovery of the lost £20 per hour for the remainder of the project. Dr Pollard's defence was that the agreement had been terminated on 7 June 2006, by his acceptance of Dr Auld's repudiatory breach in persistently paying late and every prospect of doing so in the future.

13.22.7. The court found in favour of Dr Pollard, in that it considered the late payments were substantial, persistent and cynical. There had been repeated complaints made by Dr Pollard and the likelihood was that late payment would continue over the remaining year of the project. The decision of court was the subject of an appeal. However, the Court of Appeal agreed with the decision of the lower court.

13.22.8. Where there are provisions set out in the contract which give either or both parties the right to terminate the contract, disputes can arise as to whether those rights exist alongside the right to repudiate at common law or as a replacement. The decision in the case of *Stocznia Gdynia AS* v *Gearbulk Holdings Ltd* (2009) provides the answer to the question. It was held by the Court of Appeal that the existence of a contractual right to terminate the contract did not of itself deprive the contracting parties of the right to terminate the contract under the general law. Clear words are required before a court will reach the conclusion that contracting parties have agreed to give up a valuable right which the law confers upon them. Further, the Court of Appeal held that, despite the claimant having given notice to terminate in accordance with the terms of the contract, it did not prevent it from treating the contract as having been discharged at common law. The decision in this case is in keeping with the earlier case of *Architectural Installation Services Ltd* v *James Gibbons Windows Ltd* (1989).

SUMMARY

A breach of contract occurs when a party to a contract fails to perform one of its obligations under the terms of the contract. Breach of contract may provide the innocent party with alternative remedies. There will always be a right to claim financial damages where there is a failure to comply with an obligation under the terms of the contract. However, if the obligation which is breached is sufficiently serious it gives the innocent party an option to bring the contract to an end on the grounds that it has been repudiated. The word 'repudiated' is ambiguous in that it has more than one meaning. However, if one party so acts or so expresses itself as to show that it does not intend to accept the obligations imposed by the terms of the contract any further, this constitutes repudiation and, if accepted by the innocent party, both parties are released from further performance of the contract.

Every breach of contract entitles the innocent party to damages as compensation for the loss sustained as a consequence of the breach. However, when a repudiatory breach occurs, the innocent party has the option of accepting the breach, thus bringing the contract to an end and claiming damages, or treating the contract as subsisting and claiming damages.

13.23. What are the legal responsibilities of a project manager?

13.23.1. The title 'project manager' gives the impression that the role of the holder of the position is to manage the project; however, in the construction industry the boundaries of responsibility of the project manager are blurred. Whilst the roles and responsibilities of architects and quantity surveyors are well-established, the project manager's role has been developing over a relatively few years. The ideal situation is for the duties and responsibilities of the project manager to be provided in some detail in the terms of engagement, but unfortunately this is often not the case. Sometimes a standard set of conditions is used, or they are specially drafted for the project; but, all too often, an ill-defined set of duties is provided in a letter of appointment.

13.23.2. The codes of practice and standard conditions of appointment provided by professional governing bodies are useful, but do not provide insurance against disputes arising as to the responsibilities of the project manager. The CIOB's Code of Practice for Project Management for Construction and Development, Third edition, 2010, describes project management in the following terms:

> as an established discipline which exclusively manages the full development process, from the client's idea to funding, co-ordination and acquirement of planning and statutory

controls approval, sustainability, design delivery, through to the selection of the project team, construction commissioning, handover, review, and facilities management co-ordination.

13.23.3. The RICS has produced a Project Manager's Services document for use with the RICS Short Form of Consultant's Appointment, which comprises schedules containing tick boxes. The danger of a tick-box system is that, whilst it indicates the scope of the services to be provided, it does not avoid the legal arguments as to nuances which can arise from the brief description of the services to be provided.

13.23.4. There have been a number of legal cases relating to the provision of project management services. In the case of *Pozzolanic Lytag Ltd* v *Bryan Hobson Associates* (1999), Bryan Hobson was appointed by Pozzolanic to act as project manager in respect of a construction project. The contractor was appointed to provide a design and construct service, which included having professional indemnity insurance in place. A design fault occurred involving the collapse of a dome, but, unfortunately, the contractor did not have adequate insurance to recover the rectification costs. Pozzolanic commenced an action against Bryan Hobson for failing to ensure that the contractor had proper insurance cover. It was held that Bryan Hobson owed a duty of care to Pozzolanic, which including ensuring that the contractor had taken out adequate insurance. Bryan Hobson's conditions of appointment required them to establish Pozzolanic's wishes regarding insurance.

13.23.5. In the case of *Chesham Properties Ltd* v *Bucknall Austin Management Services Ltd* (1996), the court held that the project manager had an implied duty to the employer to report any deficiencies in the performance of other professional companies who were appointed with regard to the project whose roles the defendant was responsible for co-ordinating.

13.23.6. In the case of *Pride Valley Foods Ltd* v *Hall and Partners (Contract Management) Ltd No1* (2000), Hall and Partners were appointed to manage the construction of a new factory. A fire destroyed the factory and an action was commenced against Hall and Partners because of their failure to advise about the combustibility of expanded polystyrene panels. Had proper advice been provided and acted upon, the spread of fire might have been prevented. Hall and Partners were found to have a liability for failing to provide proper advice, even though, if given, it would not have been heeded by Pride Valley Foods. The conditions of appointment of Hall and Partners required them to specify the materials to be used, but unfortunately they failed to advise on the fire risk of using expanded polystyrene panels.

13.23.7. Another fire, this time at a restaurant, resulted in the case of *Six Continents Retail Ltd* v *Carford Catering Ltd* (2003) being brought before the courts. Carford was engaged as project manager for the design and installation of kitchen equipment. A problem developed with the spit roaster after the restaurant had reopened, which caused a fire. Carford had received a letter from the rotisserie manufacturers, which included recommendations which should have been followed and, if so, would have prevented the fire. The court held that Carford had a duty to ensure that the installation was not susceptible to fire and that this was not fulfilled by merely sending the letter which included recommendations. A more proactive approach was required, which included assessing the fire risk likely to be caused by the installation.

SUMMARY

There is little history with regard to the appointment of project managers and, therefore, they have no traditional role. To ensure that the employer and project manager are in accord with regard to the duties and responsibilities which go with the position, it is essential that proper comprehensive conditions of appointment are agreed before commencement of the provision of the service; a brief letter of appointment is totally insufficient. The RICS has published terms for the appointment of a project manager, which could provide a good starting point. However, the services to be provided are indicated by a tick-box list, which requires care when being employed.

13.24. Are project managers, when performing duties relating to the Engineering and Construction Contract (NEC 3), required to act impartially, or do they merely act as agents for the employer?

13.24.1. It is well established that an architect, when issuing certificates and dealing with applications for extension of time, is obliged to act impartially; this was established in the decision in *Sutcliffe* v *Thackrah* (1974). Does this obligation extend to project managers? This issue was raised in the case of *Costain Ltd* v *Bechtel Ltd* (2005). The case arose out of a contract to refurbish St Pancras station. The contract used was the NEC 3 contract, which provided for payment on a cost-reimbursable basis with a pain/share: gain/share mechanism. The main contractor was a consortium comprising Costain Ltd, O'Rourke Civil Engineering Ltd, Bachy Soletanche Ltd and Emcor Drake and Scull Ltd, referred to as 'Corber'. A consortium of Bechtel Ltd, Ove Arup and Sir William Halcrow comprised the project manager, referred to as 'RLE'.

13.24.2. Work had been ongoing for some considerable time before the dispute arose which led to the case coming before the court. Under the terms of the contract the companies which formed the project manager consortium would benefit if the target cost exceeded the outturn cost, but would have a financial liability if the target cost was less than the outturn cost.

13.24.3. Mr Fady Bassily was the leading person who represented the project manager. Concern began to arise with regard to the level of projected outturn costs when compared with the projected target costs. It was the project manager's task to certify the sums to be paid by the employer on a monthly basis. The sums certified included the value of work properly executed, less any disallowed costs, which included the cost of remedying defective work. By the time a certificate of payment was issued on 5 February 2005, £1.4 m had been deducted in respect of disallowed costs, out of a total certified of approximately £268 m. This sum for disallowed costs increased to £5.8 m by the time a payment certificate was issued, on 8 April 2005. A meeting of the Bechtel staff was held on 15 April 2005, at which Fady Bassily made it clear to the Bechtel staff present that there was a need to be more vigilant in identifying disallowed costs. He stressed the financial benefits to Bechtel and the employer of ensuring that the certified outturn costs were kept to the minimum.

13.24.4. News of what was said at this meeting soon filtered through to the consortium of contractors, who were very concerned. They had seen the sum for disallowed costs shoot up in the month preceding the meeting and expected more of the same. It was the view of the contractors that Bechtel, in performing its project management duties, was acting in the interests of the employer and not that of their own. The contractors' view was that, when issuing a payment certificate, the project manager should act impartially between the interests of the contractor and employer.

13.24.5. The contractors' consortium applied to the court for an injunction to prevent the project manager from acting in the interests of the employer and failing to act impartially. In support of its case, the contractors' consortium drew attention to the well-known decision of the House of Lords in *Sutcliffe* v *Thackrah* (1974). It was held in this case that an architect, when issuing a certificate, is under an implied duty to act impartially between the contractor and employer. Bechtel argued that this case dealt with the duties of an architect under a JCT contract and, as the duties of a project manager under an NEC 3 contract differ from the duties of an architect under a JCT contract, there is no implied duty to act impartially. The contract provided a dispute resolution mechanism which could be used by the contractors if there was any disagreement with the decisions of the project manager.

13.24.6. The court agreed with the reasoning of the contractors' consortium. The principles laid down in *Sutcliffe* v *Thackrah* (1974) applied equally to a project manager operating under an NEC 3 contract. When exercising some of its duties the project manager would be acting in the interests of the employer, for example, when changing the design; however, when issuing payment certificates, the project manager must act impartially.

13.24.7. Despite agreeing with the argument put forward by the contractors' consortium, the court refused to issue an injunction, the reason being that if it did so there would be a need for the court to supervise the project manager to ensure that the terms of the injunction were being observed. It was considered that this was outside any service the court could offer. The decision, however, laid down a principle which the project manager would be obliged to follow.

SUMMARY

It is well-established that an architect acting under a JCT contract is obliged to act in an impartial manner when issuing payment certificates and making decisions concerning extensions of time. This can be compared with the architect's duty to act as the employer's agent when issuing variations and drawings. The decision in the case of *Costain* v *Bechtel* (2005) established that a project manager, appointed in relation to the NEC 3 contract, must act impartially, in the same manner as an architect who operates in accordance with a JCT contract.

13.25. Can a quantity surveyor who is employed by the employer be liable for any losses incurred by a contractor who successfully tenders for a project, where the bill of quantities contain an error which results in the tender being lower that it would have been if there had been no error?

13.25.1. Some standard forms of contract make specific provision for what will happen if the bill of quantities contains an error. JCT 2011 With Quantities, which provides for the bill of quantities to be a contract document, states in clause 2.14.1:

> if in the Contract Bills . . . there is . . . any error in description or in quantity or any omission of items . . . the error or omission shall not vitiate this Contract but shall be corrected.

Any such correction may affect the price which will become payable to the contractor. In the case of *Co-operative Insurance Society Ltd* v *Henry Boot (Scotland) Ltd* (2002), there was held to be an error in the bill of quantities, in that it failed to stipulate the ground water level. It was possible for the contractor to have calculated the ground water level from other information included with the tender documents; it was held that the error had to be corrected and dealt with as a variation.

13.25.2. Many contracts are employed where there is no wording similar to that contained in JCT 2011 With Quantities, where the contractor will be at risk in the event of there being an error in the bill of quantities. In *Williams* v. *Fitzmaurice* (1858) and *Patman and Fotheringham* v. *Pilditch* (1904) the court held that, where items were obviously required, such as floor boards, then the contractor should have made provision in the price, even though, because of an error, they were not included in the bill of quantities.

13.25.3. The case of *Priestly* v. *Stone* (1888) dealt with the principle as to whether a quantity surveyor could be liable to a contractor appointed by the employer for errors in the bill of quantities. It was held by the Court of Appeal that the quantity surveyor who was employed by the employer or architect could not be liable to the contractor for errors in the bill of quantities. In preparing the quantities, the quantity surveyor owed a duty in contract to the employer or architect to whom he provided the service. The quantity surveyor had no contract with the contractor and did not owe a duty to the contractor, as he wasn't making any representation that they were accurate.

13.25.4. The only cases which deal with the matter of errors in the bill of quantities and possible liability of the quantity surveyor to the contractor are very old. In view of modern-day thinking, it is debatable as to whether a court would be prepared to consider that a quantity surveyor owed a contractor a duty of care when preparing a bill of quantities. The House of Lords, in the case of *Hedley Byrne and Co Ltd* v *Heller and Partners Ltd* (1964), dealt with a duty of care in respect of a negligent misstatement owed by a banker concerning a favourable financial reference given in respect of one of its customers. It is possible that a court may decide that, as the contractor relies on the accuracy of the bill of quantities when submitting a tender and as this is known to the quantity surveyor, a duty of care arises.

SUMMARY

Some standard forms of contract, such as JCT 2011 With Quantities, provide for errors in the bills of quantities to be corrected and, hence, there is no financial risk to the contractor if errors occur. In the absence of this type of provision, the legal cases dealing with this matter have established that the quantity surveyor employed by the employer owes no duty in either contract or tort to the contractor to ensure that the quantities are correct. The cases dealing with this matter are old and it remains to be seen if, in the light of modern thinking, a court would be prepared to consider that a quantity surveyor employed by the employer owed a duty of care to the contractor when preparing the bill of quantities. The House of Lords, in the case of *Hedley Byrne and Co Ltd* v *Heller and Partners Ltd* (1964) dealt with a duty of care in respect of a negligent misstatement owed by a banker concerning a favourable financial reference given in respect of one of its customers. It is possible that a court may decide that, as the contractor relies on the accuracy of the bill of quantities when submitting a tender, and as this is known to the quantity surveyor, a duty of care arises.

13.26. Can an architect who recommends a contractor to undertake a project be liable to the employer, who incurs additional cost because the contractor is incompetent to undertake the work?

13.26.1. Employers who are desirous of having construction work undertaken often, as a first port of call, appoint an architect. The architect will seek planning approval, produce a detailed design and secure the appointment of a contractor to undertake the work. On most projects, the work is completed to the reasonable satisfaction of the employer. There are occasions, however, when the work is sub-standard and, before any redress can be sought by the employer, the contractor becomes insolvent and goes out of business. Would the employer have any right to bring an action against the architect, to recover any costs incurred as a result of the shortcomings of the contractor?

13.26.2. In the case of *Valerie Pratt* v *George J Hill* (1987), Miss Pratt retained the services of the defendant to design a bungalow. The defendant recommended two contractors, whom he described as 'very reliable'. As a result, Miss Pratt entered into a contract with one of them. Unfortunately, the contractor's performance in every respect was unsatisfactory. Miss Pratt terminated the contracts of both the defendant and the contractor. The contractor became insolvent and went out of business before an arbitration, which had been started, was completed. Miss Pratt incurred substantial costs as a result of the contractor's unsatisfactory performance, which she sought to recover from the defendant.

13.26.3. The court held that the defendant, in referring to the contractor as 'very reliable', was making a representation as to its competence. As the contractor proved not to be 'very reliable', the court held that the defendant was liable in damages to Miss Pratt for a negligent misrepresentation and hence had to pay her the additional costs she had incurred.

13.26.4. It is not clear whether the architect would have been found liable, had he not described the contractor as being 'very reliable'. However, before recommending a contractor to be included on a tender list, it is suggested that a reasonable amount of due diligence should be undertaken by the architect. The appointment of a contractor who would not have been included on the tender list if proper due diligence had been undertaken by the architect, would, if additional costs were incurred by the employer due to the contractor's incompetence, leave him exposed to a claim for failing to apply reasonable skill and care in the performance of his duties. Architects are advised to warn employers if contractors about whom the architect has little knowledge are included on a tender list.

SUMMARY

Architects need to take care before including a contractor on a tender list. If the architect describes the contractor in a certain manner, such as 'very reliable', he could find himself liable to reimburse the employer who incurs additional costs because of the contractor's incompetence. Architects are recommended to undertake a certain amount of due diligence before including a contractor on a tender list. Failure to do so could leave the architect exposed to a claim from the employer, should additional cost be incurred due to the incompetence of the contractor. If a contractor about whom the architect has little knowledge is included on a tender list, the architect should warn the employer that the contractor's level of competence is unknown and therefore constitutes a risk.

Chapter 14
Adjudication

Contractual Problems Nos 14.1 to 14.31 arise out of adjudication under Part II of the Housing Grants, Construction and Regeneration Act 1996, referred to as the 'Construction Act 1996', and the Local Democracy, Economic Development and Construction Act 2009, which apply to all contracts entered into after 1 October 2011.

14.1. Will an adjudicator's decision be enforced by the courts?

14.1.1. Prior to an enforcement case coming before the courts, it was considered by some eminent authorities that the courts would not, and in fact could not, enforce an adjudicator's award by a summary procedure. The reasons were based upon a number of cases, including *Halki Shipping Corporation* v. *Sopex Oils Ltd* (1997).

14.1.2. The first court case concerning the enforcement of an adjudicator's decision was *Macob Civil Engineering Ltd* v. *Morrison Construction* (1999). In this case, the parties entered into a contract under which the claimant, Macob, was to carry out groundworks. A payment dispute arose, which was referred to Mr E. Mouzer, an adjudicator. His decision was that the subcontract provided a payment mechanism which did not comply with the Act; consequently the Scheme for Construction Contracts applied. Mr Mouzer concluded that the defendant had served a notice indicating an intention to set off from payments due, out of time, and therefore immediate payment should be made to Macob. The adjudicator's decision was in the form of a pre-emptory order, under which either party could apply to the court for enforcement. The defendant applied to the court for a stay to arbitration on the grounds that the decision was wrong on its merits, and also that there had been a breach of the rules of natural justice. The judge rejected the defendant's arguments, holding that to refuse enforcement would substantially undermine the effectiveness of adjudication. He considered that the provisions of the Act should be construed positively. The decision of the adjudicator was therefore held to be binding.

14.1.3. In *Outwing Construction Ltd* v. *H. Randall* (1999), a dispute under DOM/1 was referred to Mr Talbot, an adjudicator appointed by the Chartered Institute of Building. He considered that, as the terms of the subcontract did not comply with the Act, the Scheme for Construction Contracts applied. It was the decision of Mr Talbot that the defendant should pay the claimant, plus his fees and expenses. He ordered his decision to be made pre-emptorily. The claimant issued an invoice for the amount included in the

200 Contractual Problems and their Solutions, Third Edition. Roger Knowles.

adjudicator's decision. In the absence of payment, solicitors acting for the claimant provided a deadline for payment, threatening to apply for summary judgment in the event of non-payment. After the date had passed, the defendant responded, to the effect that it intended to seek a court stay to arbitration. A writ was issued by the claimant for payment, plus interest and costs. The defendant made payment, but refused to pay the claimant's costs. The court found in favour of the claimant. It was the view of the judge that the intention of Parliament is clear: disputes may be referred to adjudication and the decision of the adjudicator has to be complied with.

14.1.4. It was not long before the Scottish court became involved with regard to adjudication. In the case of *Rentokil Allsa Environmental Ltd* v. *Eastend Civil Engineering Ltd* (1999), the adjudicator found in favour of the pursuer. A cheque for the full amount was provided by the defendant, who simultaneously lodged an arrestment, freezing the payment they were obliged to make to the pursuer. This is a mechanism used in Scotland which does not apply in England, whereby, to put pressure on a reluctant payer, the payee seeks to arrest or 'freeze' payments due to the payer by third parties, in this case freezing payment from itself to the pursuer. The court found in favour of the pursuer, refusing to allow the use of the arrestment mechanism to circumvent and negate the effect of the adjudicator's decision.

SUMMARY

It has been made clear by the courts that in passing the Construction Act the intention of Parliament is clear, in that a decision of an adjudicator appointed under the Act can be enforced by summary procedure.

14.2. Will a court enforce part only of an adjudicator's award?

14.2.1. The original orthodoxy on severability, as expressed in *Keating on Construction Contracts*, Eighth Edition, where at 17.045 it states:

> It seems probable that if there is a breach of natural justice, the whole decision is unenforceable and it is not possible to sever the good from the bad.

14.2.2. There have been a number of cases where this view has been challenged. It has been argued that if part of an adjudicator's decision is unenforceable because it is tainted, for example where the adjudicator has exceeded his jurisdiction, this can be severed and the remainder enforced. Mr Justice Akenhead considered this matter in the case of *Cantillon* v. *Urvasco* (2008). He reviewed a number of cases, including *Shimizu Europe* v. *Automajor* (2002), where Judge Seymour considered two unconnected disputes presented to an adjudicator for a decision; for example, one regarding extensions of time and the other additional cost. If the additional cost decision could not be enforced because of lack of jurisdiction, it would be open to have the part relating to an extension of time enforced. Mr Justice Akenhead considered that, for this to apply, there must be more than one dispute which has been referred to the adjudicator.

14.2.3. This was the problem faced by the court in the case of *R. Durtnell and Sons* v. *Kaduna Ltd* (2003). Kaduna and Durtnell entered into a JCT 80 standard form of contract for work to be undertaken at Kaduna's property in Hampshire. The parties were in dispute concerning the sum which Durtnell was due to be paid under the terms of the contract and their entitlement to an extension of time.

14.2.4. The adjudicator made a declaration concerning Durtnell's entitlement to an extension of time and an award of £1.2m in respect of the payment claim. For some strange reason, Kaduna paid half of the £1.2m and then contested the adjudicator's decision on the basis that, under the terms of the contract, the architect had 12 weeks to deal with applications for extensions of time and the period had not expired. Durtnell applied to the court to have the balance of the adjudicator's decision paid.

14.2.5. The court agreed with Kaduna that the adjudicator should not have included an entitlement to an extension of time in his decision, as the 12-week period available to the architect for granting extensions of time had not elapsed. However, the court ordered Kaduna to pay to Durtnell the balance of the £1.2m.

14.2.6. A decision would not be severed, for example, if it related to an entitlement to an extension of time where part only of the decision was tainted, leaving the remainder of the decision intact. The court would not sever the parts of the extension of time which were tainted from the remainder. This is to be distinguished from the situation which occurred in *Cantil*, where there were two distinct periods of time, one of 13 weeks and the other of 16 weeks, where each period related to different facts and were presented and dealt with separately. Mr Justice Akenhead indicated that these two disputes were severable.

SUMMARY

The original orthodoxy was that adjudicators' disputes were not severable. If part was tainted, say for lack of jurisdiction, then no part of the decision can be enforced. This view came under review and the general consensus of opinion which prevails now is that, if the referral relates to two separate disputes, then the good part can be severed and enforced. In the case of *R. Durtnell and Sons* v. *Kaduna Ltd* (2003), the adjudicator in his decision awarded a payment of £1.2m and a declaration concerning an extension of time. Unfortunately, the adjudicator acted outside his jurisdiction concerning the extension of time. Nonetheless, the court ordered the payment to be made.

14.3. When can it be said that a dispute has arisen which gives rise to an entitlement for the matter to be referred to adjudication?

14.3.1. Section 108(1) of the Construction Act states that:

> A party to a construction contract has the right to refer a dispute arising under the contract for adjudication under a procedure complying with this section.

For this purpose 'dispute' includes difference.

14.3.2. If there is no obvious dispute, there is nothing to be referred to adjudication. This may seem obvious, but, nonetheless, in the case of *Sindall Ltd* v. *Sollard* (2001), a matter was referred to an adjudicator before a dispute was alleged to have taken place. Sindall was the main contractor for the refurbishment of Lombard House in Mayfair. The employer was Sollard and the contract administrator Michael Edwards. Work was delayed and the contractor requested an extension of time. A period of 12 weeks was awarded, which did not meet with Sindall's minimum requirements, and the matter was referred to adjudication. The adjudicator decided that an extension of 28 weeks was appropriate. Further delays occurred and the employer threatened to determine Sindall's employment. Sindall drew attention to delays because of the issue of 123 instructions by the contract administrator and requested a further extension of time. They sent a box of files to the contract administrator and gave him seven days for a response. Michael Edwards asked for more time to consider the submission, but the request was ignored and Sindall commenced adjudication. The court considered that, in view of the short period of time given to Michael Edwards to reach a decision, there could be no dispute which was referable to adjudication. The judge said:

> It must be clear that a point has emerged from the process of discussion, or the negotiations have ended, and that there is something which needs to be decided.

14.3.3. Whether or not a dispute has in fact occurred will often depend upon the facts of the case. Remarks of judges in the following cases are very relevant:

(1) *Fastrack Contractors* v. *Morrison Construction* (2000), in which Judge Thornton QC stated:

> A dispute can only arise once the subject matter of the claim, issue or other matter has been brought to the attention of the opposing party and that party has had an opportunity of considering and admitting, modifying or rejecting the claim or assertion.

(2) *Edmund Nuttall* v. *R.G. Carter* (2002), where Judge Seymour QC said:

> 'For there to be a dispute there must have been an opportunity for the protagonists each to consider the position adopted by the other and to formulate arguments of a reasoned kind.

14.3.4. Other cases brought before the courts where the question of whether a dispute had arisen include:

- *Cowlin Construction Ltd* v. *CFW Architects* (2002);
- *Costain Ltd* v. *Westcol Steel Ltd* (2003);
- *Beck Peppiatt Ltd* v. *Norwest Holst* (2003);
- *Orange EBS Ltd* v. *ABB* (2003); and
- *Collins (Contractors) Ltd* v. *Baltic Quay Management* (2004).

14.3.5. The Court of Appeal in the case of *Amec Civil Engineering Ltd* v. *The Secretary of State for Transport* (2005), in considering all the authorities which dealt with the meaning of 'dispute', considered there were seven propositions on what constitutes a dispute, namely:

1. The word 'dispute' does not have a special meaning and should be given its normal meaning;
2. There is no hard-edged legal meaning to the word 'dispute', but judicial decisions have produced helpful guidance;
3. The fact that one party has notified the other of the existence of a claim does not in itself give rise to a dispute: it requires a claim not to be admitted for a dispute to arise;
4. There may be several circumstances which may lead to a claim not being admitted: the claim may be rejected, or one of the parties may prevaricate, giving rise to the inference that the claim is not admitted; or the responding party may simply remain silent for a period of time, again giving rise to the same inference;
5. The period of time a party may remain silent depends upon the facts: where the claim is notified through an agent such as an architect, more time will be required than if the claim is submitted direct to the other party; and
6. If the claimant imposes a deadline for responding, it does not mean that a dispute arises merely because the deadline has expired.

Should a claim be so nebulous and ill-defined that the respondent cannot sensibly respond to it, a non-admission is unlikely to give rise to a dispute.

14.3.6. The court, in the case of *VGC Construction Ltd* v. *Jackson Civil Engineering Ltd* (2008), had to decide whether a submission by a claimant was too nebulous and ill-defined for it to be regarded as a dispute. It arose out of a subcontract relating to the provision of ducts and cabling on the M3. VGC Construction Ltd, the subcontractor, overran the subcontract period by 26 weeks. In September 2007 VGC Construction submitted Application for payment no 13 to the main contractor Jackson Civil Engineering Ltd. The application included a one-line claim, which stated:

> Delay and disruption – £300,000

Jackson Civil Engineering Ltd was not impressed and made no payment at all against this item. This was followed by VGC Construction Ltd submitting a request for a 26-week extension of time accompanied by four pages of details which provided reasons for the delays. Several meetings took place between the contractor and subcontractor, but no agreement was reached. A further Application No 14 was submitted by VGC Construction Ltd and also a detailed as-built programme. No further information was provided to support the delay and disruption claim. VGC Construction Ltd commenced adjudication proceedings for a decision to be made concerning the sums due and their extension of time entitlement. Jackson argued that the claim was nebulous and ill-defined and therefore did not constitute a dispute. The judge recognised that, if the claim was nebulous and ill-defined, there could be no dispute. However, he did not consider that the claim for delay and disruption fell into that category and, therefore, a dispute existed. In view of the total lack of detail which accompanied the delay and disruption claim, it is difficult to see what the judge would consider to be ill-defined. The adjudicator included in his decision a sum in respect of the delay and disruption, which Jackson Civil Engineering Ltd was obliged to pay.

SUMMARY

The Court of Appeal recognised that the authorities had, when taken together, laid down seven propositions which would constitute a dispute, namely:

1. The word 'dispute' does not have a special meaning and should be given its normal meaning;
2. There is no hard-edged legal meaning to the word 'dispute', but judicial decisions have produced helpful guidance;
3. The fact that one party has notified the other of the existence of a claim does not in itself give rise to a dispute: it requires a claim not to be admitted for a dispute to arise;
4. There may be several circumstances which may lead to a claim not being admitted: the claim may be rejected; one of the parties may prevaricate, giving rise to the inference that the claim is not admitted; the responding party may simply remain silent for a period of time, again giving rise to the same inference;
5. The period of time a party may remain silent depends upon the facts: where the claim is notified through an agent such as an architect, more time will be required than if the claim had been submitted direct to the other party;
6. If the claimant imposes a deadline for responding, it does not mean that a dispute arises merely because the deadline has expired; and
7. Should a claim be so nebulous and ill-defined that the respondent cannot sensibly respond to it, a non admission is unlikely to give rise to a dispute.

14.4. To comply with the Construction Act 1996 and be subject to adjudication, the contract must be 'in writing or evidenced in writing'. What is meant by 'in writing or evidenced in writing'? Has this been amended by the Local Democracy, Economic Development and Construction Act 2009?

14.4.1. Section 107 of The Construction Act 1996 stipulates that the provisions of the Act only apply to contracts which are in writing or evidenced in writing. This requirement has been repealed by the Local Democracy, Economic Development and Construction Act 2009, which applies to all contracts entered into from 1 October 2011. Construction contracts which are entered into orally are now catered for by the legislation. The answer to this problem therefore differs in respect of contracts let before, from those entered into from 1 October 2011.

14.4.2. It was obvious from the outset that there would be disagreement as to the meaning of 'evidenced in writing'. The first of a line of cases to deal with the question as to what is meant by 'evidenced in' was *RJT Consulting Engineers Ltd* v. *DM Engineering (Northern Ireland) Ltd* (2002), when the court had to decide whether a contract between a consulting engineer and a specialist engineering company complied with the requirement to be 'in writing or evidenced in writing'. The contract between RJT Consulting Engineers and DM Engineering was essentially oral. Both the parties to the dispute were involved with the refurbishment of the Holiday Inn in Liverpool. RJT's representative verbally agreed with a representative from DM to undertake some design work for a fee of £12,000. DM was the mechanical and electrical contractor and RJT the consulting engineer. A dispute arose, whereby DM levied a claim for negligence in the sum of £858,000 against RJT. The matter was referred to adjudication by DM. RJT applied to the court for a declaration that the agreement was not in writing and therefore not covered by the Construction Act. DM argued that, whilst the contract was not in writing, it was 'evidenced in writing'. RJT's case was that to be evidenced in writing the evidence must recite the terms of the agreement. Judge Mackay disagreed. He noted that the material concerning the contract was extensive. RJT had submitted fee accounts which identified the nature of the work, the names of the parties and the place of work. There were minutes of a meeting, which again referred to parties and the nature of the work undertaken. The judge concluded that the evidence in support of the agreement would not be required to identify its terms and that, because of the extensive number of documents, the agreement came within the ambit of the Construction Act.

14.4.3. On appeal, the Court of Appeal took a different view, as expressed by Lord Justice Ward when he made the following observations:

> On the point of construction of section 107, what has to be evidenced in writing is literally the agreement, which means all of it, not part of it. A record of the agreement also suggests a complete agreement, not a partial one.

14.4.4. The Court of Appeal, in coming to this decision, seemed to be taking the view that there is little difference in meaning between a contract 'in writing' and one 'evidenced in writing'. There are three propositions arising from this case, which His Honour Judge Bowsher in *Carillion Construction Ltd* v. *Devonport Royal Dockyard* (2003) considered must be satisfied before it can be said that a contract has been 'evidenced in writing':

- A contract is not evidenced in writing merely because there are documents which indicate the existence of a contract;
- All the terms of the oral agreement must be evidenced in writing; and
- Alternatively, the material terms of the agreement must be evidenced in writing.

14.4.5. In the case of *Rok Building Ltd* v. *Bestwood Carpentry Ltd* (2010), the court had to decide on the situation where all the terms of the contract had been agreed except for price. This would, on its face, seem fatal, but it was argued that in the absence of a written term in the contract one could be implied. Rok was the main contractor on a project in London involving the construction of residential units. In early February 2006, Rok was having difficulties with its joinery subcontractor and, in an effort to correct the situation, requested Bestwood to supply operatives on a labour-only basis to undertake the joinery work. Rok and Bestwood attended a meeting in the week commencing

6 February 2006, which was followed up by a fax referring to the rates being agreed. However, the fax did not provide details of the rates which had been agreed at the meeting. Bestwood undertook the work but got into a dispute with Rok over the sums due to be paid. Bestwood referred the matter to adjudication, but it was argued by Rok that statutory adjudication was not applicable, as the contract was not in writing or evidenced in writing. In the earlier case of *Murray Building Services* v. *Spree Development* (2004), it was held that it was not necessary for the actual contract price to be recorded in writing. In the current case, the judge elaborated by stating that it was open to the parties to have a written agreement as to a mechanism for arriving at a price, which would constitute a written contract, or one evidenced in writing.

The judge went further, by drawing attention to the situation where there was no agreement about price, oral, in writing or otherwise. He explained that it would still constitute a written contract under the Construction Act. Where this situation occurs, there will normally be a term implied at law that payment will be on a fair and reasonable basis. It was held by the judge that, under these circumstances, the contract could be said to be 'in writing or evidenced in writing'. However, where there has been an oral agreement as to price which is referred to in writing without providing the details, then the Construction Act will not apply. As Rok and Bestwood had agreed the rates at a meeting during the week commencing 6 February 2006 and referred to the agreement in writing, but not how much would be paid, there was no contract in writing. Bestwood was therefore prevented from referring the dispute to statutory adjudication.

14.4.6. Since the *RJT* case there have been a number of cases where the court had to decide whether there was a contract which had been 'evidenced in writing'. However, these decisions turned upon the facts of the case, but with regard to legal principles, those laid down by the Court of Appeal in the *RJT* case were followed in every case.

14.4.7. The repeal of the need for contracts to be in writing or evidenced in writing will solve one problem, but there is a likelihood that it will be replaced by another one. Adjudications are usually conducted on a documents-only basis. There is rarely a hearing and if one takes place it is normally in the form of a meeting to give the adjudicator an opportunity to ask questions of the parties by way of clarification of matters referred to in the documents. Oral contracts, by their nature, are rarely evidenced in writing; therefore, the only acceptable evidence to the effect that an oral contract has been entered into is in the form of oral evidence. Some adjudicators may be satisfied with a witness statement to authenticate the oral agreement, but there will be others who will insist upon hearing witnesses to prove that the oral contract exists. Questions may then be raised as to whether the evidence should be given under oath.

SUMMARY

There are three matters which, it has been held, must be satisfied before it can be said that a contract has been 'evidenced in writing':

- **A contract is not 'evidenced in writing' merely because there are documents which indicate the existence of a contract;**

- All the terms of the oral agreement must be evidenced in writing; and
- Alternatively, the material terms of the agreement must be 'evidenced in writing'.

The repeal of the need for contracts to be 'in writing or evidenced in writing' by the Local Democracy Economic Development and Construction Act 1998, which applies to all contracts entered from 1 October 2011 will solve one problem, but there is a likelihood that it will be replaced by another one. Adjudications are usually conducted on a 'documents-only' basis. There is rarely a hearing and if one takes place, it is normally in the form of a meeting to give the adjudicator an opportunity to ask questions of the parties, by way of clarification of matters referred to in the documents. Oral contracts, by their nature, are rarely 'evidenced in writing'; therefore, the only acceptable evidence to the effect that an oral contract has been entered into is in the form of oral evidence. Some adjudicators may be satisfied with a witness statement to authenticate the oral agreement, but there will be others who will insist upon hearing witnesses to prove that the oral contract exists. Questions may then be raised as to whether the evidence should be given under oath.

14.5. Can a dispute concerning oral amendments to a construction contract be referred to adjudication?

14.5.1. Section 107 of the Construction Act requires a contract to be 'in writing or evidenced in writing' for it to be referable to adjudication. Where the contract itself is in writing and statute compliant, does this provision apply to amendments to the contract? The Local Democracy, Economic Development and Construction Act 2009, which applies to all contracts entered into from 1 October 2011, repealed the need for construction contracts to be in writing or evidenced in writing, and hence construction contracts which have been entered into based upon an oral agreement are now governed by the legislation.

14.5.2. A dispute concerning this matter arose in the case of *Carillion Construction* v. *Devonport Royal Dockyard* (2003). Carillion was contracted to undertake refurbishment at Devonport Royal Dockyard. The contract was in writing, under which Carillion was to be reimbursed its costs and a fee. A gain-share /pain-share arrangement applied, whereby any underspend compared with the target was shared by the parties and, in like manner, the parties contributed to any overspend. The target cost was, over the period of contract, increased from £56m to £100m. Carillion argued that an oral agreement had been reached to abandon the gain-share/pain-share arrangement, with payment being made on a fully cost-reimbursable basis. An application for payment to accord with achieving milestone 33 was made by Carillion, based upon its costs and a fee. The employer refused to make the payment and the dispute was referred to adjudication.

14.5.3. The adjudicator decided that a binding agreement had been entered into to amend the contract and that the employer was due to pay to Carillion the sum of £7,451,320 plus VAT. It was argued by the employer that there was no binding oral agreement and, if one existed, as it had not been evidenced in writing, it fell outside the adjudicator's jurisdiction. It was held by the court that the Construction Act does not provide for adjudication in respect of an oral variation to a written construction contract.

14.5.4. The Local Democracy, Economic Development and Construction Act 2009, which applies to all contracts entered into after 1 October 2011, repeals the need for contracts to be in writing or evidenced in writing. This being the case, an oral variation to a written contract would be governed by the legislation.

SUMMARY

There is no provision in the Construction Act 1996 which would provide for adjudication in respect of an oral variation to a written construction contract. The Local Democracy, Economic Development and Construction Act 2009, which governs all contracts entered into from 1 October 2011, repealed the requirement that all contracts had to be in writing or evidenced in writing to be governed by the statute. An oral variation to a written contract would, if entered into after the revision came into force, be enforceable.

14.6. Where a mediator is appointed in relation to a dispute in connection with a construction contract and the dispute is not resolved, but referred to adjudication, is the mediator barred from being appointed as adjudicator?

14.6.1. In the case of *Glencot Development & Design Co Ltd* v. *Ben Barratt & Son (Contractors) Ltd* (2001), the court was asked to decide whether there was evidence of bias on the part of the adjudicator which would prevent the enforcement of his decision. The claimant, Glencot, was a subcontractor to the defendant, Barratt, for the provision of 1,200 mild steel wind posts which were provided as part of a brickwork subcontract. It was agreed by the parties that the value of the work was £390,000 plus VAT. The dispute, however, concerned whether Barratt was entitled to a 3% discount. Mr Peter Talbot was appointed as the adjudicator. In the first instance, at the request of the parties, he acted as a mediator. The mediation did not resolve the dispute and Mr Talbot reverted to his role of adjudicator. He wrote to both parties offering to withdraw from the adjudication, if either party felt that his ability to make an impartial decision had been affected by his presence during the settlement negotiations. Barratt was of the opinion that Mr Talbot should withdraw. Mr Talbot took legal advice and decided to continue with the adjudication. The decision of Mr Talbot was that the final account should be £431,616 plus VAT, giving a balance due of £160,016 plus VAT, having taken into account amounts already paid.

14.6.2. The court had to consider whether to enforce the decision. It was argued that the decision should not be enforced because of bias on the part of the adjudicator. Bias can come in one of two forms: actual bias is where it can be demonstrated that a judge, arbitrator or adjudicator is actually prejudiced, in favour of or against one of the parties; apparent bias may occur where circumstances exist which give rise to a reasonable apprehension that the judge, arbitrator or adjudicator may have been biased. The mediation process involved the mediator holding private discussions with each party without the other party being informed of what was said. It would appear that, in the process, heated discussions took place, which may have given the impression that Mr Talbot had already formed a view. The court was also influenced by Mr Talbot's writing to each party offering to withdraw, which may lead to a conclusion that there was doubt in his mind. Under the circumstances, the court refused to enforce the adjudicator's decision.

SUMMARY

Where a dispute is referred to a mediation process which fails to resolve the matter and the mediator is subsequently appointed as the adjudicator, it is unlikely that a court will enforce any decision the adjudicator may make. However, the decision of the court may depend upon the circumstances.

14.7. Will a court enforce an adjudicator's award which is clearly wrong?

14.7.1. Adjudicators are human and, as such, are likely to make errors. Where an error occurs in an adjudicator's decision, will the court refuse to enforce the decision, refer it back to the adjudicator for correction, or order enforcement?

14.7.2. The case of *Bouygues UK Ltd* v. *Dahl & Jensen Ltd* (2000) addressed the question as to whether an adjudicator's award which contains an obvious error will be enforced by the court. A dispute arose in connection with claims for additional works, delay and disruption. There was a purported determination of the contract under an express term, which led to a counterclaim. The adjudicator found in part for both parties. In his decision, he set out the original tender amount of £5,450,000, to which he added amounts for additional works and delay, to give a total of £7,626,542. From this he deducted sums in respect of the counterclaim to arrive at a final sum of £6,979,912. This was a gross amount and included 5% retention. He then deducted the net amount paid, which was £6,732,171, leaving a balance due of £247,741. It was agreed that, in deducting a 'net paid' figure with no retention from a gross figure which included retention, the adjudicator had made an error. No payment was made and enforcement proceedings were commenced. The judge in the case said he had to decide how a mistake should be characterised. In providing a decision regarding a matter which was not referred to him, the adjudicator acts outside his jurisdiction and any decision would not be enforced. If, however, he decides on a matter which is referred to him, but his decision is mistaken, then the decision would be enforced. In other words: if the adjudicator is doing what he is asked to do and is answering the right questions, but giving the wrong answer, his decision would be enforced by the court. In this case the adjudicator answered the right question, but gave the wrong answer; and therefore it was enforced by the court. Mr Justice Dyson, in arriving at his decision, explained it in the following terms:

> Mr Furst submits that if Dahl-Jensen is permitted to enforce the decision, which is plainly erroneous, Bouygues will suffer an injustice, and this will bring the adjudication scheme into disrepute. But as I said in Macob, the purpose of the scheme is to provide a speedy mechanism for settling disputes in construction contracts on a provisional interim basis, and requiring the decisions of adjudicators to be enforced pending final determination of disputes by arbitration, litigation or agreement, whether those decisions are wrong in points of law or fact. It is inherent in the scheme that injustices will occur, because from time to time adjudicators will make mistakes.

14.7.3. The case of *Shimizu Europe Ltd* v. *Automajor Ltd* (2001) also dealt with an adjudicator's error. Automajor contracted with Shimizu for the design and construction of an office under a JCT 98 With Contractor's Design contract. A dispute arose concerning set-off, extensions of time, the status of the agreement and release of retention. A further matter in dispute related to variations to the smoke ventilation works. The adjudicator ruled that there had been no variation to the smoke ventilation works, but nonetheless included a sum of £161,996 in his decision in respect of this matter. This was an obvious error, but nonetheless the decision was enforced by the court.

14.7.4. Judge Thornton, in the case of *Sherwood and Casson Ltd* v. *MacKenzie* (2000), summarised the position on adjudicator's errors in the following terms:

> 1. A decision of an adjudicator whose validity is challenged as to its factual or legal conclusions or as to procedural error, remains a decision that is both enforceable and should be enforced.
> 2. A decision that is erroneous, even if the error is disclosed by the reasons, will still not ordinarily be capable of being challenged and should, ordinarily still be enforced.
> 3. A decision may be challenged on the ground that the adjudicator was not empowered by the HGCRA to make the decision, because there was no underlying construction contract between the parties, or because he had gone outside his terms of reference.
> 4. The adjudication is intended to be a speedy process in which mistakes will inevitably occur. Thus the court should guard against characterising a mistaken answer to an issue, which is within an adjudicator's jurisdiction, as being in excess of jurisdiction. Furthermore, the court should give a fair, natural and sensible interpretation to the decision in the light of the disputes that are the subject of the reference.
> 5. An issue as to whether a construction contract ever came into existence, which is one challenging the jurisdiction of the adjudicator, so long as it is reasonably and clearly raised, must be determined by the court on the balance of probabilities with, if necessary, oral and documentary evidence.

14.7.5. There have been a number cases relating to the enforcement of an adjudicator's decision on the grounds that it is factually or legally wrong, but in each case the court has held firm to the five principles laid down by Judge Thornton in the *Sherwood and Casson* case. In *Rok Building* v. *Celtic Composting Systems* (2010), Judge Coulson held that the mere fact that there was an error, even if it were a glaring and serious error, it should not affect the enforceability of the decision. He held that, provided the adjudicator has the necessary jurisdiction and has not acted in a manner which is contrary to the rules of natural justice, his decision will be enforced.

14.7.6. There have, however, been a few cases where judges, whilst not contesting the five principles, have refused to enforce adjudicators' decisions which were patently wrong. In

Geoffrey Osborne Ltd v. *Atkins Rail* (2009), the judge allowed a CPR Part 8 application for a declaration that the adjudicator's decision was wrong in law. The part of the decision to which this error related was severed and the remainder enforced. His Honour Mr Justice Ramsay, in the case of *Forest Heath District Council* v. *ISG Jackson Ltd* (2010), confirmed that the court might exercise its discretion by making a declaration regarding an alleged legal error, if there are no disputed factual issues.

SUMMARY

Whether a court will enforce an adjudicator's award which contains an error will depend upon the type of error involved. If a matter is referred to an adjudicator who, in his decision, arrives at the wrong answer, the decision will be enforced. Should the adjudicator's decision include a matter which was not referred to him, it will be regarded as being outside his jurisdiction and will not be enforced. An adjudicator's decision will not be enforced if he does not have jurisdiction because of a defective appointment or he acts outside the rules of natural justice. There has been a move by the courts to allow a CPR Part 8 application for a declaration that the adjudicator's decision is wrong in law, where the facts are not in dispute.

14.8. If a dispute is the subject of ongoing litigation, can one of the parties, whilst the litigation is in progress, refer the matter to adjudication?

14.8.1. Litigation can be a slow process, particularly when it involves complex construction matters. It is understandable, therefore, if one of the parties, usually the claimant, becomes impatient and considers commencing adjudication proceedings. Would a court be prepared to intervene to stop the adjudication? If both judge and adjudicator were to make a decision with regard to the same matter, there is a distinct possibility that their decisions would, in some marked way, differ.

14.8.2. In the case of *Herschel Engineering Ltd* v. *Breen Properties* (2000), county court proceedings were commenced for the recovery of sums alleged to be due under a construction contract. Despite the existence of court proceedings, the dispute was also referred to adjudication. It was the defendant's opinion that adjudication was inappropriate, as the matter was already the subject of litigation. The adjudicator made an award in the claimant's favour and, as the defendant refused to pay, the matter was referred to court. It is well-established law that the same dispute cannot be heard before two legal forums, such as a court and arbitration. The court, however, decided that this rule does not apply to adjudication under the Construction Act.

14.8.3. The judge explained that Parliament's intention in enacting the statute was to produce a speedy mechanism for resolving disputes arising out of construction contracts, on a provisional and interim basis. Decisions of adjudicators are to be enforced pending final determination by arbitration, litigation or agreement. He considered that, if Parliament had intended that a party should not be able to refer a dispute to adjudication once litigation or arbitration had commenced, it would have been expressly stated. The claimant was therefore successful in recovering the amount of the adjudicator's decision. If the defendant was successful in the county court action, the claimant would have to repay the sum awarded by the adjudicator.

14.8.4. In the case of *DGT Steel and Cladding Ltd* v. *Cubitt Building Interiors* (2007), proceedings were commenced in court with regard to a claim for payment in the sum of £250,000. An application was made to the court to have proceedings stayed whilst the dispute was resolved by adjudication. The judge agreed to stay proceedings, on the basis that the parties had contractually agreed to have disputes referred to adjudication and the court would ensure that this obligation was fulfilled. Whilst in this case the contract between the parties provided for adjudication as a contractual entitlement, Judge Coulson made it clear that the same entitlement would exist if there was no such provision in the contract, but that one of the parties wished to exercise the right conveyed by the Construction Act to have disputes referred to adjudication.

SUMMARY

Where litigation and adjudication are in running in parallel with regard to the same dispute, the court on application of one of the parties will stay the proceedings to allow the adjudication to take place. It is of no relevance whether the adjudication relates to a contractual entitlement or to adjudication commenced in accordance with the Construction Act.

14.9. Can an adjudicator withhold a decision from the parties until his fees are paid?

14.9.1. Most adjudicators operate on a commercial basis and require payment for their services. It is common practice in the commercial world for a supplier to require payment before delivery, which helps to ensure that, at the end of the financial year, write-off for bad debts is kept to a minimum. Adjudicators know the uncertainties attached to receiving remuneration for their efforts and some insist upon payment of their fees before releasing a decision.

14.9.2. In the case of *St Andrew's Bay Development* v. *HBG Management and Mrs Janey Milligan* (2003), a dispute concerning the construction of a leisure centre was referred to adjudication. To ensure that the statutory time-scale of 28 days for reaching a decision was adhered to, the adjudicator's decision was required not later than 5 March 2003. Sometime after 5.00 pm on that day, HBG contacted the adjudicator to enquire when a decision could be expected. They were advised that the adjudicator had reached a decision, but it would not be released until her fee had been paid. A fee account was issued later that day and on the following day, HBG undertook to pay the whole of the fee.

14.9.3. The adjudicator conveyed her decision by fax on 7 March 2003, which was two days late. The court had to decide whether the decision, because it was two days late, was invalid and whether the adjudicator is entitled to withhold the release of a decision until the fees have been paid. The statute makes no reference to communicating the decision, but the conditions of contract required the decision to be issued 'forthwith'. This, explained the judge, meant immediately. However, it was held that, whilst the two-day delay was serious, it was not sufficient to render the decision a nullity.

14.9.4. With regard to the payment of fees, the judge ruled that an adjudicator is not entitled to withhold the decision until the fees have been paid. This difficulty could be overcome if the parties entered into an agreement whereby the adjudicator could withhold fees. Most adjudicators now include such a clause in their terms of appointment. Even when agreement is reached on this matter, the adjudicator is not entitled to withhold the issue of the decision beyond the 28-day statutory period, whether or not the fees have been paid.

14.9.5. The requirement that the adjudicator be paid before releasing the decision, often referred to as a lien, has come in for some criticism by the courts since the *St Andrews Bay* case. In the case of *Cubitt Building and Interiors Ltd* v. *Fleetglade Ltd* (2006), the parties agreed to the lien clause. However, the adjudicator made the decision a day before it was due and notified the parties of his requirement to be paid before releasing the decision. Midnight on the due day arrived, but payment had not been made and so on the next day the adjudicator sent his decision to the parties. The judge was not impressed with the lien clause. He considered that it was in conflict with the timing for making the decision as included in the Construction Act. The judge concluded that the adjudicator was unable to exercise the lien, as the decision which was due for delivery, in accordance with para 19(3) of the Scheme for Construction Contracts, which applied in this case, was to be issued as soon as possible after the decision had been made.

14.9.6. In the case of *Mott MacDonald* v. *London Regional Properties Ltd* (2007), the court expressed disapproval of the lien clause on the basis that, as the adjudicator had to release the decision as soon as possible after it had been made, this was in conflict with the time which would elapse whist awaiting payment of the fees.

14.9.7. In real life, if neither party wished to pay the adjudicator's fees before receiving his decision, they could just allow time to pass until the deadline was reached, at which time the adjudicator would be required to release his decision.

SUMMARY

An adjudicator can withhold the issue of a decision until the fees have been paid only if the parties agree. Most adjudicators now include such a term in their conditions of appointment. However, even when agreement has been reached, the adjudicator cannot withhold the decision beyond the statutory period. The courts have shown their disapproval of this type of clause on the ground that if it were operated strictly, the statutory time period for issuing a decision could pass without a decision being sent to the parties because of the non-payment of the adjudicator's fees. This would invalidate the decision. If the parties were intent on not complying with the lien, they could allow time to pass without making payment until the adjudicator was legally obliged to issue his decision.

14.10. If an adjudicator issues a decision late, can it be enforced?

14.10.1. The Scheme for Construction Contracts provides for an adjudicator to make a decision within 28 days of the referral notice. In the case of *Ritchie Brothers (PWC) Ltd* v. *David Phillips (Commercials)* (2004), it was held that, as the adjudicator was late in making a decision, it was unenforceable. The courts seem to have lacked consistency, however, in deciding what constitutes 'late'.

14.10.2. In deciding whether an adjudicator's decision is 'late', it is necessary to establish the date when the 28-day period begins to run. The court had to make a decision in the *Ritchie Brothers* case as to when does the period 'within 28 days after the referral' commence? It was decided that the period begins to run on the date the referral notice is dispatched. However, in *Aveat Heating Ltd* v. *Jarram Falkus Construction Ltd* (2007), His Honour Justice Harvey considered that the period could not begin to run until the adjudicator had received the referral notice, on the basis that something could not be referred to a person until he had received it. This reasoning was followed in *Epping Electrical Company Ltd* v. *Biggs Forrester (Plumbing Services) Ltd* (2007).

14.10.3. In the *Ritchie Brothers* case, the court took the view that the adjudication process does not end 28 days after the referral. Provision is made in the Scheme for the adjudicator to be replaced if the one originally appointed does not produce a decision. This being the case, a late decision would still be enforceable, provided another adjudicator had not been appointed. It was the view of the court that the 28-day period is directory only and not mandatory. On appeal, this decision was overturned.

14.10.4. A different approach to the matter was taken in the earlier case of *Barnes & Elliott Ltd* v. *Taylor Woodrow Holdings and George Wimpey Southern Ltd* (2003). The court in this case was influenced by the decision in *St Andrew's Bay Development Ltd* v. *HBG Management Ltd and Mrs Janey Milligan* (2003), which drew attention to the provisions of the Scheme for Construction Contracts, under which the decision is to be made within 28 days, with no reference as to the time for its communication. The judge considered that a decision made two days after the 28-day period had expired would not invalidate that decision. In *Barnes & Elliott*, the decision was communicated two days outside the 28-day period. The court enforced the adjudicator's decision, which was considered to have been made within the 28-day period, although delivered two days late. The court considered it to be a two-stage arrangement: the decision is made in stage one and communicated in stage two.

14.10.5. The court, in *Simons Construction Ltd* v. *Aardvark Developments Ltd* (2003), followed a similar line to that in the lower court in *Ritchie Brothers (PWC) Ltd* v. *David Phillips (Commercials) Ltd* (2004). It was held that a late adjudicator's decision was not invalid unless a fresh notice of referral had been issued before the release of the adjudicator's decision. On appeal to the Inner House of the Court of Session in Scotland in the *Ritchie Brothers* case, it was held that the decision had to be made and communicated in the 28-day period. In *Hart Investments Ltd* v. *Fidler* (2006), His Honour Mr Justice Coulson said:

> As the requirement that an adjudicator must produce his decision within 28 days (unless an extension is agreed by the parties) and not thereafter, I consider that the decision in *Ritchie*

> *Brothers* is a correct statement of the law. It seems to me that a decision reached outside the 28-day period is a nullity unless there is an agreed extension of the period.

A similar view was expressed by the judge in the case of *Cubitt Building and Interiors Ltd* v. *Fleetglade Ltd* (2006), where a decision was reached on 24 November and communicated shortly after noon on 25 November and held by the judge to be 'forthwith' and therefore valid. The judge considered that the time to be allowed for the communication of a decision should at the latest be only a few hours after it was reached. His Honour Mr Justice Coulson explained that the following principles applied when deciding if a decision was on time:

1. There is a two-stage process involved in an adjudicator's decision, which is expressly identified in clause 41A. Stage 1 is the completion of the decision. Stage 2 is the communication of that decision to the parties, which must be done forthwith: see *Bloor* and *Barnes & Elliott.*
2. An adjudicator is bound to reach his decision within 28 days or any agreed extension date: see *Barnes & Elliott* and *Ritchie.*
3. A decision which is not reached within 28 days or any agreed extended date is probably a nullity: see *Ritchie.*
4. A decision which is reached within 28 days or an agreed extended period, but which is not communicated until after the expiry of that period, will be valid, provided always that it can be shown that the decision was communicated forthwith: see *Barnes & Elliott.*

14.10.6. His Honour Mr Justice Coulson was in action again in the case of *Dalkia Energy and Technical Services Ltd* v. *Bell Group UK Ltd* (2009) and remained consistent, when saying:

> A decision has to be reached in the mandatory time limits, and so I would regard *Epping* as being entirely correct. But it is settled law that, provided that a decision is reached within the statutory time limits the adjudicator has a short additional period in which to issue that decision.

14.10.7. The CIC Adjudication Rules allow an extension of time for the delivery of an adjudicator's decision, up to the time of appointment of any replacement adjudicator. These conditions applied in the case of *Epping Electrical Company Ltd* v. *Briggs and Forrester (Plumbing Services) Ltd* (2007), but the court held that, as these conditions were at variance with the Construction Act, they were invalid. The adjudicator's decision was due to be reached on 21 November but, due to the adjudicator trying to secure his fees before releasing the decision, it was not sent until 23 November. It was held by the court to be out of time and hence unenforceable.

14.10.8. It would appear that the GC/Works/1 contract does not comply with the legal requirements, as decided in the *Ritchie Brothers* case. The provisions of condition 59(5) state that 'the adjudicator's decision shall nonetheless be valid if issued after the time allowed'.

SUMMARY

The courts have not been very consistent on the matter of late decisions of adjudicators. However, His Honour Mr Justice Coulson appears to have summed up the law accurately, when he outlined the following principles which should apply, in making a decision as to whether an adjudicator's decision was on time:

1. There is a two stage process involved in an adjudicator's decision, which is expressly identified in clause 41A. Stage 1 is the completion of the decision. Stage 2 is the communication of that decision to the parties, which must be done 'forthwith'.
2. An adjudicator is bound to reach his decision within 28 days or any agreed extension date.
3. A decision which is not reached within 28 days or any agreed extended date is probably a nullity.
4. A decision which is reached within 28 days or an agreed extended period, but which is not communicated until after the expiry of that period, will be valid, provided always that it can be shown that the decision was communicated forthwith.

14.11. The Construction Act states that 'a party to a construction contract has the right to refer a dispute arising under the contract for adjudication'. As this suggests disputes can only be referred one at a time, does this mean that a dispute regarding variations and also delays will have to be the subject of separate references?

14.11.1. In the case of *Fastrack Contractors Ltd* v. *Morrison Construction Ltd* (2000), Judge Thornton held that the 'dispute' is whatever is disputed at the time a referral is made to adjudication, when he said:

> During the course of a construction contract, many claims, heads of claim, issues, contentions and causes of action will arise. Many of these will be, collectively or individually disputed. When a dispute arises, it may cover one, several or all of these matters.

This line of reasoning was followed in the decisions in *KNS Industrial Services (Birmingham) Ltd* v. *Sindall Swansea Housing* (2001), *Sindall Ltd* v. *Solland* (2001) and *David McClean Housing Contractors Ltd* v. *Swansea Housing Association* (2002), *Benfield Construction Ltd* v. *Trudson (Hatton) Ltd* (2008) and *Witney Town Council* v. *Beam Construction (Cheltenham) Ltd* (2011).

In *Construction Adjudication*, by HH Judge Peter Coulson QC, at para 2.82, it states:

> Thus although those responsible for the 1996 Act probably did not envisage it being used for this purpose, a contractor with a complex final account claim is entitled to argue that his claim is, in essence, one single claim for an unpaid sum and that therefore he is entitled to adjudicate his final account claim no matter how large the claim might be and how voluminous the supporting documentation.

14.11.2. Lord MacFadyen, however, in *Barr Ltd* v. *Law Mining* (2001), sounded a word of caution; that it may be necessary for the adjudicator to consider whether there is sufficient connection between the disputed matters for them to be dealt with as a single dispute.

14.11.3. The very wide definition of 'dispute' made by Judge Thornton may be subject to review, in the light of the comments made by Lord MacFadyen. In the case of *David and Teresa Bothma (In Partnership)T/A DAB* v. *Mayhaven Healthcare Ltd* (2006), the notice of adjudication included four disputed matters: namely, the date for completion; the non-withdrawal of the notice of non-completion; the sum of valuation No 9; and the scope and validity of the architect's instructions. The judge considered that the extension of time part of the referral was unconnected with the financial part of the claim and, as a result, there were two disputes.

SUMMARY

Judge Thornton held that claims, heads of claim, issues, contentions or causes of action which are in dispute and included in a reference to the adjudicator will all be classified as 'a dispute'. This view, however, needs to be tempered by comments and later decisions in the lower courts. It would seem that, if the matter in dispute relates to the sum of money to which the contractor is entitled, even though made up of various items of cost, this would be regarded as one dispute. Where an extension of time and associated delay costs were the subject of dispute, again this would be regarded as one dispute. Where, however, there were disputes regarding an extension of time and the evaluation of variations, then, as these two matters are unconnected, for the purposes of the Construction Act they would be regarded as two disputes.

14.12. Where a compromise agreement relating to a dispute on a construction contract is itself the subject of a dispute, can it be referred to adjudication?

14.12.1. Whether a dispute arising out of a settlement agreement is a construction contract has been the subject of several disputes which have been referred to the courts. In *Lathom Construction Ltd* v. *AB Air Conditioning* (2000), a dispute arose between the employer and contractor in relation to the construction of a factory in Skelmersdale. The matter was referred to an adjudicator appointed by the RICS. Before the adjudicator commenced work, the parties reached a compromise agreement which was committed to writing. Unfortunately, a dispute arose out of the compromise agreement because of a disagreement as to its meaning. The contractor requested the RICS to appoint an adjudicator and the original one was reappointed. It was argued by AB Air Conditioning, the employer, that the compromise agreement was not a construction contract and therefore not catered for by the Construction Act. The adjudicator disagreed and proceeded to make a decision. The employer refused to comply with the adjudicator's decision and the contractor referred the matter to the court for enforcement. The matter before the court was whether or not the adjudicator had jurisdiction to act. The court refused to enforce the adjudicator's decision, as it was considered the adjudicator had no jurisdiction to deal with a compromise agreement.

14.12.2. In the case of *Shepherd Construction Ltd* v. *Mecright Ltd* (2000) there was a dispute about a settlement agreement, which it was contended had been reached under duress. The court decided that the settlement agreement was not a construction contract. It was not a dispute under the original construction contract and, therefore, the matter could not be the subject of adjudication.

14.12.3. This matter, however, is not always straightforward. From the decision in the case of *L Brown and Sons Ltd* v. *Crosby Homes (North West) Ltd* (2005), it would seem that what has to be decided is whether the compromise agreement is a variation of the original construction contract. If this is the position, then any dispute arising under the second agreement could be the subject of a referral to adjudication. In *Quarmby* v. *Larraby* (2003), the judge decided that, as a matter of construction disputes under the original contract which were not caught by the compromise agreement could still be the subject of a referral to adjudication.

14.12.4. The judge in the case of *McConnell Dowell Constructors* v. *National Grid* (2007) likened the two contracts, namely the original construction contract and the compromise agreement, to two buckets. In this case, the compromise agreement was arrived at before the work was completed. The claims which were settled by the compromise agreement were to go in one of the buckets and the remainder remained in the original contract bucket. Those claims which remained in the original bucket, it was held, could still be the subject of adjudication proceedings.

SUMMARY

The basic rule is that a compromise agreement which arises out of a dispute on a construction contract is not, in itself, normally classed as a construction contract. Any dispute, therefore, which relates to the compromise agreement cannot be referred to adjudication. However, if the dispute has not been caught by the terms of the compromise agreement it could still be the subject of adjudication proceedings.

14.13. A matter in dispute can only be referred to adjudication once. Where an adjudicator's decision has been received relating to a dispute over sums included in an interim certificate in respect of variations, can a dispute relating to the value of those variations when included in the final account be referred to adjudication?

14.13.1. The Scheme for Construction Contracts applies in respect of an adjudication under the Act where the terms of the contract are not in compliance with the Act. Paragraph 9(2) of the Scheme stipulates that an adjudicator must resign where the dispute referred to him is substantially the same as one that has previously been decided by another adjudicator. There have been a number of cases dealing with this matter. The courts are reluctant to investigate the circumstances surrounding the dispute in an effort to ascertain whether or not they are the same; consequently, they seem to err on the side of finding in favour of the disputes being different.

14.13.2. A dispute concerning this paragraph was referred to the court in the case of *Sherwood & Casson Ltd* v. *MacKenzie* (1999). Sherwood was a cladding contractor for the construction of a grandstand at Barrow RUFC. MacKenzie was the main contractor. The non-payment for variations in interim payments became a dispute and was referred to an adjudicator, who awarded £6,550 plus VAT to Sherwood. The claim was repeated in the final account, backed up by substantially more supporting information. Sherwood also submitted a claim for loss and expense. MacKenzie argued that the matter had already been referred to adjudication and therefore, under paragraph 9(2) of the Scheme for Construction Contracts, the second adjudicator should resign. The adjudicator disagreed and rejected the loss and expense claim, but allowed a further £12,000 in respect of variations. MacKenzie refused to pay and the matter was referred to the court. Judge Thornton considered that, when reviewing the question of the adjudicator's jurisdiction, the court should give considerable weight to the adjudicator's decision on the matter. The two disputes were not substantially the same; one dealt with an interim application for payment, whilst the other related to the final account. The judge was also influenced by the loss and expense claim and the information supporting the variation claim. The judge, in passing, said that his decision would have been the same if the adjudication rules issued by one of the institutions applied.

14.13.3. In the case of *Holt Insulation Ltd* v. *Colt International Ltd* (2001), a subcontractor's claim was rejected. The subcontractor reformulated its claim, which produced a smaller sum and referred it for adjudication a second time, when it was decided that it should be paid. It was held by the court that, whilst the same matter was the subject of the referrals, they represented two distinct and separate disputes. A similar situation arose in the case of *Skanska Construction UK Ltd* v. *The ERDC Group Ltd* (2003), where a claim failed in adjudication due to lack of supporting information, following which the referring party resubmitted the same claim but with proper substantiation. It was held that they were not the same dispute and, therefore, the adjudicator's decision was enforceable.

14.13.4. The case of *Quietfield Ltd* v. *Vascroft Contractors Ltd* (2006) dealt with an extension of time. It was the adjudicator's decision that the contractor had no entitlement and, as a result, it was exposed to a claim for liquidated damages. The payment of the liquidated damages, in respect of the same period, was the dispute in a further adjudication. This time, the contractor produced much more detail to support his entitlement to an extension of time. The adjudicator refused to consider the contractor's submission, on the grounds that it had been the subject of an earlier adjudication. The court held that the adjudicator was wrong, in that the submission regarding the extension of time included far more detail than was provided in the earlier adjudication and should therefore have been accepted. The decision of the lower court was referred to the Court of Appeal, which upheld the judge's decision.

14.13.5. Since the decision in the *Quietfield* case, the attitudes of the courts seem to have hardened and they appear more inclined to look more closely when asked to decide whether a dispute had been decided in an earlier adjudication. Such a case is *HG Construction Ltd* v. *Ashwell Homes (East Anglia) Ltd* (2007). This case involved a dispute arising from a contract which was let using the JCT With Contractor's Design 1998 standard form. The contract provided for partial possession and a dispute arose relating to the employer's right to deduct liquidated damages where the contractor failed to complete by the sectional completion dates. The contractor's argument was that the application of liquidated damages relied upon it being possible to allocate a value of the work in each section and that it was impossible to make such allocation. In the first adjudication, the adjudicator decided that it was possible to allocate the value of work to each section and that liquidated damages were enforceable. Following the deduction of liquidated damages amounting to £184,627, the contractor made a further referral to adjudication, arguing that it was not possible to value the work in each section and therefore the liquidated damages should be repaid. The adjudicator, in his decision, accepted the argument of the contractor and ordered that the liquidated damages should be repaid. It was the view of the adjudicator in the second adjudication that the first adjudicator dealt with the legal principles, whereas he had applied the facts of the matter. The court held that the second adjudicator had dealt with the same dispute and under the Construction Act, a dispute once decided by adjudication could not be heard a second time. In addition, the wording in the contract, stating that the decision of the adjudicator shall be binding until finally decided by arbitration or legal proceedings meant that the decision of the second adjudicator could not be enforced.

14.13.6. In the case of *Benfield Constructon* v. *Trudson* (2008), the court had to decide whether a dispute concerning partial possession was the same as one relating to practical completion. The court decided that there were different legal concepts involved, but nonetheless they amounted to the same dispute. His Honour Mr Justice Coulson summed up the situation succinctly when he said:

> Adjudication is supposed to be a quick one-off event; it should not be allowed to become a process by which a series of decisions by different people can be sought every time a new issue or a new way of putting a case occurs to one or other of the parties.

SUMMARY

A matter which has been referred to adjudication once cannot be referred a second time. However, if a dispute has been based upon factual differences it may be the subject of a further adjudication. For example, the referral of an extension of time claim, based upon limited supporting information, which does not meet favour with the adjudicator, may be the subject of a second adjudication if full details are provided. Another example would be a claim for interim payment purposes that may be resubmitted with the final account, where more supporting details are provided. It would seem, however, that the courts are latterly taking a more restricted view as to what constitutes a separate dispute from those cases heard earlier.

14.14. Will a clause in a construction contract which states that all disputes are subject to the exclusive jurisdiction of the Austrian Courts and Austrian Law, or the courts and law of another jurisdiction outside the UK, result in disputes falling outside the UK courts' power to enforce an adjudicator's award?

14.14.1. There have been many attempts to circumvent the effects of the adjudication process under the Construction Act. In the case of *Comsite Projects Ltd* v. *Andritz AG* (2003), a clause in the contract stated that all disputes are subject to the exclusive jurisdiction of the Austrian courts and Austrian law.

14.14.2. A dispute arose between the parties which was referred to adjudication. Enforcement proceedings were commenced in respect of the adjudicator's decision. Despite the wording of the clause, the court ordered that the parties must comply with the adjudicator's decision. However, having complied with the adjudicator's decision, either party could refer the dispute to the Austrian court for a final decision.

SUMMARY

Attempts have been made to circumvent the effects of the adjudication process under the Construction Act. A clause written into the contract, which states that all disputes will be subject to the exclusive jurisdiction of the Austrian courts and Austrian law, or the courts and law of any other jurisdiction outside the UK, will not affect the enforcement of an adjudicator's decision by the UK courts.

14.15. Are the parties entitled to challenge an adjudicator's fees on the grounds that they are unreasonably high?

14.15.1. Most adjudicators ensure that the basis on which their fees are to be calculated are agreed by the parties as soon as possible. It is usual for the adjudicator to charge fees on the basis of an hourly rate or a day rate. Whilst the rates may be agreed, the parties may consider that the number of hours or days included in the adjudicator's fee account is excessive.

14.15.2. The case of *Stubbs Richards Architects* v. *WH Tolley & Son Ltd* (2001) is an example of the parties challenging an adjudicator's fees. Tolley & Son carried out construction work for Torridge District Council and, as a result, disputes arose. The matter was referred to adjudication and a partner of Stubbs Richards, a firm of architects, was appointed as the adjudicator. Two disputes were referred to the adjudicator, who delivered his decision along with his fee account in the sum of £1,500 plus VAT. The fee was paid, but Tolley & Son alleged that it was excessive, in that the number of hours which the adjudicator claimed to have worked could not be justified. An action was commenced to recover some of the fees which Tolley & Son had paid. The district judge who heard the case considered that an adjudicator does not have immunity from a claim that the hours and remuneration are unreasonable. He ordered that some of the fees should be repaid to Tolley & Son. This decision was the subject of an appeal. On appeal, a different view was expressed in finding for the adjudicator. The court drew attention to the wording of the Construction Act, where it is stated that the adjudicator is not liable for anything done or omitted in the discharge of his or her functions, unless carried out in bad faith. It was considered that the adjudicator's fee formed an integral part of his agreement and was therefore caught by the wording in the Act. The district judge was criticised for basing his judgment on what he considered would be the amount of time likely to have been spent by the solicitor in presenting a case before a court. This was felt to be inappropriate, as an adjudicator would spend a lot more time reading through files, interviewing the parties and visiting the site than would a solicitor in presenting a case in court.

14.15.3. The matter of disputed adjudicator's fees arose in the case of *Jerram Falkus* v. *Fenice* (2011), where the adjudicator's fees, amounting to 56 hours at £350 per hour, were challenged. Judge Waksman expressed the view that any party who intends to challenge an adjudicator's fees will need carefully to consider whether there is a realistic basis for disputing the fees and that in the usual run of cases, there will be no grounds for mounting a challenge.

SUMMARY

The parties to an adjudication cannot challenge the level of an adjudicator's fees. Under the Construction Act the adjudicator is not liable for anything done or omitted in discharge of his duties, unless carried out in bad faith.

14.16. Does a draft adjudicator's decision constitute a final decision?

14.16.1. There seems to be a growing trend among some adjudicators to issue a draft decision to the parties and invite comments. This may show a lack of confidence on the part of those adjudicators who indulge in this practice; or perhaps they like to err on the side of caution.

14.16.2. In the case of *Simons Construction Ltd* v. *Aardvark Developments Ltd* (2003), Simons were engaged as the main contractor under a JCT main contract. A dispute arose, which was referred to adjudication. The 28-day period within which the adjudicator is required to produce an award was extended by agreement of the parties to a decision required by 17 June 2002. A draft decision was issued on that day, inviting the parties to offer their comments. A final decision was given a week later. It was argued by Aardvark, the successful party, that the draft decision was a valid decision under the Act. Simons argued that the draft decision was not a final decision and as the final decision was given late it was invalid. The court held that the draft decision was not intended to be a final decision and, as such, did not comply with the Act. However, as no fresh notice of referral had been issued by either party, the final decision stood as valid, even though it was issued late. In view of more recent cases, such as *Ritchie Brothers* v. *David Phillips*

Commercial Ltd (2005), it is unlikely that an adjudicator's decision which is given late would be enforceable.

14.16.3. In the case of *Lanes Group plc* v. *Galliford Try Infrastructure* (2011), the adjudicator issued a document entitled 'Preliminary Views and Findings of Fact', which resembled a draft judgment, on which the parties were invited to comment. Judge Waksman expressed the view:

> in the normal run of an adjudication I would not have thought that documents expressing provisional views on which parties were then invited to comment were likely to be helpful or appropriate.

SUMMARY

Where an adjudicator issues a draft award, the court will look at his intentions when deciding whether or not it can be considered to be a final award. It would seem unlikely, except in the most unusual of cases, for an adjudicator to issue a draft award which he considers to be a final award. Judge Waksman considered that documents expressing provisional views are unhelpful. Adjudicators are therefore to be encouraged to issue one decision only and that is the final decision.

14.17. Does an adjudicator who communicates with one party without disclosing the details of the communication to the other party risk his decision being nullified by the courts on the grounds that his conduct amounted to bias?

14.17.1. The rules of natural justice provide that all persons who are accused are entitled to know the charges made against them and must be given a proper opportunity to be heard. These rules have been examined and re-examined, interpreted and misinterpreted by judges and legal commentators down the years. Arbitrators have been almost paranoid in ensuring that these rules are not contravened by giving parties and their advisers every opportunity to be heard and seeing that arbitrators do not hear evidence from one party without giving the other party a proper opportunity to respond. Are the decisions of adjudicators also vulnerable from a failure to comply with the rules of natural justice?

14.17.2. In the case of *Discain Project Services* v. *Opecprime Developments* (1999), the court had to decide whether an adjudicator had contravened the natural justice rules. A payment dispute arose, which was referred to adjudication. The adjudicator's decision was that the respondent should pay the referring party a sum of £55,552. No payment was made, on the grounds that there had been a breach of the rules of natural justice, and the matter was referred to court. It appears that the adjudicator had a private telephone conversation concerning the adjudicator's jurisdiction, the validity of the application for payment and the withholding notice. This was followed by a fax from the adjudicator which stated his intention to proceed to a decision, without the benefit of further comment from the parties. A second telephone conversation between the adjudicator and the applicant took place after the fax had been sent. The court refused to enforce the adjudicator's decision. In the opinion of the judge there was a very serious risk of bias and there were clear failures to consult with one of the parties on important submissions made by the other party. The adjudicator should also have made sure he involved both parties in discussions concerning his jurisdiction, which he failed to do.

14.17.3. A similar situation occurred in the case of *Woods Hardwick* v. *Chiltern Air-Conditioning Ltd* (2000). In this case, the adjudicator consulted the representatives of Woods Hardwick and Chiltern's subcontractors without informing Chiltern that information was being obtained from those sources or what the information comprised. It was held by Judge Thornton that the adjudicator had not acted impartially and therefore he declined to enforce the adjudicator's decision.

14.17.4. In *Glencot Development and Design Co Ltd* v. *Ben Barratt and Son (Contractors)* (2001), the adjudicator also acted as the mediator, which involved holding meetings separately with the parties. It was held by the court that this amounted to a breach of natural justice and the court declined to enforce the adjudicator's decision. It has been held that the fact that an adjudicator has acted as a mediator in an earlier dispute involving one of the parties would not bar him from acting as adjudicator in a subsequent one. It all comes down to whether a fair-minded and informed observer, having considered all the circumstances which have a bearing on the suggestion that the decision-maker was biased, would conclude that there was a real possibility that he was.

14.17.5. Adjudicators under the rules included in The Scheme for Construction Contracts are allowed to make their own enquiries to ascertain the facts concerning the matters in dispute. There is, however, a fine line to be drawn between making those enquiries and breaching the rules of natural justice.

SUMMARY

Adjudicators are allowed to make their own enquiries to ascertain the facts relating to the matters in dispute. However, if ascertaining the facts involves communicating with one of the parties, an adjudicator runs the risk that his decision may not be enforced.

14.18. Can an adjudicator employ the services of an expert to assist in making a decision?

14.18.1. Adjudicators, in their terms of appointment, are usually given power to engage the services of an expert in arriving at a decision. When taking a decision to appoint an expert, however, the adjudicator's path does not always run smoothly.

14.18.2. In the case of *Try Construction Ltd* v. *Eton Town House Group Ltd* (2003), an adjudication was commenced, following which a meeting took place between the parties and the adjudicator. At this meeting the parties agreed that the adjudicator could appoint a programming expert, a Mr Lowsley. During a second meeting, both parties agreed that Mr Lowsley could contact their respective programming experts independently.

14.18.3. The adjudicator in his decision awarded an extension of time to Try and the payment of £269,916, plus interest. Eton refused to pay on the grounds that the adjudicator, in using the information provided by Mr Lowsley, had adopted a methodology which they had not had a chance to consider and comment upon. Try argued that Eton had agreed to the appointment of the expert and they could not therefore raise an objection after the adjudicator had reached a decision.

14.18.4. The court, in upholding the adjudicator's decision, held that the evidence showed that the defendant had participated in the adjudication process without demur and knew that Mr Lowsley, as an expert programmer, would investigate the claim on his terms. The defendant had not requested a preliminary or draft finding as the basis for further submissions. The court accepted evidence that it had been agreed at the parties' first meeting with the adjudicator that Mr Lowsley would use his own expertise in the delay analysis.

14.18.5. A different outcome occurred in the case of *RSL (South West) Ltd* v. *Stansell Ltd* (2003). This case involved a dispute between a contractor and subcontractor relating to the fabrication and erection of structural steelwork and staircases for a project in Union Street, Bristol. The dispute included delay issues and the adjudicator required someone with planning experience to assist him with the programming matters. Both parties agreed to the adjudicator seeking assistance and Stansell requested sight of any report which might be prepared by the expert and reasonable time to pass comment. The expert prepared a report, which was passed to the parties for comment. RSL made a series of comments and submitted further information, as they did not find the report very favourable. Stansell, obviously pleased with the report, didn't offer any comments. Following receipt of RSL's comments, the expert produced a final report. Unfortunately, the adjudicator did not show the final report to the parties and RSL therefore contested the decision. Judge Richard Seymour QC considered the complaint by RSL justified. The adjudicator should have disclosed the final report to the parties and in failing to do so was in breach of the rules of natural justice.

14.18.6. In *Balfour Beatty Construction Ltd* v. *The Mayor and Burgesses of the London Borough of Lambeth* (2002), the court held that the adjudicator had failed to comply with the rules of natural justice. It seems the adjudicator was not impressed with the details concerning an extension of time entitlement submitted by Balfour Beatty Construction Ltd. He decided to employ the help of two of his colleagues to construct a collapsed as-built analysis to assist him in arriving at a decision. This information was not passed to the parties for comment and, as a result, the court refused to enforce the adjudicator's decision on the grounds of bias and breach of the rules of natural justice.

SUMMARY

If an adjudicator wishes to use the services of an expert, he must seek the agreement of the parties. In addition, any report produced by the expert must be passed to both parties to allow them to pass comment.

14.19. Where an adjudicator seeks legal advice to assist him in reaching a decision, is he obliged to reveal the advice to the parties involved in the adjudication?

14.19.1. Many adjudicators are neither legally qualified nor legally trained. Despite this lack of knowledge and training, adjudicators are regularly required to deal with legal matters as part of their duties. It is not uncommon, therefore, for adjudicators to seek independent legal advice to assist them in reaching a decision. Most adjudicators will include a term in their conditions of appointment which gives them authority to acquire legal opinions, or there may be provision within the contract between the parties which deals with the matter.

14.19.2. This was the situation in the case of *BAL* v. *Taylor Woodrow Construction Ltd* (2004). A dispute arose out of a contract Taylor Woodrow undertook on behalf of Nationwide Anglia Property Services Ltd, which involved BAL as one of the subcontractors. Nationwide issued a letter of intent giving Taylor Woodrow an instruction to commence work. The letter of intent placed a cap of £750,000 on the value of work to be carried out, without any further instructions. The subcontract included a clause which allowed any adjudicator who might be appointed to obtain advice from specialists, provided one of the parties requests or consents and agrees to be responsible for the costs and expenses incurred. A dispute arose between Taylor Woodrow and BAL, which was referred to adjudication. The adjudicator indicated to the parties that he might require to seek legal advice concerning the effect of the cap in the letter of intent. BAL agreed and Taylor Woodrow did not object. The adjudicator did not inform the parties as to what information he provided to his legal adviser or the advice he was given. Taylor Woodrow argued that there had been a breach of the rules of natural justice, as the adjudicator had not disclosed to the parties the legal advice he had received. The court agreed and refused to enforce the adjudicator's decision.

14.19.3. A similar situation arose in the case of *Costain Ltd* v. *Strathclyde Builders Ltd* (2004), where the adjudicator took legal advice, which was not shown to the parties. It was held by the court that the adjudicator's decision would not be enforced.

SUMMARY

Where an adjudicator receives legal advice to enable him to reach a decision, he should disclose it to the parties. If he fails to do so, the courts are likely to consider this failure to be a breach of the rules of natural justice and refuse to enforce the adjudicator's decision.

14.20. Have the courts laid down any general guidelines as to how the rules of natural justice should be applied in respect of adjudication?

14.20.1. The rules of natural justice have been applied over many years to ensure the parties get a fair hearing. Often, courts appear to bend over backwards to see that fairness applies.

Basically, the rules of natural justice are aimed at ensuring that both parties know fully the case which is being made against them and are given a proper opportunity of providing answers. All courts of law, arbitrations, tribunals and adjudications are required to apply the rules of natural justice.

14.20.2. In the early case of *Discain Project Services Ltd* v. *Opecprime Development Ltd* (1999), it was held that an adjudicator's decision would not be enforced because of a private telephone conversation which took place between the adjudicator and one of the parties, as it was held to be a breach of natural justice. His Honour Judge Bowsher, in arriving at a decision, said:

> That scheme makes regard for the rules of natural justice more rather than less important. Because there is no appeal on fact or law from the adjudicator's decision, it is all the more important that the manner in which he reaches his decision should be beyond reproach . . . I do not think that the court should enforce a decision reached after substantial breach of the rules of natural justice.

14.20.3. In the case of *Glencot Development and Design Co Ltd* v. *Ben Barratt and Sons (Contractors) Ltd* (2001), the judge said that it was accepted that the adjudicator 'has to conduct the proceedings in accordance with the rules of natural justice or as fairly as the limitations imposed by Parliament permit'.

14.20.4. In the case of *Costain Ltd* v. *Strathclyde Builders Ltd* (2004), Strathclyde Builders refused to comply with an adjudicator's decision on the grounds that, during the procedure, there had been a breach of the rules of natural justice. The adjudicator arrived at a decision following discussions with his legal adviser. However, he failed to disclose to the parties the details received from the lawyer. Like the decision in *BAL* v. *Taylor Woodrow Construction Ltd* (2004), it was held by the court that the actions of the adjudicator amounted to a breach of the rules of natural justice.

14.20.5. The court then took the opportunity to formulate guidelines which should be applied by adjudicators, to ensure they do not fall foul of the rules of natural justice:

(1) Each party must be given a fair opportunity to present its case. That is the overriding principle and everything else is subservient to it.
(2) Subject to (1) above, together with any express provisions in the parties' contract, procedure is entirely under the control of the adjudicator.
(3) In considering what is fair, it must be remembered that adjudications are conducted according to strict time limits; thus time limits will be severely restricted.
(4) The adjudication procedure is designed to be simple and informal; all that it will normally require is that each party should be given an opportunity to make comments at any relevant stage of the adjudication process.
(5) Each party must be given an opportunity to comment on the other's submissions and additional new submissions.
(6) An adjudicator is normally given power to use his own knowledge and experience in deciding the issues in dispute. If he raises new points that have not been canvassed by the parties, it will normally be appropriate to invite submissions.
(7) If the adjudicator obtains more information or carries out tests, it will be appropriate to make the results known to the parties and call for their comments.

(8) An adjudicator may be given power to obtain from other persons such information and advice as he considers necessary on technical or legal matters. If he does so, the situation is similar to (6) above.

(9) In this connection no distinction can be drawn between issues of fact and issues of law.

14.20.6. The Court of Appeal has to some extent put the brakes on with regard to the development of natural justice as it applies to adjudication. In the case of *Amec Capital Projects* v. *Whitefriars City Estates* (2004), the Court of Appeal overturned the decision of the lower court, which refused to enforce an adjudicator's decision on the grounds of breach of natural justice. The lower court considered that a private telephone conversation with one of the parties and a failure to give the parties an opportunity to comment on advice the adjudicator had taken made the adjudicator's decision unenforceable, due to a breach of the rules of natural justice. The Court of Appeal disagreed, saying:

> It is easy enough to make challenges of breach of natural justice against an adjudicator . . . The intention of Parliament . . . will be undermined if allegations of breach of natural justice are not examined critically, when they are raised by parties who are seeking to avoid complying with adjudicator's decision. It is only where the defendant has advanced a properly arguable objection . . . that he should be permitted to resist summary proceedings.

The Court of Appeal showed a level of consistency in the case of *Carillion* v. *Devonport* (2005), where it was said:

> To seek to challenge the adjudicator's decision on the ground that he had exceeded his jurisdiction or breached the rules on natural justice (save in the plainest of cases) is likely to lead to waste of time and expense.

14.20.7. His Honour Peter Coulson, in his book *Construction Adjudication*, has this to say concerning the application of the rules of natural justice in the case of an adjudicator's decision:

> Accordingly it is safe to conclude that, whilst an argument that the adjudicator has failed to comply with the rules of natural justice will be considered with a certain amount of scepticism by the court, where elementary and basic principles of natural justice have not been observed, with a resulting serious effect upon the decision in question, the courts will be prepared to refuse to enforce summarily that decision.

14.20.8. An adjudicator must consider all submissions made to him if he is to avoid allegations that there has been a breach of natural justice. In the case of *Pilon Ltd* v. *Breyer Goup plc* (2010), the court refused to enforce an adjudicator's decision on the grounds that the adjudicator had failed to consider the defence put forward by the responding party. A similar situation occurred in the case of *Quartzelec Ltd* v. *Honeywell Systems Ltd* (2008), where the adjudicator declined to consider a defence and in doing so had failed to act in accordance with the rules of natural justice.

SUMMARY

The rules of natural justice apply to adjudication. To ensure compliance, the adjudicator must ensure that both parties know fully the case which is being made against them and are given a proper opportunity of providing answers. Arguments put forward by the losing party that a breach of natural justice has occurred, however, will often be met with some scepticism by the courts. Where elementary and basic principles of natural justice have not been observed, the courts will be prepared to refuse to enforce the adjudicator's decision. Case law has established that, if the adjudicator fails to consider submissions made by the parties, he risks breaching the rules of natural justice.

14.21. Where an adjudicator issues a decision involving the payment of a sum of money by the winning party to the losing party, can the amount paid be reduced or eliminated in compliance with a clause in the contract?

14.21.1. When coming to a decision as to whether a set-off can be made from an adjudicator's award, the courts have to decide whether the enforcement of an adjudicator's decision is based upon a separate statutory force or no more than a contractual obligation, which is backed by statute. When the question as to whether an adjudicator's award could be reduced or eliminated as a result of contractual entitlement came before the courts, the view taken was that an adjudicator's decision creates a contractual obligation which is backed by statute.

14.21.2. In *Parsons Plastics* v. *Purac Ltd* (2002), a contract between a main contractor and subcontractor contained a clause which expressly preserved the contractor's common law rights of set-off. The Court of Appeal held that an adjudicator's decision was subservient to this clause, allowing the contractor to set off entitlements against the sum included for payment to the subcontractor in the adjudicator's decision.

14.21.3. This line of argument was followed in the case of *Bovis Lendlease* v. *Triangle Developments* (2003), which dealt with the payment provisions where a contractor's employment is terminated. Clauses which deal with this situation in most standard forms of contract state that, following termination, no further payment needs to be made until the final account and claims have been resolved. In this case an adjudicator's decision was not immediately honoured, but made subject to the provisions of the termination clause.

14.21.4. The Court of Appeal made an about-turn in the case of *Levolux* v. *Ferson* (2003). In a similar fashion to the *Bovis Lendlease* case, the decision involved a termination, but in this case it was a subcontractor whose employment was terminated. A dispute arose concerning the subcontractor's right to payment. The dispute was referred to adjudication and a decision made in favour of the subcontractor. Due to the fact that the subcontractor's employment had been terminated, the contractor considered that, following the provisions of the termination clause in the contract, he was under no obligation to make payment. The Court of Appeal disagreed, in holding that, to allow the contractor to withhold payment of the adjudicator's decision would defeat the intended purpose of adjudication under the Construction Act. The contract must be interpreted to achieve this purpose and any offending clause would be struck down.

14.21.5. A slight twist to the situation occurred in *Shimizu Europe Ltd* v. *LBJ Fabrications Ltd* (2003). In this case, a dispute arose in connection with cladding work at an Oxford Science Park. The adjudicator, in his decision, provided for Shimizu to pay LBJ the sum of £47,000 plus VAT 'without set off'. Payment was to be made by Shimizu not later than 28 days after LBJ had delivered a VAT invoice. After the issue of the VAT invoice, but before the payment was required to be made, Shimizu issued a withholding notice for set-off in respect of alleged defective work. The court allowed Shimizu to withhold the money, on the grounds that the withholding notice complied with the contract and the money was not due for payment until 28 days after the issue of the VAT invoice.

14.21.6. In the case of *JPA Design and Build Ltd* v. *Sentosa (UK) Ltd* (2009), it was held that, where there had been two adjudications between the parties, the adjudicator's decision in one of the adjudications could be set off against the sum awarded in the other adjudication.

SUMMARY

The Court of Appeal has changed its position in respect of whether there is a right of set-off from an adjudicator's decision in compliance with a clause in the contract. In *Parsons Plastics* v. *Purac Ltd* (2002), the Court of Appeal held that an adjudicator's decision is subservient to a clause under the contract which allows for set-off. A shift of position occurred in the decision of *Levolux* v. *Ferson* (2003), when the Court of Appeal stated that to allow a set-off from an adjudicator's decision would defeat the intended purpose of adjudication under the Construction Act.

14.22. Where an adjudicator's decision provides for a sum of money to be paid by the employer to the contractor, can the employer deduct from the sum awarded liquidated and ascertained damages due under the contract?

14.22.1. The case of *The Construction Centre Group* v. *The Highland Council* (2003) dealt with an adjudicator's decision arising out of a design, construction and maintenance contract on the West Coast of Scotland. Disputes arose which were referred to an adjudicator, whose decision provided for the employer to pay the contractor a sum in the region of £250,000. No payment was made, as the employer claimed that the contractor owed liquidated damages in the sum of £420,000.

14.22.2. Within seven days of the adjudicator's decision, the employer served a withholding notice under section III of the Construction Act in respect of the liquidated damages. There had been no earlier set-off notice, as the previous valuation was nil and therefore no sum of money was due from which the liquidated damages could be deducted. The Outer House of the Court of Session in Scotland did not accept the employer's arguments. It was considered that the employer could have raised the matter of liquidated and ascertained damages in its submission to the adjudicator, which would have provided an opportunity of having them dealt with in the decision. The employer was not at liberty to make a deduction from the adjudicator's award.

14.22.3. In *A* v. *A* (2002), the adjudicator awarded an extension of time for 46 weeks out of 112 claimed. The adjudicator was implicitly ruling that there was a delay of 66 weeks to completion, which would attract liquidated and ascertained damages of £75,000 per week. The employer sought to set off the liquidated damages from the sum awarded, but the court disagreed. He was obviously in error in not including a claim for liquidated and ascertained damages as part of the submission to the adjudicator and therefore the court enforced the adjudicator's decision in full.

14.22.4. There is an exception to the general rule that liquidated damages which are due cannot be set off against an adjudicator's decision. This applies where the deduction of liquidated damages follows logically from an adjudicator's decision. In the case of *Balfour Beatty Construction* v. *Serco* (2004), the adjudicator in his decision ordered that a sum of money be paid to the referring party. The adjudicator also made a decision with regard to the referring party's entitlement to an extension of time. Following the decision and before making payment, the employer deducted liquidated damages, having applied the extension of time included in the adjudicator's decision. Judge Jackson considered that liquidated damages could be deducted if the entitlement has been determined by the adjudicator, either expressly or by implication. He also expressed the view that there may be a right to set off liquidated damages, dependent upon the terms of the contract and the circumstances of the case.

14.22.5. The decision in the case of *Avoncroft Construction* v. *Sharbu Homes* (2008) follows the general line of legal thinking that liquidated damages are not usually deductible from the sums payable in respect of an adjudicator's decision.

SUMMARY

An employer cannot deduct liquidated damages due under a contract from an adjudicator's award. If the employer considers he has an entitlement to liquidated damages, reference to the entitlement should be included in the employer's submission to the adjudicator. An exception, however, arises where, within the adjudicator's decision, there is expressly or by implication an entitlement to deduct liquidated damages. Whether there is a right to deduct liquidated damages from an adjudicator's decision depends upon the terms of the contract and the circumstances of the case.

14.23. Can a party to a construction contract who is reluctant to have a dispute referred to adjudication successfully argue that the adjudicator's decision should not be enforced as, because of the very restricted nature of the adjudication process, it is contrary to the European Convention on Human Rights?

14.23.1. There has been a great deal of speculation that Article 6 of the European Convention on Human Rights, which came into force in the UK on 1 October 2000, has the effect of outlawing statutory adjudication. Article 6 states that each party must be afforded a reasonable opportunity to present his case, including his evidence, under conditions that do not place him at a substantial disadvantage vis-a-vis his opponent. An action was commenced in the case of *Elanay Contracts Ltd* v. *The Vestry* (2000) to enforce an adjudicator's award. The defendant's case was that enforcement should not be ordered, as the adjudication procedure was unfair. This was partly due to the fact that, for much of the time, the defendant's key witness was attending hospital, visiting his dying mother. The shortness of the proceedings, which were over in 35 days, added to the sense of unfairness. Judge Richard Havery took a different view. He considered that Article 6 does not apply to statutory adjudication. The reason he gave was that, whilst proceedings before an adjudicator determine questions of civil rights, they do not provide for a final determination. An adjudicator's decision is provisional, in that the matter can be reopened for a final decision in a subsequent court hearing or before an arbitrator.

14.23.2. In the case of *Austin Hall Building* v. *Buckland Securities* (2001), the losing defendant refused to comply with the decision of the adjudicator on the grounds that there had been a breach of the Human Rights Act. The basis of the case was that the adjudication process under the Construction Act was unfair and therefore contravened the European Convention on Human Rights. The defendant's line of argument was in respect of the 28-day period within which the adjudicator made his decision. This did not get very far with the judge. Whilst the Human Rights Act states that it is unlawful for a tribunal to act in a way which is incompatible with a Convention right, it goes on to add that this requirement does not apply if, under a provision of primary legislation, the tribunal could not have acted differently. The adjudicator was obliged under the Construction Act to give a decision within the 28-day period and therefore it fell within the exception to the requirement of the Convention right. The judge went on to consider whether an adjudicator is in fact a tribunal. He felt that, as an adjudicator is not involved in legal proceedings, he is not therefore a tribunal.

SUMMARY

It is clear from the decisions in *Elanay Contracts Ltd* v. *The Vestry* (2000) and *Austin Hall Building* v. *Buckland Securities* (2001) that the adjudication process does not contravene the European Convention on Human Rights.

14.24. Can the party that is successful in adjudication recover its costs from the losing party?

14.24.1. Where disputes are referred to litigation or arbitration, the judge or arbitrator has power to include in the award that the losing party is to pay the winning party's costs. Experience has shown that the costs incurred by a party in referring a dispute to adjudication are significantly less than would be the case if the matter were referred to arbitration or litigation. Nonetheless, the costs incurred can be substantial.

14.24.2. A party whose dispute when referred to adjudication is successful will, no doubt, be looking to the losing party for reimbursement of the costs incurred. Does an adjudicator have the same powers as a judge or arbitrator when it comes to awarding costs? A starting point may be by reference to the paragraph 25 of the Scheme for Construction Contracts, which makes the parties jointly and severally responsible for the payment of the adjudicator's fees and gives the adjudicator power to apportion them between the parties. There is no specific power given to him to order that one of the parties will pay the other's costs. Most of the adjudication provisions in the majority of the standard forms in common use make no provision for the adjudicator to order, in his decision, that one of the parties pays the other's costs. One exception is GCWorks/1, where condition 59(5) indicates that the adjudicator's decision may indicate 'whether one party is to bear the whole or part of the reasonable legal and other costs and expenses of the other'. The ORSA Adjudication Rules, which apply in Scotland, provide another example of adjudication rules which give the adjudicator the power to order that one of the parties pays the costs incurred by the other. The parties will be free to enter into an agreement which gives the adjudicator to include in his decision that one of the parties will pay the costs of the other.

14.24.3. What is the position where the contract is silent on the matter? In the case of *Northern Developments (Cumbria) Ltd* v. *J. & J. Nichol* (2000), Northern Developments Ltd was the main contractor for the construction of an outlet for M. Sport Ltd at Doventry Hall in Cumbria. J. & J. Nichol was a subcontractor appointed to construct the steel frame, roofing, cladding and associated works. The terms of the subcontract were DOM/2 and the contract was governed by the Construction Act 1996. As the terms of the subcontract did not comply with the provisions of the Act, the Scheme for Construction Contracts applied. On 13 July 1999, the subcontractor made an application for payment and a set-off notice was issued by the main contractor on the same day. The notice referred to defective work and delays. No payment was made to the subcontractor, as the contractor argued that the set-off for defective work exceeded the value of the work executed. The subcontractor withdrew from the site on 6 August 2000. This was treated by the contractor as a repudiatory breach, which it accepted and appointed another contractor to finish off the work. The matter was referred to adjudication by the subcontractor, who claimed an entitlement to a sum of £237,120 plus VAT, together with £11,456 repayment of cash discount. The main contractor claimed set-off with regard to:

(1) Defective work;
(2) Delays; and
(3) Damages due to the subcontractor's repudiation of the contract.

No reference was made in the set-off notice to the damages relating to repudiation, as the withdrawal from site by the subcontractor took place after the issue of the notice. The adjudicator had to decide, in the first instance, whether he was empowered to deal with the matter of repudiation. His decision was that he had no power to deal with the repudiation issue, but was limited to deciding matters which were under the contract. The repudiation issue did not, in his opinion, constitute a matter under the contract.

The judge considered that the adjudicator was wrong in not dealing with the repudiation. Acceptance of repudiation brings to an end the performance of the contract. The contract still exists and rights arising under it can be enforced. Nonetheless, the damages associated with the repudiation could not be set off as they had not been referred to in the notice. The adjudicator's decision was correct, but for the wrong reasons.

Costs were awarded by the adjudicator in favour of the subcontractor, but this was challenged unsuccessfully by the contractor. In the case of *John Cothliff Ltd* v. *Allen Build (North West) Ltd* (1999), it was held in Liverpool County Court that an adjudicator had power to award costs under the Scheme Rules. The judge in the *Northern Developments* case disagreed with this reasoning, as he considered that the Scheme gave the adjudicator power to deal with case management matters only. If Parliament had intended to give power to award costs, it would have said so. However, the judge considered that the parties could agree to give the adjudicator the power to award costs. In the *Northern Developments* case the parties had in fact given the adjudicator the power to award costs, as both of them had asked for costs in the referral document and response.

14.24.4. The Local Democracy, Economic Development and Construction Act 2009, which applies to all contracts entered into from 1 October 2011, deals with this matter in Section 108A and provides for an agreement to be reached on the allocation of costs as between the parties. However, to be valid, this agreement must not be made before the notice to refer.

SUMMARY

An adjudicator has power to provide in his decision that one of the parties pays the other party's costs, if the power is granted to him by the conditions of the contract. Such power is provided for in the GC/Works/1 (1998) Conditions and the ORSA Adjudication Rules, which apply in Scotland. There is, however, no such provision in the Scheme for Construction Contracts. In the absence of express powers in the contract, an adjudicator has no power to provide in his decision that the losing party is to pay the winning party's costs. However, both parties may agree to vest in the adjudicator such powers. To be effective, however, the agreement must not pre-date the notice to refer a dispute to adjudication, as provided for in the Local Democracy, Economic Development and Construction Act 2009. The provisions of GC/Works/1 and the ORSA Adjudication Rules, in the light of the requirements of the Act, will cease to be effective in respect of all contracts entered into after the Act comes into force on 1 October 2011.

14.25. Where a contractor includes in its subcontract terms a clause which states that if a dispute is referred to adjudication the subcontractor, even if successful, will be liable to pay the contractor's costs, is such a clause enforceable?

14.25.1. It would seem very unfair for a main contractor to include a clause in its subcontract conditions to the effect that, if a dispute between the parties is ever referred to adjudication, the subcontractor will always be obliged to pay the costs of employing the adjudicator and the main contractor's costs in addition to its own. The provision is expressed as applying even if the adjudicator's decision goes in favour of the subcontractor. Would a court enforce such a clause?

14.25.2. In the case of *Bridgeway Construction* v. *Tolent Construction* (2000), the contractor included in the subcontract conditions a clause which required the subcontractor to pay his own and the main contractor's costs, even if the subcontractor wins the case. The wording of the clause states:

> the party serving the Notice to Adjudicate shall bear all the costs and expenses incurred by both parties in relation to the adjudication, including but not limited to, all legal and expert fees.

It was argued on behalf of the subcontractor that the clause, by its very nature, inhibited parties from pursuing their remedies under the Act and was therefore void. This being the case, the adjudicator had exceeded his jurisdiction by applying the clause. The court held that the clause was valid. The mere fact that, in this particular case, the subcontractor was disgruntled did not entitle it to successfully argue that the clause was unfair and should not apply.

14.25.3. The decision in the *Tolent* case was considered by His Honour Mr Justice Edwards Stuart, when arriving at a decision in the case of *Yuanda (Co) (UK)* v. *WW Gear Construction* (2010), to have been wrongly decided. The case arose out of a dispute relating to the contract for the provision of curtain walling at the Plaza Hotel on Westminster Bridge. The contract conditions included a *Tolent*-type clause, which required the contractor to be 'fully responsible for meeting and paying both his own and the Employer's legal and professional costs in relation to Adjudication'. The judge considered that this clause was in conflict with section 108 of the Construction Act, which makes the adjudicator's decision binding on the parties. His view was that a situation could arise whereby the amount of the Employer's costs were so great that they were more than the adjudicator's decision, leaving the contractor to make a payment to the employer instead of the employer paying a sum to the contractor. This meant that the contractor was deprived of the remedy provided by Section 108 up to the employer's costs.

14.25.4. Subsequent to the decision in *Tolent*, a similar matter was referred to the Outer House of the Court of Sessions in Scotland, in the case of *Profile Projects Ltd* v. *Elmwood Glasgow Ltd* (2011). In this case the subcontract between the parties included the following provision:

> the referring party shall bear the whole costs of the adjudication in their entirety and both parties' legal expenses (on a solicitor and client basis and upon the scale of charges applicable to Court of Session business) in and incidental to the adjudication.

Lord Menzies agreed with the decision in *Tolent*. He considered that the courts should uphold the parties' right to freedom of contract and that such a clause was not in conflict with the provisions of the Act.

14.25.5. *Tolent*-type clauses are, in any event, ineffective, as a result of the provisions of the Local Democracy, Economic Development and Construction Act 2009, where in section 108A it states:

> (1) This section applies in relation to any contractual provision made between the parties to a construction contract which concerns the allocation as between those parties of costs relating to the adjudication of a dispute arising under the contract.
> (2) The contractual provision referred to in (1) is ineffective unless it is made in writing after the giving of notice or intention to refer the dispute to adjudication.

It has been suggested by some legal commentators that the wording in the statute is insufficiently robust to render a *Tolent* clause ineffective. Time alone will tell, following the Act coming into force on 1 October 2011.

SUMMARY

A clause which appeared in a subcontract to the effect that in the event of an adjudication between the parties, irrespective of who wins, the subcontractor would pay the main contractor's legal costs, was recognised as valid in the case of *Bridgeway Construction* v. *Tolent Construction* (2000). This decision was said by the judge to be unsound in the case of *Yuana (UK) Co Ltd* v. *WW Gear Construction Ltd* (2010), as he considered such a clause to be in conflict with Section 108 of the Construction Act. The courts in Scotland took a different view in the case of *Profile Projects Ltd* v. *Elmwood (Glagow) Ltd* (2011), where it was held that a *Tolent*-type clause was valid.

This type of clause has in any event been rendered ineffective by the provisions of the Local Democracy, Economic Development and Construction Act 2009, which applies to all contracts entered into from 1 October 2011. There is still uncertainty, however, as expressed by some legal commentators, as to whether the provisions of the Act will affect the enforceability of a *Tolent*-type clause.

14.26. It has been argued that adjudication in accordance with the Construction Act 1996, because of the short time scale involved, should not be used in complex cases, as it is likely to result in a breach of natural justice; is this correct?

14.26.1. Disputes arising out of construction contracts often involve extensive detail, because of the complexity of the matters in dispute. Delays to progress and completion, involving claims for extensions of time and liquidated damages, where many delaying factors have to be considered, employing a very detailed delay analysis, are commonplace. Argument concerning a contractor's entitlement to payment can involve enquiry into hundreds and sometimes thousands of variations and changes. Is it possible for an adjudicator to consider all these matters and arrive at a decision within 28 days? Will the responding party have adequate time to reply to the great amount of detail provided in the referral?

It has been argued in a few cases that, because of the complexity of some disputes, there is a danger of a breach of natural justice.

14.26.2. Due to the short timescales involved in the adjudication process, the complexity of cases and the amount of detail to be considered, it has also been suggested that the adjudicator may not have fully understood the case or had time to read all the documentation submitted to him.

14.26.3. The case of *CIB Properties Ltd* v. *Birse Construction Ltd* (2004) dealt with a second adjudication relating to quantum; the first adjudication had decided that CIB were entitled to determine Birse Construction's contract. A claim for £16.6m was levied by CIB against Birse Construction and the referral included 50 lever arch files which, during the adjudication process, extended to 150. The adjudicator's timescale for arriving at a decision was extended by the parties on a number of occasions. The decision was reached by the adjudicator on 24 February 2004, when CIB Properties was awarded £2,164,892. One of the grounds for challenging the adjudicator's decision was that the size and complexity of the dispute made it impossible for it to be resolved fairly within the timescale.

14.26.4. Judge Tomlin said the complexity of the case was not the test to be applied, but rather whether the adjudicator was able to reach a fair decision within the timescale allowed by the parties. The judge decided that the adjudicator was able to reach a fair decision, despite the complexity of the case and the amount of detail involved. In the judge's opinion, the adjudicator was able to deal fairly with all matters raised by the parties, probably because of the additional time which they allowed. It was noted by the judge that the parties were not obliged to allow more time. This left open what the situation would be if the responding party refused to agree to the adjudicator being allowed more time and subsequently contested the adjudicator's decision on the grounds that there was insufficient time available for the adjudicator to make a fair decision.

14.26.5. In the case of *Carillion Construction Ltd* v. *Devonport Royal Dockyard* (2005) it was considered that there would be difficulties in successfully preventing the enforcement of an adjudicator's decision by simply arguing that the matters in dispute were too complex for adjudication.

14.26.6. The adjudicator, in the case of *The Dorchester Hotel Ltd* v. *Vivid Interiors Ltd* (2009), was faced with a referral which ran to 92 pages, accompanied by 37 lever arch files. There were six witness statements and two experts' reports; the net amount claimed totalled £1.6m and there was also a 16 weeks' extension of time submission to be considered. The adjudicator indicated that he would not be prepared to take up the appointment unless Vivid agreed to an extension of time to take account of the Christmas holiday period. There may have been some suspicion of the motives of Vivid in commencing an adjudication during the run-up to Christmas. However, Vivid offered a longer period of time, up to 28 January 2009, by which time the decision should be made. Dorchester Hotel sought a declaration from the court, part way through the adjudication, that there was a serious risk of a breach of natural justice. It was argued that there might not be a fair and reasonable opportunity for them to review the documentation and submit a response, together with factual and expert evidence. The decision of the adjudicator, it was claimed, would therefore be unenforceable because of a breach of natural justice.

14.26.7. Judge Coulson considered that it was not always easy to reconcile the rules of natural justice with the necessity for speed in an adjudication. He considered that it was for the adjudicator to decide whether he could reach a fair decision within the time available and that this was the most important factor the court would take into account when deciding matters related to a possible breach of natural justice. He considered that it would only be in the most obvious cases that the court would intervene. The adjudicator had stated that he would be able to deal with the matter by 28 January, which seemed to satisfy the judge. In addition, the claimant was not left without a remedy, as the adjudicator's decision could be challenged through the courts.

14.26.8. Challenges to the adjudicator's decision on the grounds of complexity and the likely breach of natural justice are becoming a regular method of attempting to avoid paying the sum decided by the adjudicator. In the case of *HS Works* v. *Enterprise Managed Services* (2009), an adjudicator was required to decide a dispute relating to the amount of a final account. The subcontractor claimed £30,963,763, whereas the main contractor considered that £24,707,783 was the correct amount. Enterprise had a term contract on behalf of Thames Water Infrastructures, with the many carriageway repairs being subcontracted to HS Works. Remarkably, there were 51,000 job orders which had to be considered when arriving at a final sum. The adjudicator decided that the correct value of the final account was £23,253,931. Rather unusually, it was the main contractor who served the adjudication notice, the reason being that in an earlier adjudication the subcontractor had been awarded the sum of £1.8m in respect of a dispute relating to a deduction from earlier payment in respect of *contra* charges. No doubt the main contractor was hoping to recover this amount by way of an adjudication concerning the amount due on the final account. The court was asked to decide whether both of these adjudicator's decisions should be enforced, or whether one could be offset from the other and only the net amount paid.

14.26.9. One issue raised by HS Works related to a breach of natural justice. It was argued by HS Works that they would be seriously and unfairly disadvantaged in having to answer the submission made by Enterprise in a few days. Mr Justice Aikenhead looked at how the adjudicator arrived at his decision. He also took into account the dialogue which occurred between the parties before the adjudication commenced. The adjudicator's method comprised spot-checking items in the final account and weighing up which side was the more credible. The adjudicator followed this process and decided which of the two parties' evidence he preferred. It was held by the judge that the information which was available to HS Works before the adjudication commenced was adequate. The judge, in arriving at a decision, considered that there had been no breach of natural justice. He also decided that both the adjudicator's decisions should be enforced and did not allow setting one decision off against the other.

SUMMARY

It is not uncommon for an adjudication to involve a major dispute regarding variations, extensions of time, loss and expense and liquidated damages, referred to by Judge Coulson as a kitchen-sink final account adjudication. He warned of the unsuitability of adjudication for dealing with such a final account. Parties to adjudication have therefore, on a few occasions, tried to resist enforcement of an adjudicator's decision on the grounds that because of the complexity, there is a danger of a breach of natural justice. So far, parties resisting enforcement on the grounds of complexity have been unsuccessful, and judges in arriving at their decisions have made the following observations:

- It is not always easy to reconcile the rules of natural justice with the necessity for speed in an adjudication.
- The complexity of the case is not the test to be applied, but rather whether the adjudicator is able to reach a fair decision within the timescale allowed by the parties.
- It is for the adjudicator to decide whether he can reach a fair decision within the time available and this is the most important factor for the court to take into account when deciding matters related to a possible breach of natural justice.
- The procedure adopted by the adjudicator for reaching a decision will be taken into consideration when deciding if there has been a breach of natural justice.
- The dialogue which occurred between the parties, before the adjudication commenced, is relevant when deciding whether there has been a breach of natural justice.

14.27. Where one party claims that the adjudicator has no jurisdiction, but, whilst maintaining this position, continues to take part in the proceedings, can that party avoid paying the adjudicator's fees on the grounds that he claimed the adjudicator had no jurisdiction?

14.27.1. Disputes with regard to the payment of adjudicators' fees are unusual, whereas challenges to his jurisdiction are commonplace. The case of *Christopher Michael Linnet* v. *Halliwells LLP* (2009) is a rare example of a case which involves both a challenge to the jurisdiction of the adjudicator and a dispute with regard to his fees.

14.27.2. ISG entered into a contract with Halliwells to fit out offices in Manchester, using a JCT 1998 form of contract. A dispute arose; a notice of adjudication was served by ISG on Halliwells on 22 May 2008, and Mr Linnet was nominated as the adjudicator by the RICS on 28 May 2008. His terms and conditions were agreed by ISG, but not by Halliwells. The Referral Notice was sent by fax to both the adjudicator and Halliwells on 29 May 2008. The attached files were sent by post and, whilst they were received by Halliwells on 30 May 2008, those intended for the adjudicator went astray in the post and the adjudicator did not receive his copy until 4 June 2008. The JCT Adjudication Rules require the Referral Notice to be served not later than 7 days after the adjudication notice. Halliwells argued that, as the supporting files were part of the Referral Notice and served on the adjudicator outside the seven days, they were out of time. This being the case, the adjudicator did not have jurisdiction to continue and should withdraw. In addition, Halliwells argued that the contract was not wholly in writing and that therefore the dispute could not be referred to adjudication.

14.27.3. The adjudicator took the view that the late receipt of supporting files had not in any way inhibited him in the exercise of his duties. Halliwells were not affected, as they received the files on time. The adjudicator argued that, as he intended to commence the adjudication period on 4 June 2008, when he received the supporting files, the net result was to give Halliwells an additional five days in which to prepare their response. Further, as Halliwells had not adduced any evidence that the contract was not in writing, this was no longer an issue.

14.27.4. The adjudicator found in favour of ISG and, in his decision, required Halliwells to pay his fees, which amounted to £2,436. Halliwells refused to pay, on the grounds that they had challenged the adjudicator's jurisdiction and had never agreed to pay his fees. The matter was referred to the TCC, which held that an adjudicator's entitlement to the payment of his fees was a contractual matter. If a party had not specifically agreed to pay the adjudicator's fees, there would be an implied obligation to pay his reasonable fee. If a party raises a jurisdictional challenge, but then continues to argue the merits of the case without prejudice, it was usually receiving a service from the adjudicator for which it would be obliged to pay a reasonable fee. The fee was payable because there was a contract; alternatively, payment could be justified on the ground of undue enrichment. With regard to the issue relating to the contract not being in writing, the court held that the adjudication was governed by the JCT Adjudication Rules and not the Construction Act 1996 and these rules did not required the contract to be in writing.

14.27.5. The court agreed with the adjudicator that the late delivery of the supporting files did not affect his jurisdiction.

14.27.6. The court held that, if a party had raised a jurisdictional challenge, it was left with two options. Having made the challenge, it can withdraw from the proceedings. Alternatively, having made the challenge, it may decide to participate in the adjudication on a 'without prejudice' basis. However, by continuing to take part in the adjudication, that party is nevertheless requesting the adjudicator to undertake some work and make a decision, albeit one which he has no jurisdiction to make. The party then comes under a contractual obligation to pay the adjudicator any fee which has been agreed, or, if there is no agreement, a reasonable fee. In the event, Halliwells had not challenged the quantum of the adjudicator's fee and they were obliged to make payment of the sum as claimed.

SUMMARY

If one of the parties challenges an adjudicator's jurisdiction, he has to make a choice. He can withdraw from the proceedings, in which case there will be no obligation to pay any fees. Alternatively, having raised the challenge, he can continue to take part in the adjudication, in which case there will be a contractual obligation to pay the fees.

14.28. What happens if the adjudication provisions as set out in the contract are at variance with the provisions of the Housing Grants, Regeneration and Construction Act 1996?

14.28.1. The Housing Grants, Construction and Regeneration Act 1996 clearly sets out, under Section 108, requirements which relate to the adjudication process. Section 114 indicates that the minister shall by regulation make a scheme which contains provisions about the matters referred to in the Act which relate to adjudication, referred to as the Scheme for Construction Contracts.

14.28.2. The parties to a construction contract are free to agree rules which will apply to any adjudication relating to a dispute which may arise between them. However, if the adjudication rules which the parties agree are at variance with any of the requirements of the Act, Section 108(5) indicates that the provisions of the Scheme for Construction Contracts will apply.

14.28.3. Drafters of standard forms of contract and other learned bodies lost no time in drafting adjudication rules intended to apply where their standard conditions of contract are employed. Examples of this are the JCT, GC/Works/1, CIC and ICE. Unfortunately, the adjudication rules drafted by some of these organisations do not comply with the Act. Reference to the courts has been made in respect of divergences between some of these standard conditions and the provisions of the Act.

14.28.4. The Act stipulates, in Section 108(2)(c) that the adjudicator must reach a decision within 28 days of the referral or such longer period as is agreed by the parties after the dispute has been referred. Subsection 108(2)(d) allows the adjudicator to extend the 28-day period by a further 14 days with the consent of the referring party. Paragraph 25 of the CIC standard conditions states, despite this requirement, that the decision of an adjudicator will be valid, provided he does so before the dispute is referred to a replacement adjudicator. This provision is obviously at variance with the requirements of the Act as it could give the adjudicator licence to delay the decision for a significant period.

14.28.5. In the case of *Epping Electrical Company Ltd* v. *Briggs and Forrester (Plumbing Services) Ltd* (2007), a dispute arose between the parties where the contract provided for the CIC adjudication rules to apply. Epping served its referral notice on 4 October 2006 and a decision to comply with the 28-day period provided by the Act was due by 1 November 2006. Epping agreed to a 14-day extension, to take the date for a decision up to 14 November 2006, following which the parties agreed to a further 7 days' extension, which gave a date required for the decision of 21 November 2006.

14.28.6. When 21 November 2006 arrived, the adjudicator refused to issue a decision until his fees had been paid, but he subsequently relented and sent the decision on 23 November 2006. Once the adjudicator reaches a decision it must be sent forthwith, as explained in *Cubitt Building and Interiors Ltd* v. *Fleetglade* (2006). This did not happen, which would have rendered the decision unenforceable if the Scheme for Construction Contracts had applied.

14.28.7. The court then had to decide the effect of paragraph 25 of the CIC adjudication rules, as to whether the decision would still be effective if made before the referral of the dispute to a replacement adjudicator. Judge Harvey, in deciding that paragraph 25 was at variance with the provisions of the Act, stated:

> I now come to consider the effect of paragraph 25 of the CIC procedure. On the face, that unambiguously renders the adjudicator's decision unenforceable . . .

It is clear that the CIC provisions do not comply with the requirements of the Act, and therefore the Scheme for Construction Contracts applied.

14.28.8. Judge Harvey had to decide the effect of the application of the adjudication rules included in the GC/Works/1 contract, in the case of *Aveat Heating Ltd* v. *Jerram Falkus*

Construction Ltd (2007). A dispute arose out of a subcontract for plumbing and heating work, which was referred to adjudication. The adjudicator decided that a sum of £385,000 should be paid by Jerram Falkus to Aveat Heating. Jerram Falkus refused to make payment for a number of reasons, one of which was that the adjudicator's decision was out of time. The GC/Works/1 adjudication rules include the following relating to the timing of the adjudicator's decision:

> The adjudicator may extend the period of 28 days by up to 14 days with the consent of the party by whom the dispute was referred. The adjudicator's decision shall nevertheless be valid if issued after the time allowed . . .

The judge held that, as this provision was not compliant with the Act, the Scheme for Construction Contracts applied. It was argued that the offending clause in the contract would be substituted by a compliant clause in the Scheme for Construction Contracts. This was rejected by the judge, who explained that adjudication rules contained in a contract cannot co-exist with the Scheme for Construction Contracts. The effect of the GC/Works/1 conditions being non-compliant was to make them void, to be substituted by the Scheme for Construction Contracts.

SUMMARY

The Housing Grants, Construction and Regeneration Act 1996 includes basic requirements relating to adjudication. Detailed provisions for the working of adjudication have been produced by the minister and incorporated into the Scheme for Construction Contracts. The Scheme is not mandatory and, therefore, most of the organisations responsible for publishing standard conditions of contract and other learned bodies have produced their own adjudication rules. However, the Act states that if the contract does not comply with the Act, then the Scheme for Construction Contracts will apply. The adjudication rules which have been drafted for the CIC and GC/Works/1 contracts provide for the timescales within which the adjudicator must reach a decision which can be extended without the agreement of the parties. It has been held in the case of both of these sets of standard conditions that this does not comply with the Act and therefore the Scheme for Construction Contracts applies. Where this has occurred, it has been held that the adjudication rules in the contract are void in their entirety and should be replaced with the Scheme for Construction Contracts.

14.29. Is it possible for a Referring Party to restrict or exclude part of the defence and are the parties entitled to introduce evidence which was not disclosed before the adjudication commenced?

14.29.1. There is a generally held view that the manner in which the adjudication notice is worded will set the boundaries of the dispute. This being the case, the responding party will be restricted in its response to the subject matter included in the adjudication notice. Even more restricting was the view, based upon early case law, that the facts and matters brought to the attention of the adjudicator were limited to those exchanged between the parties prior to the commencement of adjudication proceedings.

14.29.2. These views need to reviewed in the light of more recent decisions, which have resulted in a move away from these restricted ideas. In the case of *Quartzelec Ltd* v. *Honeywell Control Systems Ltd* (2008), the referring party submitted a claim for additional costs and loss and expense in relation to extensions of time in the sum of £465,280. In the response, Honeywell claimed a sum of £36,578 in respect of work omitted. Quartzelec objected to this item, as it had not been raised by Honeywell prior to the service of the notice of adjudication. The adjudicator agreed with Quartzelec and did not take it into account when deciding that Honeywell should make a payment of £134,708. The court refused to enforce the adjudicator's decision, on the grounds that the omission had not been taken into account. Judge Stephen Davies, in his judgment, stated:

> Where the dispute referred to adjudication by a claimant is one which involves a claim to be paid money, it is difficult to see why a respondent should not be entitled to raise any defence open to him to defend himself against the claim, regardless of whether or not it was raised as a discrete ground of defence in the run-up to the adjudication.

14.29.3. In the case of *Cantillon Ltd* v. *Urvasco Ltd* (2008), Mr Justice Aikenhead was of the same opinion when he stated:

> It is, I believe, accepted by both parties, correctly in my view, that whatever dispute is referred to the adjudicator, it includes and allows for any ground open to the responding party which would amount in law or in fact to a defence of the claim with which it is dealing.

14.29.4. It may be concluded from these cases that the following principles have been established:

- An overly legalistic analysis of what was in dispute should be avoided;
- The disputed claim should be determined in broad terms;
- It is necessary to look at the essential claim and the fact that it has been challenged, rather than the precise grounds of rejection; and
- The disputed claim, or assertion, is not necessarily limited by the evidence or arguments submitted by the parties to each other before the notice of adjudication is served.

14.29.5. The difficulty in allowing new items to be introduced into the response is that the referring party will need adequate time to respond, which may in turn involve further new items arriving. Will this alter the original dispute? Whilst the court has in recent cases taken a liberal view of what may be introduced into the response, it is a subject to which the courts are likely to return from time to time.

SUMMARY

It is the intention that the adjudication notice fixes the boundaries of the dispute. Having said that, case law has decided that it is open to the responding party to raise any arguments by way of defence in response to the claim, when it is referred to adjudication. This applies even if the defence had never been previously raised. Further, the parties are not limited to the arguments, contentions and evidence put forward before the dispute crystallised.

14.30. Can a losing party refuse to comply with an adjudicator's decision on the grounds that the Referral Notice was served late?

14.30.1. With regard to the timing of the service of the referral notice, section 108 of the Housing Grants, Construction and Regeneration Act 2006 states:

> The contract shall:
>
> - enable a party to give notice at any time of his intention to refer a dispute to adjudication;
> - provide a timetable with the objective of securing the appointment of an adjudicator and referral of the dispute to him within seven days of such notice.

The Scheme for Construction Contracts, which expands the statutory provisions, states with regard to timescale:

> Where an adjudicator has been selected in accordance with paras 2, 5 or 6, the referring party shall, not later than seven days from the date of the notice of adjudication, refer the dispute in writing (the Referral Notice) to the adjudicator.

14.30.2. In the case of *Hart Investments* v. *Fidler and Another* (2006), there were a number of disputes between the parties which were referred to adjudication, where the adjudicator found in favour of Larchpark, the joint defendant, in the sum of £145,192. Hart resisted enforcement of the adjudicator's decision on the grounds that the referral notice was served outside the timescale provided by section 108 of the Act.

14.30.3. The judge was persuaded by the wording in the Scheme for Construction Contracts where it uses the term 'not later than seven days'. He considered that, if the seven days was allowed to be qualified in some way, how is the qualification to be formulated, let alone assessed? Would two or more days be allowed, or perhaps a period which seems just and equitable? The judge went for the easy solution by stating that 'not later than seven days' meant what it said and, as the referral was late, the adjudicator lacked jurisdiction. The responding party could have waived the requirement, but declined to do so.

14.30.4. A different result occurred in the case of *Cubitt Building and Interiors Ltd* v. *Fleetglade Ltd* (2006), where again the court had to deal with the late service of a referral notice. This case involved a project where Cubitt carried out superstructure work at Fleetglade's site in Hampton Wick. A dispute arose in connection with the amount due in respect of the final account. A notice to refer the dispute to adjudication was served by Cubitt on Fleetglade on 20 September 2006. Cubitt applied to the RICS on

21 September 2006 for the nomination of an adjudicator. The nomination wasn't made until 27 September 2006. Later that day, the adjudicator accepted the appointment. Following the adjudicator's appointment, but still on 27 September, solicitors acting for Cubitt offered Fleetglade's solicitors the referral notice without the supporting documentation, which was declined. It was returned on 28 September 2006 accompanied by 12 lever arch files of supporting documents. It was argued that, as the time span between the adjudication notice and the service of the referral was eight days, it was out of time.

14.30.5. The contract incorporated a JCT Standard Form of Contract 1988 edition, with bespoke amendments. Clause 41A sets out the adjudication rules, which state:

> If an adjudicator is agreed or appointed within 7 days of the notice, then the party giving the notice shall refer the dispute or difference to the Adjudicator (the referral) within 7 days of the notice. If the adjudicator is not agreed, or appointed within 7 days of the notice, the referral shall be made immediately on such agreement or appointment.

14.30.6. With regard to the service of the adjudication notice, the judge considered it had been properly served on 20 September 2006 and, with regard to the timing of the referral notice in relation to the wording of clause 41A, he said:

> Firstly clause 41A has to be operated in a sensible and commercial way. It endeavours to cover the two alternative scenarios that will arise, namely the appointment of an adjudicator within the seven days and the appointment of the adjudicator beyond the seven days. But it does not – it cannot – expressly provide for everything that might happen. It does not therefore expressly provide for what should happen if (as occurred here) through no fault of the referring party, the appointment does not occur until very late on the seventh day. Plainly a sensible interpretation of clause 41A is that, if it happens late on Day 7, the referral notice must be served as soon as possible thereafter; and if that means that it is served on Day 8, then service on Day 8 would be in accordance with clause 41A.

As Cubitt had served the referral notice on Day 8, it was not out of time.

SUMMARY

Where the Scheme for Construction Contracts applies, the appointment of the adjudicator must be within 7 days of the date of the adjudication notice. In the event of the referral notice being served late, the adjudicator will have no jurisdiction. Where a JCT contract applies, the wording regarding the service of the referral notice differs from the wording in the Scheme for Construction Contracts. The JCT adjudication rules provides for the appointment of the adjudicator within 7 days of the adjudication notice. If the adjudicator is appointed later than the 7 days, then the referral notice may be served later. The judge in the *Cubitt* case indicated that the provisions of the JCT adjudication rules must be operated in a sensible manner, which may involve the referral being served on Day 8.

14.31. Section 105(1) of the Housing Grants, Construction and Regeneration Act 1996 defines what are construction operations and provided for by the Act, whereas Section 105(2) defines what operations are not construction operations and excluded from the Act. Does the Act apply if on a project some of the operations are covered by the Act, whilst others are not?

14.31.1. Section 105(1) of the Act defines what is meant by a construction operation, for the purposes of the Act, and includes construction, alteration, repair, maintenance, extension, demolition or dismantling of buildings, or structures forming, or to form, part, of the land (whether permanent or not). Site clearance, earth moving, excavation, tunnelling and boring, laying of foundations, erection, maintenance or dismantling of scaffolding, site restoration, landscaping and the provision of roadways and other access works are all examples of operations governed by the Act.

14.31.2. Section 105(2) defines operations which are not construction operations and therefore not governed by the Act, which include drilling for the extraction of oil or natural gas; assembly installation or demolition of plant or machinery, or erection or demolition of steelwork for the purposes of supporting or providing access to plant or machinery on a site where the primary activity is nuclear processing, power generation or water or effluent treatment.

14.31.3. There have been several examples of projects which comprise operations defined as construction operations, covered by the Act; and construction operations which are excluded from the Act. This has resulted in the main contractor being engaged on an operation which is an excluded one and not governed by the Act, whilst one of its subcontractors is involved in an operation which is classed as a construction operation and subject to the provisions of the Act.

14.31.4. In the case of *Palmers Ltd* v. *ABB Power Construction Ltd* (1999), ABB argued that the work undertaken by Palmers was excluded, as the prime activity of the project was the generation of power. Palmers, who were subcontractors, on the other hand, argued that as the operations they undertook, being scaffolding to support the structural frame within which ABB was installing boilers and temporary access, were defined operations and governed by the Act. His Honour Judge Thornton agreed with Palmers that, whilst the assembly of the steelwork in connection with the installation of the boilers was excluded from the Act, no such exclusion applied to the scaffolding. This was in later cases referred to as the 'narrow view'.

14.31.5. A different approach was taken by Judge Lloyd in the case of *ABB* v. *Norwest Holst* (2000). ABB was the main contractor for the construction of a power station, which is an excluded operation. Norwest Holst was a subcontractor for the fabrication and installation of insulation to the boiler ducting, silencer pipes, drums and tanking. ABB argued that as the primary work on the site was power generation, all operations related to the project were excluded.

14.31.6. Judge Lloyd, in finding in favour of ABB, considered that on the one project operations should be dealt with consistently and either all be covered by the Act, or all excluded. His words were:

> It is in my judgment clear from the language used in Section 105(2) that it was intended that, if the regimes were not to apply, it would be invidious if they applied to some but not all construction contracts on site or for a project. Defining the exempt construction operations by reference to the nature of the project or by reference to a site should minimise the possibility that, for example one contractor or sub-contractor would think that it was better or worse off than another working alongside it, or preceding or following it.

This was, in later cases, referred to as the 'broad view'.

14.31.7. In the case of *Homer Burgess Ltd* v. *Chirex (Annan) Ltd* (2000), the court had to decide whether pipework connected to machinery is a construction operation or one which is exempt. 'Assembly, installation or demolition of plant and machinery' are excluded operations. The court held that, without the pipework, the individual piece of machinery would be unable to operate and hence, like the machinery, the pipework was exempt from the operation of the Act. This decision would appear to support the broad view.

14.31.8. In arriving at a decision the court may conclude that the main function of the project is the deciding factor. In the case of *ABB Zantingh Ltd* v. *Zedal Building Services Ltd* (2001) the main purpose of the project was the construction of a printing facility. A dispute arose in respect of field wiring to the electricity generator, which provided the necessary power. The court held that, as the main purpose of the project was printing, which is not an exempt operation, the Act applied to the field wiring to the power generators. This case again appears to support the broad view.

14.31.9. The narrow view was preferred by Judge Ramsey, in *North Midlands Construction PLC* v. *AE&E Lentjes UK Ltd* (2009). AE&E were engaged by Scottish and Southern Energy Ltd to install flue gas desulphurisation units at two coal-fired power stations. The enabling and civil engineering works were sublet to North Midlands Construction, who got into dispute with AE&E and wished to have the dispute referred to adjudication. The matter was referred to court for a declaration as to whether the enabling and civil engineering works were construction operations or exempt. Judge Ramsey, in applying the narrow view and finding in favour of North Midland Construction, said:

> As I have observed, if the intention had been to exclude all construction operations on a site where the primary activity was power generation, then that could easily have been done, or if it had been intended to exclude all preparatory activities, then a sub-section similar to s.105(1)(e) could have been added.

14.31.10. Judge Ramsey was again called upon to decide an issue relating to the application of the Act in the case of *Cleveland Bridge (UK) Ltd* v. *Whessoe-Volker Stevin Joint Venture* (2010). The facts were a little different from those in earlier cases. The project involved work at a liquid natural gas terminal. It was the judge's opinion that the production of fabrication drawings, off-site fabrication and delivery to site of the fabricated steelwork was a construction operation, but the erection of the steelwork for the piperacks and pipebridges were excluded operations. The adjudicator had found in favour of the referring party, on the basis that all the work was subject to the Act. However, as it was not possible to sever the part of the decision which was governed by the Act from the excluded operation, the adjudicator's decision was held to be unenforceable.

SUMMARY

The drafting of the Act, which provides for some operations, such as power generation, to be excluded from the Act, but leaving unanswered what the position is regarding operations on the project which are defined as construction operations: for example, scaffolding, has presented some difficulties for the courts. Judge Thornton has taken what has been referred to as the 'narrow view', in allowing the Act to apply to scaffolding, which is a construction operation but forms part of a power generation project, which is an exempt operation. A 'broad view' has been taken by Judge Lloyd, who was strident in explaining that it would be invidious to allow adjudication as a dispute resolution process for a part of a project, but excluding it from other parts. Judge Ramsey, in a case heard some ten years after the decisions of judges Thornton and Lloyd, adopted the narrow view.

The situation is still unsettled, but the most recent case supports the 'narrow view'. Adopting the narrow view, it is possible for the Act to apply with regard to construction operations as defined by the Act but where the main purpose of the project is an exempt operation, where the Act does not apply.

Table of Cases

ABB Engineering and Construction Pty. Ltd. *v.* Abigroup Contractors Ltd., [2003] NSWSC 665; [2003] 34 BLISS 7 72
Abigroup Contractors Pty. Ltd. *v.* Peninsula Balmain Pty. Ltd., [2001] NSWSC 752; [2002] 18 BCL 15 119
A.C. Controls *v.* British Broadcasting Corporation (BBC), [2003] 89 ConLR 52 72, 74
ACT Construction Ltd. *v.* E. Clarke & Son (Coaches) Ltd., [2002] 28 BLISS 4; [2003] 85 ConLR 1 179
A.L. Barnes Ltd. *v.* Time Talk (UK) Ltd., [2003] BLR 331 179
Aldi Stores Ltd. *v.* Galliford, [2000] 11 BLISS 7 257
Alexander Corfield *v.* David Grant, (1992) 59 BLR 102; [1992] 29 ConLR 58. . . 357
Alfred McAlpine Homes North Ltd. *v.* Property and Land Contractors Ltd., (1995) 76 BLR 59; [1996] 47 ConLR 74 207
Alghussein Establishment *v.* Eton College, [1991] 1 All ER 267; [1988] 1 WLR 587 123
Aluminium Industrie Vaasen BV *v.* Romalpa Aluminium Ltd., [1976] 1 WLR 676 346
Amalgamated Building Contractors Ltd *v.* Waltham Holy Cross Urban District Council, [1952] 2 All ER 452 127
Amec Building Ltd. *v.* Cadmus Investments Co. Ltd., [1996] 51 ConLR 105 . . . . .104, 171, 271
Amec Process and Energy Ltd. *v.* Stork Engineers & Contractors BV, [1999] 21 BLISS 4; [2000] 68 ConLR 17 194
Amec Process and Energy Ltd. *v.* Stork Engineers & Contractors BV (No. 4), [2002] 17 BLISS 4 218
Amiri Flight Authority *v.* BAe Systems plc and Crossair Limited Company for Regional European Air Transport, [2003] 2 Lloyd's Reports 767 355
Aoki Corporation *v.* Lippol and (Singapore) Pte Ltd., [1994] SLR 609. 145
Archivent Sales & Developments Ltd *v.* Strathclyde Regional Council, (1985) 27 BLR 98. 348
Ascon Contracting *v.* Alfred McAlpine, [1999] 43 BLISS 5. 86
Attorney General for the Falklands *v.* Gordon Forbes Construction (Falklands) Ltd. (No. 2), [2003] 14 BLISS 8; [2003] BLR 280 188
Auriema Ltd *v.* Haigh and Ringrose Ltd., (1988) 4 Const. LJ 200 259

200 Contractual Problems and their Solutions, Third Edition. Roger Knowles.

Babcock Energy Ltd. *v.* Lodge Sturtevant Ltd. (Formerly Peabody Sturtevant Ltd.), [1994] 41 ConLR 45 212
Bacal Construction (Midlands) Ltd. *v.* The Northampton Development Corporation, (1975) 8 BLR 88. 36
Bacardi-Martini Beverages Ltd. *v.* Thomas Hardy Packaging, [2002] 2 Lloyd's Reports 379; [2002] 2 All ER (Comm) 335 355
BAL *v.* Taylor Woodrow Construction Ltd., [2004] 7 BLISS 1 . . . . 410, 411
Balfour Beatty Building Ltd. *v.* Chestermount Properties Ltd., (1993) 62 BLR 1; [1993] 32 ConLR 139 112
Balfour Beatty Civil Engineering Ltd. *v.* Docklands Light Railway Ltd., (1996) 78 BLR 42; [1996] 49 ConLR 1. 327
Balfour Beatty Construction Ltd. *v.* The Mayor and Burgesses of the London Borough of Lambeth, [2002] BLR 288; [2002] 84 ConLR 1 409
Banque Paribas *v.* Venaglass Ltd., 1 February 1994 unrep. 176
Barnard Pipeline Technology Ltd. *v.* Marston Construction Co. Ltd., 12 June 1992 354
Barnes & Elliott Ltd. *v.* Taylor Woodrow Holdings and George Wimpey Southern Ltd., [2003] BLR 111 398
Beck Peppiatt Ltd. *v.* Norwest Holst, [2003] BLR 316 386
Beechwood Development Company (Scotland) Ltd. *v.* Stuart Mitchell (t/a Discovery Land Surveys), 2001 SLT 1214. 210
Belcher Food Products Ltd. *v.* Miller & Black and Others, 1999 SLT 142 336
Bell (A.) & Son (Paddington) Ltd *v.* CBF Residential Care and Housing Association, (1989) 46 BLR 102; [1989] 16 ConLR 62 146
Berhards Rugby Landscapes Ltd. *v.* Stockley Park Consortium, (1997) 82 BLR 39 . 105
BFI Group of Companies Ltd *v.* DCB Integration Systems Ltd., (1987) CILL 348 . 140
Big Island Contracting (HK) Ltd. *v.* Skink Ltd., (1990) 52 BLR 110 304
Biggin Co. *v.* Permanite Ltd., [1951] 1 KB 422; [1950] 2 All ER 859. 317
Blackpool and Fylde Aero Club Ltd. *v.* Blackpool Borough Council, [1990] 1 WLR 1195; [1990] 3 All ER 25. . . . . 10, 41, 42, 327
Blue Circle Industries plc *v.* Holland Dredging Co (UK) Ltd., (1987) 37 BLR 40 . 182
Blyth & Blyth Ltd. *v.* Carillion Construction Ltd., [2002] 79 ConLR 142; 2002 SLT 231 69
Bolam *v.* Friern Hospital Management Committee (1957), [1957] 2 All ER 118. . . . . 49, 229, 332
Bonnells Electrical Contractors *v.* London Underground Ltd., 20 October 1995 unrep. 170
Bouygues UK Ltd. *v.* Dahl-Jensen UK Ltd., [2000] BLR 522; [2001] 73 ConLR 135; [2001] 1 All ER (Comm) 1041 393
Bovis Construction (Scotland) Ltd. *v.* Whatlings Construction Ltd., (1995) 75 BLR 1; [1996] 45 ConLR 103 140
Bovis Lend Lease Ltd. *v.* RD Fire Protection Ltd., [2003] 89 ConLR 169 316
BP Chemicals Ltd. *v.* Kingdom Engineering (Fife) Ltd., (1994) 69 BLR 113; [1994] 2 Lloyd's Rep 373; [1994] 38 ConLR 14 271
Bracken *v.* Billinghurst, [2003] TCLR 4 259
Bradley (D.R.) (Cable Jointing) Ltd. *v.* Jefco Mechanical Services, 1998 unrep. . . 252

Bremer Handelsgesellschaft mBH *v.* Vanden Avenne-Izegem PVBA, [1978] 2 Lloyd's Reports 109 117
Brenner *v.* First Artists Management Pty. Ltd., (1993) VR 221 180
British Airways Pension Trustees Ltd. (formerly Airways Pension Fund Trustees Ltd.) *v.* Sir Robert McAlpine & Sons Ltd., (1994) 72 BLR 26; [1995] 45 ConLR 1 . . . 103
British Eagle International Air Lines Ltd. *v.* Compagnie Nationale Air France, [1975] 1 WLR 758. 284
British Steel Corporation *v.* Cleveland Bridge & Engineering Co. Ltd., (1984) 24 BLR 94; [1999] 1 All ER 504. . . . . .40, 72, 77, 78, 179
British Sugar plc *v.* NEI Power Plant Projects Ltd., (1997) 87 BLR 42 235
British Westinghouse Electrical Manufacturing Co. Ltd. *v.* Underground Electric Railways, [1912] AC 673 234
Britvic Soft Drinks Ltd. and Others *v.* Messer (UK) Ltd. and Others, [2002] 2 Lloyd's Reports 368 355
Brogden and Others *v.* The Directors of the Metropolitan Railway Company, [1877] 2 App Cas 666 31
Bush *v.* The Trustees of the Town and Harbour of Whitehaven, (1888) 52 JP 392 . 182

Cable & Wireless *v.* IBM United Kingdom Ltd., [2002] 2 All ER (Comm) 1041; [2003] BLR 89 326
Cambridge Construction Ltd. *v.* Nottingham Consultants QBD, 28 March 1996 unrep. 229
Carillion Construction *v.* Devonport Royal Dockyard, [2003] T.C.L.R. 2; [2003] BLR 79. . . . . 389, 391
Carillion Construction Ltd. *v.* Felix (UK) Ltd., [2001] BLR 1; [2001] 74 ConLR 144 278
Carr *v.* Berriman Pty Ltd., (1953) 89 CLR 327; (1952) 70 WN (NSW) 23 110
Carter (R.G.) Ltd. *v.* Edmund Nuttall Ltd., [2002] BLR 359 386
Chaplin *v.* Hicks, [1911] 2 KB 786 227
Charles Brand Ltd. *v.* Orkney Islands Council, 2001 SLT 698. 272
Charlotte Thirty Ltd. and Bison Ltd. *v.* Croker Ltd., [1990] 24 ConLR 46 354
Chester Grosvenor Hotel Co. Ltd. *v.* Alfred McAlpine Management Ltd., (1991) 56BLR 115 353
City Inn Ltd. *v.* Shepherd Construction Ltd., [2001, 2003, 2007, and 2010] 28 BLISS 8 93
City Polytechnic of Hong Kong *v.* Blue Cross (Asia Pacific) Insurance HCA10750/1999 47
Clarke *v.* Nationwide Building Society, [1998] EGCS 47 259
Clusky (t/a Damian Construction) *v.* Chamberlain. 1994 unrep. 255
Clydebank Engineering and Shipbuilding Co. Ltd. *v.* Don Jose Ramos Yzquierdo y Castaneda and Others, [1905] AC 6 139
Commissioner of Public Works *v.* Hills, [1966] AC 368 137
Comsite Projects Ltd. *v.* Andritz, [2003] 18 BLISS 1. 405
Co-Operative Insurance Society *v.* Henry Boot, [2002] 84 ConLR 164. 232
Copthorne Hotels *v.* Arup Associates and Others, [1996] 58 ConLR 138 . . . 17, 332

Costain Ltd. *v.* Strathclyde Builders Ltd., [2003] ScotCS 316; [2003] 49 BLISS 1 . 410
Costain Ltd. *v.* Westcol Steel Ltd., [2003] 22 BLISS 1 386
Cowlin Construction *v.* CFW Architects, [2002] BLR 241 386
Crosby (J.) & Sons Ltd. *v.* Portland Urban District Council, (1977) 5 BLR 121 . . 100
Croudace Construction Ltd. *v.* Cawoods Concrete Products Ltd., (1978) 8 BLR 20 235
Crown Estate Commissioners *v.* John Mowlem & Co. Ltd., (1995) 70 BLR 1; [1994] 40 ConLR 36 335
Crowshaw *v.* Pritchard and Renwick, (1899) 16 TLR 45 46

D& C Builders Ltd. *v.* Rees, [1965] 3 All ER 837 . . . . 258, 278
Dawber Williamson Roofing Ltd *v.* Humberside County Council, (1979) 14 BLR 70 . . . . 273, 349
Design 5 *v.* Keniston Housing Association Ltd., (1986) 34 BLR 92; [1987] 10 ConLR 123 155
Dillingham Construction Pty. Ltd. and Others *v.* Downs, [1972] 2 NSWLR 49; (1980) 13 BLR 97 45
Dinkha Latchin Associates *v.* General Mediterranean Holdings SA and Nadhmi S. Auchi, [2003] EWCA Civ 1786; [2003] 49 BLISS 13. 40
Discain Project Services *v.* Opecprime Developments Ltd., [2000] BLR 402 . 407, 411
Dole Dried Fruit and Nut Co. *v.* Trustin Kerwood Ltd., [1990] 2 Lloyd's Rep 309 . 266
Donoghue *v.* Stevenson; [1932] All ER 1 50
Douglas (R.M.) Construction Ltd. *v.* Bass Leisure Ltd., (1991) 53 BLR 119; [1991] 25 ConLR 38 276
Dudley Corporation *v.* Parsons Morrin Ltd., 1959 BCCB 41 256
Dunlop Pneumatic Tyre Co *v.* New Garage and Motor Ltd., [1915] AC 79 . . 137, 143
Dunnett *v.* Railtrackplc, [2002] 12 BLISS 10. 363
Durabella Ltd. *v.* J. Jarvis & Sons Ltd., (formerly known as J. Jarvis & Sons plc); [2001] 83 ConLR 145 . . . . 72, 262
Durtnell (R.) and Sons *v.* Kaduna Ltd., [2003] BLR 225; [2003] TCLR 7; [2004] 93 ConLR 36 385

East Ham Corporation *v.* Bernard Sunley Sons Ltd., AC 406; [1965] 3 All ER 619 358
Edmund Murray Ltd. *v.* BSP International Foundations Ltd., [1992] 33 ConLR 1 . 355
Ellis-Don Ltd. *v.* The Parking Authority of Toronto, (1978) 28 BLR 98 . 206, 220, 230
Elsley and Others *v.* J.G. Collins Insurance Agencies Ltd., (1978) 83 DLR (3d) 1 . 157
Elvin (C.J.) *v.* Peter and Alexa Noble, [2003] 20 BLISS 10 252
Emson Eastern Ltd. *v.* E.M.E Developments, (1991) 55 BLR 114; [1992] 26 ConLR 57. 304
Enco Civil Engineering Ltd. *v.* Zeus International Development Ltd., (1991) 56 BLR 43; [1992] 28 ConLR 25 276
English Industrial Estates Corporation *v.* Kier Construction Ltd., (1992) 56 BLR 93 89
Equitable Debenture Assets Corporation Ltd. *v.* William Moss Group, [1984] 2 ConLR 1 60
Euro Pools plc *v.* Clydeside Steel Fabrications Ltd., [2003] 20 BLISS 4 . . . . 202, 212
Expo Fabrics (UK) Ltd. *v.* Naughty Clothing Co. Ltd., [2003] EWCA Civ 1165 . . 355

Fairclough Building Ltd. *v.* Borough Council of Port Talbot, [1993] 33 ConLR 24; (1993) 62 BLR 82 10
Fairclough Building Ltd. *v.* Rhuddlan Borough Council, (1985) 30 BLR 26; [1985] 3 ConLR 368 212
Fairweather (H.) & Co Ltd *v.* The Mayor, Aldermen and Burgesses of the London Borough of Wandsworth, (1987) 39 BLR 106 . . . . 64, 95
Farr (A.E.) Ltd. *v.* Ministry of Transport, (1960) 5 BLR 94. 270
Farrans Construction Ltd. *v.* RMC Ready Mixed Concrete Ltd., [2003] 10 BLISS 15. 355
Fastrack Contractors Ltd. *v.* Morrison Construction Ltd. and Another, [2000] BLR 168; [2001] 75 ConLR 33; [2001] 75 ConLR 33 . . . . 386, 400
Fillite (Runcorn) Ltd. *v.* APV Pasilac Ltd. and APV Corporation Ltd., 22 I 1993, unrep. 355
Finnegan, (J.F.) Ltd. *v.* Community Housing Association Ltd., (1996) 77 BLR 22; [1996] 47 ConLR 25 . . . . 143, 148, 155
Finnegan (J.F.) Ltd. *v.* Ford Sellar Morris Developments Ltd., (1991) 53 BLR 38; [1991] 25 ConLR 89 247
Fischer, (George) (GB) Ltd. *v.* Multi Design Consultants Ltd., Roofdec Ltd. and Severfield Reece and Davis Langdon & Everest, [1998] 61 ConLR 85. 43
Fisher *v.* Ford, 1840 12 Ad 2 El 654. 151

Gaymark Investments Pty. Ltd. *v.* Walter Construction Group, (1999) NTSC 143; (2000) 16 BCL 449. . . . . 119, 124
Gleeson (M.J.) plc *v.* Taylor Woodrow Construction Ltd., (1989) 49 BLR 95; [1991] 21 ConLR 71 . . . . 149, 160, 166
Glencot Development and Design Co. Ltd. *v.* Ben Barrett & Son (Contractors) Ltd., [2001] BLR 207 367
Glenlion Construction Ltd *v.* The Guinness Trust, (1987) 39 BLR 89; [1988] 11 ConLR 126 . . . . 195, 198
Governor & Company of the Bank of Scotland *v.* Fuller Peiser, 2001 SLT 574. . . 355
Gray and Others (Special Trustees of the London Hospital) *v.* T.P. Bennett & Son, Oscar Faber & Partners and McLaughlin & Harvey Ltd., (1987) 43 BLR 63; [1989] 13 ConLR 22. 338
Greaves Contractors Ltd. *v.* Baynham Meikle & Partners, [1975] 1 Lloyd's Reports 31 51
Grogan *v.* Robin Meredith Plant Hire and Triact Civil Engineering Ltd., [1996] 53 ConLR 87 352

Hadley *v.* Baxendale, (1854) 9 Ex 341 156 ER 145 148
Haley (John L.) Ltd. *v.* Dumfries and Galloway Regional Council, 1988 39 GWD 1599. 125
Halki Shipping Corporation *v.* Sopex Oils Ltd., [1998] 1 Lloyd's Reports 49; [1997] 1 WLR 1268; [1997] 3 All ER 833 . . . . 276, 277, 383
Hall & Tawse Construction Ltd. *v.* Strathclyde Regional Council, 1990 SLT 775 . . 269
Halsey *v.* Milton Keynes General Hospital Trust, [2004] 20 BLISS 2. 363

Hanson (W.) (Harrow) Ltd. *v.* Rapid Civil Engineering and Usborne Developments, (1987) 38 BLR 106; [1988] 11 ConLR 119. 347
Hargreaves (B.) Ltd. *v.* Action 2000 Ltd., (1992) 62 BLR 72; [1994] 36 ConLR 74 . 266
Harvey (W.S) (Decorators) Ltd. *v.* H.L. Smith Construction Ltd., [1997] 16 BLISS 4 399
Havant Borough Council *v.* South Coast Shipping Company Ltd., (1996) CILL 1146. 90
Hawkins (Victor Stanley) *v.* Pender Bros Pty. Ltd., [1994] 1 Qd R 135. 108
Hayter *v.* Nelson and Others, [1991] 23 ConLR 88; [1990] 2 Lloyd's Rep 265. . . 276
Hedley Byrne Co. Ltd. *v.* Heller & Partners Ltd., [1963] 3 WLR 101; [1963] 2 All ER 575; [1964] AC 465 342
Hendy Lennox (Industrial Engines) Ltd. *v.* Grahame Puttick Ltd., [1984] 2 All ER 152; [1984] 1 WLR 485. 347
Henry Boot Construction Ltd. *v.* Alstom Combined Cycles Ltd., [2000] BLR 123; [1999] 64 ConLR 32 256
Henry Boot *v.* Malmaison Hotel, [1999] 70 ConLR 32 94
Herbert Construction (UK) Ltd. *v.* Atlantic Estates plc, (1993) 70 BLR 46 246
Herschel Engineering Ltd. *v.* Breen Property Ltd., 28 July 2000 395
Hersent Offshore S.A. and Amsterdamse Ballast Beton-en-Waterbouw B.V. *v.* Burmah Oil Tankers Ltd., (1978) 10 BLR 1 . . . . 192, 224
Hills Electrical and Mechanical plc *v.* Dawn Construction Ltd., [2003] 15 BLISS 1 263
Home and Overseas Insurance Co Ltd *v.* Mentor Insurance Co (UK) Ltd., [1990] 1 WLR 153; [1989] 3 All ER 74. 275
Hotel Services Ltd. *v.* Hilton International Hotels, [2000] BLR 235; [2000] 1 All ER (Comm) 750 235–6
How Engineering Services Ltd. *v.* Linder Ceilings Partitions plc and Another, [1999] 64 ConLR 59 105
Howard Marine and Dredging Co. Ltd. *v.* A. Ogden & Sons (Excavations) Ltd., (1978) 9 BLR 34. . . . . 35, 38, 45
Humber Oil Terminals Trustee Ltd. *v.* Harbour and General Works (Stevin) Ltd., (1991) 59 BLR 1; [1993] 32 ConLR 78 233
Hurst Stores & Interiors Ltd. *v.* ML Property Ltd., [2004] 94 ConLR 66; [2004] BLR 249 260

IBM United Kingdom Ltd *v.* Rockware Glass Ltd., [1980] FSR 335 108
Interfoto Picture Library Ltd. *v.* Stiletto Visual Programmes Ltd., [1988] 2 WLR 615; [1988] 1 All ER 348 328
Imperial Chemical Industries (ICI) plc *v.* Bovis Construction Ltd., GMW Partnership and Oscar Faber Consulting Engineers, [1992] 32 ConLR 90. 102
Impresa Castelli SpA *v.* Cola Holdings Ltd., [2002] 20 BLISS 5 308
Independent Broadcasting Authority *v.* EMI Electronics Limited, (1980) 14 BLR 1 50
Inserco Ltd. *v.* Honeywell Control Systems Ltd., [1996] 21 BLISS 3 . . . 104, 151, 221
Isovel Contracts Ltd. (In administration) *v.* ABB Building Technologies Ltd. (Formerly ABB Steward Ltd.), [2001] 47 BLISS 4 277

James (T.L.) and Co. Inc. *v.* Traylor Brothers, [2002] 42 BLISS 6 37
James Longley & Co. Ltd. *v.* South West Regional Health Authority, (1983) 25 BLR 56 203
JDM Accord *v.* Secretary of State for the Environment and Rural Affairs, [2004] 93 ConLR 133 254
Jeancharm Ltd. *v.* Barnet Football Club Ltd., [2003] 92 ConLR 26 . . . . . 137, 153
John Doyle Construction Ltd. *v.* Laing Management (Scotland) Ltd., [2004] 24 BLISS 10; [2004] BLR 295 98,105
John Moodie & Co. and Others *v.* Coastal Marine (Shipbuilders) Ltd., [2002] 37 BLISS 10. 355
John Mowlem & Co plc *v.* Carlton Gate Development Co Ltd., [1990] 21 ConLR 113; (1991) 51 BLR 104 274
Johnston *v.* W.H. Brown Construction (Dundee) Ltd., 2000 SLT 791; [2000] 69 ConLR 100; [2000] BLR 243. 315

Kier Construction Ltd. *v.* Royal Insurance (UK) Ltd., [1992] 30 ConLR 45. . 191, 224
Kingston-Upon-Thames (Mayor and Burgesses of the Royal Borough of) *v.* Amec Civil Engineering Ltd., [1993] 35 ConLR 39 271
Kitsons Sheet Metal Ltd. *v.* Matthew Hall Mechanical & Engineers Ltd., (1989) 47 BLR 82; [1990] 17 ConLR 116. 85
Koufos *v.* C. Czarnikow Ltd. ('The Heron II'), [1969] 1 AC 350; [1967] 3 All ER 686. 231

Lachhani Brothers *v.* Destination Canada Ltd., [1997] 1 BLISS 1 . . . . . . 178, 180
Laserbore Ltd. *v.* Morrison Biggs Wall Ltd., [1993] 37 BLISS 7 . . . . 176, 179
Leedsford Ltd. *v.* The Lord Mayor, Aldermen and Citizens of the City of Bradford, (1956) 24 BLR 45 360
Leyland Shipping Co. Ltd. *v.* Norwich Union Fire Insurance Society Ltd., [1918] AC 350 96
Lindenberg (Edward) *v.* Joe Canning and Jerome Contracting Ltd. and William Archibald Brown (t/a WAB Design Associates), (1992) 62 BLR 147 . . . . . 57, 61
Lomond Assured Properties Ltd. *v.* McGrigor Donald, 2000 SLT 189 . . . . 203, 213
London Borough of Barking and Dagenham *v.* Terrapin Construction Ltd., [2000] BLR 479; [2001] 74 ConLR 100 336
London Borough of Merton *v.* Stanley Hugh Leach Ltd., (1986) 32 BLR 51 . . . . .101, 104, 190, 201, 224
London Underground *v.* Kenchington Ford, [1998] 63 ConLR 1 . . . . . . 55, 187

MacJordan Construction Ltd. *v.* Brookmount Erostin Ltd. (In Administrative Receivership), (1991) 56 BLR 1 247
Macob Civil Engineering Ltd. *v.* Morrison Construction Ltd., [1999] 64 ConLR 1; [1999] BLR 93 383
Maidenhead Electrical Services Ltd. *v.* Johnson Control Systems Ltd., [1996] 7 BLISS 7 . . . . 118, 193

Malkins Nominees Ltd. *v.* Societe Finance, [2002] EWHC 1221 (Ch) 363
Marston Construction Co Ltd *v.* Kigass Ltd., (1989) 46BLR 109; [1989] 15 ConLR 116 39
Martin Grant & Co. Ltd. *v.* Sir Lindsay Parkinson & Co. Ltd., [1984] 3 ConLR 12 85
Matthew Hall Ortech Ltd. *v.* Tarmac Roadstone Ltd., (1997) 87 BLR 96 337
McAlpine Humberoak Ltd. *v.* McDermott International Inc., (1992) 58 BLR 1; [1992] 28 ConLR 76 180
McLaren, Murdock and Hamilton *v.* The Abercromby Motor Group, [2002] 47 BLISS 3 345
Metropolitan Special Projects Ltd. *v.* Margold Services Ltd., [2001] 43 BLISS 5 . . 72
Michael Salliss & Co Ltd *v.* E.C.A. & F.B. Calil and William F. Newman & Associates, [1988] 13 ConLR 68 . . . . 342, 344
Miller *v.* London County Council, [1934] All ER 656; (1934) 50 Times Law Reports 479 127
Minister of Public Works *v.* Lenard, 1992, unrep.. 180
Minter (F.G.) Ltd. *v.* Welsh Health Technical Services Organisation (WHTSO), (1980) 13 BLR 1. 218
Mitsui Babcock *v.* John Brown, [1996] 51 ConLR 129. 78
Mitsui Construction Co. Ltd. *v.* The Attorney General of Hong Kong, (1986) 33 BLR 1; [1987] 10 ConLR 1 170
Mondel *v.* Steel, (1841) 1 BLR 106 . . . . 235, 287
Monk Building and Civil Engineering Ltd. *v.* Norwich Union Life Insurance Society, (1992) 62 BLR 107 179
Moresk Cleaners Ltd. *v.* Thomas Henwood Hicks, (1966) 4 BLR 50 . . . . . . 53, 56
Morgan Grenfell Ltd and Sunderland Borough Council *v.* Seven Seas Dredging, (1990) 49 BLR 31 270
Morrison-Knudsen Co. Inc and Others *v.* British Columbia Hydro and Power Authority, (1978) 85 DLR (3d) 186 223
Mostcash plc (Formerly UK Paper), Fletcher Challenge Forest Industries Ltd. and Metsa-Serla (Holdings) Ltd. *v.* Fluor Ltd., [2002] 28 BLISS 13 355
Motherwell Bridge Construction (t/a Motherwell Bridge Storage Tanks) *v.* Micafil Vakuumtechnik and Another, [2002] 15 BLISS 11; [2002] 81 ConLR 44 223
Mowlem plc *v.* Newton Street Ltd., [2003] 89 ConLR 153 240

Nash Dredging Ltd *v.* Kestrel Marine Ltd., (1987) SLT 641. 269
Neodox *v.* The Mayor, Aldermen and Burgesses of the Borough of Swinton and Pendlebury, (1958) 5 BLR 34 . . . . 200, 225
Nevill H.W. (Sunblest) Ltd. *v.* William Press & Son Ltd., (1981) 20 BLR 78 . . . . .304, 311, 323, 324
Newton Moor Construction Ltd. *v.* Charlton, (1981) 13 Const. LJ 275. 258
Norta Wallpapers (Ireland) Ltd. *v.* John Sisk & Sons (Dublin) Ltd., (1977) 14 BLR 49 59
Norwest Holst Construction Ltd. *v.* Co-operative Wholesale Society Ltd. and Another, [1998] 22 BLISS 4 209

Ogilvie Builders Ltd. *v.* The City of Glasgow District Council, (1994) CILL 930. . 219
Omega Trust Fund *v.* Wright, Son & Pepper, [1997] PNLR 424; [1996] NPC 198 . 355
Orange EBS Ltd. *v.* ABB, [2003] 20 BLISS 3 386
Outwing Construction Ltd. *v.* H. Randall & Son Ltd., [1999] BLR 156; [1999] 64 ConLR 59 383
Ovcon (Pty.) Ltd. *v.* Administrator, Natal, 1991 (4) SA 71 196

Pacific Associates Inc and Another *v.* Baxter and Others, (1989) 44 BLR 33; [1989] 2 All ER 159; [1989] 16 ConLR 90 . . . . 343, 344
Paul Tsakok & Associates *v.* Engineer & Marine Services (Pte.) Ltd., [1991] SLR 942 333
P.C. Harrington Contractors Ltd. *v.* Co-Partnership Developments Ltd., (1998) 88 BLR 44. 248
Peak Construction (Liverpool) Ltd. *v.* McKinney Foundations Ltd., (1970) 1 BLR 111 . . . . 94, 111, 141, 142, 150, 156, 220
Pearce (C.J.) & Co. *v.* Hereford Corporation, (1968) 66 LGR 647 233
Pearson (S.) & Son Ltd. *v.* Dublin Corporation, [1907] AC 351. 35
Pegler Ltd. *v.* Wang (UK) Ltd., [2000] BLR 218; [2000] 70 ConLR 68 203
Penvidic Contracting Co. Ltd. *v.* International Nickel Co. of Canada Ltd., (1975) 53 DLR (3d) 748 228
People's Park Chinatown Development Pte. *v.* Schindler Lifts (Singapore) Pte. Ltd. 1993 unrep. 350
Pepsi-Cola *v.* Coca Cola, [1942] 49 RPC 131. 242
Perini Corporation *v.* Commonwealth of Australia: Supreme Court of New South Wales, (1969) 12 BLR 82 223
Philips Hong Kong Ltd. *v.* The Attorney-General of Hong Kong, (1993) 61 BLR 41 . . . . 140, 153
Pillings (C.M.) & Co. Ltd. *v.* Kent Investments Ltd., (1985) 30 BLR 80; [1986] 4 ConLR 1 274
Pioneer Shipping Ltd. Others *v.* B.T.P. Tioxide ('The Nema'), [1982] 2 All ER 1030 . . . . 13, 242
Pitchmastic plc *v.* Birse Construction Ltd., [2000] 22 BLISS 5 314
P & O Developments Ltd. *v.* Guy's and St. Thomas' National Health Service Trust, [1999] BLR 3; [1999] 62 ConLR 38. 187
Pratt Contractors Ltd. *v.* Transit New Zealand, [2003] BLR 143. 42
PSC Freyssinet Ltd. *v.* Byrne Brothers (Formwork) Ltd., 11 December 1996, unrep. 52

Quebec (Commission Hydroelectrique) *v.* Banque de Montreal, 1992 unrep.. . . . 45
Quinn *v.* Burch Bros. (Builders), [1966] 2 QB 370 99

Railtrack plc *v.* Pearl Maintenance Services Ltd., [1995] 6 BLISS 8 37
Rapid Building Group Ltd *v.* Ealing Family Housing Association, (1984) 29 BLR 5; [1985] 1 ConLR 1 . . . . 142, 156
Rayack Construction Ltd. *v.* Lampeter Meat Co. Ltd., (1979) 12 BLR 30 . . . . 245, 248–249

Reardon Smith Line Ltd *v.* Hansen Tangren, [1976] 3 All ER 570 242
Rees and Kirby Ltd. *v.* The Council of the City of Swansea, (1985) 30 BLR 1; [1986] 5 ConLR 34 . . . . 191, 218
Rees Hough Ltd. *v.* Redland Reinforced Plastics Ltd., (1984) 27 BLR 136; [1985] 2 ConLR 109 353
Regalian Properties *v.* London Docklands Development Corporation, [1995] 45 ConLR 37; [1995] 1 WLR 212 39
Rentokil Ailsa Environmental Ltd. *v.* Eastend Civil Engineering Ltd., [1999] 16 BLISS 2 384
Richard Roberts Holdings Ltd. *v.* Douglas Smith Stimson Partnership and Others, (1990) 46 BLR 50 63
Ritchie Brothers (PWC) Ltd. *v.* David Phillips (Commercials) Ltd., [2004] 16 BLISS 1 398
RJT Consulting Engineers Ltd. *v.* DM Engineering (NI) Ltd., [2002] BLR 217; [2002] 83 ConLR 99 389
Robophone Facilities *v.* Blank, [1966] 3 All ER 128 143
Rolls Royce Engineering and Another *v.* Ricardo Engineers Ltd., [2003] EWHC 2871 (TCC); [2003] 47 BLISS 14 355
Routledge *v.* Grant, (1828) 4 Bing 653; 3 C&P 267; 6 LJOS 166; 1 Moo & P 717 . 47
RSL (South West) Ltd. *v.* Stansell Ltd., [2003] 24 BLISS 2 409
Ruxley Electronics and Construction Ltd. *v.* Forsyth Laddingford Enclosures Ltd. *v.* Same, [1995] 3 All ER 307; (1995) 73 BLR 1; [1995] 45 ConLR 61; [1995] 3 WLR 118 344

Sainsbury (J.) plc *v.* Broadway Malyan, [1998] 61 ConLR 31 . . . . 54, 317
St. Albans City and District Council *v.* International Computers Ltd., [1996] 4 All ER 481 355
St. Andrew's Bay Development *v.* HBG Management and Mrs. Janey Milligan, [2003] 15 BLISS 4 . . . . 396, 398
St. Modwen Developments *v.* Bowmer Kirkland, [1996] 38 BLISS 4 209
Saint Line Ltd. *v.* Richardsons Westgate, [1940] 2 KB 99. 221
Salvage Association *v.* CAP Financial Services Ltd., [1992] FSR 654 355
SAM Business Systems Ltd. *v.* Hedley & Co. (A Firm), [2002] 1 All ER (Comm) 465 355
Scobie & McIntosh Ltd *v.* Clayton Browne Ltd., (1990) 49 BLR 119; [1991] 23 ConLR 78. 249
Seath & Co. *v.* Moore, (1886) 13 RHL 57 350
Secretary of State for Transport *v.* Birse-Farr Joint Venture, [1993] 35 ConLR 8; (1993) 62 BLR 36 270
Semco Salvage & Marine Pte. Ltd. *v.* Lancer Navigation Co. Ltd. ('The Nagasaki Spirit'), [1997] 1 Lloyd's Reports 323 175
Serck Controls Ltd. *v.* Drake & Scull Engineering Ltd., [2000] 73 ConLR 100. . . 177
Shanks and McEwan (Contractors) Ltd. *v.* Strathclyde Regional Council, 1994 SLT 172 54
Sherwood & Casson Ltd. *v.* Mackenzie, (2000) 2 TCLR 418 394

Shimizu Europe Ltd. *v.* Automajor Ltd., [2002] 4 BLISS 6 394
Shimizu Europe Ltd. *v.* LBJ Fabrications, [2003] BLR 381 414
Shirayama Shokusan Co. Ltd. and Others *v.* Danova Ltd., [2001] 10 BLISS 11 . . 363
Sim and Associates *v.* Tan, 1997 unrep. 358
Simkins Partnership *v.* Reeves Land & Co. Ltd.,18 July 2003 unrep. 235
Simons Construction Ltd. *v.* Aardvark Developments, [2003] TCLR 2; [2004] BLR 117; [2004] 93 ConLR 114 . . . . 398, 406
Sindall Ltd. *v.* Solland and Others, [2001] 37 BLISS 1 400
Sir Lindsay Parkinson & Co. Ltd. *v.* Commissioners of Works and Public Buildings, [1949] 1 All ER 208 179
Skanska Corporation *v.* Anglo-Amsterdam Corporation, [2002] 84 ConLR 100. . 307
Slater (James) and Hamish Slater (A Firm) and Others *v.* Finning Ltd., [1996] 2 Lloyd's Reports 353 51
Smith & Smith Glass Ltd. *v.* Winstone Architectural Cladding Systems Ltd., [1993] 2 NZLR 473 261
Stent Foundations Ltd. *v.* Carillion Construction (Contracts) Ltd., [2000] 29 BLISS 13 6
Stephenson (WA) (Western) Ltd. *v.* Metro Canada Ltd., (1987) 27 CLR 113 (BCSC) 223
Stewart Gill Ltd. *v.* Horatio Myer Co. Ltd., [1992] 2 All ER 257; [1993] 31 ConLR 1; [1992] 2 WLR 721. 354
Stour Valley Builders (A Firm) *v.* Stuart and Another, 21 December 1993, unrep. 259
Strachan & Henshaw *v.* Stein Industrie (UK) Ltd., (1997) 87 BLR 5; [1999] 63 ConLR 160 193
Stubbs Rich Architects *v.* W.H. Tolley & Son Ltd., [1997] 1 BLISS 1. 405
Sutcliffe *v.* Thackrah, [1974] 1 All ER 319; (1977) 4 BLR 16 331
Sweatfield Ltd. *v.* Hathaway Roofing Ltd., [1997] 9 BLISS 8 361

Tate and Lyle Food and Distribution Ltd. *v.* Greater London Council, [1982] 1 WLR 149 202
Team Services plc *v.* Kier Management and Design Ltd., (1993) 63 BLR 76; [1994] 36 ConLR 32 241
Temloc Ltd. *v.* Errill Properties Ltd., (1987) 39 BLR 30 . . . . 145, 165
Tennant Radiant Heat Ltd. *v.* Warrington Development Corporation, (1988) 4 Const. LJ 321 97
Terrell *v.* Maby Todd & Co., (1952) 69 RPC 234 108
Tesco Stores Ltd. *v.* Costain Construction Ltd., [2003] 27 BLISS 11 72
Thorn *v.* Mayor and Commonality of London, [1876] 1 AC 120 181
Tinghamgrange Ltd. (t/a Gryphonn Concrete Ltd.) *v.* Dew Group and North West Water Ltd., [1996] 47 ConLR 105 173
Trolex Products Ltd. *v.* Merrol Fire Protection Engineers Ltd. 1991 355
Try Construction Ltd. *v.* Eton House Group Ltd., 20 November 1991, unrep.. . . 408
Turner Corporation Ltd. (Receiver and Manager Appointed) *v.* Austotal Pty. Ltd., (1998) 13 BCL 378. . . . . 119, 123

Turner Page Music *v.* Torres Design Partnership, (1997) CILL 1263 99
Turriff Ltd. *v.* Welsh National Water Development Authority and McCreath Taylor & Co. Ltd. and Trocoll Industries Ltd., 1979 unrep.. 35
Twintec Ltd. *v.* GSE Building & Civil Engineering Ltd., [2003] 13 BLISS 9 78

University Court of the University of Glasgow *v.* William Whitfield and John Laing (Construction) Ltd. (Third Party), (1988) 42 BLR 66 60
University of Keele *v.* Price Waterhouse (A Firm), [2004] EWCA Civ 583 355

Vascroft (Contractors) Ltd. *v.* Seeboard plc, (1996) 78 BLR 132; [1997] 52 ConLR 1 309
Victoria Laundry Windsor Ltd. *v.* Newman Industries Ltd., [1949] 2 KB 528; [1949] 1 All ER 997 . . . . 148, 221
Victoria University of Manchester *v.* Hugh Wilson & Lewis Womersley (A Firm) and Pochin (Contractors) Ltd., [1984] 2 ConLR 43. . . . . 60, 62

Wall *v.* Rederiaktiebolaget Luggude, [1915] 3 KB 66 157
Walter Lawrence and Son Ltd. *v.* Commercial Union Properties (UK) Ltd., [1984] 4 ConLR 37 . . . . 131, 132
Wates Construction (London) Ltd. *v.* Franthom Property Ltd., (1991) 53 BLR 23 . . . . 245, 247
Wells *v.* Army and Navy Co-operative Society Ltd., (1902) LT 764; (1902) KB 346. 94
Weldon Plant Ltd. *v.* Commission for the New Towns, [2000] BLR 496; [2001] All ER (Comm) 264; [2001] 77 ConLR 1 . . . . 177, 179
Wescol Structures Ltd. *v.* Miller Construction Ltd., [1998] 22 BLISS 14 243
West Faulkner Associates (A firm) *v.* The London Borough of Newham, (1992) 61 BLR 84; [1993] 31 ConLR 105. 341
Westminster (The Lord Mayor, Aldermen and Citizens of the City of) *v.* J. Jarvis & Sons Ltd. and Another, (1978) 7 BLR 64. . . . . 303, 305
Wharf Properties Ltd. and Another *v.* Eric Cumine Associates Others, (1991) 52 BLR 1; [1992] 29 ConLR 113 . . . . 101, 226
Whittall Builders Co. Ltd. *v.* Chester-le-Street District Council, [1989] 11 ConLR 40 . . . . 206, 226
William Hill Organisation Ltd.v Bernard Sunley & Sons Ltd., (1982) 22 BLR 1 . . 338
William Tomkinson and Sons Ltd. *v.* The Parochial Church Council of St. Michael's and Others, unrep. 1990 312
Williams *v.* Roffey Bros. & Nicholls (Contractors) Ltd., (1990) 48 BLR 69; [1990] 1 All ER 512; [1990] 2 WLR 1153 278
Wimpey Construction UK Ltd *v.* D.V. Poole, (1984) 27 BLR 58. 50
Wood *v.* Grand Valley Railway Co., (1913) 16 DLR 361 227
Wraight Ltd *v.* P.H. & T. Holdings Ltd., (1980) 13 BLR 26 221

Yorkshire Water Authority *v.* Sir Alfred McAlpine & Son (Northern) Ltd., (1985) 32 BLR 114 89
Young and Marten *v.* McManus Childs, [1969] 1 AC 454 51

Index

200 Contractual Problems and their Solutions, Third Edition. Roger Knowles.

Printed and bound by CPI Group (UK) Ltd, Croydon, CR0 4YY
18/07/2024
14530251-0001